D1199869

Essential Excel 2019

A Step-By-Step Guide

Second Edition

David Slager
Annette Slager

Apress®

Essential Excel 2019: A Step-By-Step Guide

David Slager
Fort Wayne, IN, USA

Annette Slager
Fort Wayne, IN, USA

ISBN-13 (pbk): 978-1-4842-6208-5
https://doi.org/10.1007/978-1-4842-6209-2

ISBN-13 (electronic): 978-1-4842-6209-2

Managing Director, Apress Media LLC: Welmoed Spahr
Acquisitions Editor: Joan Murray
Development Editor: Laura Berendson
Coordinating Editor: Jill Balzano

Cover image designed by Freepik (www.freepik.com)

Distributed to the book trade worldwide by Springer Science+Business Media New York, 1 New York Plaza, Suite 4600, New York, NY 10004. Phone 1-800-SPRINGER, fax (201) 348-4505, e-mail orders-ny@springer-sbm.com, or visit www.springeronline.com. Apress Media, LLC is a California LLC and the sole member (owner) is Springer Science + Business Media Finance Inc (SSBM Finance Inc). SSBM Finance Inc is a **Delaware** corporation.

For information on translations, please e-mail booktranslations@springernature.com; for reprint, paperback, or audio rights, please e-mail bookpermissions@springernature.com.

Apress titles may be purchased in bulk for academic, corporate, or promotional use. eBook versions and licenses are also available for most titles. For more information, reference our Print and eBook Bulk Sales web page at http://www.apress.com/bulk-sales.

Any source code or other supplementary material referenced by the author in this book is available to readers on GitHub via the book's product page, located at www.apress.com/9781484262085. For more detailed information, please visit http://www.apress.com/source-code.

Printed on acid-free paper

We are dedicating this book to our parents, Andrew and Carrie (Sterk) Slager and Russell and Florence Carson. We would like to thank them for all their love, guidance, and the sacrifices they made for us.

We are also dedicating this book to our wonderful children, Rhianna and Marten.

Table of Contents

About the Authors..xvii

About the Technical Reviewer xix

Acknowledgments xxi

Introduction xxiii

Chapter 1: Becoming Acquainted with Excel..1
   What Is Excel?...1
   Excel Versions 2
   History of Spreadsheets..4
   This Book 4
   Excel Navigation Basics 5
   Creating, Saving, and Opening Workbooks...7
   Getting to Know the Ribbon 12
      Ribbon Contextual Tabs 13
      Resizing the Ribbon..14
      Using Dialog Box Launchers 15
      Minimizing and Hiding the Ribbon...16
      Using Ribbon Shortcuts 18
   Quick Access Toolbar 19
      AutoSave 21
      Switch Between Touch and Mouse Mode..21
   Identifying the Current Cell 24
   Entering Data into a Worksheet 25

Getting Help 27

    Screen Tips 28

    Excel's Tell Me What You Want to Do Feature 28

    Help Tab 33

Summary 39

Chapter 2: Navigating and Working with Worksheets 41

Moving Between Cells Using the Keyboard 42

Selecting Cells 46

    Selecting Cells Using a Mouse 46

    Deselecting Cells Using a Mouse 47

    Selecting Cells Using a Keyboard 50

    Select Cells by Using Their Cell References in the Name Box 53

    Going Directly to Any Cell 55

Worksheets 60

    Naming Worksheets 60

    Adding and Removing Worksheets 61

    Changing a Worksheet Tab Color 63

    Selecting Multiple Worksheets 64

    Hiding and Unhiding Worksheets 64

    Reordering and Copying Worksheets 65

    Using Tab Buttons to Move Through the Worksheets 67

Summary 73

Chapter 3: Best Ways to Enter and Edit Data 75

Data Types 76

Inserting Special Characters 79

How to Change Column Widths 83

    Automatically Resize Column Widths to Fit Number of Characters in the Cell 87

    Changing the Column Width for Multiple Columns 87

How to Change Row Heights 91

Correcting Typing Mistakes 93

    Changing Specific Characters 93

    Returning a Cell to Its Original Value 94

    Clearing the Contents of a Cell That Has Already Been Accepted 95

Shortcuts for Entering and Correcting Data 96

    Using the AutoCorrect Feature 96

    Using AutoComplete to Enter Data 102

    Pick from Drop-down List 103

    AutoFill 105

Creating, Viewing, Editing, Deleting, and Formatting Cell Notes 116

    How to Add a Note 118

    How to View a Note 118

    How to View All the Notes at the Same Time 119

    How to Change the Default Name for Notes 119

    Editing and Deleting Notes 119

    Printing a Note 120

The New Collaborative Comments 124

Summary 128

Chapter 4: Formatting and Aligning Data 131

Formatting Your Text Using the Font Group 132

    Using Bold, Italics, Underline, and Double Underline 132

    Changing the Font and Its Size 132

    Using the Font Group's Dialog Box Launcher 135

    Formatting with Color 136

    Check Which Formats Have Been Applied to the Current Cell 140

    Cell Borders 144

Formatting Numeric Data Using the Number Group 155

    Using Default Formats 156

    Formatting Monetary Values 158

    Converting Values to Percent Style 159

Converting Values to Comma Style... 160

Changing the Number of Decimal Places 160

Accessing the Format Cell Dialog Box 161

Aligning Data Using the Alignment Group 165

Fitting More Text into a Cell 166

Aligning and Indenting Text in a Cell 170

Align Text Vertically and Horizontally 171

Rotating Text 172

Using Format Painter to Copy Formatting 178

Using the Mini-Toolbars and the Context Menu 182

Inserting, Deleting, Hiding, and Unhiding Rows and Columns 184

Hiding and Unhiding Columns and Rows 185

Inserting Columns and Rows 190

Deleting Columns and Rows 191

Inserting and Deleting Cells 192

Summary 195

Chapter 5: Different Ways of Viewing and Printing Your Workbook 197

Views 198

Page Break Preview 199

Page Layout View 205

Printing 212

Creating a Print Area 212

Adding Additional Cells to the Print Area 213

Removing the Print Area 213

Using Paste Special for Printing 216

Dividing the Excel Window into Panes 218

Freezing Rows and Columns 221

Synchronizing Scrolling 223

Custom Views How to Create, Show, and Delete 228

Summary 232

Chapter 6: Understanding Backstage **233**

Backstage Overview 234

Info Group—Viewing, Adding, and Editing Information About the Workbook 235

Properties Pane 236

Protect Workbook Options 239

Check for Issues 242

New Group—Creating a New Workbook 248

Open Group—Open a Workbook 251

Opening an Existing Workbook 251

Options Affecting the Open Group 254

Save and Save As Groups—Saving a Workbook Using Save or Save As 256

Document Recovery 260

Saving Workbooks with Protections: Backups and Limiting Changes 267

Print Group—Printing a Workbook 272

Selecting a Printer 273

Printer Settings 274

Share Group—Sharing Workbooks 282

Sharing Online with OneDrive 282

Sharing Files Using Email 285

Account Group 287

Summary 288

Chapter 7: Creating and Using Formulas **289**

Formulas 290

Introducing Formulas 290

Entering Formulas 292

Copying Formulas 298

AutoCalculate Tools 301

AutoSum 301

Average, Count Numbers, Max, Min 306

Viewing Formulas 311

Creating Named Ranges and Constants 312

    Naming Ranges 312

    Naming Noncontiguous Ranges 315

    Naming Constants 315

    Name Manager 316

    Using Column or Row Headings for Range Names...................................... 318

    Selecting Named Ranges Rather Than Typing Them into Formulas 320

Absolute Cell References 332

Mixed Cell References 338

    Order of Precedence... 342

Summary... 344

Chapter 8: Excel's Pre-existing Functions 347

Excel's Built-In Functions.. 348

Function Construction 349

Functions That Sum Values 350

    SUM Function 351

    Using the Insert Function Option 355

    SUMIF—Adds the Cells That Meet a Specified Criteria............................. 360

    SUMIFS—Adds the Cells That Meet Multiple Criteria............................... 364

IF—Returns Different Values Depending upon If a Condition Is True or False 369

AND—Returns TRUE if All of Its Arguments Are TRUE................................ 373

OR—Returns TRUE If Any Argument Is TRUE 374

Nested Functions 375

New Excel 2019 Functions... 381

    IFS—Returns Value Based on Criteria.. 382

    MAXIFS and MINIFS—Returns Highest or Lowest Value Based on Criteria 390

Date Functions 398

    TODAY Function—Returns the Current Date 399

    NOW Function—Returns the Current Date and Time 402

    DATE Function—Returns the Serial Number of the DATE 404

MONTH, DAY, and YEAR Functions 405

DAYS—Returns the Number of Days Between Two Dates 408

Summary 411

Chapter 9: Auditing, Validating, and Protecting Your Data 413

Validating Your Data and Preventing Errors 414

Data Validation 414

Evaluating Formulas 431

Using IFERROR 432

Correcting Circular References 434

Formula Auditing 436

Tracing Precedents and Dependents 437

Using the Watch Window 440

Using the Evaluate Formula Feature to Evaluate a Nested Function
One Step at a Time 443

Proofreading Cell Values—Have Excel Read Back Your Entries 447

Spell Checking 449

Thesaurus 451

Protect Worksheets and Cells from Accidental or Intentional Changes 453

Protect Your Data at the Worksheet Level 453

Protect Your Data at the Cell Level 457

Summary 459

**Chapter 10: Using Hyperlinks, Combining Text, and Working with the
Status Bar 461**

Working with Hyperlinks 462

Concatenation and Flash Fill 475

Using the Status Bar 485

Cell Mode 487

Flash Fill Blank Cells and Flash Fill Changed Cells 487

Caps Lock, Num Lock 487

Scroll Lock 488

Fixed Decimal 488

Overtype Mode 488

End Mode 488

Macro Recording 488

Accessibility Checker 489

Selection Mode 489

Page Number 489

Average, Count, Numerical Count, Minimum, Maximum, Sum 489

View Shortcuts 489

Zoom and Zoom Slider 489

Summary 495

Chapter 11: Transferring and Duplicating Data to Other Locations 497

Moving and Copying Data 498

Moving and Copying Cells Using the Drag-and-Drop Method 498

Moving and Copying Cells Using the Cut and Copy Buttons 501

Moving and Copying Cells Using the Keyboard 502

Paste Button Gallery 504

Copy Data to Other Worksheets Using Fill Across Worksheets 512

Copy Data from One Workbook to Another 516

Paste Special 518

Using Paste Special 519

Using Paste Special to Transpose Rows and Columns 522

Using Paste Special to Perform Calculations 524

Inserting Copied or Moved Cells 530

Insert Copied Cells 531

Insert Cut Cells 533

Using the Microsoft Office Clipboard 534

Entering Data into Multiple Worksheets at the Same Time 536

Summary 541

Chapter 12: Working with Tables...**543**

Creating and Formatting Tables 544

Sort and Filter a Table 546

Adding to the Excel Table 548

Filtering Data with a Slicer...558

Using Themes 560

Applying and Defining Cell Styles 565

Conditional Formatting..569

Summary..581

Chapter 13: Working with Charts **583**

Chart Types 584

Creating and Modifying Charts..587

Pie Charts..607

The Standard Pie Chart...608

Pie of Pie Subtype 612

Combination Chart 620

Hierarchical Charts 625

Treemap Chart 625

Sunburst Chart 633

Funnel Chart 642

Map Chart 644

Using Bing Maps 649

Sparklines 656

Summary..664

Chapter 14: Importing Data **665**

Importing Data into Excel..665

Importing Text Files...666

Delimited Text Files 666

Fixed-Width Text Files 679

Importing Data from an Access Database..683

Importing Data from a Website 686

Importing Data Using Transform 690

Summary...705

Chapter 15: Using PivotTables and PivotCharts................................707

Working with PivotTables..708

Creating a PivotChart 741

Creating PivotTable on a Relational Database..752

Summary...763

Chapter 16: Geography and Stock Data Types.................................765

Use Geographic Data Types in a Table 770

Using Geographic Location Data in Formulas 777

Stocks 786

Summary...797

Chapter 17: Enhancing Workbooks with Multimedia 799

Adding Pictures to the Worksheet..800

Using Screenshot..809

Working with WordArt 813

Adding and Modifying Shapes 819

Using SmartArt...838

Inserting Sound into a Worksheet..845

Inserting Video into a Worksheet 848

Summary...849

Chapter 18: Icons, 3D Images, and Object Grouping 851

Grouping and Ungrouping Icons, Images, Shapes, and WordArt............................855

3D...863

Summary...873

Chapter 19: Automating Tasks with Macros.. **875**

Creating (Recording) a Macro 876

The Problem with Absolute Cell References 880

Saving a Macro-Enabled Workbook 882

Creating a Macro Using Relative Cell References............................... 884

Adding Macros to the Quick Access Toolbar and Other Objects........... 886

Sharing the Personal Workbook with Others 894

Looking at VBA Code... 894

Creating Macros from Code 897

Summary.. 901

Index.. **903**

About the Authors

David Slager has been a computer programmer for four decades, with a focus on Excel. He was also the computer department head of a college for many years. David has worked with spreadsheets since their introduction. As a consultant, he developed major e-learning training projects for agriculture and steel businesses and designed a simulation program that trained feed market managers to use analytics to improve their market position. He enjoys working with analytics and solving problems and has taught learners of all ages and levels. David holds an MS in Education, specializing in Instructional Media Development, a BA in Organizational Management, and an associate's degree in Accounting as well as many certifications. He is currently a software manager.

Annette Slager has been involved in data management and employee training in the nonprofit and higher education sectors. She has been responsible for coordinating donor stewardship events and processes and overseeing information entered into the donor/alumni data system. She has transitioned systems from manual accounting and processing to shared databases and created learning manuals and training for employees. Annette has a bachelor's degree in English and a post-baccalaureate certificate in Technical Writing.

About the Technical Reviewer

 A native to Northern Indiana and former student of Mr. Slager, **Jake Halsey** has over a decade of experience working with various Microsoft products, services, and development tools in the IT industry. Working in the Fort Wayne and Chicago areas as a senior-level software developer and application administrator, he's made frequent use of the skills learned in Mr. Slager's classroom, many of which are now expounded upon in this very book, to perform complex data analysis and prepare professional reports using the many powerful features built into Microsoft Excel. He's excited to be a part of Mr. Slager's latest endeavor to bring knowledge and technical skill to the next wave of business professionals and to pick up a few new tricks for himself in the process.

Acknowledgments

We would like to thank Jake Halsey, the Technical Reviewer for this book, for all his helpful comments and suggestions.

We would also like to thank Jill Balzano for guiding this book through the writing, editing, and publication process.

Introduction

This is a practical hands-on working book. It features numerous screenshot examples providing extensive visual reinforcement of the examples and exercises. It is for those who want to impress their supervisor and others with their Excel skills. It is designed for users who want their presentations to stand out and want to create impressive spreadsheets and charts that clearly convey data to a targeted audience. It is for problem solvers who want to go the extra mile. This is a reference book to keep on your desk. When you need to solve a problem in Excel or learn a new feature, you can go directly to the chapter that solves that problem. The book doesn't need to be read straight from beginning to end. Chapters build on each other; however, you can go directly to any chapter to find help. Readers should have a basic knowledge of computers and Excel.

Chapter 1 covers the basics such as saving and opening workbooks, using the Ribbon and Quick Access Toolbar, and how to use Excel's help features.

Chapter 2 covers various ways of selecting cells and moving between them. It shows how to work with worksheets such as adding and removing them, hiding and unhiding them, and copying them and changing their order.

Chapters 3 and 4 cover Excel's techniques to improve and speed up data entry, documenting your data through notes and comments, and how to format and align different types of data.

Chapter 5 covers how to print only what you want from a workbook and various methods of viewing your data including how to create your own custom views.

Chapter 6 covers what Excel refers to as the Backstage. The Backstage contains information about your workbook as well as options that control what happens behind the scenes of your workbook including handling files, printing, sharing and exporting workbooks, account information, and how Excel works.

Chapters 7 and 8 cover Excel's built-in functions as well as formulas that you create using different types of cell addressing as well as named ranges.

Chapter 9 covers validating, auditing, and protecting your data.

Chapter 10 covers linking to web pages, files, emails, different locations in the current workbook, and links that will even create and open a new workbook. This chapter also shows how to automate combining text and how to use the functions available on the status bar.

Chapter 11 covers the many ways of moving and copying data into one worksheet or multiple worksheets at the same time.

Chapter 12 covers tables which are grids that let you format, sort, and filter your data as well as apply conditional formatting. The chapter also shows how to use slicers to provide a quick and visual method for organizing and filtering data.

Chapter 13 shows how to create, format, and modify many different chart types. It also covers displaying geographical data on maps and creating Sparklines which are charts contained within a cell.

Chapter 14 covers getting just the data you want from different locations into your workbooks and how to create a connection between a data source and your workbook.

Chapter 15 shows how to group, rearrange, and summarize your data using Pivot Tables. It also covers using the Slicer and Time Interval tools. You will learn how to create a relational database.

Chapter 16 covers the new Geography and Stock data types added in Excel 2019. The geography data type is for gathering data about countries, provinces, regions, states, counties, cities, and towns. The stock data type lets you get the latest stock market data, which you can display in charts or use in formulas.

Chapter 17 deals with adding images, icons, shapes, WordArt, SmartArt, audio, and video to your workbooks.

Chapter 18 covers splitting an image into pieces that can be worked with individually, using 3D objects, changing the order and visibility of objects, and grouping and ungrouping objects.

Chapter 19 covers automating Excel tasks by using Macros (repeatable steps created by recording the tasks as you do them or by writing code).

CHAPTER 1

Becoming Acquainted with Excel

Excel is a powerful and versatile spreadsheet program that can be used for both business and personal needs. It has amazing capabilities that you can use to make any type of data you record more streamlined and productive. In the first chapter, you'll learn the basics of creating worksheets and how to use the Ribbon, a feature which drives the user-friendly resources in Excel.

After reading and working through this chapter, you should be able to

- Know what Excel is and know some of its capabilities

- Create, save, and open a workbook

- Identify the current cell

- Use the Ribbon

- Use and customize the Quick Access Toolbar

- Enter data in a worksheet

- Get Help by using Screen Tips and the "Tell Me What You Want to Do" feature

What Is Excel?

Excel is an electronic spreadsheet program. A spreadsheet is a grid of cells organized into rows and columns in which you enter and store your data. Excel can meet both your personal and professional needs.

© David Slager and Annette Slager 2020
D. Slager and A. Slager, *Essential Excel 2019*, https://doi.org/10.1007/978-1-4842-6209-2_1

Using Excel, you can do all the following:

- Create, edit, sort, analyze, summarize, and format data as well as graph it.

- Keep budgets and handle payroll.

- Track investments, loans, sales, inventory, and so on.

- Perform What-If Analysis to determine such things as "if the price of gas went up 20 cents per gallon," by how much would that decrease my profit, or "if I extend my loan payments from 15 years to 20 years," by how much will that affect my monthly payments, total payments, and total interest.

You can improve the appearance of a spreadsheet or better convey what you want a spreadsheet to say by adding pictures, clip art, shapes, SmartArt, video, and audio.

Microsoft Office is Microsoft's most profitable product. Microsoft devoted most of their effort in Microsoft Office 2016 to updating Excel and since then have added many new features.

Excel Versions

Note   There are differences between the PC and Mac versions. This book covers only the PC version.

There are basically three ways of getting Excel. You can use the free online version, you can subscribe and pay a monthly fee, or you can make a one-time purchase.

Office online is free to everyone. It is accessed through your Internet browser. Because it is accessed through a browser, it will look the same for both Windows and Mac users. It is a stripped-down version of office products. Go to www.office.com.

Microsoft breaks down their Microsoft Office products by "Home" and "Business".

They have three Home versions:

Office 365 Home—With this subscription, you get updates to the latest features that Microsoft has added. The subscription can be used by up to six users on all your different devices at the same time. Your subscription includes 1 TB of cloud storage for each user and Skype access. You will get Excel, Word, PowerPoint, and Outlook. If you have a PC (not a Mac), you will also get Publisher and Access.

Office 365 Personal—This is the same as Office 365 Home except that the subscription is for only one individual.

Office Home & Student 2019—There is a one-time fee to purchase this product. It can only be installed on one PC. You will only get Excel, Word, and PowerPoint. Microsoft will provide you with new security updates but not any new features. You will have to buy the next version to get any new features. New versions come out about every three years.

Note   A difference between the Office 365 subscriptions and the one-time fee for Office Home & Student 2019 is the length of time you can use the product. If you have an Office 365 subscription, once you stop paying your subscription, you will no longer be able to use it, but if you pay the one-time fee for Home & Student 2019, you will be able to use it forever.

There are four "Business" Office types. They are all subscriptions, and the fee is based on a "per/user" basis. The four different types have different products and services.

If you are a subscriber, there might be new additions or changes since this book was published. As a subscriber, you might not notice when a new feature has been added. So how do you know when a feature is added? You can click the Help tab and then What's New Or File ➤ Account ➤ What's new.

How can you be sure that you have the latest updates? Office 365 is updated automatically by default. If you want to make sure you have the latest updates, click the File menu, then click Account in the left pane, click the Office Updates button, and select Update Now. (Remember there are only security updates for Excel 2019.)

History of Spreadsheets

VisiCalc (short for Visible Calculator) was the first computerized spreadsheet available to the public. It was created by Dan Bricklin and Bob Frankston in 1979 for the Apple IIe and then released in 1981 for the newly created IBM PC. Up until this point, sales of personal computers had been slow because there wasn't a lot you could do with them. Early PCs were very expensive, and there weren't any prewritten applications. They were mostly purchased by computer programmers who thought they were fascinating and sought the ability to practice programming at home. At this point, programmers worked on large-scale computers called mainframes. At that time, you couldn't go to a store and buy software like you can today. Back then, company programmers wrote all the programs that the company needed themselves. Each company wrote their own payroll program, their own inventory program, and so on. They didn't share the software with each other. With VisiCalc, businesses now had a product that could be of great benefit. Sales of personal computers took off. VisiCalc became the world's first Super App. VisiCalc also started a revolution in businesses being started for the sole purpose of creating software to be sold to the public.

The Lotus 1-2-3 spreadsheet program was released in 1983. It was made specifically for the IBM PC. It was faster and had better graphics than VisiCalc and soon replaced it. Lotus 1-2-3 greatly increased the sales of the IBM PC.

Microsoft Excel has dominated the spreadsheet market since the 1990s.

This Book

An Excel book that taught you every possible option would be too large for you to carry. Throughout this book, you will be reading about an Excel topic followed by a practice. You can learn by reading, but to fully comprehend the different topics, you should do the exercises. Many illustrations are included to make it easier to follow along and comprehend.

As you go through the material in the book, exploring the different options and trying different things, think about how you could use what is being taught in different environments. Remember to think consciously about each step in the exercises and not to just click places and enter text only because the book instructs you to. Excel is extremely powerful, capable of many things, so allow yourself to experiment. The goal is to use Excel to solve your own problems beyond the book's exercises, so think about how you can leverage what you're learning as you read and work through the exercises. Be innovative and think about how you could use Excel to solve various problems.

Excel Navigation Basics

Before we do anything with Excel, let's get to know the main parts of the program. Figure 1-1 shows an Excel workbook. The arrows have been added to highlight the purpose of the different areas of the workbook.

Note Your Excel program might not match perfectly with this workbook. Microsoft is constantly making changes to Office 365 through the Internet. For example, your Ribbon might look a little different. The Ribbon for Excel 2019 looks a little different than the Ribbon for Excel in Office 365.

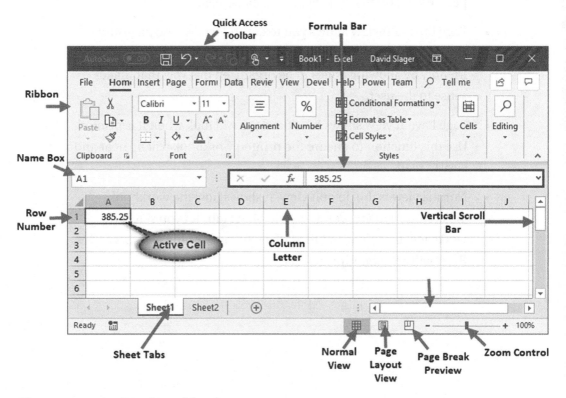

Figure 1-1. *An Excel workbook*

Figure 1-1 shows essential components of the workbook and worksheet. I'll work clockwise around the sheet starting with the Quick Access Toolbar (QAT):

- The QAT is a shortcut tool for storing the commands you use most often and want quick access to.

- The formula bar shows the formulas for the current selected cell. Excel displays the result of the formulas, not the formula itself, in each cell. This bar lets you see the formula that is producing the cell results.

- The vertical scroll bar and horizontal scroll bar allow you to move through the worksheet page.

- The Zoom control, Page Break Preview, Page Layout view, and Normal view are buttons that allow you to control how you are viewing the worksheet:

  - The Zoom control lets you increase or decrease the size (Zoom percentage) of the worksheet on your screen.

  - Page Break Preview allows you to control where one page ends and another begins. This helps make the worksheet more user-friendly by allowing pages to be organized in a way that makes sense to the user.

  - Page Layout view shows how the page will look when it is printed. Use this function to ensure the printed workbook will be neat and easy to read.

  - Normal view is the default view. It shows how the workbook looks while you are working on it. Sheet tabs let you select the worksheet that you want to work on or view. Many workbooks in Excel will have multiple sheets.

- The row number tells you what row you are on in the workbook. Excel has a potential of 1,048,576 rows. Columns are identified by letters. There are 16,384 columns in an Excel spreadsheet. This means that a single worksheet contains more than 17 billion cells. Each cell can hold 32,767 characters. How many worksheets you can have in a workbook depends upon your computer's available memory. Each cell is identified by an address which consists of the column letter and the row number.

- The Name box displays the address of the cell where you are at the moment. The Name box in Figure 1-1 displays **A1** which is the address of the current active cell.

- The Ribbon provides access to all of Excel's capabilities. The Ribbon will be discussed in much greater detail later in this chapter and in subsequent chapters.

Creating, Saving, and Opening Workbooks

The first step is to create a workbook. You should consider what you want in the workbook and how you will use the data. Select a name for the workbook that reflects its purpose. This will make it easy to identify as you open and use again. Next, make sure to save your work as you go along adding new data. Getting into the habit of frequently saving your work will prevent wasted time.

We'll start our Excel journey by creating a new workbook and then examine the different parts of the workbook. How you start Excel depends upon your operating system. Excel starts just like any other application you use.

EXERCISE 1-1: CREATING AND SAVING A WORKBOOK

In this exercise, we'll create a simple blank workbook and save it.

1. Start your Excel program. If you are using Windows 10, choose Start, then scroll to Excel, and click it. Figure 1-2 shows the opening window.

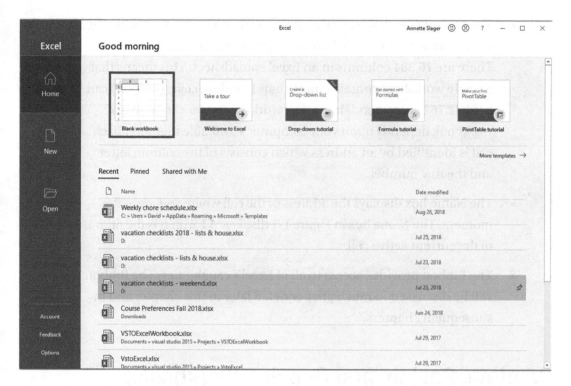

Figure 1-2. *Excel starting window*

Note   The file names on your window will not be the same as those shown in Figure 1-2 because they are the names of the files that I have opened.

---

2.   When you start a new workbook, you have two choices:

- You can start with a blank workbook by clicking the Blank workbook button.

- You can click one of the many template buttons to create a new workbook based on the templates you selected.

Click Blank workbook for this exercise.

3.   Click any cell, type any value you want, and then press the Enter key.

4.   Click the Save button 🖫 located on the Quick Access Toolbar at the top left of your window (see Figure 1-1). The first time you save the workbook Excel will display the File tab with Save As highlighted. See Figure 1-3.

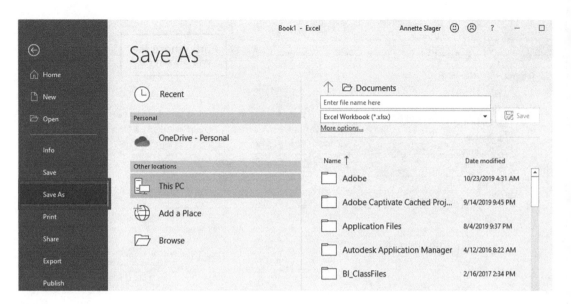

Figure 1-3. *Places to save your workbook*

Note   You will learn a lot more about saving workbooks from the File tab (known as the Backstage) in Chapter 6.

You can save your file to many different locations. If you are using this book in a school, you should ask your instructor where to save your files. Clicking the Browse button will bring up the Save As window.

    5.   Click Browse.

Your Save As window may look slightly different than the one in Figure 1-4 depending upon your version of Windows. If you have used File Explorer before, this window works the same way. Click a folder and then, if need be, click a folder within that folder as you build the path to where you want the file to be saved. The path is the drive and folders that you must go through to get to the file. The workbook being saved in Figure 1-4 is set to be saved in the Documents folder. You may want to store the file directly in the My Documents folder or you may want to create a folder under your My Documents folder and then store your files in it. The File name is Book1.xlsx by default. You should change the name to something more relevant to what you are working on. The File name can be changed by dragging across the word Book1 and then typing a new name.

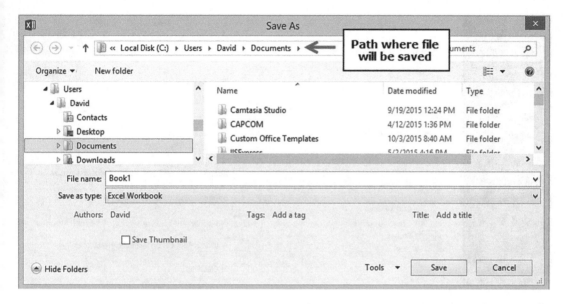

Figure 1-4. *Save As window*

6. Create the path to where you want your workbook saved by clicking the folders in the left pane of the Save As window until you are at the location where you want to store your files.

7. Change the File name from Book1 to MyFirstWorkbook. Excel adds an extension of .xlsx to the file name. Make sure the **Save as type** is Excel Workbook(*.xlsx).

8. Click the Save button.

9. Enter any value you want in another cell and then press Enter.

10. Click the Save button located on the Quick Access Toolbar. Since you previously saved the file, the Save As window doesn't appear. Excel saves the file with all the changes you made to it.

11. Close Excel by clicking the X in the upper right corner.

This exercise showed you the basics of creating a workbook. Next, you'll practice opening the same workbook to continue working on it.

EXERCISE 1-2: OPEN A WORKBOOK AND CREATE A NEW ONE

In this exercise, we'll open the file we created in the last exercise, make some changes, and then save with a new name. This will create a new workbook.

1. Start Excel.

The window in Figure 1-5 displays the most recently opened workbooks. The MyFirstWorkbook file you created in the previous exercise is at the top of the list.

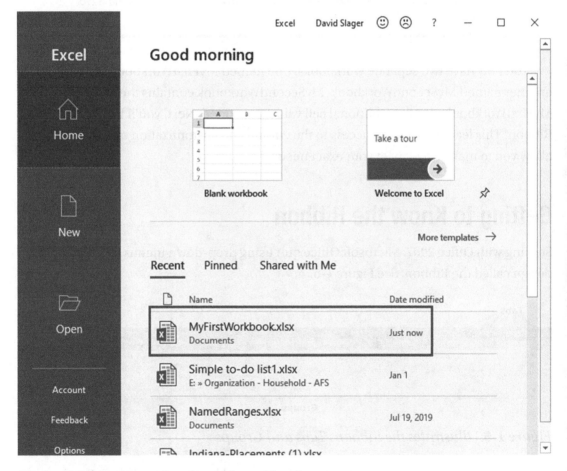

Figure 1-5. *Open a recently used workbook*

2. Click MyFirstWorkbook. The workbook opens.

Next, we will add additional cell values to this workbook and then save it under a different name.

11

<u>Create Another Workbook Under a Different Name</u>

1.   Enter any value you want in a blank cell.

2.   Click the Ribbon's File tab.

3.   In the left pane, click Save As.

4.   Click Browse.

5.   You can save this file in the same Documents folder where you saved
     MyFirstWorkbook. Change the name to MySecondWorkbook and click the Save
     button.

You now have two separate workbooks: one named MyFirstWorkbook and
another named MySecondWorkbook. MySecondWorkbook contains the same data as
MyFirstWorkbook plus the additional cell value you added. Next, you'll learn about the
Ribbon. This feature gives you access to the editing and customization options which
allow you to make Excel meet your exact needs.

Getting to Know the Ribbon

Starting with Office 2007, Microsoft Office quit using drop-down menus in favor of a tab
design called the Ribbon. See Figure 1-6.

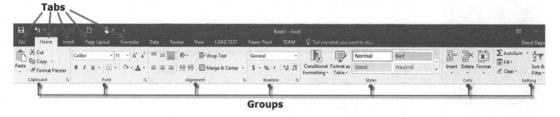

Figure 1-6. *Illustrates the Ribbons Tabs and Groups*

The Ribbon consists of tabs, groups, and command buttons. The default Excel
Ribbon contains the tabs File, Home, Insert, Page Layout, Formulas, Data, Review, View,
and Power Pivot. Your Ribbon may include additional tabs depending upon your setup.
Each tab is broken up into groups. The buttons are organized within those groups. Excel
lets you alter the Ribbon to meet your own needs. You can create your own tabs or add

new groups within your tabs. You can place the commands you use most often in your own groups. There have been many complaints about Microsoft constantly changing the look of the Ribbon and its tabs. Your Ribbon may look a little different than this.

Ribbon Contextual Tabs

In addition to the tabs that you see when you start Excel, there are many other tabs that appear and disappear depending on what you are working on. These are called **context-sensitive tabs** because the tabs are displayed based on the *context* in which you are using them. These context-sensitive tabs will appear when you are working on such things as charts, drawings, pictures, pivot tables and pivot charts, SmartArt graphics, header or footers, and so on. Contextual tabs have an additional label that appears above the tab. The labels have different background colors. Figure 1-7 shows the contextual Format tab that appears when you are working with pictures. It has a label of **Picture Tools** above it. Figure 1-8 shows the two additional tabs that appear when you click a chart in your worksheet: a Design tab and a Format tab.

Figure 1-7. *Additional tab displayed when an image is selected*

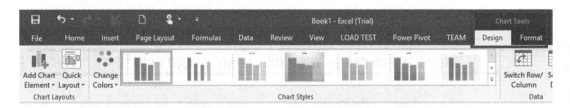

Figure 1-8. *Additional tabs displayed when a chart is selected*

These additional tabs appear under a Chart Tools label. These tabs appear only as long as the object that caused them to appear is active. Clicking off of the object to something else removes the tabs.

Resizing the Ribbon

Resizing the Excel window resizes the Ribbon. As you shrink the size of the window, the buttons start to align vertically as shown in Figure 1-9.

Figure 1-9. *Buttons aligning vertically*

Shrinking the size of the Ribbon further as shown in Figure 1-10 makes the buttons disappear. Clicking an arrow in the group will make that group's buttons display below the Ribbon.

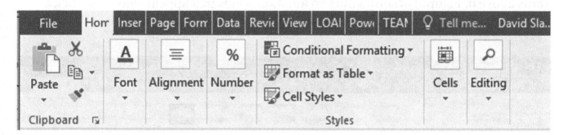

Figure 1-10. *Resized Ribbon may not show buttons*

As you work, you may need to adjust the size of the Ribbon to accommodate your working space. The next exercise shows you how.

EXERCISE 1-3: RESIZE THE EXCEL WINDOW AND RIBBON

If you see the Restore Down button (Figure 1-11) in the upper right-hand corner of your window, that means that your window is currently at its maximum size. You can't shrink the size of the window while your screen is maximized.

Figure 1-11. *Restore Down button*

1. If the Restore Down button is displayed, click it.

2. Move your cursor to the right edge of the window. The cursor will change to a double arrow. Drag the right edge toward the left to shrink the window. As you drag the window, notice how the buttons start aligning vertically, and as you drag farther to the left, the buttons in the group start disappearing.

3. Click the Maximize button (Figure 1-12).

Figure 1-12. *Maximize button*

Your window should now be maximized, and the Ribbon should be displaying all of its command buttons.

Using Dialog Box Launchers

At the bottom right corner of some Ribbon groups are boxed arrows. See Figure 1-13. They are called dialog box launchers. Dialog box launchers present a set of options to select from. A dialog box is a window that has options to select from that you must respond to before you can return to another window. It usually has an OK button and a Cancel button.

Figure 1-13. *Dialog box launchers*

Clicking the dialog box launcher for the Font group, Alignment group, or Number group will bring up the Format Cells dialog box in Figure 1-14.

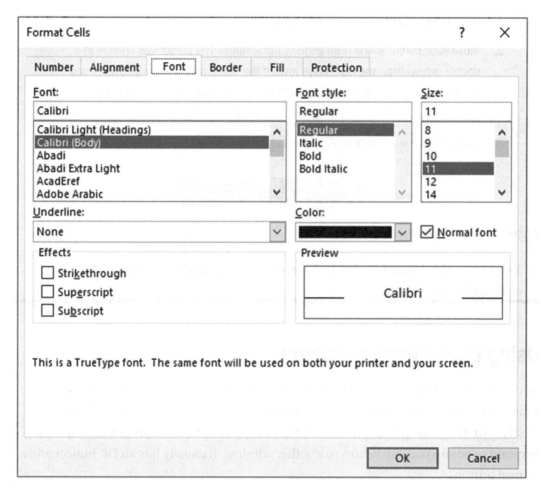

Figure 1-14. *Format Cells dialog box started from the Font dialog box launcher*

If you click the Font group's dialog box launcher, the Font tab will be selected. If you click the Alignment group's dialog box launcher, the Alignment tab will be selected. We will work with dialog box launchers in later chapters.

Minimizing and Hiding the Ribbon

If you think the Ribbon is taking up too much of your window space, you can either minimize it so that it only displays the tab names or you hide it completely. Clicking the Ribbon display button in the upper right-hand corner of the Excel window displays three options. See Figure 1-15.

Figure 1-15. *Three Ribbon options*

The options are

- Auto-hide Ribbon—Puts your Excel workbook in full-screen mode and hides the Ribbon completely. When your Ribbon is in Auto-hide mode, you will see this ⋯ ⊡ ✕ in the top right corner of your window. Clicking the three dots or anywhere to the left of them at the top of the screen will bring back the Ribbon. When you click the spreadsheet, the Ribbon will disappear again.

- Show Tabs—Shows only the Ribbon tabs. Clicking a tab will display the groups with their buttons. Clicking anywhere on the spreadsheet will hide the groups and their buttons again. Pressing Ctrl + F1 works like a toggle switch while in this mode by hiding and unhiding the groups and buttons.

- Show Tabs and Commands—This option makes the Ribbon display in full at all times.

You can also collapse the Ribbon by clicking the up arrow ⌃ at the far right side of the Ribbon.

Using Ribbon Shortcuts

Pressing the Alt key on your keyboard brings up the shortcut keys as shown in Figure 1-16 for each of the Tabs as well as the Quick Access Toolbar. Keying one of the shortcut keys for one of the Quick Access Toolbar buttons will perform that command.

Entering the shortcut key for a Ribbon tab will make that tab active. As you can see in Figure 1-16, pressing the F key will make the File tab active, and pressing the H key will make the Home tab active. Notice that the N key is used for the Insert tab. Most of the letters have no relation to the names of the options.

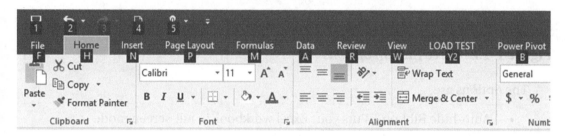

Figure 1-16. Shortcut keys for Ribbon tabs and the Quick Access Toolbar

When you press the letter key for the tab you want to use, shortcut keys appear for every option on that tab. Figure 1-17 shows the shortcut keys for all of the options on the Home tab. Notice you may need to enter more than one letter for the shortcut. It is interesting to note that even though the Home tab is selected by default, you still have to press H for the tab in order for it to show all the shortcuts under it. It will not just show the commands under the selected tab on its own when you first press Alt.

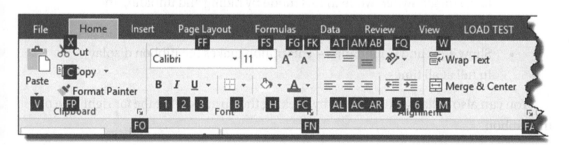

Figure 1-17. Shortcut keys for Home tab commands

Pressing a shortcut key from a tab performs that command, or it will display more shortcut keys if the command has more options available. For example, using the keyboard to apply the Merge & Center option, you would do the following:

1. Press the Alt key and then press the **H** key to select the Home tab.

2. Press **M** to select the Merge button.

3. Then press **C** to select the Merge & Center option. See Figure 1-18.

Figure 1-18. *Shortcuts for commands under the Merge & Center category*

Note   Even if the tab is active for the command you want to use, you must still press the Alt key and then the shortcut key for the tab. In other words, if the shortcut key isn't displayed, you can't use it.

You should now be able to use the Ribbon to move around and enter data into the worksheet. The Ribbon drives the functionality of the Excel program.

Besides using commands from the Ribbon, you can select them from a Quick Access Toolbar (QAT). This is what we'll cover in the next section.

Quick Access Toolbar

The Quick Access Toolbar provides a quick and convenient place for you to store and access

- The tools that you use most often

- Tools that are not normally found on the Ribbon

- Macros that you create

By default, the Quick Access Toolbar shown in Figure 1-19 is located above the Ribbon in the upper left-hand corner of the Excel window.

Figure 1-19. *Quick Access Toolbar (QAT)*

By default, the Quick Access Toolbar (QAT) displays

- The AutoSave button

- The save button which uses a diskette for an icon

- The undo and redo buttons

- A drop-down button from which you can select other tools to be displayed on the QAT

Clicking the drop-down button on the right side of the QAT displays the Customize Quick Access Toolbar from which you can select other buttons to be added to your QAT. See Figure 1-20.

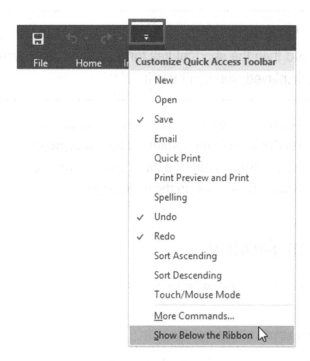

Figure 1-20. *Click the drop-down button to select Items to Add to or Remove from the QAT*

The QAT can be moved below the Ribbon by selecting the Show Below the Ribbon option from the drop-down menu. This may be a better place for it since it will provide more room for additional tools.

AutoSave

The AutoSave stays grayed out unless you are working with a workbook that is stored on OneDrive. When AutoSave is on, every time you make a change to your workbook, it is immediately saved to your workbook on OneDrive. AutoSave should only be turned on when you are sharing the workbook on OneDrive with others and you want to give them immediate access to any changes you are making.

Switch Between Touch and Mouse Mode

Because many monitors today are touch screen, Microsoft has added a Touch/Mouse Mode button. This button can be added to the QAT by selecting it from the QAT drop-down menu. See Figure 1-20. Clicking the down arrow of the Touch/Mouse Mode button displays the two options shown in Figure 1-21.

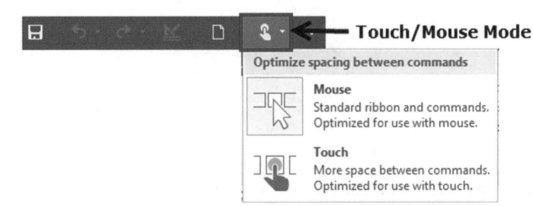

Figure 1-21. Options for optimizing Ribbon for using the Mouse or Touch monitor

The Touch option is for those users who are using touch monitors. Selecting the touch option places more space between the Ribbon buttons as shown in Figure 1-22, making it easier to select the correct button with your finger.

Figure 1-22. *Ribbon set up for Touch screen monitors*

Changing the Touch and Mouse mode in any of the Microsoft Office products changes it for all office products.

You can easily remove a button from the Quick Access Toolbar by either right-clicking the button you wish to remove and selecting Remove from Quick Access Toolbar or you can click the drop-down button, and then click the checked item you wish to remove.

Your QAT is not limited to the items appearing in the Customize Quick Access Toolbar menu. Buttons that are on the Ribbon can be added to the QAT by right-clicking a Ribbon button and then selecting Add to Quick Access Toolbar from the menu. See Figure 1-23.

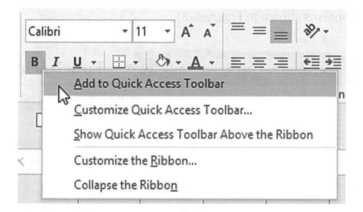

Figure 1-23. *Adding button from Ribbon to the QAT*

The order in which the buttons appear on the QAT can be rearranged. You can save your QAT customization to a file and then later import it into another workbook.

Three ways to get to the QAT customizations in the Excel options window are

- Click the drop-down arrow on the QAT and then select More Commands....

- Right-click the Ribbon and then select Customize Quick Access Toolbar....

- Click the File tab of the Ribbon. Select Options. Select Quick Access Toolbar from the left side of the Excel Options window.

EXERCISE 1-4: USING THE QUICK ACCESS TOOLBAR

In this exercise, you will add command buttons to your QAT.

1. Click the drop-down button of the Quick Access Toolbar and then select **Print Preview and Print** from the Customize Quick Access Toolbar menu.

2. Click the drop-down button of the Quick Access Toolbar and then select **New**.

3. Click the drop-down button of the Quick Access Toolbar and then select Open.

   The Print Preview and Print, New, and Open buttons have been added to the end of your Quick Access Toolbar. See Figure 1-24.

Figure 1-24. *Quick Access Toolbar*

Notice that the tools appear in the order that they were selected.

4. Right-click the Print Preview and Print button 🖺 on the QAT and select **Remove from Quick Access Toolbar**.

5. Right-click the QAT and then select **Show Quick Access Toolbar Below the Ribbon**.

6. Click the **Review** Tab on the Ribbon. In the Proofing group, right-click the Spelling button and select **Add to Quick Access Toolbar**. Your QAT should now appear as follows. See Figure 1-25.

Figure 1-25. *Quick Access Toolbar*

Identifying the Current Cell

Columns are represented by letters. Rows are represented by numbers. A combination of a column letter and a row number gives each cell a unique address. The first cell in a worksheet would have an address of A1. A cell that is at the intersection of column G and row 5 would have a cell address of G5. The cell address is also called a cell reference. Individual cells contain text, numbers, or formulas. The result of a formula is displayed in the same cell where you inserted the formula.

The current (active) cell in Figure 1-26 is B6. The current cell can be identified by the following:

- Its border is bolded.

- Its column head and row head are highlighted.

- The address appears in the Name Box.

- The cell's value or formula is displayed in the formula bar.

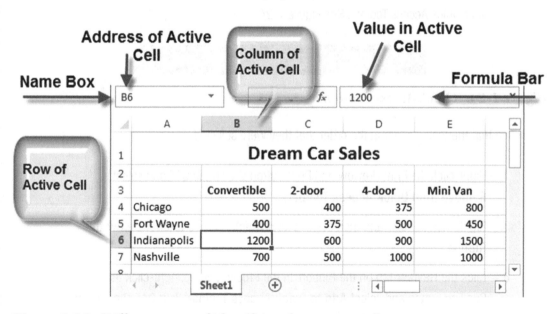

Figure 1-26. *Different ways of identifying the current cell*

Once you've identified the current cell, you are ready to start entering your data!

Entering Data into a Worksheet

The data you enter in a cell is not accepted until you do one of the following:

- Press the Tab key—Cursor moves to the next cell.

- Press the Enter key—Cursor moves to the next cell.

- Any of the arrow keys—Cursor moves to the next cell in the direction of the arrow.

- Click the check mark icon on the formula bar—Cursor remains in the cell.

- Pressing Ctrl + Enter—Cursor remains in the cell.

Note   You can't format the data in a cell until the data has been accepted.

If you want to overwrite all the data in a cell, you can click the cell and type the new data. If you only want to change part of the data in a cell, you need to be in Edit mode. Double-clicking a cell puts it in Edit mode; pressing F2 will do the same. As you are typing data in a cell, the data appears in both the active cell and the formula bar. Because the data appears in both the cell and the formula bar, making changes in either location will update the cell data.

EXERCISE 1-5: ENTERING AND ACCEPTING A CELL ENTRY

In this exercise, enter data in cells and use different options for accepting the entries.

1. First, enter some column headings and use the Tab key to accept them:

    a. Type **Assets** in cell A1. Notice that as you're typing the text in cell A1, it is also being typed into the formula bar. Press the Tab key.

    b. Type **Cash** in cell B1. Press the Tab key.

    c. Type **Supply** in cell C1. Press the Tab key.

    d. Type **Land** in cell D1. Press the Enter key. Cell A2 becomes Active.

2. Next, type **Liability** in cell A2, but don't press the Tab key. Move your cursor over the check mark in the formula bar. If the data in the cell hasn't been accepted, it will change color. See Figure 1-27. Click the Enter button (the check mark). The data is accepted and the button becomes grayed out. See Figure 1-28.

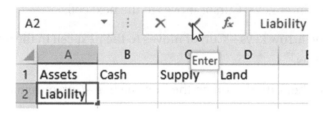

Figure 1-27. *Click the Enter button to accept the data*

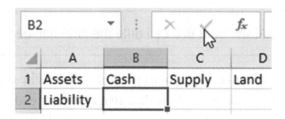

Figure 1-28. *Once the data is accepted, the button becomes grayed out*

3. Press the Tab key.

4. Type **Loan** in cell B2, but don't press the Tab key. Click the Cancel button ✕ on the formula bar. The entry is cleared. Type **Loan** in cell B2 again. Press Ctrl + Enter. Cell B2 remains the active cell. The cursor is still in cell B2, but you can't see it.

Note   Another way to cancel the text you are entering or to clear it even after it has been accepted is to move your cursor over the square at the bottom right of the cell (see the cell on the left). The cursor will change to crosshairs. Drag the cursor toward the center of the cell. The text will fade, as in the cell on the right, and when you let go of the mouse button, the text will be gone.

---

5. Press the Tab key. Notice the word **Ready** in the bottom left corner of the status bar. This means that the cell is ready for you to enter data into.

Note   If you don't see the word Ready, then right-click Excel's Status bar at the bottom of the window and select Cell Mode.

---

6. Type **Wages** into cell C2. When you start typing text in the cell, the word Ready on the status bar changes to **Enter**. Press the Tab key.

7. Double-click cell C1. Looking at the bottom left side of the status bar, you should see that you are in **Edit** mode. Change Supply to Supplies. It doesn't matter if you make the change in cell C1 or in the formula bar. Press Enter when you are done.

8. Click cell A2. Press the F2 key. This is another way of placing the cell in Edit mode. Change Liability to Liabilities. Press Ctrl + Enter.

9. Click once in cell D1. Since you didn't double-click, you are not in Edit mode. The status bar still shows Ready. Notice that the cursor does not display. Type the letter R. The word Land is cleared from the cell. Finish typing the word Replace. Press Ctrl + Enter.

10. Type **Use** in cell C5. Press the down arrow key.

11. Type **the** in cell C6. Press the left arrow key.

12. Type **arrow** in cell B6. Press the up arrow key.

13. Type **keys** in cell B5. Press the right arrow key.

Getting Help

Excel provides help to the user within the program. Screen Tips and the "Tell Me What You Want to Do" feature answer questions about formatting and entering data into your worksheet while you are working on it. Smart Lookup enables you to search on cell contents. One of the Tabs on the Excel Ribbon is the Help tab that contains additional help features.

Screen Tips

Screen tips are a quick way of getting snippets of information about the commands on the Ribbon and some other objects on the screen. Screen tips can be viewed by merely moving your cursor over an object. The screen tip shows the name of the object and what it does. Some of the screen tips also provide graphic illustrations to illustrate what the command does. Figure 1-29 shows a screen tip for the Merge & Center button, which is located on the Home tab in the Alignment group. If you wanted additional information, you could click the Tell me more link.

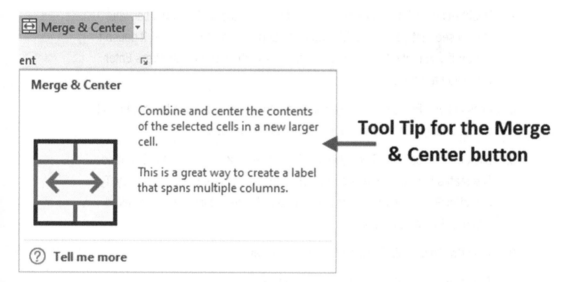

Figure 1-29. *Screen Tip for the Merge & Center button*

Excel's Tell Me What You Want to Do Feature

This feature was added to Office 2016. You will also find it in Word and PowerPoint. It is located in the tab area of the Ribbon. When you click the text box, it displays your most recent requests. You can select one of the requests or you can start typing into the text box. Excel displays what is available for the characters you have entered so far.

Let's say that I want help creating a chart. I start by entering a C into the text box. You are probably thinking why did the list in Figure 1-30 display items that don't all start with a C? The reason is that Math & Trig, More, and Financial are groups of formulas, some of which start with a C. New from Template appears because some templates start with a C. I can refine what appears in the list by entering more text.

Figure 1-30. *Suggested choices for Tell Me What You Want to Do feature*

After I have finished entering the word "Chart," the list has changed to items about charts. See Figure 1-31.

If a range of data was already selected like shown in Figure 1-32 when I clicked Create Chart in Figure 1-31, then an Insert Chart window as shown in Figure 1-33 would display from which I could pick the chart I wanted.

If I didn't have any data selected when I clicked Create Chart, then I would get a message saying how to create a chart.

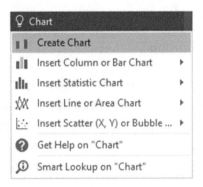

Figure 1-31. *Provided suggestions after entering the word Chart for Tell Me What You Want to Do feature*

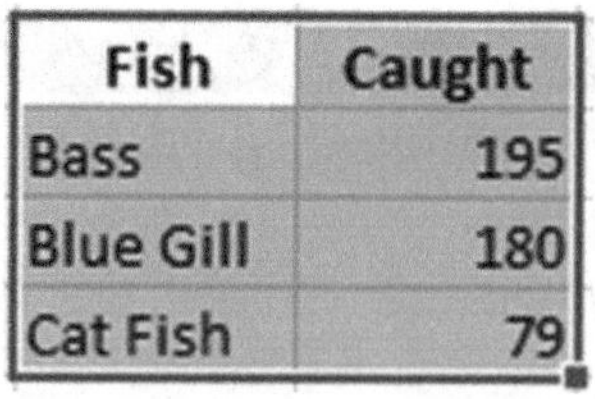

Fish	Caught
Bass	195
Blue Gill	180
Cat Fish	79

Figure 1-32. *Data selected before entering the word Chart for Tell Me What You Want to Do feature*

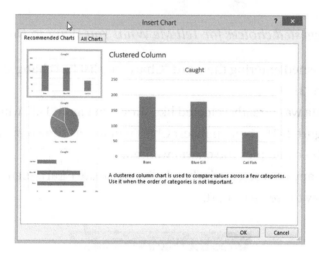

Figure 1-33. *Insert Chart window*

The last item in the list in Figure 1-31 is Smart Lookup on "Chart". Clicking Smart Lookup performs an Internet search for what you have entered in the text box.

You can also use Smart Lookup to do a search for text that you have entered into a cell by right-clicking the cell and selecting Smart Lookup. Figure 1-34 shows the results of having selected Smart Lookup from a cell that contained the text **mammoth**. There are two tabs: Explore, which it shows by default, and Define. Clicking Define gives you the origin of the word and its various meanings as shown in Figure 1-35.

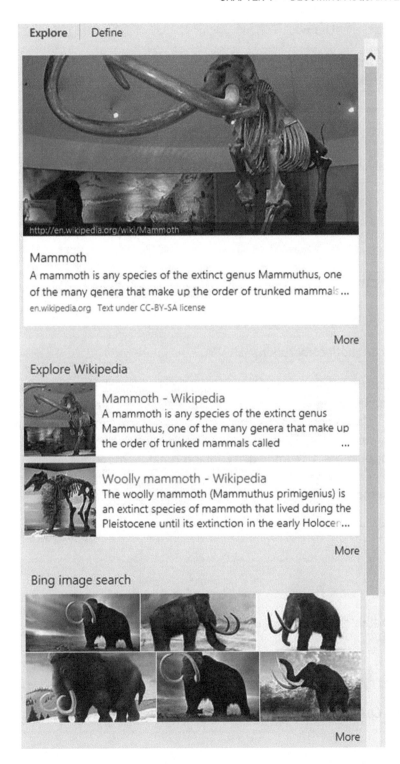

Figure 1-34. *Result of using Smart Lookup for the word Mammoth*

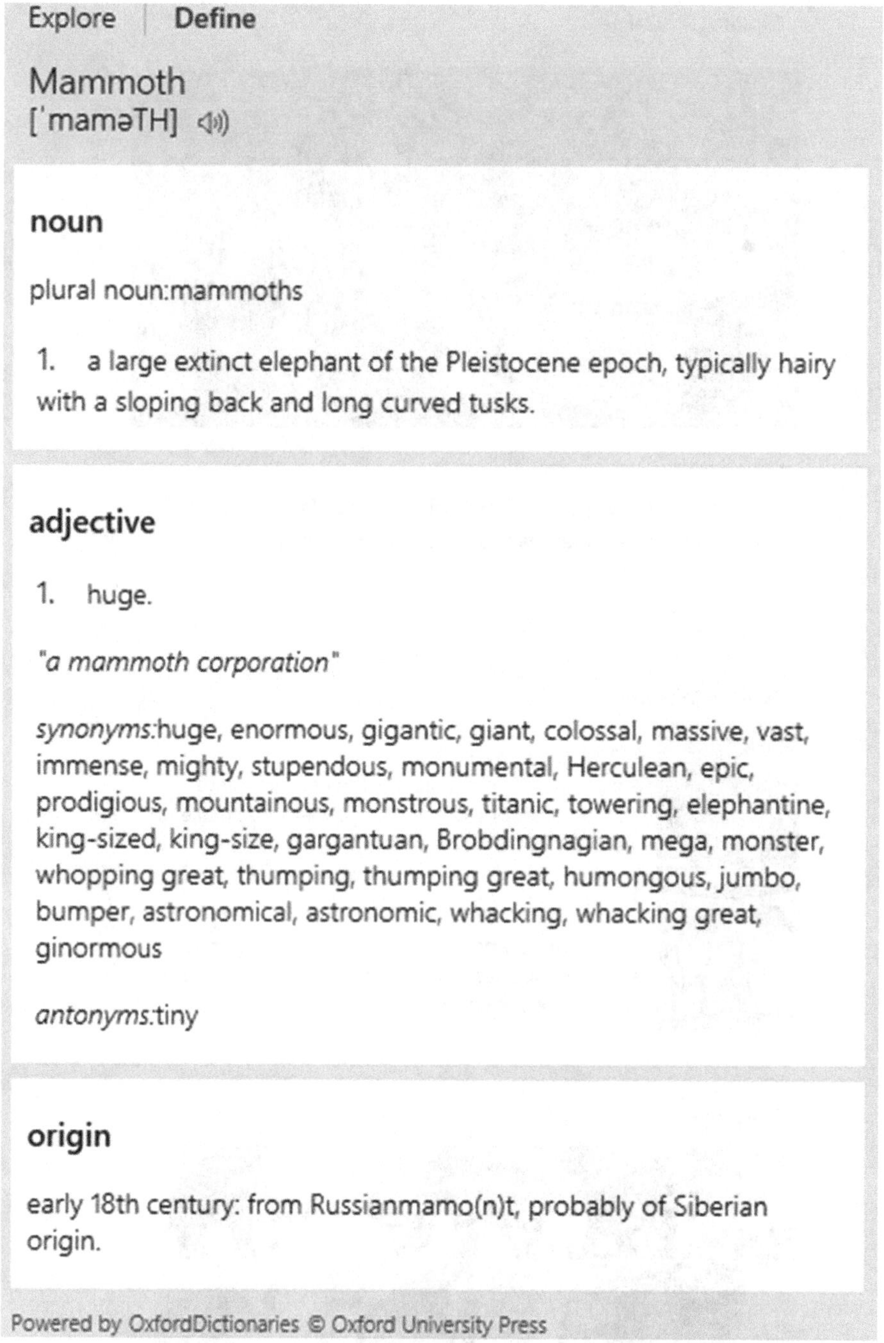

Explore | **Define**

Mammoth
['maməTH] ◁))

noun

plural noun:mammoths

1.   a large extinct elephant of the Pleistocene epoch, typically hairy with a sloping back and long curved tusks.

adjective

1.   huge.

"a mammoth corporation"

*synonyms:*huge, enormous, gigantic, giant, colossal, massive, vast, immense, mighty, stupendous, monumental, Herculean, epic, prodigious, mountainous, monstrous, titanic, towering, elephantine, king-sized, king-size, gargantuan, Brobdingnagian, mega, monster, whopping great, thumping, thumping great, humongous, jumbo, bumper, astronomical, astronomic, whacking, whacking great, ginormous

*antonyms:*tiny

origin

early 18th century: from Russianmamo(n)t, probably of Siberian origin.

Powered by OxfordDictionaries © Oxford University Press

Figure 1-35. *Origin of the word Mammoth and its various meanings*

Help Tab

Excel's latest update includes a tab dedicated to just help and support. You can also provide feedback to Microsoft telling them what you like and don't like and suggestions for improving Excel. The Help tab contains only one group. It is shown in Figure 1-36.

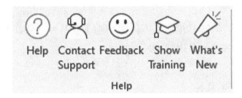

Figure 1-36. *Help tab*

What's New

Clicking the What's New button shows Office 365 users the newest features that have been added to Excel. See Figure 1-37.

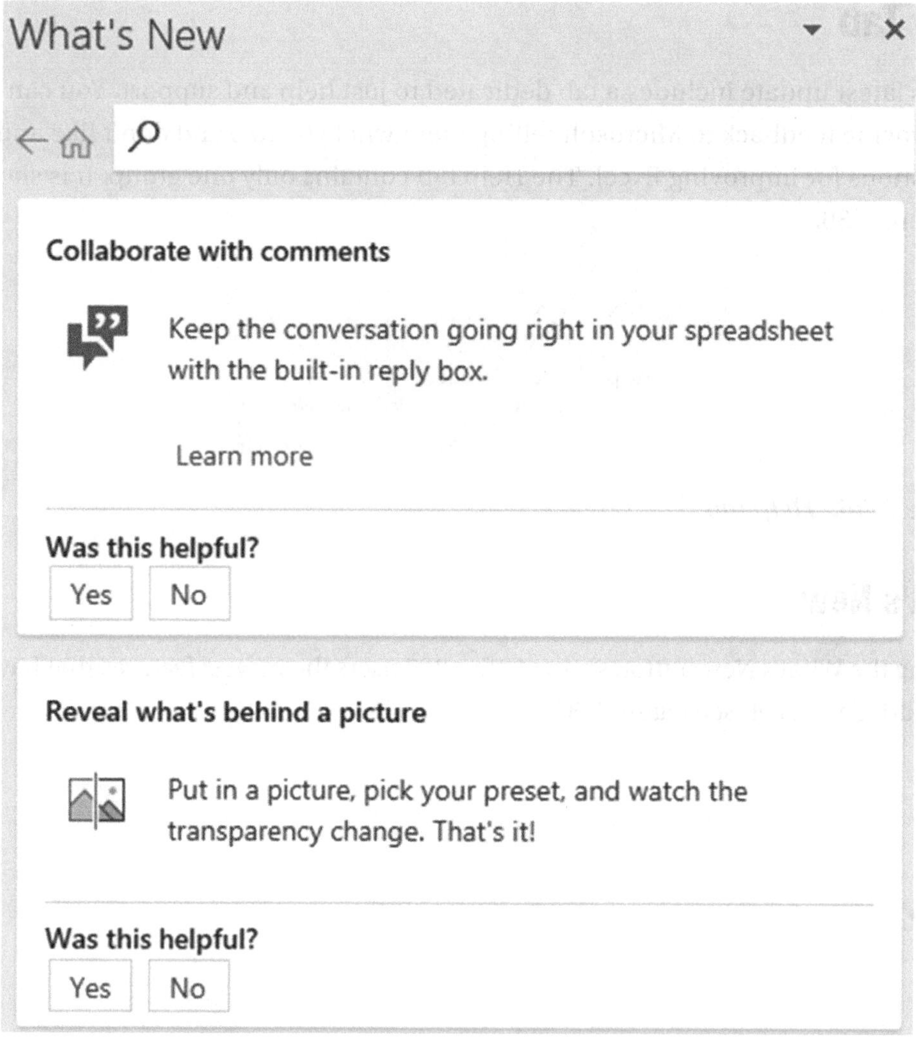

Figure 1-37. *What's New*

Show Training

The Show Training option brings up a series of training categories. See Figure 1-38.

Figure 1-38. *Training Options*

When you click one of the categories, you will see a list of help topics for that category. Clicking the Quick start category displays the topics shown in Figure 1-39. Clicking any of these topics gets you help for that topic.

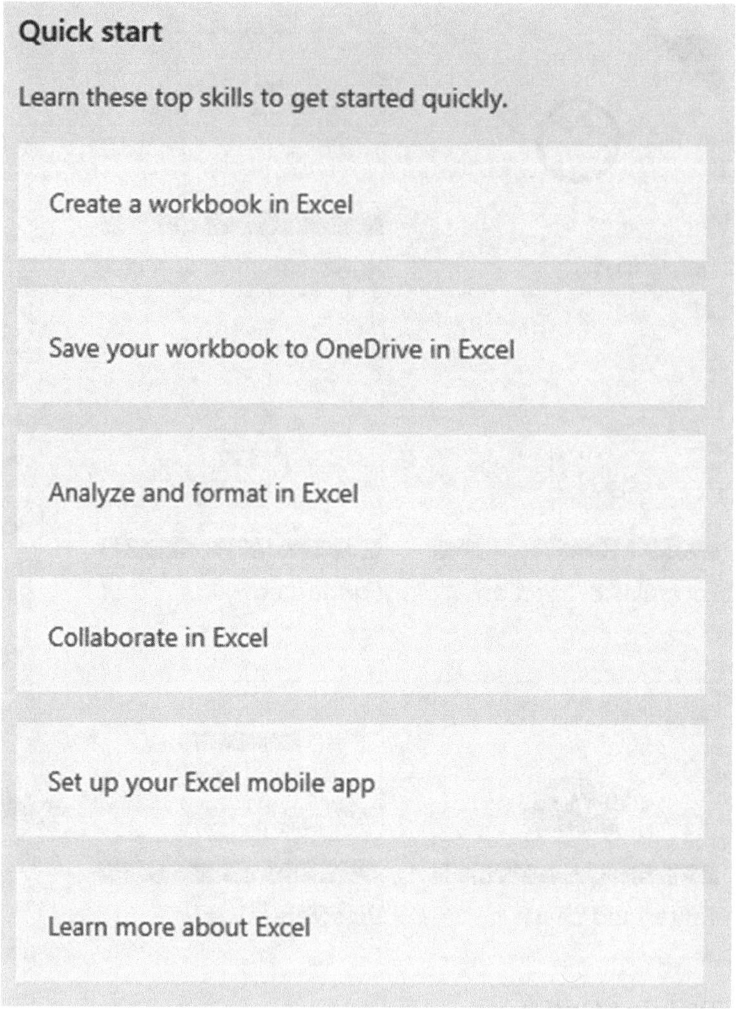

Quick start

Learn these top skills to get started quickly.

Create a workbook in Excel

Save your workbook to OneDrive in Excel

Analyze and format in Excel

Collaborate in Excel

Set up your Excel mobile app

Learn more about Excel

Figure 1-39.   *Quick start topics*

Contact Support

The Contact Support option brings up a text box in which you enter the problem you are having. Clicking the Get Help button after you have entered your problem will display some self-help solutions for you to try. If this still hasn't resolved your problem, you can click Live Chat at the bottom of the window to get help from a Microsoft support person. See Figure 1-40.

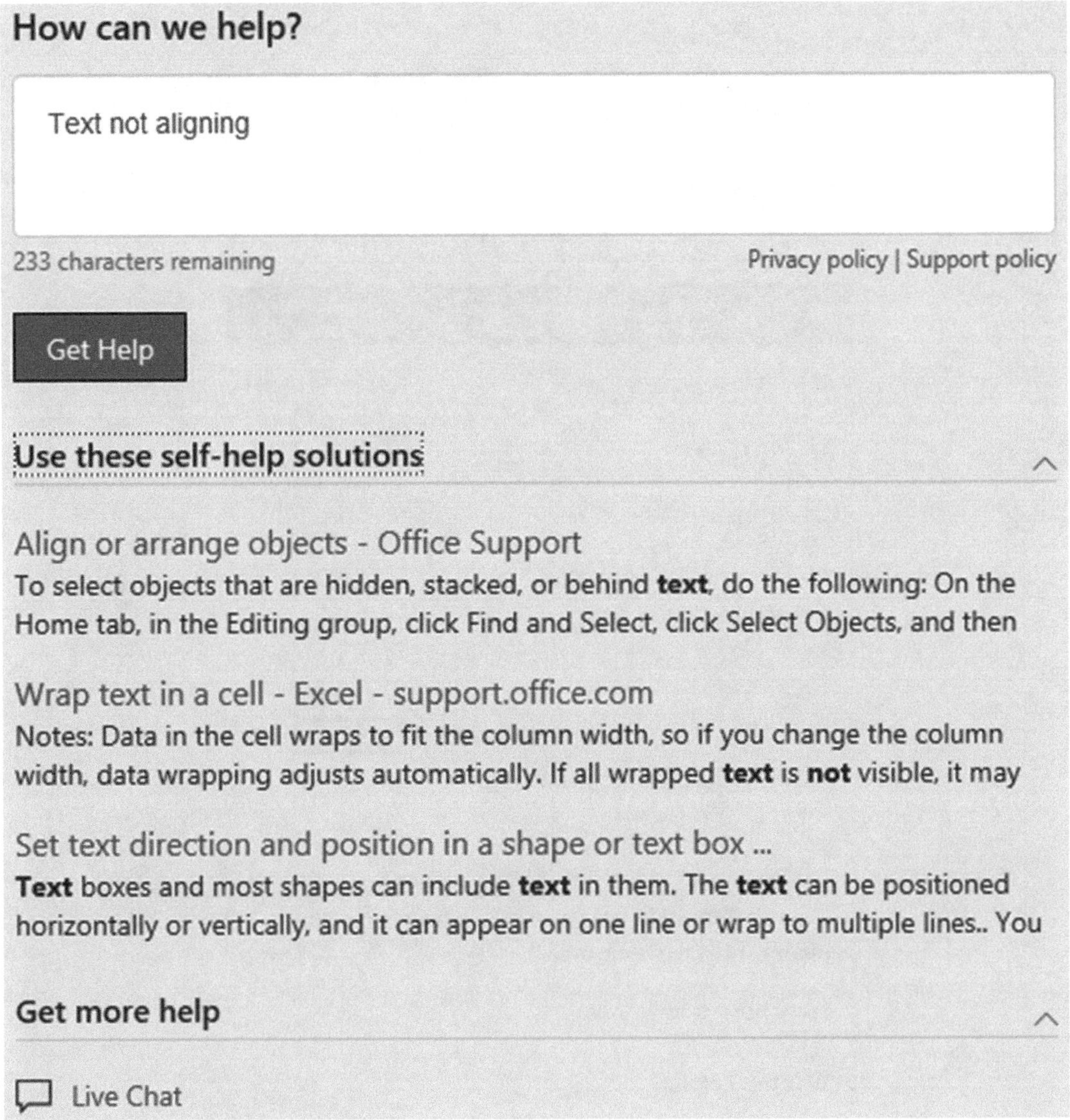

How can we help?

Text not aligning

233 characters remaining   Privacy policy | Support policy

Get Help

Use these self-help solutions   ∧

Align or arrange objects - Office Support
To select objects that are hidden, stacked, or behind **text**, do the following: On the
Home tab, in the Editing group, click Find and Select, click Select Objects, and then

Wrap text in a cell - Excel - support.office.com
Notes: Data in the cell wraps to fit the column width, so if you change the column
width, data wrapping adjusts automatically. If all wrapped **text** is **not** visible, it may

Set text direction and position in a shape or text box ...
Text boxes and most shapes can include **text** in them. The **text** can be positioned
horizontally or vertically, and it can appear on one line or wrap to multiple lines.. You

Get more help   ∧

▭ Live Chat

Figure 1-40. *Contact support*

Help

Clicking the Help button on the Help tab provides another way of selecting a help topic
or entering a search topic. See Figure 1-41.

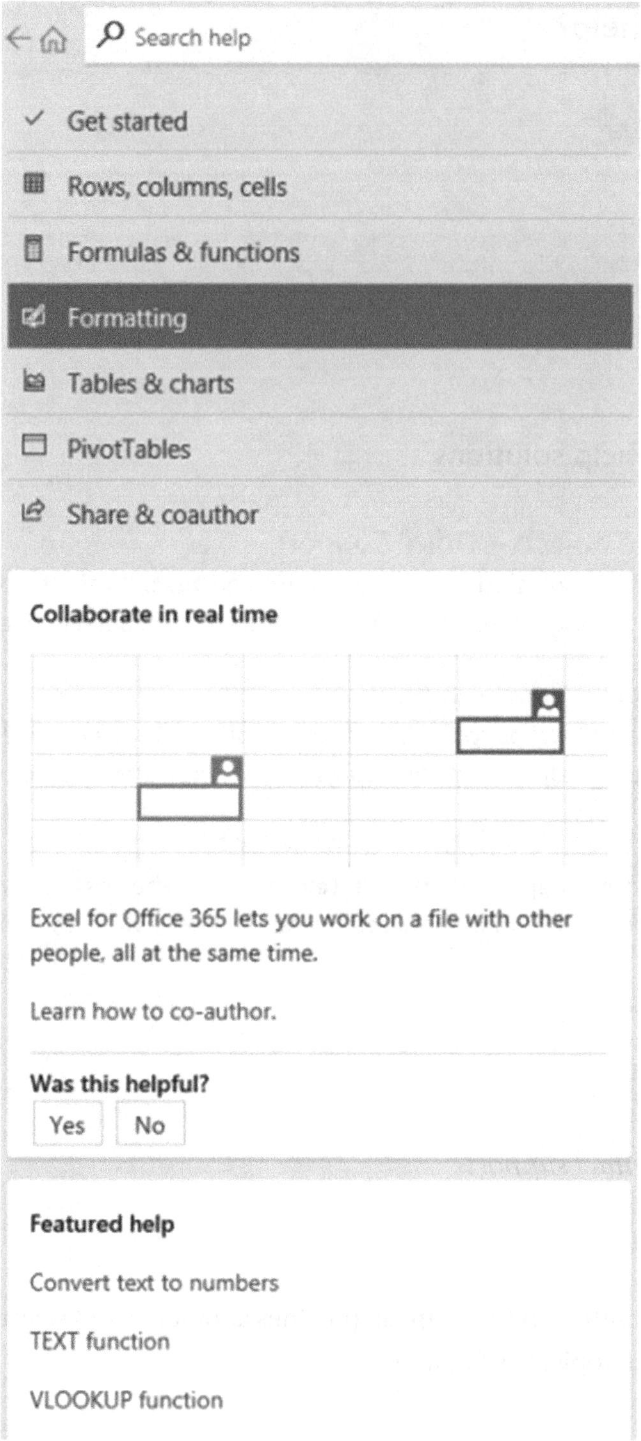

Figure 1-41. *Help*

Summary

You've learned how to create, save, and open a worksheet and how to use the Ribbon. You've practiced customizing the Ribbon and creating a Quick Access Toolbar to meet your specific needs. These are the basic skills you need to begin utilizing the powerful features in Excel. Excel has an abundance of help features. It provides multiple methods of getting to the same help topic. You can ask Excel for help on any topic. Screen Tips and the "Tell Me What You Want to Do" feature are two of the most convenient help features in Excel. You get intuitive help answering your questions about formatting and entering data into your worksheet while you are working on it. Smart Lookup lets you use Internet resources without leaving your worksheet. You can use any of the additional features available on the Help tab to get additional help or Microsoft support. In Chapter 2, you will practice moving around the worksheet. You'll learn useful shortcuts to help you move efficiently and easily around the worksheet, reducing the time you spend on repetitive tasks.

Summary

CHAPTER 2

Navigating and Working with Worksheets

In this chapter, you'll learn how to navigate the worksheet using the mouse and keyboard commands. You'll also become familiar with shortcuts that help you move quickly between cells and begin practicing working with multiple worksheets within the same workbook.

After reading and working through this chapter, you'll be able to

- Easily move around the worksheet

- Use shortcuts to get to different cells

- Reference an individual cell, a column, a row, or a range of cells

- Select cells using a mouse

- Select cells using a keyboard

- Select cells using the Name Box

- Name worksheets

- Add and remove worksheets

- Change a worksheet's tab color

- Select multiple worksheets

- Hide and unhide worksheets

- Reorder and copy worksheets

- Move through the worksheets

© David Slager and Annette Slager 2020
D. Slager and A. Slager, *Essential Excel 2019*, https://doi.org/10.1007/978-1-4842-6209-2_2

Excel is designed for the beginner to learn and use very quickly. The exercises guide you as you discover how to create, enter, and manipulate data. It is easy to make changes as you become familiar with Excel's features.

Moving Between Cells Using the Keyboard

Because there are so many cells, you need a quick way to move around the worksheet. Knowing your movement keys saves time by getting to the desired cell in the shortest amount of time. Click a cell to make it active or use one of the key combinations in Table 2-1 to jump to a particular cell.

Table 2-1. *Shortcut Movement Keys to Jump to Cells*

To Change the Active Cell	Press
Down to the next cell	Enter
Up to the next cell	Shift + Enter
Next cell	Tab
Previous cell	Shift + Tab
First cell in the row	Home
Next cell in the direction of the arrow	Up, down, left, or right arrow keys
Last cell in worksheet that contains data	Ctrl + End
First cell in worksheet	Ctrl + Home
Cell in next screen	Page Down
Cell in previous screen	Page Up
Moves to the last cell in the direction of the arrow, provided there are no blank cells between	Ctrl + Arrow

---

EXERCISE 2-1: MOVING AROUND THE WORKSHEET

This exercise will introduce you to the basic techniques for moving around using the keyboard.

Note   As you are going through this exercise, if you are on a computer where the arrow keys also work as other keys, you may need to click the Numeric Lock to turn it on or off depending upon what you want to do.

---

1. Start your Excel program.

2. Select File ➤ New, and then select Blank workbook.

3. Enter the data on the Sheet1 tab as shown in Figure 2-1.

⬚	A	B	C	D	E	F
1	Inventory					
2						
3	City	SUV	2-door	4-door	Mini Van	Totals
4	Denver	500	400	375	800	2075
5	Boston	400	375	500	450	1725
6	Chicago	1200	600	900	1500	4200
7	Nashville	700	500	1000	1000	1000
8						
9	Totals	2800	1875	2775	3750	11200

Figure 2-1. *Enter these values into the worksheet*

4. Click the Save button 💾 on the Quick Access Toolbar. Since you haven't previously saved this file, the Save As window displays.

    You can save the file in OneDrive which is a free storage location that Microsoft provides on the Web. The benefit of saving your workbook to OneDrive is that you can access the workbook anywhere in the world as long as you have access to the Internet. Another benefit is that you can allow other people access to it without having to email it to them. If you give others permission, they can make changes to the workbook or enter comments in the workbook.

5. Click **Save As** in the left pane. (1) Click **This PC**. Excel displays the Documents folder at the top and the folders on your drive below. (2) Enter Chapter2 for the name of your file that will be stored in the Documents folder. (3) Click the Save button. Excel displays locations you have used most recently as shown in Figure 2-2. Click **Documents**, which should be displayed because you used it in Chapter 1. Excel opens the Save As dialog box.

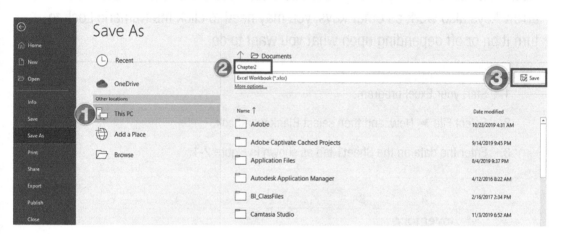

Figure 2-2. *The File tab, also called Backstage*

6. Enter *Chapter 2* in the File Name text box of the **Save As** window and click the Save button.

<u>Use the Tab Key</u>

Pressing the Tab key moves the cursor one cell to the right of the current cell.

1. Click cell C6 (sixth row in column C).

2. Press the Tab key to move to cell D6.

3. Press the Tab key to move to cell E6.

<u>Use the Shift+ Tab Key</u>

Holding down the Shift key while pressing the Tab key moves the cursor one cell to the left.

1. Hold down the Shift key while you press the Tab key to move to cell D6.

2. Hold down the Shift key while you press the Tab key to move to cell C6.

3. Hold down the Shift key while you press the Tab key to move to cell B6.

Use the Arrow Keys

Use Arrow keys to move to the next cell in the direction of the arrow key.

1. Click the up arrow key to move to cell B5.

2. Click the right arrow key to move to cell C5.

3. Click the down arrow key to move to cell C6.

4. Click the left arrow key to move to cell B6.

Ctrl + End Key

Holding down the Ctrl key while pressing the End key will move the cursor to the last cell that contains data.

1. Hold down the Ctrl key and press the End key to go to cell F9.

Ctrl + Home Key

Holding down the Ctrl key while pressing the Home key will always move the cursor to cell A1.

1. Hold down the Ctrl key and press the Home key to go to cell A1.

Page Down Key

Pressing the Page Down key moves the cursor down one screen.

1. Press the Page Down key twice to go down two screens.

Page Up Key

Pressing the Page Up key moves the cursor up one screen.

1. Press the Page Up key twice to go up two screens.

2. Click the Save button 💾 on the Quick Access Toolbar.

In this section, you've learned to use the keyboard to move between cells while entering data. Next, you'll learn how to select cells, first by using a mouse and then by using a keyboard.

Selecting Cells

There are several ways to select individual cells, a range of cells, a combination of individual cells and cell ranges, and even cells that are nonadjacent to each other. Selecting multiple cells is useful for applying formatting to multiple cells at the same time or when creating formulas or deleting or inserting a complete row or column.

While you are dragging across a block of adjacent cells (see Figure 2-3), Excel displays in the Name Box how many rows and columns are in that block. The Name Box in Figure 2-3 shows that the block of cells spans three rows and two columns.

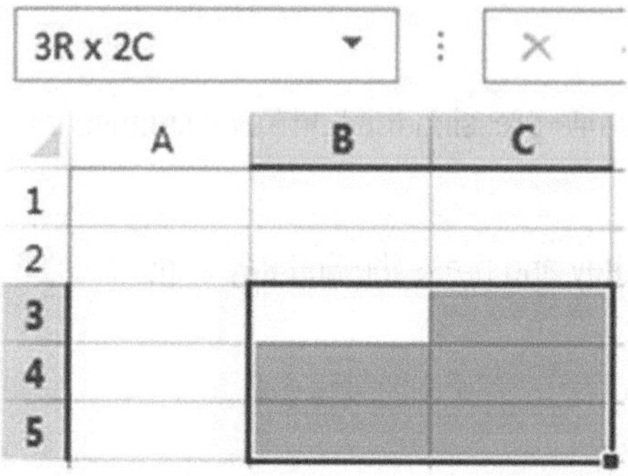

Figure 2-3. *Three rows and two columns*

Selecting Cells Using a Mouse

Table 2-2 tells how to select different worksheet ranges using a mouse.

Table 2-2. *How to Select Different Worksheet Ranges Using a Mouse*

To Select This	Do This
Column	Position the cell pointer on the column header (a letter) and then click the left mouse button.
Row	Position the cell pointer on the row header (a number) and then click the left mouse button.
Adjacent cells	Drag with mouse to select the cells in the range—or click the first cell of the range, hold down the Shift key, and click the last cell in the range.
Nonadjacent cells	Hold down the Ctrl key while clicking a column header, row header, or specific cells.
All cells in worksheet	Click the Select All button. 

Deselecting Cells Using a Mouse

Deselecting cells using a mouse for some reason doesn't work on laptop computers. If you are working on a desktop, try these: if you have a selected cell range and you wish to deselect some of the cells from that range, you can hold down the Ctrl key while you drag across the cells you want to remove from the selection. If you need to reselect any of those cells you have deselected or you want to add additional cells, continue holding the Ctrl key and select those cells.

EXERCISE 2-2: SELECTING CELLS USING A MOUSE

In this exercise, you select different ranges of cells. Although we'll cover formatting in detail later, you can get an idea of how to use them in this simple exercise.

Selecting Entire Rows or Columns

1. Open the workbook named Chapter 2, which you created in the previous exercise.

2. Position your cursor on column header B. You will see a down arrow; hold down the left mouse button and drag across column head C so that columns B and C are both selected. As you are dragging, you will see a box that displays the number of rows and columns that are currently selected. See Figure 2-4.

⯅	A	B	C	1048576R x 2C	E	F
1	Inventory					
2						
3	City	SUV	2-door	4-door	Mini Van	Totals
4	Denver	500	400	375	800	2075
5	Boston	400	375	500	450	1725
6	Chicago	1200	600	900	1500	4200
7	Nashville	700	500	1000	1000	3200
8						
9	Totals	2800	1875	2775	3750	11200

Figure 2-4. *When selecting column heads, Excel displays the number of rows and columns selected*

Click in a blank area on the worksheet to clear the selection.

3. Position the cell pointer on row header 4. The cursor changes to a right-facing arrow, hold down the left mouse button, and drag down so that row headers 4, 5, 6, and 7 are selected. As you are dragging, you will see a box that displays the number of rows and columns that are currently selected. See Figure 2-5.

	A	B	C	D	E	F
1	Inventory					
2						
3	City	SUV	2-door	4-door	Mini Van	Totals
4	Denver	500	400	375	800	2075
5	Boston	400	375	500	450	1725
6	Chicago	1200	600	900	1500	4200
7	Nashville	700	500	1000	1000	3200
4R x 16384C						
9	Totals	2800	1875	2775	3750	11200

Figure 2-5. *When selecting row heads, Excel displays the number of rows and columns selected*

Selecting Adjacent Cells

1. Click inside cell A3 and then drag to the right through cell F3.

2. On the Ribbon's Home tab, in the Font group, click the B (this will bold the selected cells).

Note   You can also select adjacent cells by clicking the cell that you want to be in the upper left cell of the range and then hold down the Shift key while clicking the cell you want to be in the bottom right cell of the range.

---

3. Click inside cell B4, hold down the Shift key, and click inside cell F9.

4. On the Ribbon's Home tab, in the Number group, click the $ (this will add dollar signs, commas, and decimals to the selected cells).

Selecting Nonadjacent Cells

1. Click a cell that is not part of the selection to clear the previous selection.

2. Click column head C. Make the following selections while holding the Ctrl key down:

   a. Drag across row heads 3 and 4.

   b. Click cell E6.

   c. Click cell F9.

   d. Click cell A1.

3. Choose a format for the selection such as changing the font color or changing the font size.

Selecting All the Cells in the Worksheet

1. Click the Select All button ◢ at the top left corner of the worksheet. It is above the row heads and to the left of the column heads.

Deselect and Reselect Some of the Cells in the Worksheet

1. Do the following only if you have a desktop computer: hold down the **Ctrl** key. Drag across the cell range C4:E7 to deselect those cells. Keep the Ctrl key held down while you drag across cells D5 and D6 to reselect them.

2. Click the Save button on the Quick Access Toolbar.

Now that you have learned how to select cells using a mouse, the next step is to learn how to select cells using combinations of keys on the keyboard. The next section has a table *detailing how to do this*.

Selecting Cells Using a Keyboard

Table 2-3 tells how to select different worksheet ranges using the keyboard.

Table 2-3. *How to Select Worksheet Ranges Using a Keyboard*

To Select	Press
An entire column	Ctrl + spacebar
Multiple columns	Shift + left or right arrow key and then Ctrl + spacebar
An entire row	Shift + spacebar
Multiple rows	Shift + up or down arrow key and then Shift + spacebar
Cells in direction of arrow key	Shift + arrow key
Cells from current cell to beginning of the row	Shift + Home
Go to beginning of worksheet	Shift + Ctrl + Home
Go to last cell in worksheet containing data	Shift + Ctrl + End
Entire worksheet	Ctrl + A or Ctrl + Shift + spacebar

Note   Ctrl + A and Ctrl + Shift + spacebar selects the entire worksheet if you select a blank cell or a cell that doesn't have an adjacent cell containing data. If you use one of these key combinations with a cell selected that does have adjacent cells with data, it selects the block of adjacent cells with data. Don't worry if this isn't clear. We'll try both in the following exercises, and you'll see how they work.

EXERCISE 2-3: SELECTING CELLS USING A KEYBOARD

Continue using the same workbook from the previous practice.

Selecting an Entire Column

Pressing Ctrl + spacebar selects all the cells in the column of the active cell.

1. Click cell D3.

2. Hold down the Ctrl key and press the spacebar.

Selecting an Entire Row

Pressing Shift + spacebar selects all the cells in the row of the active cell.

1. Click inside cell C3.

2. Hold down the Shift key and press the spacebar. Notice that the B in the Font group on the Ribbon is highlighted. This is because you previously bolded this row. Click the B again. This turns off the bolding for the row.

Selecting Multiple Columns

1. Move your cursor to cell D1.

2. Hold down the Shift key while you press the right arrow key twice. This will select cells D1, E1, and F1. Release the Shift key.

3. Next, hold down Ctrl key while you press the spacebar. This will select all of columns D, E, and F.

Selecting Multiple Rows

1. Click cell A1.

2. Hold down the Shift key and press the down arrow twice.

3. Hold down the Shift key and press the spacebar. This selects rows 1, 2, and 3.

Selecting an Area

In this exercise, you select several different ranges using different key combinations.

1. Click cell A1. Hold down the Shift key and keep pressing the down arrow key until you are down to row 7. Hold down the Shift key and keep pressing the right arrow key until you have selected columns A to G. Hold down the Shift key and press the left arrow to deselect column G. This selects the range A1–F7.

2. Click cell F3. Hold down the Shift key plus the Home key. This selects all cells from the current cell to the beginning of the row.

3. Click cell F4. Hold down the Shift key plus the Ctrl key plus the Home key. This selects everything from the current cell to cell A1.

Selecting from the Current Cell to the Last Cell Containing Data

1. Enter 380 in cell J23. Press Ctrl + Home to go to cell A1.

2. Press Shift + Ctrl + End. This selects all cells between A1 and J23.

Selecting an Entire Worksheet or All Cells

1. Click any blank cell.

2. Press Ctrl + A. This selects all the cells in the worksheet.

3. Click a blank cell again and press Shift + Ctrl + spacebar for the same result.

Selecting an Entire Block

1. Click cell B5. Press Ctrl + A. This selects all the cells in the block.

2. Click cell B5. Press Shift + Ctrl + spacebar.

3. Click the Save button on the Quick Access Toolbar.

You have learned how to select cells using a mouse and keyboard. Next, you'll be learning how to select cells by entering their cell addresses.

Select Cells by Using Their Cell References in the Name Box

Referencing a value in a cell by using its address is called a cell reference. A reference can be used to identify a single cell or a range of cells on the current worksheet or on another worksheet. You can even reference cells in other workbooks. A reference to cells in another workbook is called a link.

When referencing a range of adjacent cells, the upper left cell address is listed first, followed by a colon, and then the lower right cell's address. The address B3:D5 includes cells B3, C3, D3, B4, C4, D4, B5, C5, and D5. See Figure 2-6.

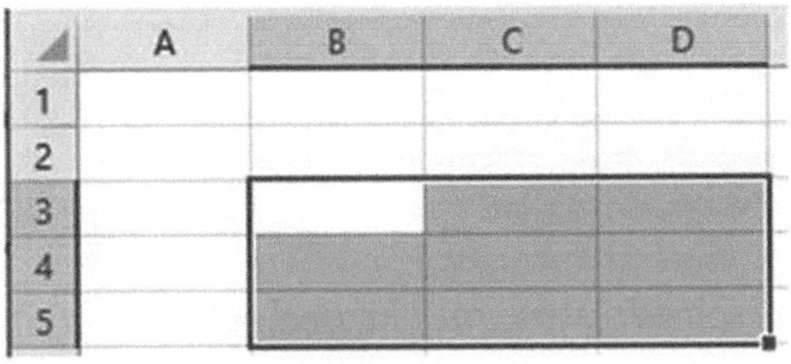

Figure 2-6. *The selected range is B3:D5*

In Table 2-4, the first column specifies the cells to reference, and the second column gives an example of how to select that reference.

Table 2-4. *Cell Selection Examples*

Reference to Cell(s)	Example
Reference a single cell. Enter column letter followed by the row number	C12
A range of cells in a single column	A1:A15 (column A, rows 1 through 15)
A range of cells in a single row	A12:E12 (row 12, columns A through E)
A range of cells in multiple rows and columns	A1:F15 (rows 1 through 15 in columns A through F)
All cells in row 8	8:8
All cells in rows 8 through 12	8:12
All cells in column C	C:C
All cells in columns C through F	C:F
Cell B3 on the worksheet named Sheet1	Sheet1!B3

The cell references can be entered into the Name Box to select those cells. You can enter any combination of cell references that are separated by commas in the Name Box and then press Enter to select those cells. Figure 2-7 shows the result of having pressed Enter after typing A2, B3:C5,E4:G5 in the Name Box. (Note that the first cell in the last range becomes the active cell.)

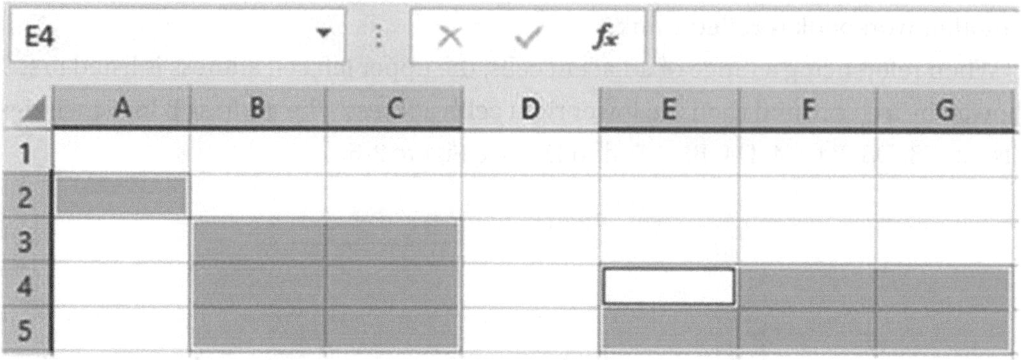

Figure 2-7. *Enter the cell addresses you wanted selected in the Name box*

EXERCISE 2-4: SELECTING CELLS USING THE NAME BOX

The previous practices have covered selecting cells using your mouse and your keyboard. In this exercise, you will practice selecting individual cells, ranges of cells, and rows and columns of cells by entering cell addresses in the Name Box.

1. Create a new worksheet in your Chapter 2 workbook.

2. Enter C12 in the Name Box and then press Enter. Cell C12 should be selected.

3. Enter A1:A15 in the Name Box and then press Enter.

4. Enter the other cell addresses in the Example column in Table 2-4.

5. Enter **A5**, **A8**, **B3:B7**, and **D4:F8** in the Name Box and then press Enter.

___

Note   If you are entering data in a selected range, the active cell stays within that range when you press the Enter key or the Tab key.

___

6. Enter **A8:C10** in the Name Box and then press Enter.

7. It is usually better to enter data across in rows because you can tab across to the next cell. When you get to the last cell, press Enter to go to the next row automatically. If you prefer to enter data going down one column and then to the next column, you can select a range of cells and press the Enter key after each cell entry.

   a. With cell range A8:C10 selected type **car** without first clicking in a cell. Press the Enter key.

   b. Type **van**. Press the Enter key.

   c. Type **truck**. Press the Enter key.

   d. Type **semi**. Press the Enter key.

   e. Type **bus**. Press the Enter key.

   f. Type **bike**. Press the Enter key.

   g. Type **trike**. Press the Enter key.

   h. Type **train**. Press the Enter key.

   i. Type **skates**. Press the Enter key. You now are back at cell A8.

8. Reenter the data you entered in each step, except this time press the Tab key after each entry.

In this section, you learned how to select cells using their cell reference. Next, you'll be learning how to use the **Go To** feature to move rapidly to individual cells. This function becomes increasingly important when worksheets become very large over time. When your worksheet contains extensive data, you have to be able to make quick jumps to work efficiently.

Going Directly to Any Cell

Excel provides two methods for jumping directly to any cell. One way is to type the cell address into the Name box and press Enter. Another way is to use the Go To feature.

Entering the cell address in the Name box is the quicker method, but the Go To feature has the benefit, in that it stores the cells that you jumped to previously, making it easy to jump back and forth between cells.

The Go To feature can be accessed by clicking the Home tab, and then, in the Editing group, click the Find & Select arrow to see the drop-down menu. See Figure 2-8. The shortcut for this feature is pressing Ctrl + G.

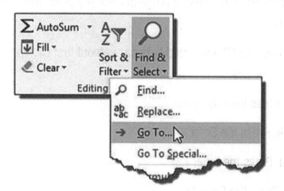

Figure 2-8. *Select Go To from the Find & Select button*

Selecting the Go To... option brings up the Go To dialog box. See Figure 2-9.

Figure 2-9. *Go To dialog box*

To jump to a particular cell, you would type the cell address for the cell you want to go to in the Reference area, and then press Enter or click the OK button.

EXERCISE 2-5: PRACTICE FOR GOING DIRECTLY TO ANY CELL

1. Use the same worksheet you created in the previous practice.

2. Click cell B5.

3. On the Ribbon, click the Home tab. In the Editing group, click the Find & Select button. Select Go To….

4. Type w35 in the Reference area. See Figure 2-10.

Figure 2-10. *Enter the cell address in the Reference area of the Go To dialog window*

5. Click the OK button.

6. You should now be at cell W35.

7. Press the shortcut Ctrl + G to bring up the Go To dialog window. See Figure 2-11. The cell address B5 is displayed in the Reference area, and its absolute address is displayed in the Go to area. Absolute addresses use dollar signs in front of the column and row. Absolute addresses will be discussed later. Excel saves the location you were at when you used the Go To feature so you can easily jump back to where you were.

Figure 2-11.   *Cell address B5 is displayed because it was the location you were at when you started the Go To feature*

8. Type IV4 over the B5 in the Reference area. Click the OK button.

9. Your cursor should be in cell IV4.

10. Press Ctrl + G. The previous cell address that you jumped to using Go To and the first cell address when starting the Go To feature are displayed in the Go to area. See Figure 2-12. Clicking an address in the Go to area takes you back to that address.

Figure 2-12. *Clicking an address in the Go to area takes you back to that address*

11. Click $W$35 in the Go to area, and then click the OK button to return to cell W35.

12. Press Ctrl + G. Click $B$5 in the Go to area. Click the OK button.

<u>Use the Name Box to Jump to a Cell</u>

13. Type F15 in the Name box and press Enter.

Next, you learn how to work with worksheets. You will learn how to select them, rearrange them, and change their tab colors—as well as adding, deleting, and hiding them.

Worksheets

Spreadsheets in Excel are referred to as *worksheets*. Individual worksheets are stored together in a *workbook*. When you save your work in Excel, you do not save individual worksheets; rather, you save the workbook. Data entered in one worksheet can be entered into other worksheets at the same time. Data can be passed between worksheets. Data can also be imported into worksheets from other workbooks or other sources.

A workbook can consist of one or more worksheets. The tabs for these worksheets appear at the bottom left-hand corner of the screen. Clicking a sheet tab makes that sheet the current worksheet. The current tab can be easily identified because its tab text is bolded and has a thick line below the text.

Figure 2-13. *Worksheet tabs*

In addition to adding and formatting data, the following are some of the ways to manipulate worksheets:

- Rename them.

- Add or remove them.

- Hide and unhide them.

- Reorder them.

- Copy them to the same workbook.

- Copy or move them to another workbook.

Naming Worksheets

If you add a second worksheet, it would be named Sheet2 by default. If you add a third worksheet, it would be named Sheet3 by default and so on. If you create three sheets, delete Sheet3, and create a new one, it will name it Sheet4, not create a second Sheet3. It remembers used names. These names are not helpful because they don't provide any clue as to what is on the worksheet, but fortunately you can rename them to something more meaningful.

There are two ways to rename a worksheet:

- Right-click a sheet tab and select Rename. This will highlight the tab you right-clicked (see Figure 2-14a) and put the tab text in Edit mode in which you can type over the current text (see Figure 2-14b).

- Double-clicking a sheet tab will also put the tab text in Edit mode.

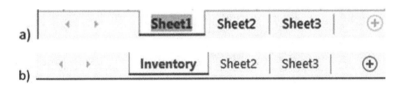

Figure 2-14. *a) The tab name is in Edit mode, and b) the tab has been renamed*

Adding and Removing Worksheets

You can add new worksheets to your workbooks or remove them.

Worksheets can be added by doing one of the following:

- Clicking the **New Sheet** button ⊕. Clicking the New Sheet button adds the new worksheet after the currently selected worksheet.

- Right-click a worksheet tab and select Insert, which displays the Insert dialog box (see Figure 2-15). Select Worksheet and click OK to add a blank worksheet.

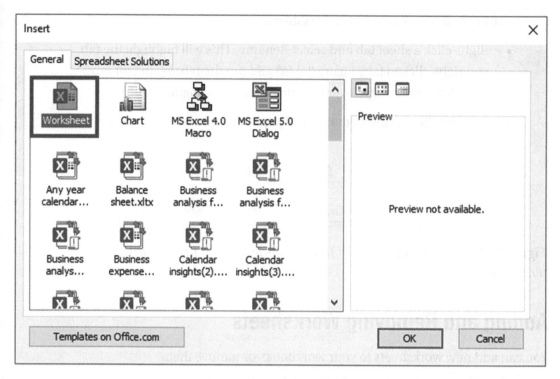

Figure 2-15. *Select Worksheet from the Insert dialog box to add a blank worksheet*

You also can use add a prebuilt worksheet known as a template. Click the Spreadsheet Solutions tab to see the available templates (see Figure 2-16). There are some very nice templates available on the Spreadsheet Solutions tab. There is a Blood Pressure Tracker, a Personal Monthly Budget template, a Loan Amortization template, and so on. Click a template to view it in the Preview area. Double-clicking one of these templates adds it as a worksheet in your workbook.

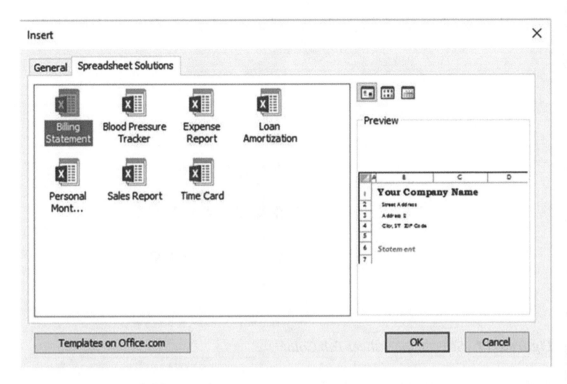

Figure 2-16. *Available templates*

A worksheet can be removed by right-clicking its tab and selecting Delete.

Changing a Worksheet Tab Color

You can make the different tabs easily distinguishable by making them different colors. You can change the background color of each tab by right-clicking a tab, selecting Tab Color, and then selecting a color. See Figure 2-17.

Figure 2-17. *Select a worksheet Tab Color*

Selecting Multiple Worksheets

You can select multiple worksheets by holding down the Ctrl key while you click the worksheet tabs. You can select multiple adjacent worksheets by clicking the first worksheet tab you want to use and then holding down the Shift key and clicking the last worksheet tab you want to use.

Hiding and Unhiding Worksheets

You can hide a worksheet by right-clicking its tab and selecting Hide from the menu. Hiding doesn't delete the worksheet or change it in any way. You can bring back any worksheets you have hidden by right-clicking any of the worksheet tabs and selecting Unhide.... The Unhide dialog box displays all of the worksheets you have hidden. To Unhide a worksheet, just select the worksheet you want to redisplay and click the OK button. See Figure 2-18.

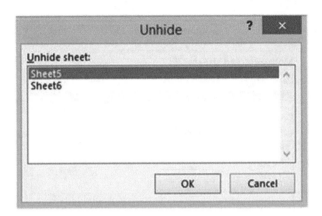

Figure 2-18. *Select worksheets to Unhide*

Reordering and Copying Worksheets

You can change the order of the worksheets by dragging and dropping their tabs to where you want them placed. When you drag a worksheet tab, the cursor displays as a document. A down arrow ▼ moves as you drag indicating where the worksheet will be placed when you let go of the mouse button. See Figure 2-19.

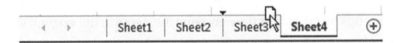

Figure 2-19. *Drag a worksheet to a new location*

Another way to change the order of the worksheets is to right-click the tab of the worksheet you wish to move and then select Move or Copy from the context menu. The Move or Copy dialog box displays all of the sheets in your workbook. You can either select which sheet you want to place it before or you can select (move to end) to make it the last worksheet in your workbook. See Figure 2-20.

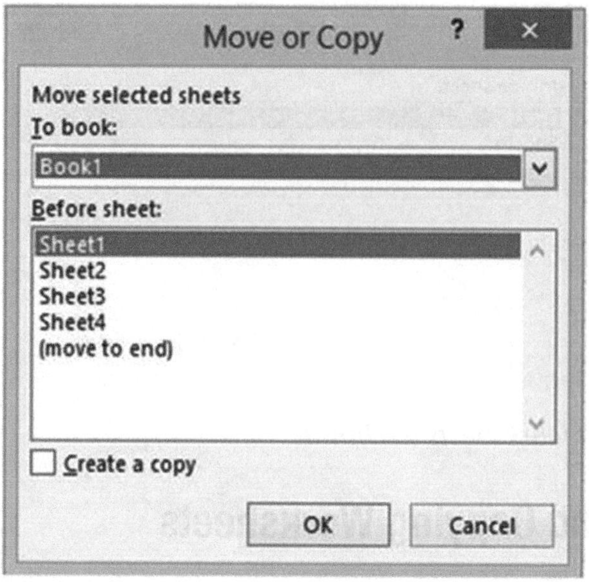

Figure 2-20. *Move current sheet to the end or before another sheet*

The **Move or Copy** dialog box has a check box option for creating a copy of the worksheet. Selecting this option will make a duplicate worksheet of the one you right-clicked, and it will place the new copy in the location you specify in the **Before sheet:** area of the dialog box. Excel names the new copy the name of the original sheet plus it adds a number of the copy in parentheses. Figure 2-21 shows that a copy of Sheet1 was moved to the end and named Sheet1 (2). If another copy was made of Sheet1, it would be named Sheet1 (3). You can also rename the copied worksheets to something more meaningful.

Figure 2-21. *Excel adds a number to the end of the name of the copied worksheet*

Note   You can also move or copy worksheets to a different workbook.

Using Tab Buttons to Move Through the Worksheets

If there are dots to the left of the worksheet tabs, this means that there are more worksheets to the left of those currently showing. These dots are called ellipsis. If there is an ellipsis to the right of the worksheet tabs, this means that there are more worksheets to the right of those currently showing. See Figure 2-22.

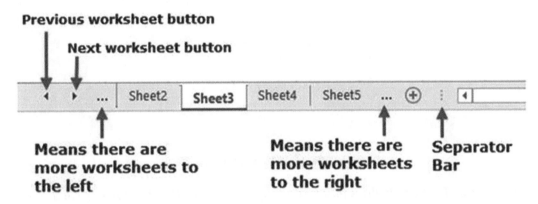

Figure 2-22. *Handling worksheets that aren't visible*

Clicking the Next worksheet button displays the tab for the next worksheet (it doesn't change the currently selected tab) giving the result as shown in Figure 2-23.

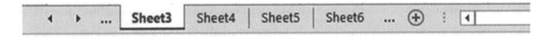

Figure 2-23. *Sheet6 appears after clicking the Next worksheet button*

Clicking the Previous worksheet button twice at this point takes me back two worksheet tabs. Since this is the first worksheet, the Previous worksheet button is grayed out and the left ellipsis is no longer displayed. See Figure 2-24.

Figure 2-24. *Previous worksheet button is grayed out and the left ellipsis button is removed*

If you want more room for displaying the worksheet tabs, you can drag the separator bar to the right.

If you move your cursor over either the Previous or Next worksheet button, a tooltip displays available options. See Figure 2-25.

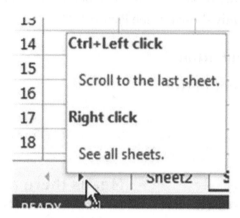

Figure 2-25. *Tooltip displays when placing cursor over the Previous or Next worksheet button*

Holding down the Ctrl key while clicking the Previous worksheet button scrolls to the first worksheet. Holding down the Ctrl key while clicking the Next worksheet button scrolls to the last worksheet.

If you right-click either the Previous or Next worksheet button, a dialog box displays all of the worksheets from which you can activate the worksheet you want. See Figure 2-26.

Figure 2-26. *All of the worksheets appear in the Activate dialog box*

EXERCISE 2-6: WORKING WITH WORKSHEET TABS

This exercise covers how to move through the worksheet tabs and how to hide and unhide worksheets.

1. Create a new Workbook named WorksheetTabs.

2. Right-click the Sheet1 tab. Select Rename from the menu. Type **Qtr1 Sales** over the text Sheet1.

3. Click the New sheet button three times.

4. Double-click the Sheet2 tab. Type **Qtr1 Expenses** over the text Sheet2. Press Enter.

5. Double-click Sheet3. Type **Qtr1 Accts Receiv** over the text Sheet3. Press Enter.

6. Double-click Sheet4. Type **Qtr1 Payroll** over the text Sheet4.

7. Insert four new worksheets. The first worksheets may no longer be visible. Don't worry, they are still there. Drag the separator bar ⁞ far to the right so that you can see all of the worksheet tabs plus extra space so that you can write longer names on the tabs.

8. Name the new worksheets the same as the existing ones, except change Qtr1 to Qtr2. Your tabs should look like those in Figure 2-27.

| Qtr1 Sales | Qtr1 Expenses | Qtr1 Accts Receiv | Qtr1 Payroll | Qtr2 Sales | Qtr2 Expenses | Qtr2 Accts Receiv | **Qtr2 Payroll** |

Figure 2-27. Your worksheet tabs should look like these

9. Drag the separator bar ⁞ to the left so that you only see the first five worksheet tabs.

10. Click the Next worksheet button ▶ twice so that you can view the next two tabs that are hidden to the right.

11. Click the Previous worksheet button ◀ so that you can view the tab that is hidden to the left.

12. Click Ctrl + the Next worksheet button ▶ to view the last tab.

13. Click Ctrl + the Previous worksheet button ◀ so you can see the first tab.

14. Drag the separator bar ⁞ to the right until you can see all eight worksheet tabs.

15. To activate a worksheet, right-click either the next Tab button or the previous Tab button. Select Qtr1 Payroll from the Activate dialog box. See Figure 2-28. Click the OK button.

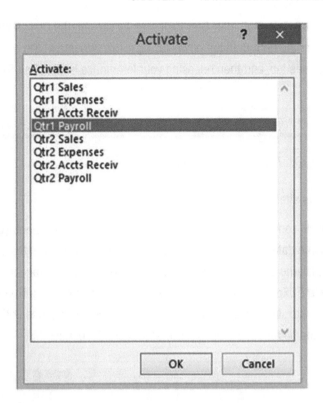

Figure 2-28. *Select Qtr1 Payroll from the Activate dialog box*

16. Next, you will hide the worksheets for both the **Qtr1** and **Qtr2 Accounts Receivable**. Click the **Qtr1 Accts Receiv** tab, hold down the Ctrl key, and click the **Qtr2 Accts Receiv** tab. Right-click the **Qtr 2 Accts Receiv** tab and then select Hide.

17. You will now bring back the two hidden worksheets. Right-click any one of the tabs and then select Unhide… from the menu. Select Qtr1 Accts Receiv and then click OK. The worksheet is displayed back in its original position. Right-click any one of the tabs and then select Unhide… from the menu. Select **Qtr2 Accts Receiv** and then click OK.

18. Let's remove the **Qtr1** and **Qtr2 Accounts Receivable** worksheets permanently from the workbook. Click the **Qtr1 Accts Receiv** tab, hold down the Ctrl key, and click **Qtr2 Accts Receiv** tab. Right-click the **Qtr2 Accts Receiv** tab and select Delete.

19. Move the **Qtr1 Sales** tab after Qtr1 Expenses by clicking the Qtr1 Sales tab, dragging it to the right until the down-facing arrow appears at the end of the **Qtr1 Expenses** tab, and then releasing your left mouse button. See Figure 2-29.

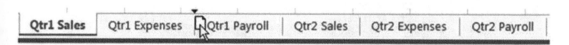

Figure 2-29. *Drag Qtr1 Sales after Qtr1 Expenses*

Move the **Qtr2 Sales** after the **Qtr2 Expenses** tab.

20. Next, you will copy a worksheet, but before you copy the worksheet, you need to enter some data into the worksheet being copied to verify that the copy worked. Click the **Qtr1 Expenses** tab. Enter some data in several cells. Right-click the **Qtr1 Expenses** tab. Select Move or Copy. This brings up the Move or Copy dialog box. In the **Before sheet:** area, select Qtr1 Sales. Check the **Create a copy** at the bottom of the dialog box. See Figure 2-30. Click OK.

Figure 2-30. *Select Create a copy*

Worksheets **Qtr1 Expenses** and **Qtr1 Expenses (2)** should contain the same data.

21. Save the workbook as **Chapter2Tabs**.

As you learn more about worksheet functions and options, think about how you will use the data. As you go through this book, you'll learn more about planning and designing workbooks and worksheets to meet your present and future needs.

Summary

You've learned a quick way to move around worksheets and how to use special keyboard shortcuts to jump to different cells. You've also worked with creating and customizing multiple worksheets within one workbook. In Chapter 3, you'll create different types of data and learn how to insert special characters and adjust the size of columns and rows. Included are shortcuts for entering the data to help you continue to use Excel in the way that best suits your needs.

CHAPTER 3

Best Ways to Enter and Edit Data

In this chapter, you'll learn how Excel worksheets use three types of data: text, numeric, and date/time. It is important to understand the different types of data. You'll also begin using special characters that you can't find on the keyboard.

Excel has several tools and features to make your data entry quick, efficient, and in the format you want. Excel automatically determines how data is aligned based on the type of data you enter. The data you enter can be formatted by using various text, numeric, and data formats. If the data you enter doesn't fit in the cell the way you want, you can adjust the column widths and row heights.

Excel's AutoCorrect feature can be useful for correcting commonly misspelled words, correcting misuse of capitalization, entering special symbols that can't be directly entered from the keyboard, and creating shortcuts for entering often used words or phrases.

The **AutoComplete** and **Pick from Drop-down List** features allow you to easily duplicate existing column values. AutoFill can be used to create duplicate cells, enter a series of values, create a custom list, or use a pattern that you taught it.

After reading and working through this chapter, you should be able to

- Create the three types of data

- Insert special characters

- Change column widths and row heights

- Correct typing mistakes

- Use AutoCorrect to make corrections and to create shortcuts for entering words or phrases

- Use AutoComplete to enter data

© David Slager and Annette Slager 2020
D. Slager and A. Slager, *Essential Excel 2019*, https://doi.org/10.1007/978-1-4842-6209-2_3

- Use Pick from Drop-down List to enter data

- Use AutoFill to create duplicate cells and create a series

- Create a custom list

- Add Notes and Comments to your spreadsheet

Data Types

There are three types of data that can be entered into a cell: text, numeric, and date and time.

- Text data—Text data is also known as string data. The data can consist of a combination of letters, numbers, and some symbols. Text data is left-justified in a cell. You can't use text data in a formula.

- Numeric data—A numeric cell can only contain the characters shown in Figure 3-1.

Numbers	Characters	Character Meaning
0	,	comma
1	.	period
2	+	positive
3	-	negative
4	( )	parenthesis
5	%	percent
6	E	exponential notation
7	e	exponential notation
8		
9		

Figure 3-1. *Only these numbers and characters can be used in a numeric value*

If any other character or even a space is included in the cell, Excel treats the data as text. Only cells that contain numeric values can be used in a formula. Numeric data is right-justified in the cell.

- Date and time data—Date and time data can be used for handling dates, time, or a combination of the two. You can enter a date as January 16, 2005 or 01/16/05 or 01-16-2005, but how the date appears in the cell depends on how the cell is formatted. Date and time data, like numeric data, is right-justified in a cell.

Note   If you enter a date as January 15 2010, Excel will display it as text rather than a date because there is nothing to identify it as a date. If you place a comma after the day, Excel will be able to identify it as a date and right-justify it.

The first column in Figure 3-2 shows exactly what text was entered in the cell; the second column shows how that entry would be displayed in Excel. You could change how the date is displayed by changing its formatting which we will look at in a later chapter.

Text Entered in Cell	Text Displayed in Cell
January 15, 2015	15-Jan-10
January 15 2010	January 15 2010
Jan 15, 2010	15-Jan-10
01/15/2010	1/15/2010
01-25-10	1/25/2010
1/15	15-Jan

Figure 3-2. *Date entered in a cell and how it's displayed*

Note   When you are entering time into a cell, be sure to place a space between the time and AM or PM. The time should be entered as 11:35 AM rather than 11:35AM. If you make the entry without the space, the time will be treated as text data and it will be left-justified in the cell. See Figure 3-3. You will not be able to perform date and time functions on dates and times that are treated as text.

◢	A	B	C
1	Date and Time Data		Date and Time Treated As Text
2	9/2/2000		10:10PM
3	9-Mar-15		9/2/2000 8:30AM
4	11:35 AM		
5	9/2/2000 8:30		
6	6/15/1954 20:30		

Figure 3-3. *Entering time without a space makes it a text entry rather than a date*

EXERCISE 3-1: PRACTICE ENTERING DIFFERENT TYPES OF DATA

In this exercise, you will create a new workbook and enter different kinds of data.

1. Start Excel program. Create a new workbook named Chapter 3.

2. Type **5 Text** in cell A1. Press the Tab key. Because it is text, it stays left-aligned.

3. Type **12.75** in cell B1. Press the Tab key. Because it is numeric, it is right-aligned.

4. Type **2/13** in cell C1. Press the Tab key. Because it is a date, it is right-aligned.

5. Type **12:15 PM** in cell D1. Press the Tab key. Because it is a time, it is right-aligned.

6. Type **5:10AM** in cell E1. Press the Tab key. Because you entered time without a space before AM, Excel left-aligned the time because Excel thinks it is text.

7. Type **March 3 2014** in cell F1. Press the Tab key. Because you entered a date without a comma after the day, Excel left-aligned the date because Excel thinks it is text.

8. Change the name on the Sheet1 tab to **Data Types**.

9. Save the workbook.

Inserting Special Characters

If you want to enter a character that isn't available on your keyboard, you can use one of Excel's symbols such as the **greater than** or **equal to sign** ($\geq$) or the **less than** or **equal to sign** ($\leq$). The special characters can be accessed by clicking the Ribbon's Insert tab, and then in the Symbols group, click the Symbol button. See Figure 3-4.

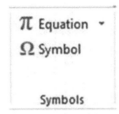

Figure 3-4. *Symbol button is located in the Ribbon's Symbols group*

The Symbol dialog box has two tabs, one for symbols and the other for special characters. The Special Characters tab displays commonly used symbols such as the copyright, trademark, different styles of closing quotes, and so on. The Symbols tab contains many more special characters, and that will be our focus in this section.

The symbols that are available on the Symbols tab depend on the selected font. Different fonts have different symbols that they can create. Some fonts have more available symbols than others. For example, the generic (normal text) option, shown in Figure 3-5, provides a very large selection of characters.

**The font you select determines
what symbols are available**

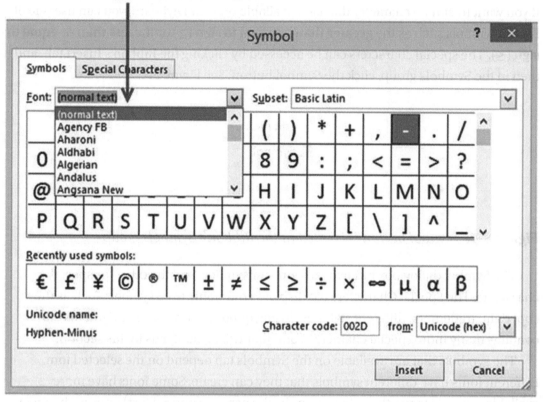

Figure 3-5. *Use this when you need to enter a character not available on your
keyboard*

The Recently used symbols list displays the last 36 symbols that you have used. You
can insert one of the symbols or special characters into a cell by clicking it and then
clicking the Insert button.

Clicking the Insert button places the selected symbol in the current cell, but it
doesn't close the Symbol dialog box. This allows you to enter multiple symbols in the
same cell without having to reopen the Symbol dialog box. You may have to move the
Symbol dialog box if it is obstructing the view of the cell in which you are entering a
symbol or special character.

Clicking the down arrow to the right of the **from:** drop-down box reveals three
categories of symbols to select from. See Figure 3-6.

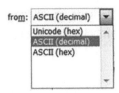

Figure 3-6. *Categories of symbols*

Note   The three options in Figure 3-6 are different ways your computer can represent data.

Unicode stands for Universal Code. Unicode code can display over 16,000 symbols. It is used to represent all the symbols from all the languages around the world.

ASCII stands for American Standard Code of Information Interchange. The ASCII format of creating symbols has been around for a long time. It is capable of creating up to 256 different symbols. It uses symbols from the United States and Western Europe.

ASCII(hex) codes are hexadecimal. They can use numbers from 0 to 15.

The **Character code** is the numeric representation for the symbol you select. The Character code will be different for the same symbol depending upon whether it is in a decimal or hexadecimal format. If you know the character code for the symbol you wish to use, you can enter it to select the symbol.

EXERCISE 3-2: ENTER SPECIAL CHARACTERS

In this exercise, you add special characters to a spreadsheet. If you have closed the Chapter 3 workbook from the last practice, reopen it.

1. Create a new worksheet. Rename the worksheet tab "Special Characters."

2. Click cell B2. On the Ribbon, click the Insert tab. In the Symbols group, click the Symbol button.

3. Select **normal text** from the **Font** drop-down box, and then select ASCII (Decimal) from the **from** drop-down list in the bottom right corner of the dialog box. Increase the size of the Symbol dialog box until you can see all the symbols without having to scroll.

4. Click the ½ symbol. Notice the Character code for the ½ symbol is 189. Click the Insert button. Move the Symbol dialog box so that you can see cell B2.

Note   Unfortunately, you can't select another cell in your spreadsheet and then insert another symbol. You must first close the Symbol dialog box, select another cell, and then reopen the Symbol dialog box.

---

5. Click the Close button on the Symbol dialog box.

6. Type **25** in cell C2. Don't press Enter. We will add a ¢ after the 25.

7. In the Symbols group on the Ribbon, click the Symbol button. Drag across (normal text) in the Font drop-down box if it isn't already highlighted. Type **Ti** and then press the Tab key. This will display the symbols for the Times New Roman font. Click the ¢ symbol and then click the Insert button. Move the Symbol dialog box so that you can see cell C2.

8. Click the down arrow of the **from** drop-down box and select Unicode (Hex). You'll see a lot more symbols. Click the different symbols; as you do, notice that some of the symbols have a letter such as A, B, C, D, E, or F in their character code.

Note   When you selected Unicode (Hex), a **Subset** drop-down box appeared in the upper right-hand corner. Click some of the different subsets.

---

9. Close the Symbol dialog box.

10. Click cell D2. In the Symbols group on the Ribbon, click the Symbol button. Click the ½ symbol in the **Recently used symbols**. Click the Insert button.

11. Click the **Special Characters** tab in the Symbol dialog box. Double-click the Copyright symbol. Double-clicking a symbol does the same thing as clicking it and then clicking the Insert button. Move the Symbol dialog box so that you can see cell D2. Cell D2 should contain ½©.

12. Click the Symbols tab. Click the down arrow for the **Font** drop-down box. Select Webdings. This font contains some images that you might find useful. Also look at the Wingdings, Wingdings2, and Wingdings3 fonts.

13. Close the Symbol dialog box.

14. Click cell E2. You can enter most of the ASCII symbols directly into a cell if you remember the character code. To enter a symbol directly into a cell, you would hold down the Alt key while you entered the character code on the numeric keypad. The character code can only be entered on the numeric keypad. Make sure the Num Lock key is on before you start entering the number.

15. Remember that the Character code for the ½ symbol is 189. Hold down the Alt key, type **0189**, and then release the Alt key. You should see the ½ symbol. The extra 0 was needed to make it a four-character number.

16. Click the Save button ▣ on the Quick Access Toolbar (QAT).

How to Change Column Widths

When the text you type is longer than the width of the column, the overflow characters are displayed in the adjacent cell(s) to the right if they do not contain any data. The characters may appear to be a part of the adjacent cells, but they are not stored there. They are stored in the cell in which you entered them. In the following example, the name Anthony Bradtmueller displays across cells A3, B3, and C3, but the entire name is actually only stored in cell A3. See Figure 3-7.

	A	B	C
1			
2	John Wilson		
3	Anthony Bradtmueller		
4	Andrew DeYoung		

Figure 3-7. *Text overflow in next cell*

If there are overflow characters and the adjacent cell to the right is not empty, then by default the overflow characters are hidden as illustrated in Figure 3-8.

	A	B	C
1	Name	Address	
2	John Wils	609 Rose Street	
3	Anthony E	E414 Lincoln Street	
4	Andrew D	309 Madison Avenue	

Figure 3-8. *Text is hidden by text in next cell*

If you are entering numeric data, the column will automatically expand to fit the size of the number. If the number that you are entering is the longest value in the column and it contains a decimal, then the decimal value may not be shown in the cell. If you change the width of a column that contains numeric data, cells in that column will no longer automatically adjust to the size of any new entries. If a cell in a column that has had its size changed contains numeric data that is longer than the width of the cell, the cell will display # signs instead of the numeric values. See Figure 3-9.

	A
1	#########
2	#########
3	279,327
4	4,715,329
5	#########

Figure 3-9. *# signs mean the column isn't wide enough to display all the data*

The sight of all these # signs always puts new users into a panic. They think that something is horribly wrong with their spreadsheet.

If you click a cell that contains # signs, the actual value will be displayed in the formula bar. Just remember when you see # signs in a column, the simple cure is to expand the width of the column.

If you enter a very large number into a cell, you may see something like 1.24312E+12. This is *exponential notation.* If you would rather view the value as a long number, you can select the cell, click the Number Format's drop-down arrow located on the Home tab, and then select which format you want the exponential format converted to. See Figure 3-10. The number will probably be too big to display in the cell; you will need to widen the column width to view the number.

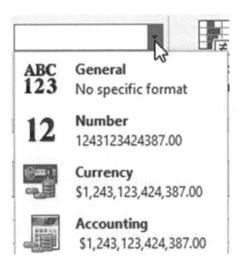

Figure 3-10. *You can convert exponential notation values to one of these formats*

You can view the column width by holding down your left mouse button while your cursor is on the line between the column headings (the cursor will turn into a double arrow). The default column width is 8.43; this allows for about 8 or 9 characters, provided you are using the default font type of Calibri and font size of 11. See Figure 3-11.

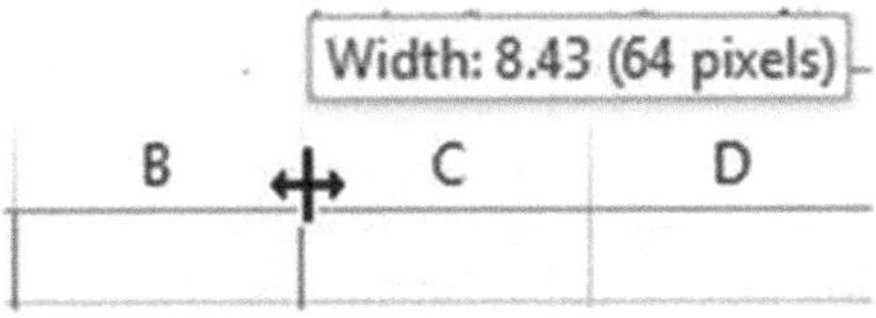

Figure 3-11. *Check column width*

Column widths can be changed by using the mouse to drag the column to the desired size. Expanding a column width can be accomplished by holding down your left mouse button while your cursor is on the line between the column heads; when the column width is displayed, drag the line to the right to the desired width. As you drag, Excel shows you the width as well as a dotted vertical line that provides a visual clue as to where the column border will be changed to. See Figure 3-12.

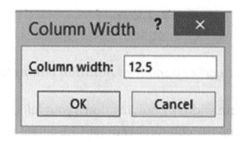

Figure 3-12. *Adjust column width*

The column width can be shrunk as well as expanded. Drag the double arrow to the left to shrink the size of the column.

You can also change a column's width by performing the following steps:

1. Right-click a column head and then select Column Width. The Column Width dialog box opens (see Figure 3-13).

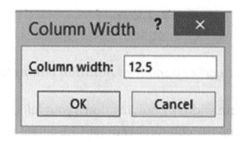

Figure 3-13. *Changing column width*

2. Enter the desired width. You can enter any value between 0 and 255. Entering a width of 0 makes the selected column(s) hidden.

3. Click OK.

The Column Width dialog box can also be accessed by clicking the Home tab, and then in the Cells group, click the Format button and then select Column Width. See Figure 3-14.

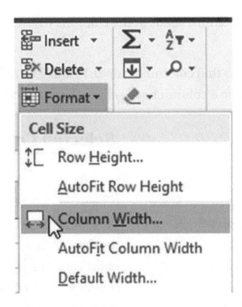

Figure 3-14. *Select column width from format button*

Automatically Resize Column Widths to Fit Number of Characters in the Cell

You can have Excel automatically adjust the size of the columns to fit the number of characters in the cell. This can be accomplished by selecting the column heads for those columns whose width you wish to change and then double-clicking between any of the column heads of the selected columns. You can also perform an Autofit by clicking the Home tab, in the Cells group, click the Format button and select **AutoFit Column Width**.

Changing the Column Width for Multiple Columns

You can change the column width for more than one column at a time. The only difference between the methods used for adjusting the width of multiple columns vs. a single column is that you need to select the column heads for all the columns whose width you want to change. The column heads you select do not need to be adjacent to each other. The following steps are an example of how to change the width of multiple columns at the same time:

1. Drag across the column heads of the columns whose width you wish to change.

   Figure 3-15 shows that column heads D, E, and F have been selected. Selecting a column head selects the entire column.

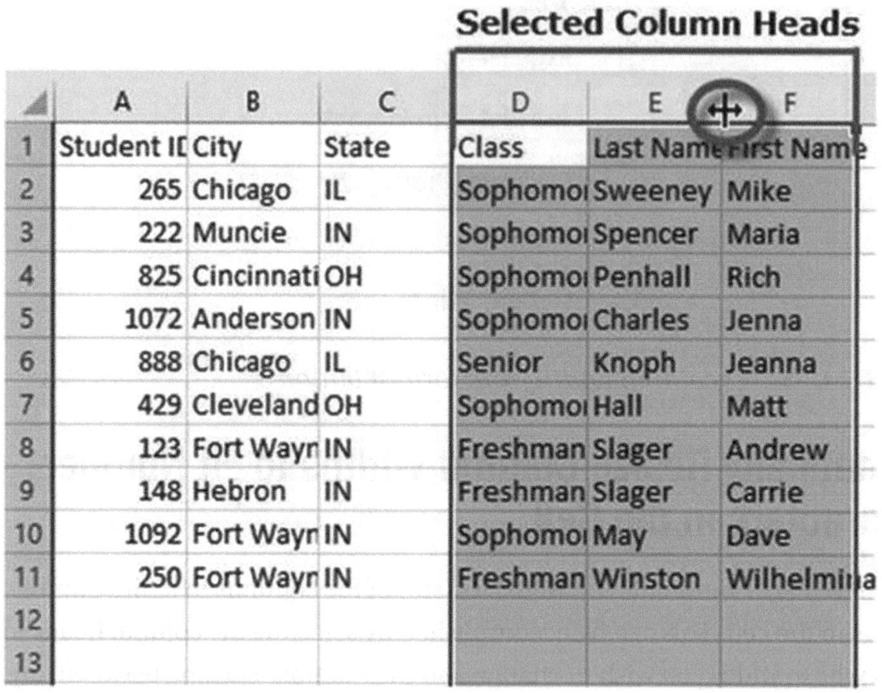

Figure 3-15. *The entire columns are selected*

2. Place your cursor between any of the selected column headings. The cursor will turn into a double arrow as shown in Figure 3-15. Hold down the left mouse button and drag it to the right to expand all of the selected columns or drag it to the left to shrink all of the selected columns.

Now, let's apply some of the things learned in this section.

EXERCISE 3-3: CHANGING COLUMN WIDTHS

In this exercise, you create a new worksheet, create columns of data, and then modify the widths of those columns. If you have closed the Chapter 3 workbook from the last practice, reopen it.

1. Create a new worksheet. Rename the worksheet tab "Column Widths."
2. Enter the Table 3-1 data into the worksheet starting in cell A1. **Do not adjust** the column widths.

Table 3-1. *Enter Data into Worksheet*

Name	Address	Profession
Pyle, Gomer	308 Marine Barracks	Marine Private
Griffith, Andy	408 Mayberry Ave.	Town Sheriff
Presley, Elvis	215 Graceland	Rock Singer
Douglas, Lisa	509 Green Acres	Unusual Housewife
Adams, Mortisha	313 Spooky Lane	Strange Housewife

After entering the data, your worksheet should look like Figure 3-16.

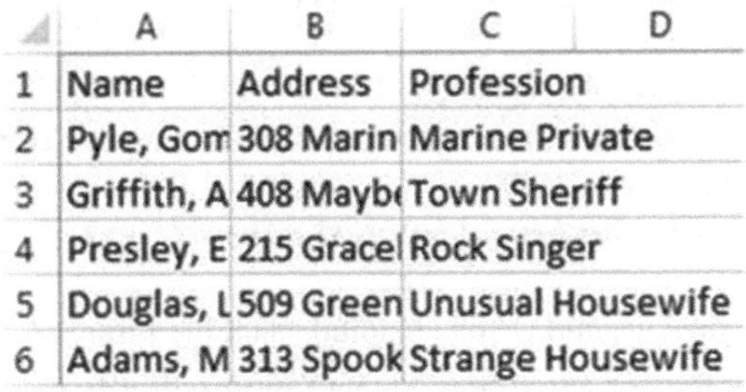

	A	B	C	D
1	Name	Address	Profession	
2	Pyle, Gom	308 Marin	Marine Private	
3	Griffith, A	408 Mayb	Town Sheriff	
4	Presley, E	215 Gracel	Rock Singer	
5	Douglas, L	509 Green	Unusual Housewife	
6	Adams, M	313 Spook	Strange Housewife	

Figure 3-16. *Enter data without adjusting column widths*

3. Place your cursor between column heads A and B. Drag your cursor to the right until the width is 16.43. See Figure 3-17.

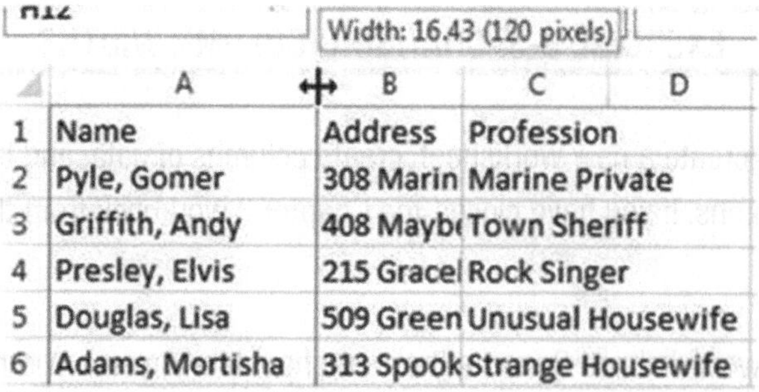

Figure 3-17. *Drag between column heads to change a column's width*

4. Place your cursor between column heads A and B. Drag your cursor back to the left until the width is 7. The text in column A is now partially hidden by the text in column B.

5. Right-click column head A and then select Column Width. Enter 8.43 for the column width. Click OK.

6. Drag across column heads A, B, and C to select all three columns.

7. Double-click between column heads B and C. Columns A, B, and C adjust to just fit the longest text in their columns as shown in Figure 3-18.

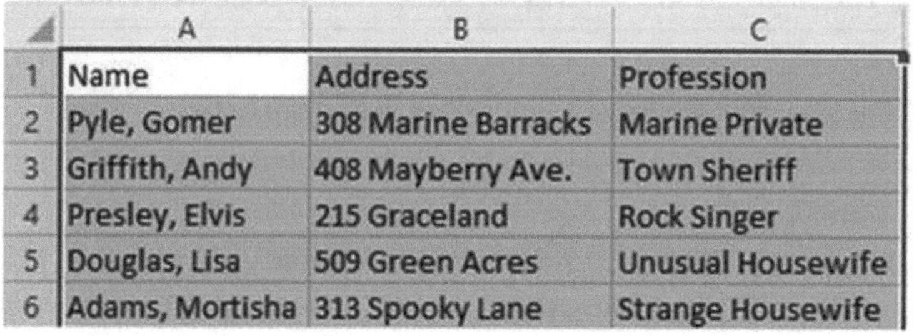

Figure 3-18. *Columns adjusted to fit longest column text*

8. With column heads A, B, and C still selected, click the Ribbon's Home tab. In the Cells group, click the Format button. Select **Column Width**.

9. In the Column Width dialog box, enter 20 and click the OK button. The three columns should each have a column width of 20.

10. Click the Format button again. Select **AutoFit Column Width**. Columns A, B, and C adjust to fit the longest text in their columns.

11. Click the Save button 🖫 on the Quick Access Toolbar.

Now that you have learned to adjust the column width, you'll learn how to change the height of the row so that you can view all of the text in a cell.

How to Change Row Heights

Changing the height of a row works the same as changing the width of a column. The row height is 15.00 by default.

Changing row heights is accomplished by selecting the row or rows whose height you want to change. You select rows the same way you select columns: click the number for a specific row, drag on the numbers to select adjacent rows, or select nonadjacent rows by holding down the Ctrl key while clicking the row numbers (see Figure 3-19). After you have made your row selections, place the cursor between the row head of one of the selected rows and another row head and then drag down to expand the height of the row or up to shrink it.

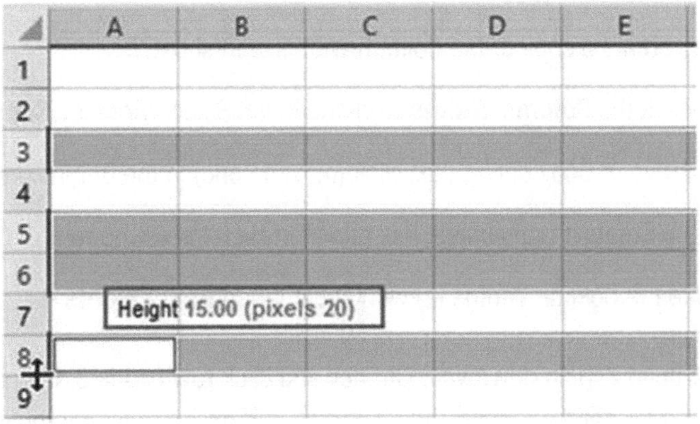

Figure 3-19. *Select rows and then drag up or down to change the height of the selected rows*

Double-clicking the line between the numbers for the selected rows adjusts the height of the row to accommodate the tallest text or object in that row.

Another way to change the row height is to follow these steps:

1. Select the row or rows that you want to change.

2. On the Home tab, in the Cells group, click Format.

3. From the menu, you can select Row Height or Autofit Row Height.

You can select all the rows in a spreadsheet by clicking the Select All button. You can then adjust the height of all rows by dragging up or down between any row headers. The Select All button is located to the left of the Column A header and above the row 1 header. See Figure 3-20.

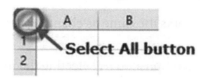

Figure 3-20. *Select All button*

EXERCISE 3-4: CHANGING ROW HEIGHTS

This exercise uses the same spreadsheet you created for practicing changing column widths. If you have closed the Chapter 3 workbook from the last practice, reopen it.

1. First, we'll create a copy of the Column Widths worksheet:

   a. Right-click the **Column Widths** worksheet tab. Select **Move or Copy**.

   b. In the Move or Copy dialog box, click (move to end) in the Before sheet list box.

   c. Click the Create a Copy check box and then click the OK button.

   d. Rename the Column Widths (2) worksheet tab to **Row Heights**.

2. Click row head 2. Hold down your Ctrl key and click row heads 5, 6, 9, and 12. Put your cursor between row heads 9 and 10 and drag down until the size is about 22.5.

3. Select cell A6. On the Ribbon, on the Home tab, in the Font group, type **22** in the Font Size text box and then press Enter. The name Adams, Mortisha is now partially hidden. Double-click between column heads A and B. This performs an AutoFit adjusting the column width to accommodate the longest text in the column.

4. Double-click between row heads 6 and 7. This performs an AutoFit adjusting the row height to accommodate the tallest text in the row.

5. Click the Select All button. Put your cursor between any two row heads and then drag down until the height is 40.50. Double-click between any row head to make all rows AutoFit.

6. With all rows still selected, right-click any cell and select Row Height from the menu. Enter **25** for the Row Height and then click OK.

7. Click any cell to unselect all of the rows.

8. Click the Save button 🖫 on the Quick Access Toolbar.

You've learned how to adjust cell sizes by changing row heights and column widths. Next, you will learn how to correct data entry errors.

Correcting Typing Mistakes

Sometimes you make errors when entering data into a worksheet. If you do, it is easy to clear or correct what you entered. There are several ways to clear or correct either some or all the data from a cell. Let's look at how to remove or change specific characters in a cell, how to clear data (and restore a cell to its previous state) before accepting your data entry, and how to clear a cell for which you have accepted the entry.

Changing Specific Characters

If you need to correct specific characters in a cell, you have a couple of options depending on which characters they are and whether you have already accepted the data entry. First, you delete the character or characters you want to change and then type in the correct characters if necessary. Here are several ways to delete specific characters in a cell:

- While still entering data in a cell, you can delete the last character or characters you entered by pressing the Backspace key.

- If you need to correct a character other than the last one(s) you typed, click just to the right of that character and press Backspace.

- If you have already accepted the data in the cell, double-click the cell to enter Edit mode and then click to the right of the character you want to change or use the left and right arrow keys to move the cursor to the character you want to delete.

You can correct typing mistakes by doing any of the following:

- Remove the latest characters that you entered by pressing the Backspace key.

- Clear an entry before it has been accepted by pressing the Esc(ape) key.

- Clear an entry before it has been accepted by clicking the Cancel button on the formula bar.

- Clear the contents of a cell that has already been accepted by right-clicking a cell and selecting Clear Contents.

If you only need to remove the latest characters that you entered, you can use the Backspace key.

Returning a Cell to Its Original Value

You can restore a cell to its original value if you have not yet accepted the data entry or if you have accepted it but have not yet done anything else.

While you are still entering data in a cell, you can clear what you've entered and restore the cell to its previous state. If it was an empty cell, you simply clear all the data. On the other hand, if you changed the value in the cell, you clear the new entry and restore the cell's original value. Use either of the following techniques to restore a cell to its previous value before the new value has been accepted:

- Press the Esc(cape) key.

- Click the Cancel button ✕ on the formula bar.

If you have already accepted the new data entry but have not yet done anything else in the workbook, you can select the Undo button on the QAT (see Figure 3-21) or press Ctrl + Z, which restores the cell to its previous state: empty or containing the previous value.

Figure 3-21. Undo button

Clearing the Contents of a Cell That Has Already Been Accepted

You can clear the contents of a cell by right-clicking the cell and then selecting Clear Contents. You can also select Clear Contents by clicking the Ribbon's Home tab and then selecting Clear ➤ Clear Contents in the Editing group.

EXERCISE 3-5: ENTERING AND EDITING DATA

In this exercise, you enter data in a new worksheet and then edit it. If you have closed the Chapter 3 workbook from the last practice, reopen it.

1. Add another worksheet and name it **Edit Data**.

2. Enter the data in Figure 3-22. For the date values in column C, let's use three different formats:

   a. Enter the data in cell C2 as 10/15/1999.

   b. Enter the data in cell C3 as 6-15-2001.

   c. Enter the data in cell C4 as Jan 31, 2005.

◢	A	B	C
1	Numeric	Text	Date/Time
2	300	street	10/15/1999
3	4000	sign	6/15/2001
4	3500.75	address	31-Jan-05
5	$620.75	313 Rdige	6:40AM
6	-300	Pay 39.95	6:40 AM

Figure 3-22. Enter this data

3. Drag across column heads A, B, and C. Right-click one of the selected column heads. Select Column Width. Enter 14. Click the OK button.

4. Double-click cell A2 to enter Edit mode. Change 300 to 550. Press Enter.

5. Click cell B5. Press F2 to enter Edit mode. Change the word Rdige to Ridge. Press Enter.

6.   Double-click cell C2. Change the date in the formula bar to 10/15/2002. Click the Enter button (check mark) ✓ in the formula bar to accept the change rather than pressing Enter.

7.   Double-click cell C3. Change the year to 1998, but don't press Enter. You decide you really didn't want to make that change. Press the Escape key to return the cell to its original value. Repeat this step again, but instead of pressing the Esc key, click the Cancel button ✕ on the formula bar.

8.   Right-click cell A6. Select **Clear Contents** from the menu.

9.   Click cell B6. Click the Home tab. In the Editing group, select Clear ➤ Clear Contents.

10. Click the Save button 💾 on the Quick Access Toolbar.

You have learned how to handle any keying errors that you have made. Now, you'll look at Excel features that not only prevent you from making data entry errors but at the same time increase the speed of entering the data.

Shortcuts for Entering and Correcting Data

Excel provides the AutoComplete, Pick data from drop-down list, AutoFill, and AutoCorrect methods to speed up data entry and ensure that your data has been entered without errors.

Using the AutoCorrect Feature

Excel's AutoCorrect feature can speed data entry as well as help prevent errors from being entered into your worksheet by automatically adjusting entries or by fixing common typing errors as they are being entered. Excel's AutoCorrect feature is useful for the four following situations:

- *Correcting commonly misspelled words.* If you spell the word "achieved" as "acheived", Excel will automatically correct it.

- *Entering special symbols that can't be directly entered from your keyboard.* If you type **(tm)**, Excel will automatically convert it to the trade symbol ™.

- *Creating shortcuts for entering words or phrases that you enter often.* You could enter your initials and Excel would convert it to your full name.

- *Correcting misuse of capitalization.* If you type **monday**, Excel corrects it to Monday.

Note   Microsoft office applications share the same AutoCorrect Options as Excel; any changes, additions, or deletions you make to the AutoCorrect options in Excel will affect those in other Microsoft Office applications and vice versa.

EXERCISE 3-6: USING AUTOCORRECT

In this exercise, you experience Excel's default AutoCorrect functionality and then look at how you can modify it for your own preferences. If you have closed the Chapter 3 workbook from the last practice, reopen it.

1. Add another worksheet and name it **AutoCorrect**.

2. In cell A1, type **abbout** and then press the Tab key. Excel corrects it to about.

3. In cell B1, type **(c)** and then press the Tab key. Excel converts (c) to a copyright symbol.

4. In cell C1, type **(r)** and then press the Tab key. Excel converts (r) to a registered symbol.

5. In cell D1, type your name starting with two capital letters such as DAvid and then press the Tab key.

Your results should look similar to those in Figure 3-23.

	A	B	C	D
1	about	©	®	David

Figure 3-23. *Results of using AutoCorrect*

Add Items to the AutoCorrect List

Now that you've seen how AutoCorrect works by default, you can add new items to
AutoCorrect.

1. On the Ribbon, click the File tab.

2. Click **Options** in the left pane. This opens Excel Options. See Figure 3-24.

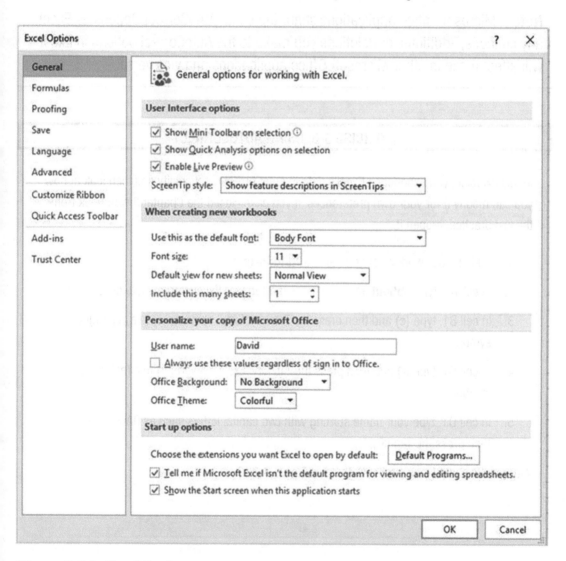

Figure 3-24. *Excel Options*

3. Click **Proofing** in the left pane.

4. In the right pane under AutoCorrect Options, click the **AutoCorrect Options** button. This brings up the AutoCorrect dialog box. See Figure 3-25.

Figure 3-25. *AutoCorrect dialog box with AutoCorrect tab selected*

You can see that there is an option for *Correct Two Initial Capitals.* This is what corrected your name entry. There are also options for capitalizing the first letter of sentences and names of days and correcting the accidental use of the Caps Lock key.

Look at the bottom half of the tab, there is a two-column table. The left column has a heading of **Replace**: and the right column has a heading of **With**: Entries in the Replace: column are replaced with those in the With: column. You used three entries in the table. You entered (c) in a cell and it was replaced with the copyright symbol ©. You entered (r) and it was replaced with the registered symbol ®. If you scroll down in the table, you can see that *abbout* will be replaced with *about* which you have already done.

5. Scroll through the table and see what other corrections are available.

6. Enter **tommorrow** in the Replace text box. Enter **tomorrow** in the With text box. Click the Add button. The new AutoCorrect item is added to the list. See Figure 3-26.

Figure 3-26. *The new AutoCorrect item has been added*

7. Type your initials in the Replace text box over the existing text. Enter your full name in the With text box over the existing text. Click the Add button.

8. Enter phoneA in the Replace text box. Enter (380)599-1275 in the With text box. Click the Add button. You could make phone numbers for your home phone and your cell phone.

9. Enter any other corrections or shortcuts you would like to make. This could be a big time-saver. If you often enter your company name, you should create a shortcut for it or any other text that you use often.

10. Click the OK button in the AutoCorrect dialog box. Click the Cancel button in the Excel Options dialog box.

11. Type tommorrow in cell A2 and then press the Enter key—Replaced with tomorrow.

12. Type your initials in cell A3 and then press the Enter key—Replaced with your full name.

13. Type phoneA in cell A4 and then press the Enter key—Replaced with (380)599-1275.

14. Put your Caps Lock on. Hold down your Shift key while typing an **E**. Let go of your Shift key and type **xcel**. Cell A5 should now contain eXCEL. Press the Enter key.

Cell A5 should now contain Excel. This was corrected by the **Correct Accidental use of cAPS LOCK key** option.

Delete Items from the AutoCorrect List

1. On the Ribbon, click the File tab.

2. Click Options in the left pane.

3. Click Proofing in the left pane of Excel Options.

4. In the right pane, click the **AutoCorrect Options** button.

5. Type **phoneA** in the Replace text box; as you do so, the list will jump to the letters you are entering. When phoneA is highlighted in the list, press the Delete button.

6. Remove the shortcut you created for your initials. Remove any other shortcuts you created that you want deleted.

7. Click the Close button on the AutoCorrect dialog box. Click the Cancel button on the Excel Options dialog box.

8. Click the Save button 💾 on the Quick Access Toolbar.

You have just used Excel's AutoCorrect feature to correct misspelled words and to create quick shortcuts for entering words or phrases. Next, you will use the AutoComplete feature to duplicate data that already exists in the column.

Using AutoComplete to Enter Data

The AutoComplete feature is automatically set up; there are no buttons to click or commands to be entered. The AutoComplete feature compares the entry you are making to entries that already exist in the same column. As you are typing your entry, Excel looks for a word that starts with the same letters as the word you are entering. If it finds an existing word in the column that starts with the same letters you have entered so far, it places that word in the active cell. See Figure 3-27.

Figure 3-27. *AutoComplete automatically added the letters rt after you typed the letters **sma***

The first two letters are the same for all words in the column. When you type the letter **a** in the second entry for the word **smart**, Excel sees that there is only one word that starts with **sma** so it places the word **smart** in the cell. You can accept the word by pressing the Enter key, the Tab key, or the down arrow key, or you can clear the cell by pressing Esc(ape).

Only after typing the second **a** in the word savant in Figure 3-28 is there a word that matches all the letters typed so far. You can accept the word by pressing the Enter key, the Tab key, or the down arrow key, or you can clear the cell by pressing the Esc key.

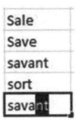

Figure 3-28. *AutoCorrect automatically added the letters nt after you typed* ***sava***

Note   AutoComplete does not work if there is an empty cell above the current cell.

Pick from Drop-down List

The **Pick from Drop-down List** feature is similar to the AutoComplete feature. The Pick from Drop-down List like the AutoComplete feature lets you use entries that already exist in the column. The Pick from Drop-down List lets you select from a list of existing entries that are automatically sorted for you. The list can be accessed by right-clicking the cell and selecting **Pick from Drop-down List** or by pressing **Alt + Down Arrow**. See Figure 3-29.

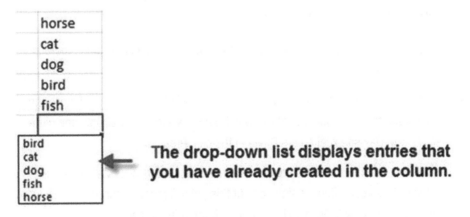

The drop-down list displays entries that you have already created in the column.

Figure 3-29. *Pick from Drop-down List*

Note   Pick from Drop-down List does not work if there is an empty cell above the current cell.

EXERCISE 3-7: AUTOCOMPLETE AND PICK FROM DROP-DOWN LIST

Let's use Autocomplete and Pick from Drop-down List to add text.

AutoComplete

Let's see AutoComplete in action in this exercise. If you have closed the Chapter 3 workbook from the last practice, reopen it.

1. Add another worksheet and name it **AutoComplete**.

2. Type the names in column A as shown in Figure 3-30.

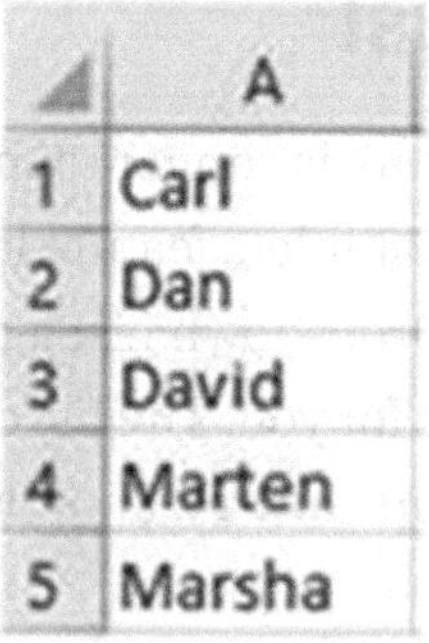

	A
1	Carl
2	Dan
3	David
4	Marten
5	Marsha

Figure 3-30. *Enter names in worksheet*

3. Type the letter **C** in cell A6. The name Carl appears in the cell because there is no other name that starts with C.

4. Type **Da** in cell A7. Since there are two names that start with a D in the column, Excel can't find a unique one at this point. Type a **v** after the Da. Only one name starts with Dav so Excel enters David in the cell. Press Enter.

5. Type **Mar** in cell A8. Since there are two names that start with Mar in the column, Excel can't find a unique one at this point. Type a **t** after the Mar. Only one name starts with Mart so Excel enters Marten in the cell.

Pick from Drop-down List

In this exercise, you use the Pick from Drop-down List feature.

1. Right-click in cell A9 and then select **Pick from Drop-down List** from the menu.

    Notice that your entries are displayed in alphabetic order and no duplicates are displayed. See Figure 3-31.

Figure 3-31. *Entries are in alphabetic order and there are no duplicate names*

2. Click David. Press the Enter key.

3. Press **Alt + Down Arrow**. Select Marten. Press the Enter key.

4. Click the Save button 💾 on the Quick Access Toolbar.

The AutoComplete feature and the Pick from Drop-down List both provide a quick way of creating duplicate data in a column. The AutoComplete feature automatically provides suggestions for you to select from as you type. The Pick from Drop-down List displays all the entries you have made thus far in the current column and then lets you pick one from the list to duplicate. Next, you will look at how the AutoFill feature not only lets you duplicate data but also creates repeated patterns of data.

AutoFill

The AutoFill feature allows you to fill cells automatically with data that follows a pattern or that is based on data in other cells. It can also be used to create duplicate cells, enter a series of values, and create a custom list.

Use AutoFill to Create Duplicate Cells

You can copy values in one cell to additional cells by dragging the AutoFill Handle in the direction that you want. The AutoFill Handle is a small black square at the bottom right corner of the active cell (Figure 3-32). Placing the cursor over the AutoFill Handle turns the cursor into a plus sign (Figure 3-33).

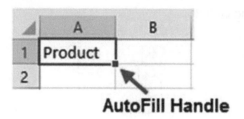

Figure 3-32. *AutoFill Handle*

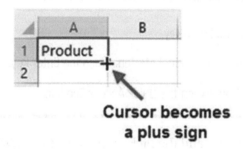

Figure 3-33. *Cursor becomes a plus sign when moved over AutoFill Handle*

A ToolTip is displayed as you drag the AutoFill Handle across other cells indicating the value that will be placed in that cell when you release the mouse button (Figure 3-34).

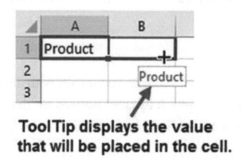

Figure 3-34. *ToolTip*

Figure 3-35 shows the result of dragging the AutoFill Handle from A1 through D1.

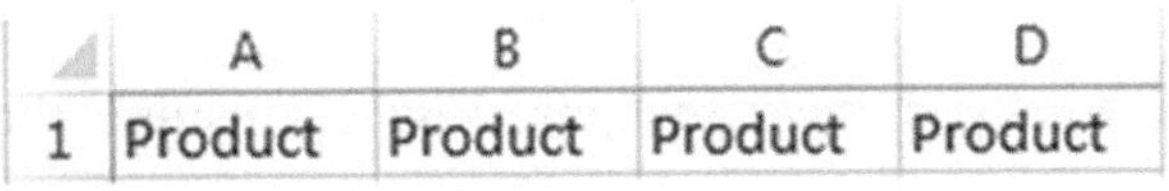

◿	A	B	C	D
1	Product	Product	Product	Product

Figure 3-35. *Results of dragging AutoFill Handle*

You can drag the AutoFill Handle across as many cells as you like, and you can drag it in any direction. Figure 3-36 shows the result of dragging the AutoFill Handle up.

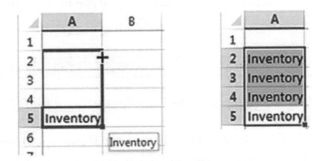

Figure 3-36. *Result of dragging Autofill Handle up*

Using AutoFill to Enter a Series of Values

AutoFill, in addition to being able to duplicate cell values, can also be used to create a series of repeating data lists. Excel has built-in data series related to days, months, years, numbers, and so on. For example, you can enter Monday into a cell and then use the Autofill Handle to automatically enter the other days of the week in adjacent cells. As you are dragging the AutoFill Handle, Excel shows you what is going to be placed in that cell when you have finished dragging (see Figure 3-37).

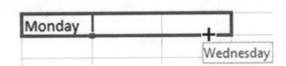

Figure 3-37. *As you drag across cells using the AutoFill Handle, Excel shows what it will place in that cell*

It doesn't matter what day of the week you start with. If you drag across more than seven cells, the days will start repeating. You can also use a three-character abbreviated day.

Dragging across the cells as shown in Figure 3-38 will result in the series shown in Figure 3-39.

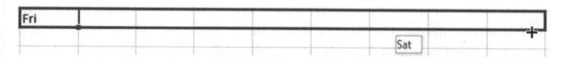

Figure 3-38. *Using the AutoFill Handle on a day of the week, it will repeat that day every seven cells*

Fri	Sat	Sun	Mon	Tue	Wed	Thu	Fri	Sat

Figure 3-39. *Result of dragging AutoFill Handle starting with Fri*

Excel knows that there are four quarters in the year. If you type **Quarter 1** and then use the AutoFill Handle to create a series, Excel will go from Quarter 1 to Quarter 4, and then it will repeat the pattern starting with Quarter 1. See Figure 3-40.

Quarter 1	Quarter 2	Quarter 3	Quarter 4	Quarter 1

Figure 3-40. *Series starting with Quarter 1*

You can also use the abbreviation Qtr for Quarter.

If you have a word followed by a number in a cell, Excel will increase the number by 1 for each cell in the series. See Figure 3-41.

Section 2	Section 3	Section 4	Section 5	Section 6

Figure 3-41. *Series starting with Section 2*

Teaching Excel to Create an AutoFill Pattern

You can teach Excel a pattern. If you want to display every third number, you can enter a value of 1 in a cell and a 4 in the next cell. You would select both cells and then, using the AutoFill Handle, drag across as many cells as you want to include in the pattern. This would produce a pattern of 1, 4, 7, 10, and so on.

If you want to display a series of adjacent cells with 15-minute intervals, select the first two cells to teach it the pattern (Figure 3-42) and then drag the AutoFill Handle for as many cells as you want to include in the pattern. When you finish dragging, the result will be as shown in Figure 3-43.

8:15	8:30					
					9:45	

Figure 3-42. *Select two adjacent cells to teach Excel an AutoFill pattern*

8:15	8:30	8:45	9:00	9:15	9:30	9:45

Figure 3-43. *Result of dragging AutoFill Handle using two adjacent times*

Using the AutoFill Options Button

When you have finished using the AutoFill Handle to copy a cell or create a series, an Autofill Options button appears to the bottom right of the last selected cell.

Clicking the Autofill Options button brings up three or more options if you used the AutoFill Handle to copy a cell that contains alpha data.

Clicking the Autofill Options button brings up three options when copying alpha data in a horizontal format. The default is Copy Cells. See Figure 3-44.

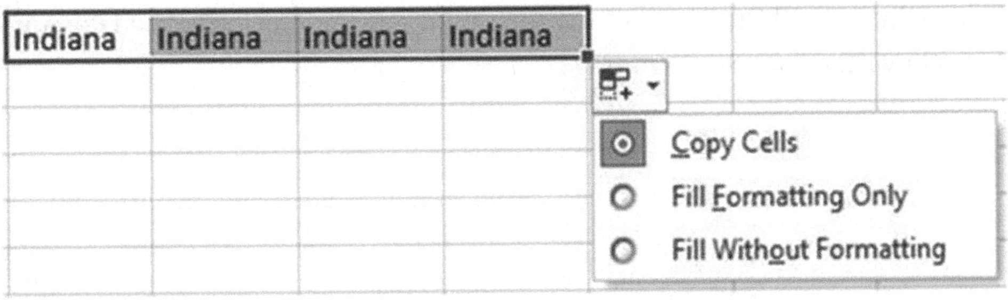

Indiana	Indiana	Indiana	Indiana	

⊙ <u>C</u>opy Cells
○ Fill <u>F</u>ormatting Only
○ Fill Witho<u>u</u>t Formatting

Figure 3-44. *Options available from AutoFill Options button*

Clicking the Autofill Options button brings up four options when copying alpha data in a vertical format. The default is Copy Cells. See Figure 3-45.

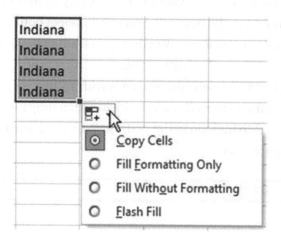

Figure 3-45. AutoFill options when copying data vertically

Clicking the Autofill Options button brings up four or more options if you used the AutoFill Handle to copy numeric data. See Figure 3-46.

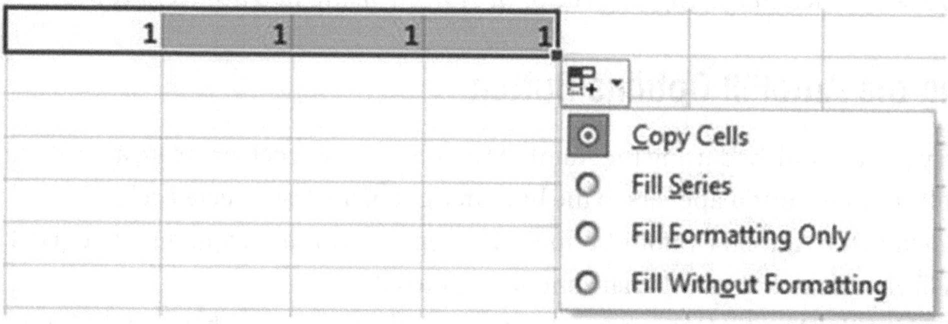

Figure 3-46. AutoFill options when copying numeric data

Clicking the Autofill Options button brings up many options if you used the AutoFill Handle to copy Date data. See Figure 3-47.

	A	B	C	D
1	10/28/2012			
2	10/29/2012			
3	10/30/2012			
4	10/31/2012			
5	11/1/2012			
6	11/2/2012			
7	11/3/2012			
8	11/4/2012			
9	11/5/2012			
10	11/6/2012			
11				
12		○ Copy Cells		
13		◉ Fill Series		
14		○ Fill Formatting Only		
15				
16		○ Fill Without Formatting		
17		○ Fill Days		
18		○ Fill Weekdays		
19		○ Fill Months		
20		○ Fill Years		
21		○ Flash Fill		
22				

Figure 3-47. *AutoFill options when copying date data*

Selecting the **Copy Cells** option would revert the series you just created back to copies of the original cell you used for making the series. The ten cells would all contain 10/28/2012.

The **Fill Formatting Only** option will only copy any formatting that you have applied to the original cell(s) to the cells you dragged across with the AutoFill Handle. It will not place any data into the cells.

The **Fill Without Formatting** option will copy the data or create a series from the original cells, but it will not apply any formatting from the original cells to the data you created using the AutoFill Handle.

The next practice, Exercise 3-8, illustrates the Autofill Options.

Create an Autofill Custom List

In addition to using Excel's built-in data series, you can create your own. If you find yourself constantly entering the same series of values, then you should create your own data series to speed up your entries. The data series can be created in the Custom Lists dialog box which can be accessed by clicking the Ribbon's File tab, selecting Options in the left pane, selecting Advanced in the left pane, scrolling the right pane until you see the General section, and then clicking the **Edit Custom Lists** button. See Figure 3-48.

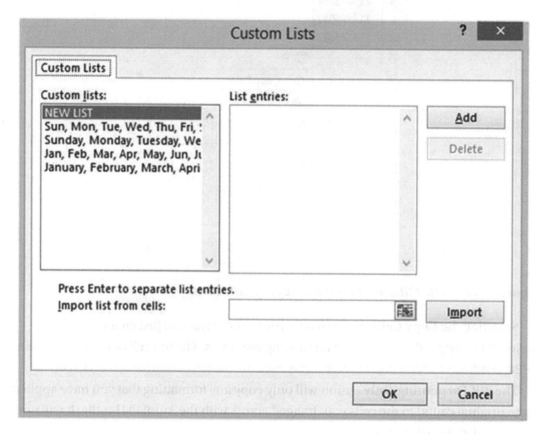

Figure 3-48. *Create your own Custom List in the Custom Lists dialog box*

You type the new data series in the List entries box. Then click Add to add the data series to the Custom lists box. You can also create a new custom list by selecting the cells on your worksheet and clicking Import in the Custom Lists dialog box.

EXERCISE 3-8: PRACTICE USING THE AUTOFILL HANDLE AND THE AUTOFILL OPTIONS BUTTON

In this exercise, you practice using AutoFill techniques. If you have closed the Chapter 3 workbook from the last practice, reopen it.

1. Add another worksheet and name it **AutoFill**.

2. Type **Tuesday** in cell A1. Using the AutoFill Handle, drag to the right. As you drag past each cell, notice what Excel is going to place in that cell. Drag to cell J1.

3. Type **Qtr1** in cell A2. Using the AutoFill Handle, drag to the right to cell F2.

4. Type **10:00** in cell A3. Using the AutoFill Handle, drag to the right to cell F3.

5. A series can be created vertically as well as horizontally. Type **Sep** in cell A4. Using the AutoFill Handle, drag down to cell A14.

6. Type **June** in cell H5. Using the AutoFill Handle, drag to the left to cell C5. You should now have a series of January through June.

7. Type **Store 3** in cell C7. Using the AutoFill Handle, drag to the right to cell G7. You should now have a series of Store 3 through Store 7.

Teach AutoFill a Pattern

You can teach AutoFill to use a consistent pattern.

1. Type **1** in cell C9. Enter a 3 in cell D9. Drag across both cells. With both cells selected, drag the AutoFill Handle through cell H9.

2. Enter a 1 in cell C10. Enter a 4 in cell D10. Drag across both cells. With both cells selected, drag the AutoFill Handle through cell H10.

3. Enter 8:30 in cell C11. Enter 9:00 in cell D11. Drag across both cells. With both cells selected, drag the AutoFill Handle through cell H11.

Use the Autofill Options Button

Try using the AutoFill Options button.

1. Type **Section 1** in cell A16 and **Section 2** in Cell B16. Select both cells. Press Ctrl + B + I to bold and italicize both cells. With both cells selected, drag the AutoFill Handle through cell F16.

2. Click the Autofill Options button. See Figure 3-49.

16	Section 1	Section 2	Section 3	Section 4	Section 5	Section 6			
17							⊞▾		
18								○ Copy Cells	
19									
20								⊙ Fill Series	
21								○ Fill Formatting Only	
22								○ Fill Without Formatting	

Figure 3-49. *AutoFill options*

3. Select Copy Cells. Instead of filling a series, the original Section 1 and Section 2 are copied to the other cells. See Figure 3-50.

16	Section 1	Section 2	Section 1	Section 2	Section 1	Section 2

Figure 3-50. *The selected two cells are copied to the other cells*

Holding down the Ctrl key as you drag the AutoFill Handle is a way of forcing Excel to copy values rather than create a pattern without having to use the Autofill Options button.

4. Press Ctrl + Z to undo the change.

5. Select cells A16 and B16. Hold down the Ctrl key while you drag the AutoFill Handle through cell F16.

6. Press Ctrl + Z to undo the change.

7. Select cells A16 and B16. With both cells selected, drag the AutoFill Handle through cell F16.

8. Click the Autofill Options button. Click **Fill Without Formatting**. The cells that were filled with the AutoFill Handle lose their formatting. See Figure 3-51.

| 16 | *Section 1* | *Section 2* | Section 3 | Section 4 | Section 5 | Section 6 |

Figure 3-51. *The values in the cells are copied but not any formatting*

9. Click the Autofill Options button. Click **Fill Formatting Only**. The data in the cells that were filled with the AutoFill Handle are cleared, but the Bold and Italicize formatting still apply to those cells. Type the word **Sales** in cell D16. The word Sales is bolded and italicized.

10. Type **2/1/2013** in cell A18. Drag the AutoFill Handle through cell F18. Click the Autofill Options button.

11. Select **Fill Weekdays**. The list skips over Saturday and Sundays.

12. Click the Autofill Options button. Select **Fill Months**. The list creates a series of months with the same day. See Figure 3-52.

| 2/1/2013 | 3/1/2013 | 4/1/2013 | 5/1/2013 | 6/1/2013 | 7/1/2013 |

Figure 3-52. *The Autofill option **Fill Month** automatically adjusts the month and year but leaves the day the same*

13. Click the Autofill Options button. Select **Fill Years**. The list creates a series of years with the same month and day. See Figure 3-53.

| 2/1/2013 | 2/1/2014 | 2/1/2015 | 2/1/2016 | 2/1/2017 | 2/1/2018 |

Figure 3-53. *The Autofill option Fill Years adjust the year but keeps the month and day the same*

14. Click the Autofill Options button. Select **Fill Without Formatting**. Excel's calendar starts with the date January 1, 1900, and since the list is unformatted, it displays the number of days since January 1, 1900. See Figure 3-54.

2/1/2013	41307	41308	41309	41310	41311

Figure 3-54. *The Fill Without Formatting option displays the number of days for that date since January 1, 1900*

15. Click the Autofill Options button. Select **Copy Cells**. The list consists of the same copied cell. See Figure 3-55.

2/1/2013	2/1/2013	2/1/2013	2/1/2013	2/1/2013	2/1/2013

Figure 3-55. *Using the option Copy Cells*

16. Click the Save button 💾 on the Quick Access Toolbar.

You have finished covering all of Excel's shortcuts for entering data. In this section, you learned how to copy data and repeat patterns. You even learned how to create your own patterns. In the next section, you'll learn how to make comments for the data in a cell. Comments serve as a memory aid for you or others to remember where the data came from or what it represents.

Creating, Viewing, Editing, Deleting, and Formatting Cell Notes

Excel has created a new type of comment that makes it easier for you to collaborate with other users. In previous versions, you would create a comment in a cell. If a user wanted to respond to that comment, they would need to create another comment. If you or other users wanted to respond to that comment, additional comments would need to be created. The new style of comments was created to resolve that issue. Now a single comment can be created in a cell in which all users can respond by placing their comments into reply boxes.

For those of you who have used Excel comments in the past, you will need to be aware of a different terminology. What was referred to in Excel 2016 and previous Excel versions as Comments are now referred to as Notes. The Ribbon's Review tab still contains the Comments group just as it did in Excel 2016, but these comment commands now refer to a new collaborative type of comment. What were previously referred to as comments are now in a new group named Notes that appear to the right of the Comments group. See Figure 3-56.

Figure 3-56. *New Comments group*

The Notes group consists of only a single drop-down command. See Figure 3-57.

We will first look at what use to be referred to as comments but now as Notes, and then we will look at the new style of collaborative Comments.

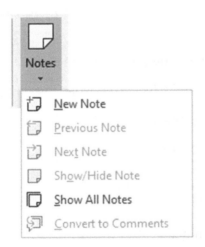

Figure 3-57. *Notes group*

You can attach a note to any worksheet cell. A note could be information that you want to share with others or just to help yourself remember—such as how or why a formula was entered, where the data came from, what data still needs to be entered, and so on. You can think of a note as a sticky note. Every place in the worksheet that you would want to place a sticky note to help you remember something, you can create a note. Notes are actually better than sticky notes because they won't fall off the sheet and they are easily editable. Cells that contain notes are easily identified by a red triangle in the upper right corner. The notes can be viewed by moving your cursor over the red triangle. See Figure 3-58.

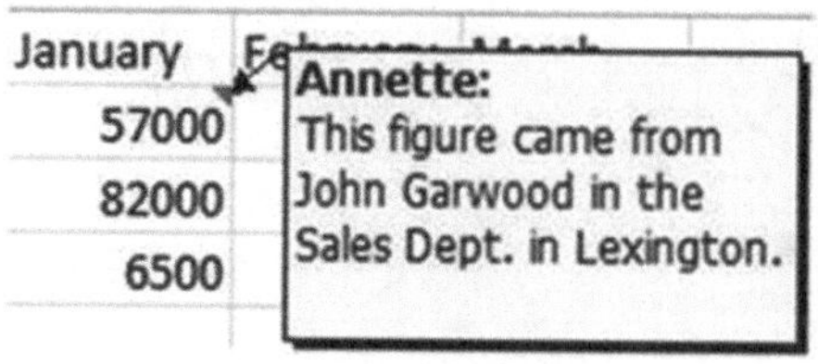

Figure 3-58. *Cells with notes have a red triangle in upper right corner*

How to Add a Note

Few people take the time to enter notes into their Excel worksheets. Adding notes is time well spent. I think it is very important to get beginners off to a good start and so I am presenting notes here in the data entry chapter because it should be a part of your normal entry routine.

1. Click the cell to which you want to attach a note.

2. Right-click the cell and select Insert Note.

3. Type the note in the note box.

4. Click outside the note box.

How to View a Note

Follow these steps to view a note:

1. Move your cursor over a red triangle.

2. If the workbook contains more than one note, you can click the **Previous** or **Next** button to move from note to note.

3. By Default, notes remain hidden until you move your cursor over the red triangle. If you want a note to remain visible even when you move your cursor away from the red triangle, then click the Show/Hide note button. The button is available from the Ribbon or by right-clicking the cell and selecting it from the context menu.

How to View All the Notes at the Same Time

You also can view all notes by doing the following:

1. Click the Review tab.

2. In the Notes group, click Show All Notes.

How to Change the Default Name for Notes

By default, Excel places your username in the note. The name can be removed and changed by doing the following:

1. To delete the username in a note, select the name and press the Delete key.

2. To change the default name for all notes, click the File tab and then click Options in the left pane.

3. Select **General** in the left pane.

4. In the **Personalize your copy of Microsoft Office** group, type the name you wish to use in the **User Name** text box.

5. Click the OK button.

Note   You might not use a worksheet for months and then things you thought you would remember you no longer do. If you are like me and write notes to yourself on a piece of paper, you will probably lose it and then spend hours looking for it. So, make life easy on yourself and use notes in Excel frequently.

Editing and Deleting Notes

If you right-click a cell that contains a note, the context menu no longer has an **Insert Note** option but instead contains an **Edit Note** and a **Delete Note** option. Clicking the Edit Note brings the note up for editing. Make the changes you wish and then click outside the note to update the note with the changes. Clicking Delete Note removes the note without asking for any verification.

Printing a Note

Notes do not print by default. The option to print notes is located on the Page Setup dialog box. The Page Setup dialog box can be accessed by clicking the Page Layout tab on the Ribbon and then clicking the dialog box launcher of the Page Setup group. It can also be accessed by clicking the File tab, then selecting Print, and then clicking the **Page Setup** link. Once you are in the Page Setup dialog box, select the **Sheet** tab. From the comment drop-down box, you can select either to have the notes printed at the end of your worksheet data or to be printed as they appear in your worksheet.

EXERCISE 3-9: HANDLING NOTES

In this exercise, you add, edit, delete, and print notes.

1. Start a new worksheet named "Notes."

2. Right-click any cell. Select **Insert Note**. Type anything you wish for a note.

3. Do the same for three more cells. You should now have four cells with red triangles in them.

View the Notes

1. Move your cursor over each note to view it. Click one of the cells that contain a note.

2. On the Ribbon, on the Review tab, click the Notes button and select **Show/Hide**. The note is displayed.

3. Select **Show/Hide** again. The note is now hidden.

4. Click the Notes button and select **Next Note**. Do this three times. Click the Notes button and select **Previous Note.** Do this twice.

5. Click the Notes button and select **Show All Notes** button. All four notes are displayed. Click the **Show All Notes** button again. The four notes are now hidden.

<u>Edit a Note</u>

1. Right-click the cell of your first note. Click **Edit Note**.

2. Type some more text in the note. Click somewhere outside the note.

<u>Delete a Note</u>

1. Right-click one of the cells that contains a note.

2. Click **Delete Note**. The note is removed. The cell no longer contains a red triangle.

<u>Print the Notes</u>

Only the notes that are currently displayed on the worksheet will be printed.

1. Click the Notes button and select **Show All Notes** to display all the notes at once.

2. Click the File tab on the Ribbon. Select Print from the left pane.

3. Click Page Setup located at the bottom of the Settings options. This brings up the Page Setup dialog box.

4. Click the Sheet tab on the Page Setup dialog box.

5. Click the down arrow of the **Comments** drop-down box. This displays three options. See Figure 3-59.

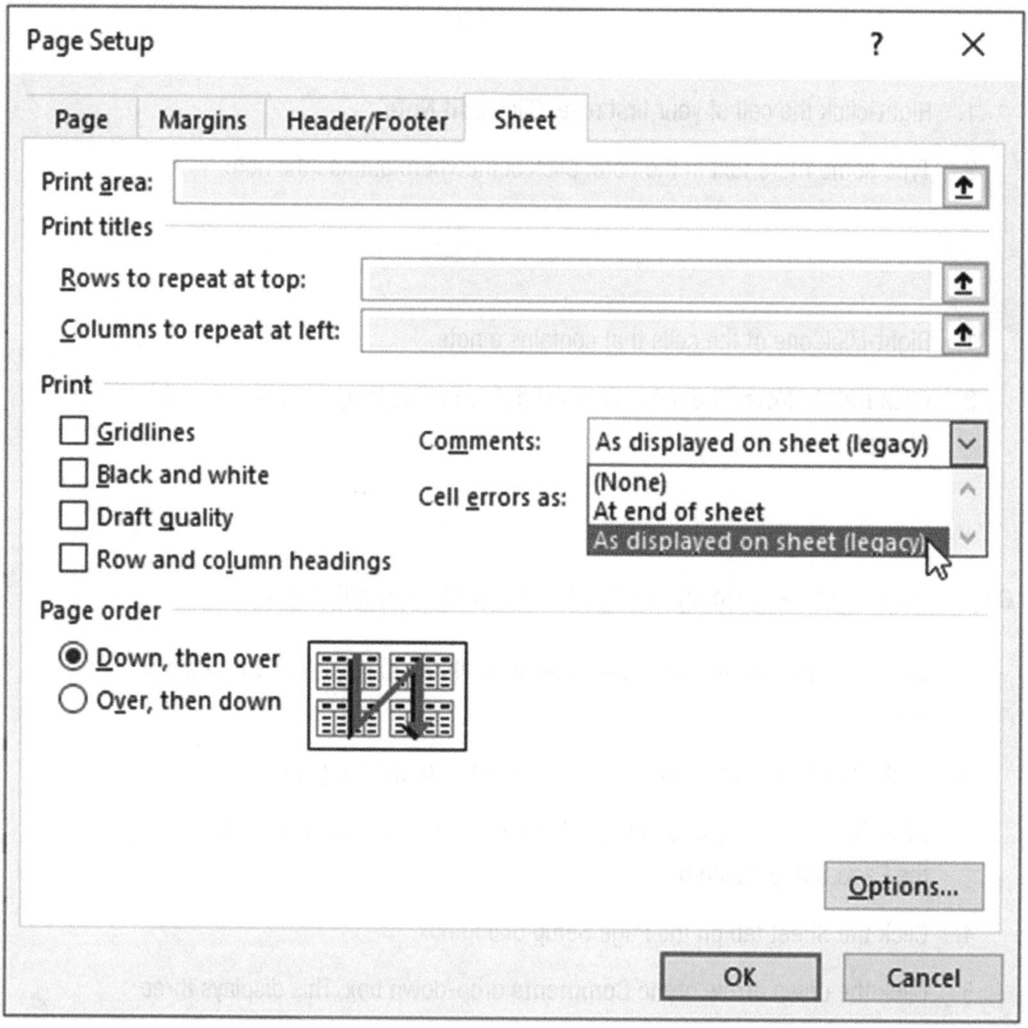

Figure 3-59. *Select where notes are to be printed*

Note   Microsoft didn't change Comments to Notes in the Page Setup dialog box. If you select **At end of sheet**, Excel will print both notes and comments. If you select **As displayed on sheet**, Excel will only print Notes, not Comments.

Selecting **None** will display no notes or comments in the printout.

Selecting **At end of sheet** will print the notes and comments after the spreadsheets have printed. The printout shows the notes along with the cell address of each note. See Figure 3-60.

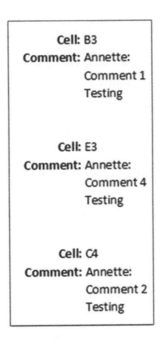

Figure 3-60. *Result of selecting At end of sheet*

Selecting **As displayed on sheet** prints the notes within the data exactly as they appear on the spreadsheet. See Figure 3-61.

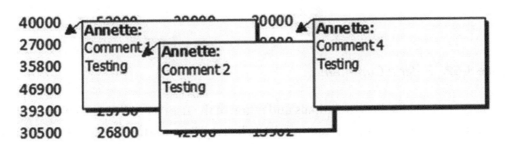

Figure 3-61. *Result of selecting As displayed on sheet*

Change the Background Color or Border Color of a Note

1. Right-click the border of a note. Select Format Comment. Select the tab Colors and Lines. You can select a color from the Fill area to change the background. You can select a color from the Line area to change the comment border color. Click the OK button.

The New Collaborative Comments

The difference between Notes and the new Comments are that the comments feature contains a reply box. With the old Comment method, each person that wanted to respond to one of your comments would have to create a new Comment, but with the new Comments, each time a user wants to respond, a new reply box is added to the Comment. This makes collaboration on spreadsheets much easier. These new comments are sometimes referred to as Thread Comments because each user is creating a new thread.

The new Comments feature is available in Word, Access, and PowerPoint as well. Figure 3-62 shows a new Thread Comment. An icon appears in the upper right-hand corner of a cell that contains a comment. Move your cursor over the icon to display the associated comment.

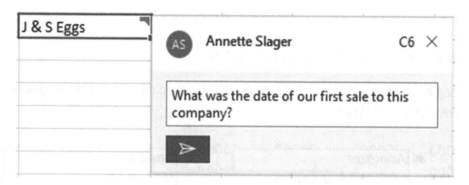

Figure 3-62. *Enter a Comment*

The Comment displays the initials and name of the user as well as the cell address. You enter your comment in the box and then click the Post button below it. Clicking the Post button displays the comment with the date and time it was posted. See Figure 3-63. If you need to make a change to the comment, click Edit. A new Reply box is displayed for another user to add a response to the comment. Unlike Notes, you can't change the size of Comments or the background color.

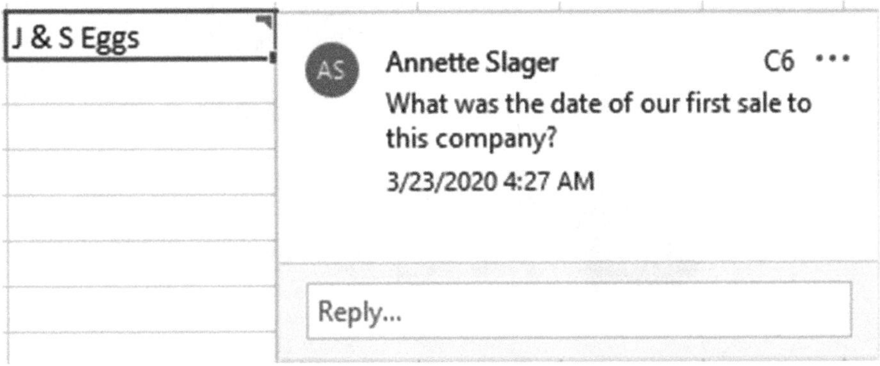

J & S Eggs
AS
Annette Slager
C6 •••
What was the date of our first sale to this company?
3/23/2020 4:27 AM
Reply...

Figure 3-63. *Posted Comment*

Once you have posted a comment, you expect a response. To get a response, you can share the spreadsheet by emailing it to someone, share it on OneDrive, or share it on a network drive. If you are sharing the document with someone who uses your same local computer, you will want to change the user information. This can be done by clicking File on the Ribbon and then selecting Account in the left pane. At the top of the right pane, you will see the current user Information. Click Switch account to switch to a different user. See Figure 3-64.

User Information
AS
Annette Slager
Sign out
Switch account

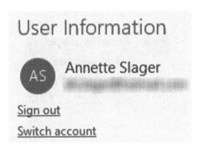

Figure 3-64. *Switch user*

Clicking Switch account brings up a dialog box at the top right of your Excel window. See Figure 3-65. You may see other available users from which you can select the user you want. If there aren't any other users to select, then click Sign in with a different account. You will then need to enter the username and password for a Microsoft account or create a new one.

Figure 3-65. *Sign in with a different user*

The second user can then move his cursor over the comment icon and then enter his response into the Reply box. See Figure 3-66.

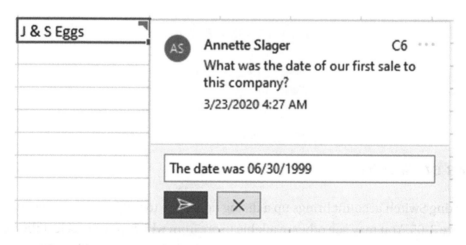

Figure 3-66. *Sign in with a different user*

Figure 3-67 shows how the comment would look after the second user clicked the Post button. Now another user or one of the same users can post another comment. This can go on and on.

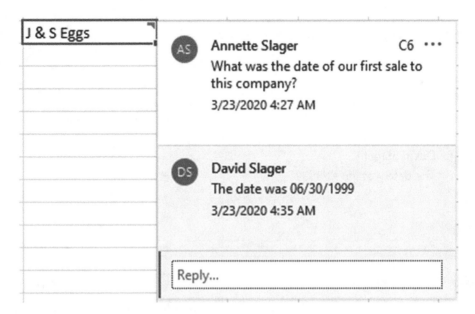

Figure 3-67. *Sign in with a different user*

The buttons for the Comments are located on the Ribbon's Review tab. See Figure 3-68.

Figure 3-68. *Comment buttons*

When you show all of the Notes in a spreadsheet, they appear next to the cell that references them, but clicking Show Comments doesn't do that; rather, it displays all comments in a pane to the right of the spreadsheet. See Figure 3-69. You can enter replies to the comments in the pane if you want instead of moving your cursor over the comment icon and then entering a reply.

Figure 3-69. *All comments can be displayed in a pane*

You can use the Next Comment and Previous Comment buttons to jump to another comment. You can delete a comment by selecting a cell that contains a comment and then click the Delete button.

Print your comments by clicking the Ribbon's File tab and then click Print. Click Page Setup. Click the Sheet tab. Click the down arrow for comments. Select **At end of sheet**. The comments will not work with **As displayed on sheet** (legacy).

Summary

This chapter covered the various data types and how those types can affect how your data looks and how it is adjusted within a cell. It also covered different ways of entering and editing data.

This chapter focused on knowing how to change the size and appearance of columns and rows in the worksheet and learning how to use special characters as well as Excel's AutoCorrect function. You can create, edit, delete, and print notes and comments that you add to a cell.

Excel's AutoCorrect feature can be useful for such things as correcting commonly misspelled words, correcting misuse of capitalization, entering special symbols that can't be directly entered from the keyboard, and creating shortcuts for entering often used words or phrases.

The AutoComplete and Pick from Drop-down List features allow you to easily duplicate existing column values.

In Chapter 4, you'll learn how to use more complex data and cell formatting features. There are also a variety of options for presenting and formatting numeric data including those for accounting purposes. You'll also begin using the mini-toolbars and context menus, which provide more shortcut options particular to what you are working on.

CHAPTER 4

Formatting and Aligning Data

You'll now learn how to refine the way your worksheets look on screen and in print. Excel provides many great options to establish a style for worksheets and the workbook as a whole that is distinctive and consistent.

After reading and working through this chapter, you should be able to

- Format data and cell backgrounds

- Create cell borders

- Check which formats have been applied to the current cell

- Align data

- Copy the formatting from one cell to other cells

- Format numeric data

- Use Context Toolbars menus

Reasons for formatting data include emphasizing certain entries (e.g., totals) and making the worksheet more visually appealing and easier to understand (e.g., by using colors to visually segment rows and columns). This helps keep information organized for both the person creating the worksheet and others who use it for analysis and reports. Formatting can also be used to change how you display numeric data (with dollar signs, commas, decimals, percent, etc.). Formatting is also used to align data. You can use Excel's Format Painter to copy all the formatting applied to a cell to other cells. You can create cell borders to separate different areas or to highlight a particular area on your worksheet.

© David Slager and Annette Slager 2020
D. Slager and A. Slager, *Essential Excel 2019*, https://doi.org/10.1007/978-1-4842-6209-2_4

Formatting Your Text Using the Font Group

The Font group located on the Home tab has options for such things as changing the font, size, color, and so on. Figure 4-1 shows the formatting command buttons in the Font group. We will discuss each of these in the following text.

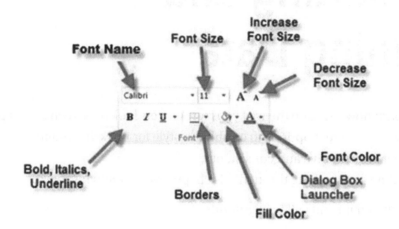

Figure 4-1. *Font formatting command buttons*

Using Bold, Italics, Underline, and Double Underline

You can use any combination of bolding, italicizing, or underlining on your text. Selecting one or any combination of these options affects the active cells. These options work like a toggle switch; clicking them turns them on, clicking them again turns them off. You can switch between underlining and double underlining by clicking the down arrow for the underline button and then making your selection. See Figure 4-2.

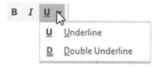

Figure 4-2. *Select Underline or Double Underline*

Changing the Font and Its Size

It's important to pick the right font and size not only to make your spreadsheet data stand out but to make it easy to read.

Fonts

The font can be changed by clicking the down arrow for the font and then picking your selection from a list. See Figure 4-3. The font list shows how the different fonts look. As you pass your mouse pointer over the various fonts in the list, Excel displays the text in the active cell using the font that your mouse pointer is passing over. Changing the font affects the text of all currently selected cells on your worksheet.

Figure 4-3. *Available fonts*

There are a lot of fonts in the list. If you want to jump to a specific font quickly, you type the first letter(s) of the font name in the Font Name Box. For example, if you wanted to use the font Bell MT, you would type the first three letters Bel (there is only one font that starts with Bel) and then press Enter.

Font Size

Clicking the down arrow for the font size shows all the available font sizes for the currently selected font. Figure 4-4 shows the top half of the available font sizes for the Calibri font.

Figure 4-4. Select font size

Clicking the Increase Font Size button **A** increases the size of the font to the next available font size. Looking at Figure 4-4 if the current font size is 11, clicking the Increase Font Size button will increase the font size to the next available size, which is 12. If the current font size is 18, clicking Increase Font Size will increase the font size to the next available size, which is 20.

Clicking the Decrease Font Size button **A** does the opposite; it changes the font size to one size less than the current size. Looking at Figure 4-4, you can see that if the current font size is 18, clicking the Decrease Font Size button will decrease the font size to 16 points. A font size is measured in points. Each point is approximately 1/72 of an inch.

EXERCISE 4-1: CHANGE THE FONT AND FONT SIZE

This exercise covers various ways of selecting fonts and changing font sizes.

1. Create a new workbook.

2. Save the workbook giving it a name of **Chapter 4**. Change the name of the worksheet from Sheet1 to Font.

3. Type your name in cell H5 and then press Ctrl + Enter to accept the value.

4. Click the Ribbon's Home tab in the Font group and click the down arrow for the Font.

5. Observe how your name changes as you move your mouse pointer over the different fonts. Select the font Arial Narrow if you have it; otherwise, select another typeface.

6. Click the down arrow for the font size.

7. Observe the size of your name as you move your mouse pointer over the different font sizes. Select Size 26.

8. Now, try the Increase and Decrease buttons:

    1. Click the Increase Font Size button. The Font Size changes to 28.

    2. Click the Increase Font Size button again. The Font Size changes to 36.

    3. Click the Decrease Font Size button. The Font Size changes to 28.

    4. Click the Decrease Font Size button. The Font Size changes to 26.

9. Double-click cell H5 to put it in Edit mode. Drag across the first three letters in your name to select them.

10. Click the down arrow for the font. Observe how only the first three letters change as you move your mouse pointer over the different fonts.

11. Type **Ti** in the **Font Name Box** to jump to the Times New Roman font. Press Enter.

You should now be familiar with all of the tools available on the menu's Font group. Next, you will see how you can access these same tools from the Font group's dialog box launcher.

Using the Font Group's Dialog Box Launcher

Clicking the Font group's dialog box launcher brings up the Format Cells dialog box with the Font tab active. See Figure 4-5. Here, you can specify most of the options that are also available on the Font group, plus you can add Effects from one convenient location.

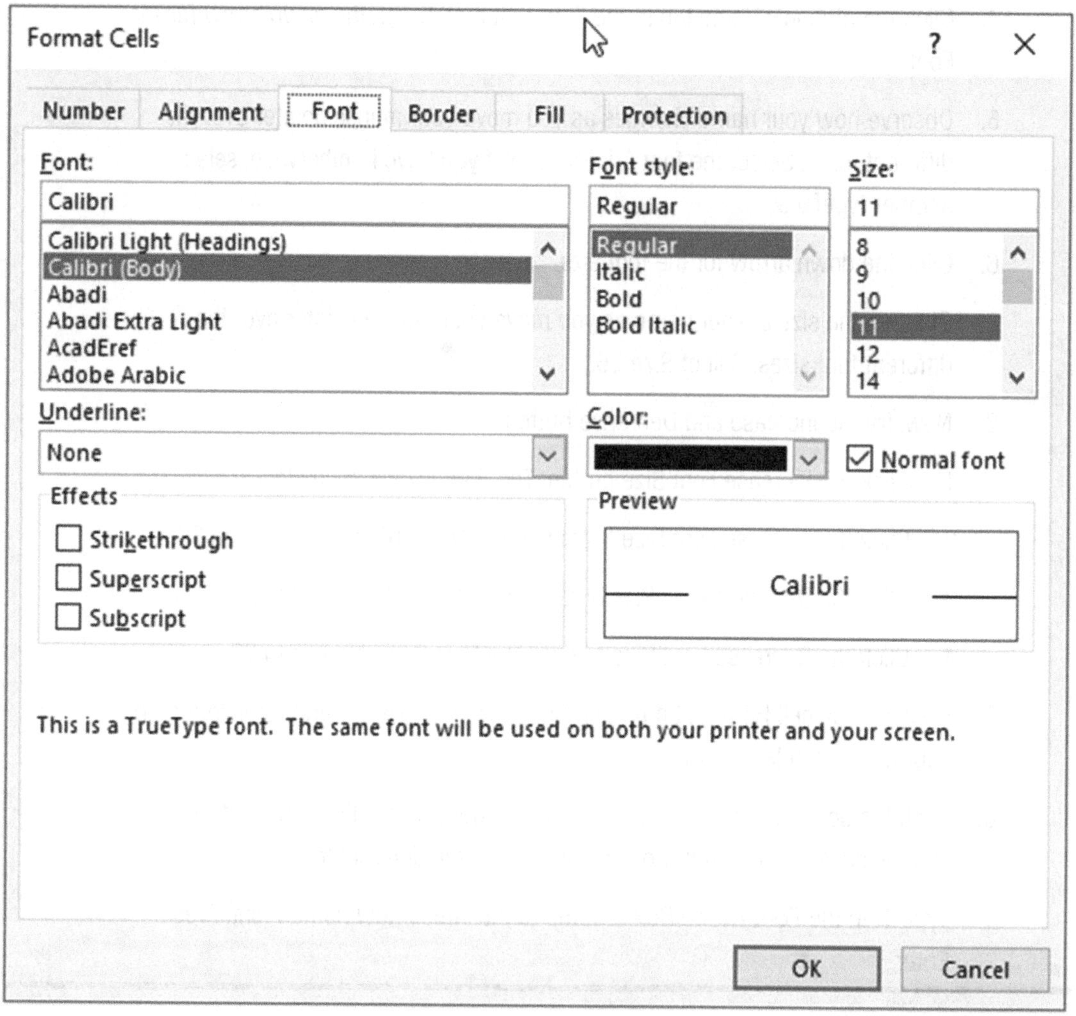

Figure 4-5. Font options

Formatting with Color

You can change your worksheets from drab to attention grabbing by adding colors. You can add colors not only to make your worksheet more stimulating but also to

- Emphasize the cells that you think are important

- Separate different areas of your worksheet

- Make your worksheet easier to read

Be careful how you use color. Using too many colors or using colors in an incorrect manner such as using dark text against a dark background can make your worksheet confusing and hard to read.

Changing the Font Color

Clicking the Font color button 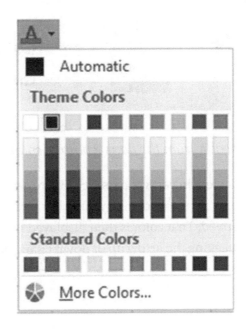 will apply the current color displayed on the icon to the text of your active cells. You can select a different color by clicking its down arrow.

You can select any color from the set of Theme Colors or from the Standard Colors as shown in Figure 4-6. If you want more colors to select from, click the More Colors option.

Figure 4-6. *Theme and Standard Colors*

Clicking the More Colors option brings up the Colors dialog box which has Standard and Custom tabs. The Standard tab lets you select color from a honeycomb type of grid. The current area in the bottom right shows what the font color is for the current cell. When you select a color from the grid, it is displayed in the New area at the bottom right of the window. See Figure 4-7. This is the color to which the font will be changed if you click the OK button.

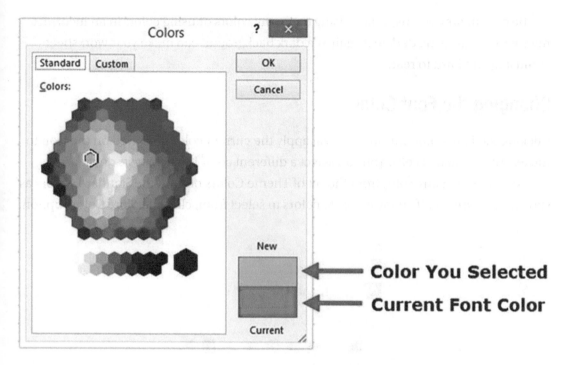

Figure 4-7. *Select a new color from the honeycomb of colors*

The Custom tab shows colors in vertical bars. See Figure 4-8. The edge of each vertical bar blends in with the adjacent color. You can click any spot in the color palette that matches the color you need. That color is then displayed in the vertical bar to the right of the color palette. Dragging the arrow up or down changes the intensity of the color. See Figure 4-8.

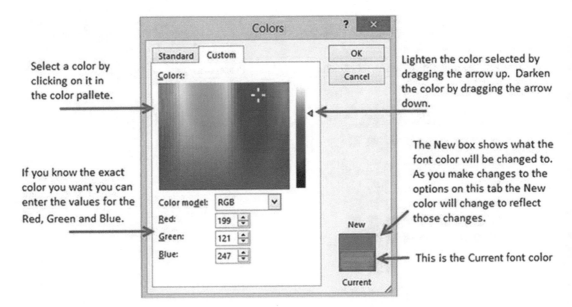

Select a color by clicking on it in the color pallete.

If you know the exact color you want you can enter the values for the Red, Green and Blue.

Lighten the color selected by dragging the arrow up. Darken the color by dragging the arrow down.

The New box shows what the font color will be changed to. As you make changes to the options on this tab the New color will change to reflect those changes.

This is the Current font color

Figure 4-8.   *Create a custom color*

Changing the Cell Background Color

Clicking the Fill Color button 🎨 ▾ will apply the current color displayed on the icon to the background of your active cells. You can select a different color by clicking its down arrow. This will bring up the same options that were available for the font color.

Clicking the dialog box launcher for the Font group brings up the Format Cells dialog box. The Fill tab in the Format Cells dialog box has options for creating Fill Effects or a patterned background. See Figure 4-9.

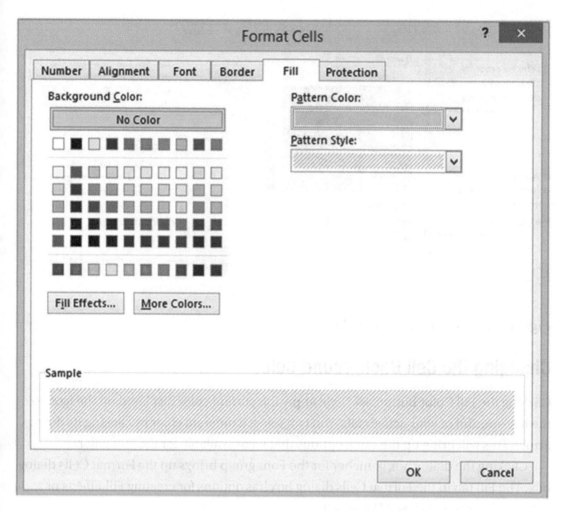

Figure 4-9. *Select a Fill Color, Pattern Color, or Pattern Style*

Check Which Formats Have Been Applied to the Current Cell

You can tell some of the formats that have been applied to a cell by just clicking the cell and then observing what is displayed in the Font group on the Ribbon. Looking at Figure 4-10, you can see that the text in cell C1 has been bolded and italicized. It has an Arial Black font and has a font size of 16.

Note   If all of the text in the cell does not have the same format, you will need to select those characters for which you want to reveal their formatting.

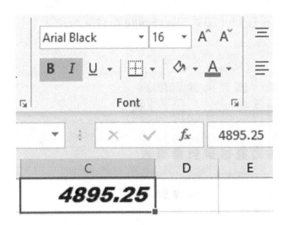

Figure 4-10. *Selected text is bolded and italicized*

The Borders, Fill Color, and Font Color options do not reflect the current cell. They reflect the selection that was made the last time they were used.

EXERCISE 4-2: CHANGING FONTS AND BACKGROUND COLORS

This exercise covers changing the font's color as well as changing the background color of a cell. The cell background can be a solid color, a gradient, or a pattern.

1. Create a new worksheet named "Colors."

2. Click cell B3. Click the Ribbon's Home tab, and in the Font group, click the down arrow for the Fill Color button ⬩ ˅. Select Black.

3. Now, type and format some text:

   a. Click the down arrow for the Font Color button **A** ˅ and select White.

   b. Type **Total Values**. Press Ctrl + Enter.

   c. Change the font type to Calibri if it isn't already.

   d. Click the Increase Font Size button **A**˄ once.

e. Press Ctrl + B to bold the text.

f. Widen the column by double-clicking between the column heads labeled B and C so that all of the text fits in the cell.

4. Click cell B4. Click the down arrow for the Fill Color button. Select the color "Green, Accent 6, Lighter 80%." See Figure 4-11.

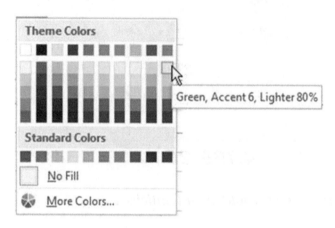

Figure 4-11. *Select Green, Accent 6, Lighter 80%*

5. Click the down arrow for the Font Color button. Select Black. Type **Total Values** in cell B4. Press Ctrl + Enter. Change the font to Cambria. Change the font size to 14. Click the bold and italics buttons. Widen the column to accommodate the changes.

6. Click cell B3 again. You can see what formatting has been applied to the cell. The Bold button is highlighted. The font shows as Calibri. The font size reflects that you clicked the Increase Font Size button. Click the down arrow for the Fill Color button. The selection box is around the color black. Click the down arrow for the Font Color. The selection box is around the color white. Click cell B4. See if you can detect what formatting has been applied to it.

Change Background to a Gradient

1. Select cells B5:D5. Click the dialog box launcher 🔲 for the Font group.

2. Click the Fill tab.

3. Click the Fill Effects button. The Gradient tab is for changing the background to a gradient color. See Figure 4-12.

4. Select any color you want for Color 1 and then select another color for Color 2.

5. Click each of the shading styles and observe the changes made to the gradient pattern.

6. When you have the colors and style you wish to use, click the Variant box that contains the pattern you want to use. The Sample shows how the background will look using the choices you have made.

7. Click the OK button for the Fill Effects dialog box. Click the OK button for the Format Cells dialog box.

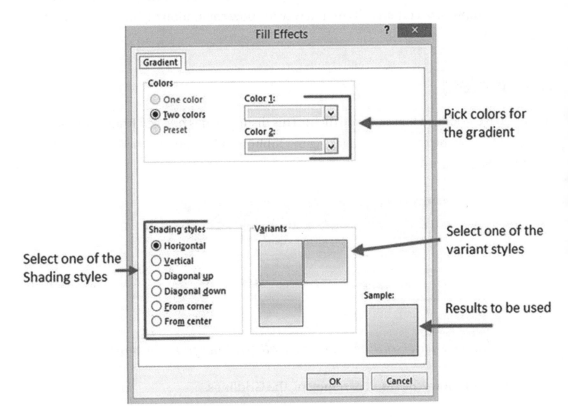

Figure 4-12. *Change cell background colors to a gradient*

<u>Change Background to a Pattern Style</u>

1. Select cells B7:D8. Click the dialog box launcher for the Font group. Click the Fill tab. Select a pattern color of your choice. Select a pattern of your choice from the Pattern Style drop-down box. Click the OK button.

2. Click the Save button 💾 on the Quick Access Toolbar.

You have learned how to alter the appearance and size of your text as well as emphasize it. You can use this knowledge to draw your user's eye to the text you want. Your company may have standards for how your worksheets should look with certain font types and colors. Next, you will see how placing borders around your cells can also be used to emphasize certain cells or as a way to show separation of different areas of your worksheet.

Cell Borders

All cells have gridlines that separate them. This distinguishes one cell from another. The gridlines can be removed by clicking the Ribbon's View tab and then unchecking Gridlines. See Figure 4-13.

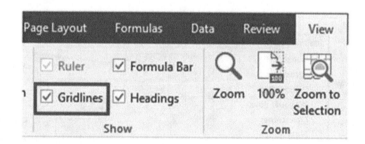

Figure 4-13. *Uncheck Gridlines to remove all gridlines from the worksheet*

Figure 4-14 shows the result of removing the Gridlines.

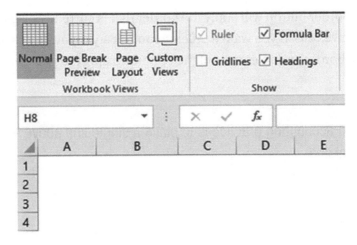

Figure 4-14. *No Gridlines*

There are times, however, when you will want to distinguish a particular cell, row, column, or group of cells. You can do this by using borders. A border can be just a line at the left, top, right, or bottom of the cell or range of cells. It can also be a combination of lines or a full box around the cell or range of cells. Figure 4-15 shows several different border types.

Barney's Burger Basket		
Store 1	**Product**	**Sales**
	Product A	3850
	Product B	1275
	Product C	419
	Product D	185
	Total	$ 5,729
Store 2	**Product**	**Sales**
	Product A	2735
	Product B	1953
	Product C	589
	Product D	672
	Total	$ 5,949

Figure 4-15. *Different border types*

Clicking the Border button will apply the border that is currently displayed on the button to your active cells. If you want a different border style, you can click the down arrow next to the border and select one of the options to apply to your selected cells. See Figure 4-16.

Note   Standard accounting practice is to place a single line above and a double line below a total line.

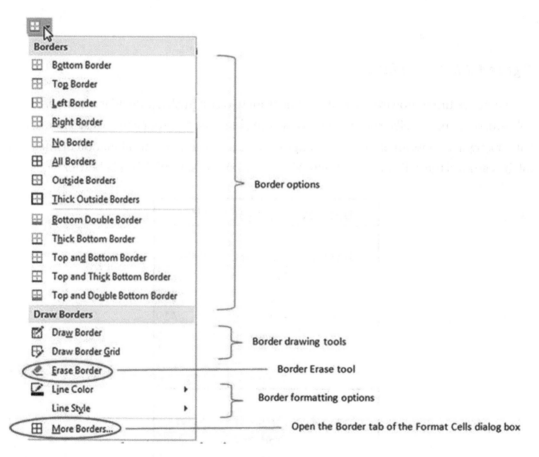

Figure 4-16. *Border options and tools*

Note   You can erase a portion of a border by using the Erase Border tool. The mouse pointer will display as an eraser when using the Erase Border tool.

Drawing Borders

You can even draw your borders by using either the Draw Border or Draw Border Grid tool. Your mouse pointer will display as a pencil when using one of the drawing border tools. The border will be drawn in the current line style.

The Draw Border tool draws a full outside border around the cells you drag across. See Figure 4-17.

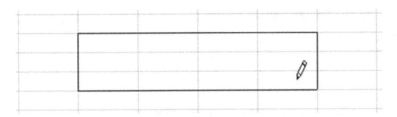

Figure 4-17. *Draw Border tool*

The Draw Border Grid tool places a full border around each cell. See Figure 4-18.

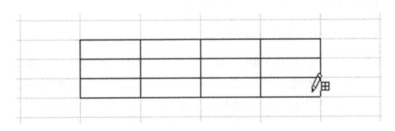

Figure 4-18. *Draws Border Grid tool*

EXERCISE 4-3: CREATING CELL BORDERS

In this exercise, you will use various methods of creating borders as well as changing border width, style, and color.

Add Borders Using the Border Button

1. Create a new worksheet named "Borders."

2. Select cells B2:C2.

3. On the Ribbon's Home tab, in the Font group, click the down arrow for the Borders button.

4. Click **Top and Double Bottom Border**. Click another cell so that you can see the results.

Draw Borders

1. Click the down arrow for the Borders button in the Font group.

2. Click **Draw Border**. The mouse pointer should turn into a pencil.

3. Draw a border around cells B4 to B8. Notice that the Border button's icon has changed to a pencil to reflect your selection.

4. Click the down arrow for the Border button.

5. Notice that there is now a shadow around the Draw Border button to indicate that the tool is currently active. Also, notice that the Line Color button has a colored line on it. This indicates the color that is currently being used.

6. When you change the Line Color or the Line Style, you are making changes to the currently active tool which is currently the Draw Border tool.

7. Click Line Color. Select the green color from the Standard Colors. See Figure 4-19.

Figure 4-19. *Change the border's line color*

8. Draw a border around cells B10:C13. The border should appear in green.

9. Click the down arrow for the Border button. Click the Draw Border button to deselect it.

10. Click the down arrow for the Border button. Click **Line Style**. Select the double line. See Figure 4-20.

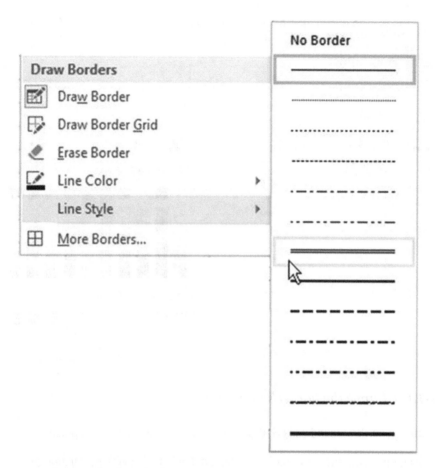

Figure 4-20. *Change the border's line style*

11.   If neither the Draw Border tool nor the Draw Border Grid tool is active when you select the Line Color or Line Style, Excel will automatically turn on the Draw Border tool.

12.   Draw a border around cells B15:C19. See Figure 4-21.

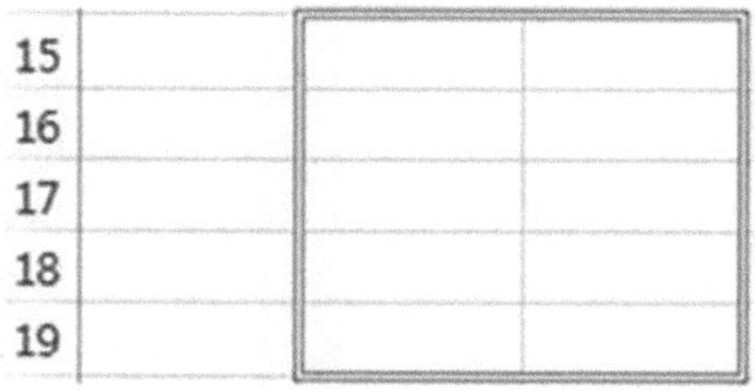

Figure 4-21. *Draw a border around the cell range B15:C19*

13. Press the Escape key to turn off the Draw Border mode.

Draw a Border Grid

1. Click the down arrow for the Border button. Notice that the Draw Border tool is no longer active. Pressing the Escape key made it inactive.

2. Click Draw Border Grid.

3. Draw a border around cells E3:F7. The border should appear as a green double line since these are the last options you selected.

4. The Border button icon has changed to reflect that the Draw Border Grid tool is currently active. See Figure 4-22.

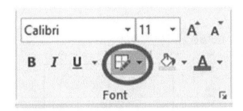

Figure 4-22. *Border button reflects that the Draw Border Grid tool is active*

5. Click the down arrow for the Border button. Notice that the Draw Border Grid tool is currently active. Select Line Color. Select the Red color from the Standard Colors.

6. Draw a border around cells E10:F12 and press the Escape key.

7. Select cells I2:J6.

8. Click the down arrow for the Border button. Select **Top and Thick Bottom Border**. Click another cell to see the result. (Notice that the lines now appear in red because this is the last line color that you used.)

Create a Border Using the Format Cells Dialog Box

1. Click cell I8.

2. Click the down arrow for the Border button. Select **More Borders**.

3. Create a border:

    a. Select the first solid line in the second column of the Style group.

    b. Select Automatic for the color.

    c. Click **Outline** in the Presets area. The Preview box in the Border area displays what the cell will look like. See Figure 4-23.

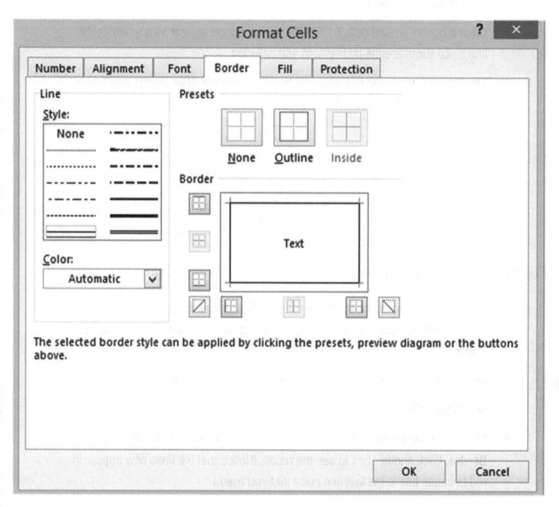

Figure 4-23. *Preview box in Border area displays what the border will look like*

4. The Inside option in the Presets area is grayed out because there is only one active cell and that is cell I8. Click **None** in the Presets area. This removes all borders from the Preview box.

5. Click the top border button in the Border area. A top border is displayed in the Text box. See Figure 4-24.

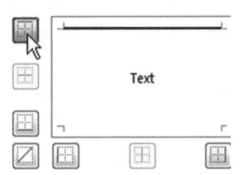

Figure 4-24. *Top border button creates a top border*

6. The Border buttons work like toggle switches. Clicking a Border button puts that border in the cell. Clicking the Border button again will remove that border. Click the top Border button to remove the border from the cell.

7. Click the left Border button and then the right Border button. See Figure 4-25.

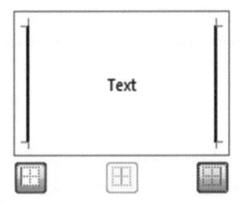

Figure 4-25. *Left and right Border buttons create left and right borders*

8. Click the OK button. Click another cell so that you can see the results. Cell I8 should now have a left and right border.

9. Select cells E15:F20.

10. Click the down arrow for the Border button. Select **More Borders**. Because multiple cells are selected, the Inside button in the Presets area and all the borders in the Border area are available.

11.   Click the **Vertical Inside** button ⊞ .

12.   Click the **None** button in the Presets area to clear the Preview area.

13.   Select the first solid line in the second column of the Style group.

14.   Select Red for the color.

15.   Click the **Outline** button in the Presets area. This activates the top, bottom, left, and right Border buttons.

16.   Select Green for the color.

17.   Click the horizontal inside button ⊟ . See Figure 4-26.

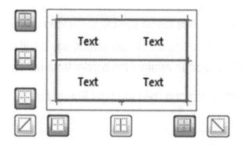

Figure 4-26. *The horizontal Border button creates horizontal border*

18.   Click the OK button.

Erase Borders

1.   Click the down arrow for the Border button.

2.   Select **Erase Border**. The mouse pointer becomes an eraser ✐ . Clicking a border line with the tip of the eraser will remove that border line.

3.   Erase the first, third, and fifth horizontal green lines. Since the horizontal lines are two cells wide, you will need to click the horizontal line for each cell. You will probably find it easier just to hold down the left mouse button and drag across the horizontal line. See Figure 4-27. If you erase too much, just click the Undo button ↺ or press Ctrl + Z and try again.

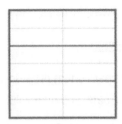

Figure 4-27. *Erased the first, third, and fifth inside border horizontal lines*

4. Press the Escape key. This turns off the Eraser tool.

5. Click the Save button 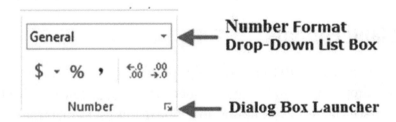 on the Quick Access Toolbar.

Formatting Numeric Data Using the Number Group

You have seen how Excel makes assumptions based on the data you entered as to how that data should be formatted and aligned in the cell, such as right-aligning numeric data. If you want to change that formatting, you will need to use one of Excel's many formatting features which are located in the Number group on the Home tab. See Figure 4-28. Table 4-1 provides a description for each of the options in the Number group.

Figure 4-28. *Ribbon's Number group on the Home tab*

Table 4-1. *Description for Each Option in the Number Group*

Buttons on Number Group	Description
General ▾	The Number Format drop-down list provides various defaults for numeric, date, and time values. See Figure 4-26.
$ ▾	Clicking the dollar sign portion of the button formats the current cell in the Accounting format (dollar signs, commas, and two decimal positions). If you want to use a different currency such as pounds or euros, you can click the down arrow to the right of the $ button.
%	Multiplies the value in the cell by 100 and adds a % sign to the end of the value.
'	Adds commas to a value. If the value doesn't already contain a decimal, it will add .00 to the value.
←.0 .00 .00 →.0	The first button (Increase Decimal) increases the number of decimal positions in the cell by one each time you click it. The second button (Decrease Decimal) decreases the number of decimal positions in the cell by one each time you click it.
⌐	Dialog box launcher brings up the Format Cells dialog box with the Number tab selected.

Using Default Formats

Let's say that the active cell currently has a value of 3859.27.

Clicking the drop-down arrow of the Number Format drop-down list displays a list of default formats (see Figure 4-29). Clicking an option in this list formats the current value as shown below each format label. What you see is what you get. There are no options to select.

For example, if you were to click the Fraction option, the format of the current cell would change and the result would be 3859 ¼.

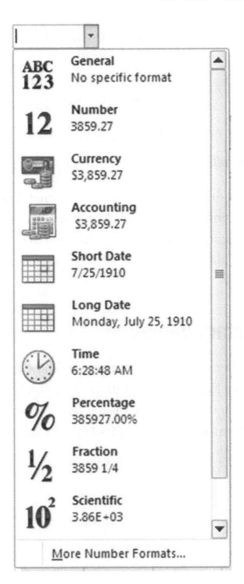

Figure 4-29. *Default formats for different data types*

Accounting and Currency Formats

The Accounting format is a special monetary format that displays values with a dollar sign, commas, and two decimal positions. The Accounting format lines up the dollar signs and decimal points in a column. See Figure 4-30. The dollar sign is always placed in the left-most position in the cell. Depending upon the width of the cell and the value you have in it, there could be many spaces between the dollar sign and the value itself.

Accounting	Currency
$ 3,857.25	$3,857.25
$ 853.29	$853.29
$ 4.75	$4.75
$ 62,895.40	$62,895.40

Figure 4-30. *Difference between Accounting and Currency formats*

The Currency format is similar to the Accounting format. It also adds dollar signs, commas, and two decimal positions to values. The difference is that the Currency format uses a floating dollar sign and the Accounting format uses a fixed dollar sign. The floating dollar sign doesn't align the dollar signs; instead, it places the dollar sign to the immediate left of the value.

Formatting Monetary Values

The $ button in the Number group performs the exact same formatting as selecting the Accounting format from the Number Format drop-down list. If you want to display the value in a different format or if you want to use a different currency such as pounds or euros, you can click the down arrow to the right of the $ button. See Figure 4-31.

Figure 4-31. *Formatted monetary values*

If you want to use something other than two decimal positions or you want to use a currency other than those shown in the drop-down box, you can select the last option **More Accounting Formats**. Selecting this option is one of the many ways of opening the Format Cells dialog box.

From here, you can select one of the many different currency options available in the Symbol drop-down list. See Figure 4-32.

Figure 4-32. *Accounting options for decimal places and monetary symbol*

Converting Values to Percent Style

The Percent Style % button multiplies the values in your selected cells by 100 and places the % after the number. The value .25 would be formatted as 25%. The % button rounds the value and then removes any decimal positions as shown in Figure 4-33.

Original	Formatted
0.254391	25%
0.25105	25%
0.255391	26%
0.257012	26%

Figure 4-33. *Formatted using percent style*

If you want to use a percentage format that displays two decimal positions, then click the drop-down arrow of the Number Format drop-down list and select **% (Percentage)**. Figure 4-34 shows the result of applying this format.

Original	Formatted
0.254391	25.44%
0.25105	25.11%
0.255391	25.54%
0.257012	25.70%

Figure 4-34. *Formatted with % option*

Converting Values to Comma Style

The Comma Style ❜ button places comma separators in the values for the cells you have selected. The value 3857495.25 would be formatted as 3,857,495.25.

Changing the Number of Decimal Places

The ⬆.00 Increase Decimal button increases the number of decimal positions by one for each time you click it. The icon shows a single decimal place changing to two decimal places.

The ⬇.0 Decrease Decimal button decreases the number of decimal places by one each time you click it. The icon shows two decimal places changing to one decimal place.

Accessing the Format Cell Dialog Box

If you want to use more options when formatting a number, then using the Number tab in the Format Cells dialog box is the way to go. Using the Number tab in the Format Cells dialog box, you can do such things as format fractions, dates, times, and other formatting that you can't do in the other sections. What options appear on the right side of the window depends upon the category selected. See Figure 4-35.

The Format Cells dialog box can be accessed from the Number group by using any of the following methods:

- Click the Dialog Box Launcher.

- Click the drop-down arrow for the Number Format drop-down box and then select **More Number Formats**.

- Click the down arrow to the right of the $ button and then select **More Accounting Formats**.

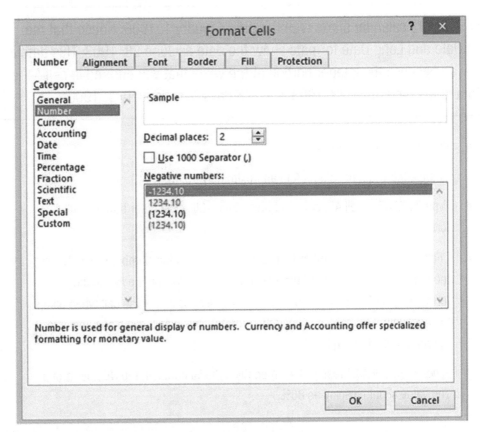

Figure 4-35. *Format Cells dialog box with Number formatting selected*

EXERCISE 4-4: FORMATTING NUMBERS

In this exercise, you use various numeric data formats. If you have closed your workbook from the last practice, you can reopen the Chapter 4 workbook.

1. Create a new worksheet named "NumberGroup."

2. Type **1275.85** in cell A1, but don't press Enter. Click the Enter button in the formula bar ✓ . Clicking the Enter button accepts the value you entered and leaves the current cell as the active cell. Once the value has been accepted, the check and delete buttons are grayed out on the formula bar.

3. On the Home tab, in the Number group, click the down arrow for the Number Format drop-down box General ⬇. Notice that every option on the drop-down box shows how the value in A1 will appear if you click that particular format.

Note   Excel's calendar starts with the date January 1, 1900. Notice that the Short Date and Long Date formats show the date as June 28, 1903. This date is displayed because the integer portion of the value that you entered in cell A1 is 1275 and the date **June 23, 1903** is 1,275 days past January 1, 1900.

---

4. Click the Currency format.

5. Change the width of column A to approximately 13.43.

6. Type **3427.35** in cell A2, but don't press Enter. Click the Enter button in the formula bar.

7. Click the Accounting Number Format button $ ⬇ in the Number group. Notice the difference between the Currency format in cell A1 and the Accounting format in cell A2. The $ appears in the left-most position of the cell when using the Accounting format. The Accounting format also leaves one blank space to the right of the last digit.

8. Type .85 in cell A3. Press Ctrl + Enter. Click the % button in the Number group. The value should display as 85%.

9. Enter 1385197 in cell A4. Press Ctrl + Enter. Click the **Comma Style** button in the Number group. The value displays as 1,385,197.00.

10. Click the Decrease Decimal button .00/→.0. One of the decimal positions is removed. Click the Decrease Decimal button a second time. The second decimal position is removed. Click the Increase Decimal button ←.0/.00 twice to change the value back to having two decimal positions.

11. Enter −385 in cell C1. Press Ctrl + Enter. Click the down arrow for the Number Format drop-down box. Notice that the Currency format displays the value as −$385.00, but the Accounting format displays it as ($385.00). In Accounting, negative numbers are displayed in parentheses.

Note   Notice that the Scientific format displays the value as **−3.85+E02**. The Scientific format when dealing with whole numbers puts the decimal after the left-most digit, rounds the number to two decimal positions, and displays the rest as an exponent.

---

12. Click Accounting to change the format in cell C1.

13. Enter 3.25 in cell C2. Press Ctrl + Enter. Click the down arrow for the Number Format drop-down box. Click **Fraction**. The Fraction format converts the decimal portion of a number to a fraction. The fraction portion .25 is changed to 1/4.

14. Enter 5/15/2012 into cell C3. Press Ctrl + Enter. Click the dialog box launcher on the Number group, shown in Figure 4-36, to bring up the Format Cells dialog box. See Figure 4-37.

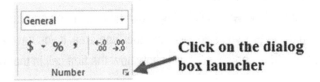

Figure 4-36. *Number group*

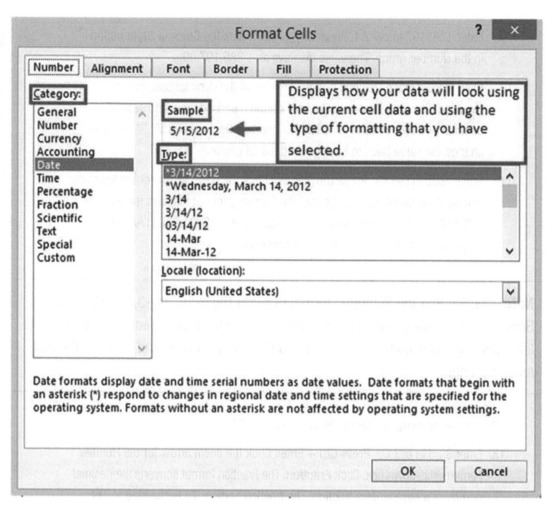

Figure 4-37. *Format Cells dialog box with Number tab selected*

15.   Excel recognizes the value as a date and so it opens the Format Cells dialog
      box with the Date category selected. After the category has been selected,
      you will need to select how you want your data to appear from the Type area.
      The Sample area of the window displays how the data in the current cell will
      appear using the option you selected from the Type area. If you have a range of
      cells selected on the worksheet, it will show how the first cell in that range will
      appear.

16.   Click Number in the Category area. The Number category has options for
      entering the number of decimal positions, for using a comma, and for letting
      you select how you want to display negative numbers.

17. Click each of the other categories to familiarize yourself with the different options for formatting your data.

18. Click the Date Category again.

19. In the Type area, select *Wednesday, March 14, 2012. The Sample area shows how this change will display the current cell.

20. Click the OK button.

21. Cell C3 may display as a number of # signs. What does this mean? That's right; the column isn't wide enough. Widen the column to see your results.

22. Click the Save button 💾 on the Quick Access Toolbar.

You have learned how to use the tools from the Ribbon's Number group to format numeric data. Formatting the data is necessary so that the user can interpret the data the way it is meant to be, such as making sure the user knows a number is a percentage and not just a numeric value. Next, you will learn how to align data within a cell.

Aligning Data Using the Alignment Group

The Alignment group is located on the Home tab. See Figure 4-38. It contains options for aligning your text vertically and horizontally, rotating text, increasing and decreasing indenting, word wrap, and merging cells. Table 4-2 provides a description for each option in the Alignment group.

Figure 4-38. *Alignment group on the Home tab*

Table 4-2. *Descriptions of Alignment Buttons*

Buttons	Description
≡ ≡ ▦	**Vertical Alignment**—The Top-, Middle-, and Bottom-Align buttons are for aligning your text to the top of the cell, the middle of the cell, and the bottom of the cell in that order.
🄰▾	**Orientation**—Clicking this button brings up options for selecting the angle and direction of the text within a cell.
🔄 Wrap Text	**Wrap Text**—Used for wrapping your data across multiple lines within a cell.
≡ ≡ ≡	**Horizontal Alignment**—The Left-, Center-, and Right-Align buttons are for aligning your text to the left side of a cell, to the middle of a cell, and to the right side of a cell in that order.
←≡ ≡→	**Indentation**—The Decrease Indent button (left) is for decreasing the amount of indentation in a cell. The Increase Indent button (right) is used for increasing the amount of indentation in a cell.
⊞ Merge & Center	-**Merge & Center**—This button provides options for merging cells, merging and centering cells, and unmerging cells.

Fitting More Text into a Cell

Excel provides three options for fitting more data into a cell:

Text Wrapping—Text automatically wraps to the next line.

Shrink to Fit—Shrinks the size of the cells contents so that they fit within the cell.

Merge Cells—Merges multiple cells into a single cell.

Text Wrapping

If your text doesn't fit within a cell, you might want to break up the text on to multiple lines. Excel has a feature called Word Wrap that will format the cell so that the text wraps automatically. You can turn the word wrap feature on by selecting the cell(s) you want to automatically word wrap and then clicking the Home tab in the Ribbon. In the Alignment group, click the Wrap Text button. You can also force a line break to occur wherever you want by pressing Alt + Enter.

Note   When word wrap is turned on, data in the cell wraps to fit the column width. When you change the column width, data wrapping adjusts automatically.

Figure 4-39 shows a before and after example of using Wrap Text.

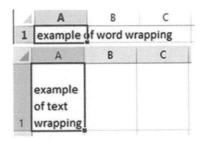

Figure 4-39.   *The upper part of the image shows text before using Wrap Text, and the lower part shows text after Wrap text has been applied*

Shrinking to Fit

Turning on the Shrink to fit option will shrink the cell's contents so that they will fit within the size of the cell. Figure 4-40 shows a before and after example.

◢	A	B		◢	A	B
1	Shrink to Fit			1	Shrink to Fit	

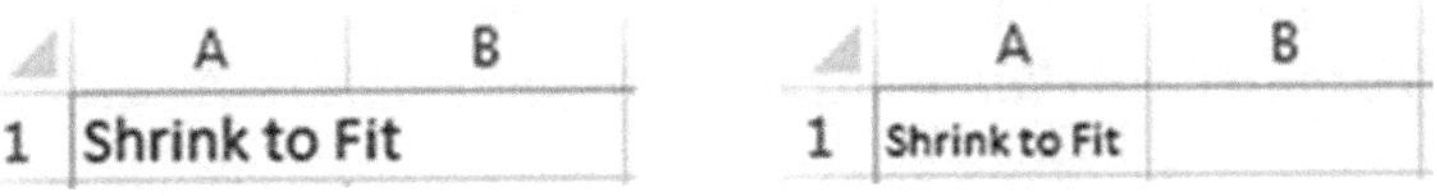

Figure 4-40.   *The image on the left shows text before using Shrink to fit, and the one on the right shows it after Shrink to fit has been applied*

If you enter a great deal of text, it may become too small to see unless you expand the width of the column. The Shrink to fit option is located on the Alignment tab of the Format Cells dialog box. See Figure 4-41. There are several ways to open this tab. One way is to click the dialog box launcher of the Alignment group and then select the Alignment tab.

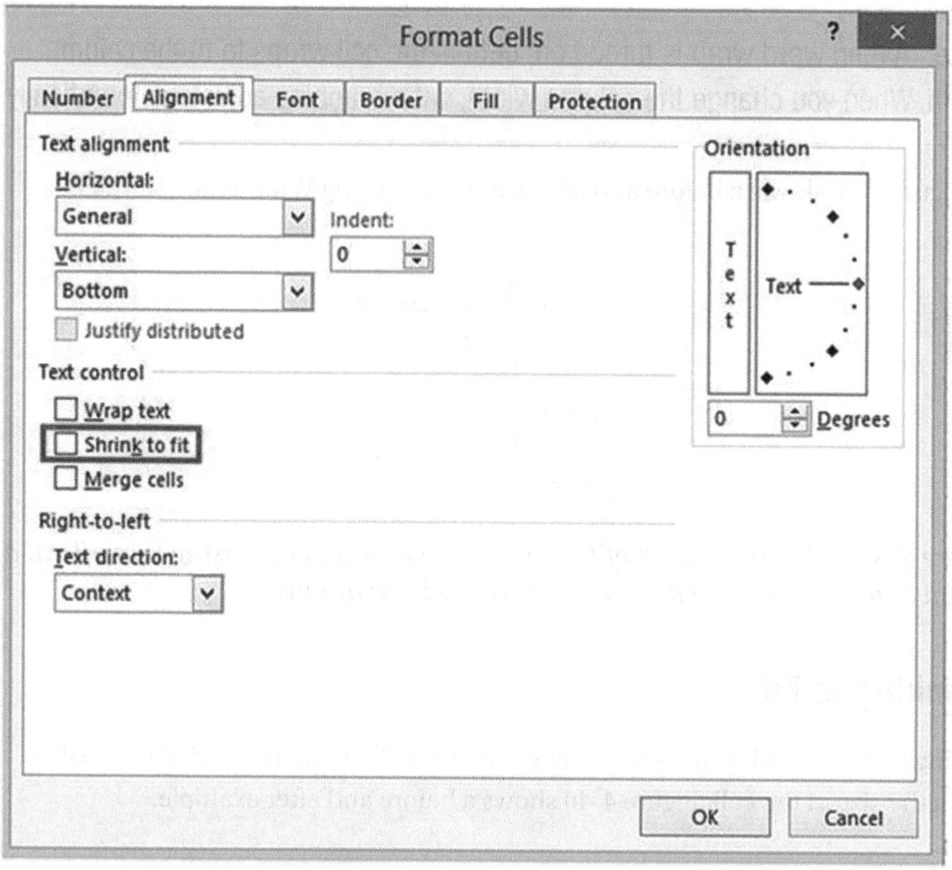

Figure 4-41. Format Cells dialog box with Alignment tab selected

The Shrink to fit option can be turned on or off by selecting or clearing the **Shrink to fit** check box. Notice that there are also options here for turning on Wrap text and Merge cells.

Merging and Unmerging Cells

You can join multiple cells to make one larger cell by using one of the three different options for merging cells:

- Merge & Center

- Merge Across

- Merge Cells

The **Merge & Center** option merges the selected cells and then centers the data horizontally in the new expanded cell. The Merge & Center button is located on the Home tab ➤ Alignment group.

You can improve the look of your spreadsheet by making your heading centered across the width of your worksheet. In Figure 4-42, the heading has been entered into cell A1. The heading would look better if it were centered across columns A through F.

	A	B	C	D	E	F
1	Thompson Boiler Sales					
2						
3	Model	Qtr1	Qtr2	Qtr3	Qtr4	Total
4	L3921	32500	27500	6000	45000	111000
5	R2929	5030	4075	5900	6201	21206
6	R5000	45802	42927	50035	50050	188814

Figure 4-42. *Heading before being centered*

To center the heading across the cells, you select the cells and select the Merge & Center option. When you merge cells A1, B1, C1, D1, E1, and F1, you end up with one cell that takes up the space of all six cells. See Figure 4-43. The single cell will have an address of A1.

	A	B	C	D	E	F
1			Thompson Boiler Sales			
2						
3	Model	Qtr1	Qtr2	Qtr3	Qtr4	Total
4	L3921	32500	27500	6000	45000	111000
5	R2929	5030	4075	5900	6201	21206
6	R5000	45802	42927	50035	50050	188814

Figure 4-43. *Cells merged and the heading centered*

The Merge & Center command lets you merge cells that are not only adjacent horizontally but vertically as well. Even though you can merge cells that are vertically adjacent, the Merge & Center command is usually used for creating labels that expand multiple columns or for spanning a heading over a report.

All the Merge options can be accessed by clicking the down arrow to the right of the Merge & Center button. See Figure 4-44.

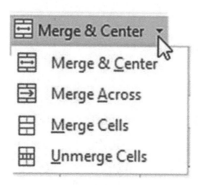

Figure 4-44. *Merge options*

The other three options do the following:

- The **Merge Across** option can only merge cells that are horizontally adjacent. It does not center the text.

- The **Merge Cells** option can merge cells that are horizontally or vertically adjacent. Like the Merge Across option, it does not center the text.

- You can unmerge cells by selecting the cells you want to unmerge and then clicking the **Unmerge Cells** option.

Aligning and Indenting Text in a Cell

Excel provides options for setting the alignment for data in a cell (see Table 4-3) and for increasing or decreasing indentation in a cell (see Table 4-4).

Table 4-3. *Vertical and Horizontal Alignment*

Buttons	Description
≡ ≡ ▦	**Vertical Alignment**—These buttons are for aligning your text to the top of the cell, the middle of the cell, and the bottom of the cell in that order. Excel aligns data at the bottom by default.
≡ ≡ ≡	**Horizontal Alignment**—These buttons are for aligning your text to the left side of a cell, to the middle of a cell, and to the right side of a cell in that order.

Table 4-4. *Increasing and Decreasing Indentation of Text in a Cell*

Buttons	Description
⇐	**Decrease Indent**—Used to decrease the amount of indentation in the cell.
⇒	**Increase Indent**—Used to increase the amount of indentation in the cell.

Align Text Vertically and Horizontally

There are other options on the Alignment tab of the Format Cells dialog box for Horizontal and Vertical alignment. See Figure 4-45. Some of these are the same as those that appear on the Alignment group of the Home tab, but here you have additional options such as the ability to set the amount of indention for the Left, Right, and Distributed horizontal alignments by using the Indent numerical text box.

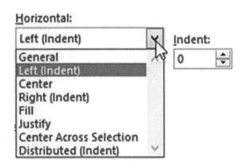

Figure 4-45. *Alignment options*

The Distributed horizontal alignment uses an equal amount of indention on both sides of the cell text. Therefore, if you enter 3 for the Indent, the cell will have three blank spaces to the left and to the right of whatever is entered in the cell.

The Justify horizontal alignment option is only useful when you are entering multiple lines of text in a cell. Excel *justifies* each line except the last line by adjusting the space between words so that the text spreads completely from the left to the right borders of the cell. See Figure 4-46.

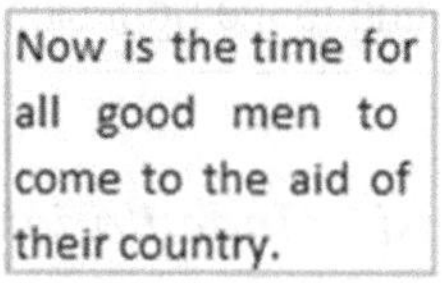

Now is the time for all good men to come to the aid of their country.

Figure 4-46. Justify alignment

Rotating Text

Clicking the Orientation button (Figure 4-47) displays a list of options for rotating your text. The first five default options are self-explanatory. If you want to specify another particular orientation, select the last option **Format Cell Alignment**. This option displays the Alignment tab on the Format Cells dialog box.

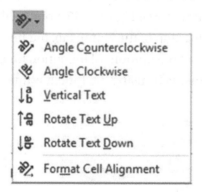

Figure 4-47. Orientation options

The Orientation section on the Alignment tab of the Format Cells dialog box is for setting the angle of your text. See Figure 4-48.

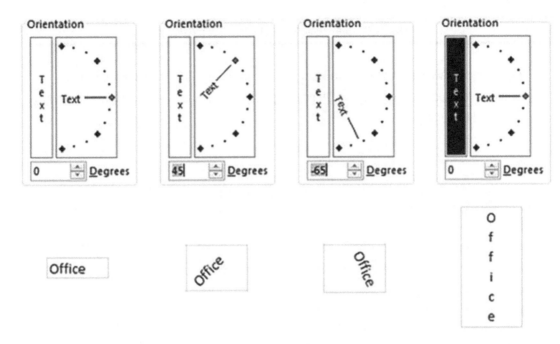

Figure 4-48. *Options for setting the angle of your text*

You can set the angle that your text will be displayed by dragging the line up or down or by setting the degrees. If you enter the number of degrees, Excel automatically adjusts the line and vice versa. Moving the line up from 0 degrees makes the text display from bottom to top. Moving the line down from 0 degrees makes the text display from top to bottom. If you are entering the degrees and you want your text to display top to bottom, you will need to enter a negative value. Clicking the Text button on the left side of the Orientation box puts the text in a completely vertical format.

Note   When you rotate text, the size of the cell automatically adjusts to fit the rotated text.

EXERCISE 4-5: USE THE ALIGNMENT GROUP

For this exercise, we will create the worksheet shown in Figure 4-49. The worksheet shows information for students attending the Office Integration and Financial Management classes at Wonderland University.

	A	B	C	D	E
1		Wonderland University			
2		Student ID	Dorm Room	Summer Session	Parking Permit
3		1357	123A	X	X
4	Office Integration	1276	125B		X
5		11295	143A	X	
6		709	302C		
7		23105	4A		
8		3057	57C	X	
9		29	105C		
10		325	204A		X
11					
12		14175	209B	X	X
13	Personal Financial Strategies	1085	46C		
14		51	177A	X	
15		2092	311C	X	
16		30195	303C	X	X
17		1495	209A		
18		1776	198B	X	
19		20215	198A		X
20		1091	200B	X	X

Figure 4-49. Worksheet you will create

1. If you have closed your workbook from the last practice, you can open the Chapter 4 workbook and then create a new worksheet named "Align."

2. Type Wonderland University in cell A1. Click the Enter button on the formula bar.

3. Select cells A1:E1. Click the Merge & Center button.

4. Select the Lucida Calligraphy font. Change the font size to 18.

5. Type the column head Student ID in cell B2 and then press the Tab key.

6. Type the rest of the column heads in row 2.

7. We want to place the column headings at a 45-degree angle. Before we do that, we need to widen the columns.

Note   If you do not widen the columns, you will end up with multiple column heads in the same column as shown here.

8. Drag across column heads B through E. Click the Ribbon's Home tab, and in the Cells group, click the Format button. Select **Autofit Column Width**.

9. Select cells B2:E2.

10. In the Alignment group, click the Orientation button ✎ ▾ and then select **Angle Counterclockwise**. Each of the column heads should now appear at a 45-degree angle in their separate columns.

11. Type **1357** in cell B3 and then press the Tab key. Since the value is a number, it should appear right-justified in the cell.

12. Type **123A** in cell C3 and then press the Tab key. Since the value contains a letter, it is considered a text or string value and is left-justified in the cell.

13. Type a capital **X** in cell D3 and then press the Tab key. X is text so it is left-justified.

14. Type a capital **X** in cell E3 and then press the Enter key.

15. Continue typing the rest of the data in cells B4:E10.

16. Select cell range A3:A10 by dragging across the cells or by typing A3:A10 in the Name box and then pressing Enter. Click the Merge & Center button.

17. Type `Office Integration`. Don't worry about how the text looks at the moment. Press Ctrl + Enter.

18. Click the Orientation button 🔲 ▾ and then select `Rotate Text Up`.

19. Click the Bold button **B** or press Ctrl + B.

20. Click the Middle Align button ≡ in the top row of the Alignment group.

21. Next, we want to create a defining space between the two classes:

    a. Select cells A11:E11. Click the Merge & Center button.

    b. Increase the height of row 11 to approximately 22.5 by clicking between row heads 11 and 12 and then dragging down.

22. Type the data in cells B12:E20.

23. Select cell range A12:A20 by dragging across the cells or by typing A12:A20 in the Name box and then pressing Enter.

24. Click the Merge & Center button.

25. Type `Personal Financial Strategies`. Don't worry about how the text looks at the moment. Click the Enter button on the formula bar.

26. Click the Orientation button 🔲 ▾ and then select `Rotate` Text Up.

27. Click the Bold button **B** or press Ctrl + B.

28. Click the Middle Align button ≡. The text extends slightly beyond its border. We will use **Shrink to fit** so that it fits completely in the cell.

29. Click the dialog box launcher on the Alignment group. In the Text Control area, click **Shrink to fit**. Click the OK button.

30. Click cell A1, hold down the Ctrl key, and then click cells A3 and A12. Click the down arrow for the Fill Color button located in the Font group. Select the Blue-Gray, Text 2, Lighter 80% color. See Figure 4-50.

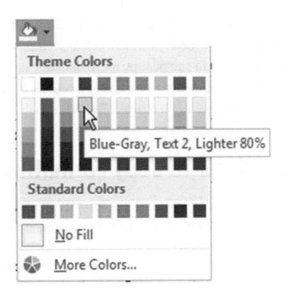

Figure 4-50. *Color selections*

31. We could improve the look of our spreadsheet by having the data appear just a little to the right of the left cell border:

    a. Select cells B3:E10. Click the Increase Indent button ⊫ in the Alignment group.

    b. Select cells B12:E20. Click the Increase Indent button ⊫.

32. Type your full name in cell B22 and then press Ctrl + Enter. Right-click cell B22. Select Format Cells. On the Alignment tab, check **Shrink to Fit**. Click the OK button. Your name should fit completely within the cell.

33. Right-click cell B22. Select Format Cells. On the Alignment tab, check **Wrap Text**. Click the OK button.

34. Type your last name in cell C23 and then press Ctrl + Enter. Right-click cell C23. Select Format Cells. Click the vertical word **Text** on the left side of the Orientation box. Click the OK button.

35. With cell C23 still selected, click the following buttons on the Alignment group: click the Align Left button, then click the Align Right button, and then the Center button.

36. Click the Save button 💾 on the Quick Access Toolbar.

You have learned how to use the commands from the Ribbon's Alignment group to align data both horizontally and vertically within a cell as well as rotating cell data. You practiced Merging Cells so that data can occupy more than one cell. The Word Wrap and Shrink to fit features allowed you to place more text in a cell. Next, you will see how you can copy all the formatting that you applied to a cell to other cells.

Using Format Painter to Copy Formatting

The Format Painter can be used for copying all the formatting in one cell to other cells. The Format Painter button has an icon of a paintbrush. It appears on the Clipboard group of the Home tab. See Figure 4-51.

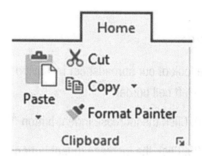

Figure 4-51. *Clipboard group*

The Format Painter not only speeds up the process of applying the same formatting to other cells but also makes it easy to give your workbook a consistent look.

Single-clicking the Format Painter button allows you to copy the formatting of a selected cell to a single range of cells, and then it is automatically turned off. Double-clicking the Format Painter button keeps the button turned on so you can keep applying the formatting to as many cell ranges as you want until you click the button again to turn it off.

EXERCISE 4-6: USING THE FORMAT PAINTER TO COPY FORMATTING

This exercise takes you through the process of copying formatting that has been applied to one cell to other cells.

1. If you have closed your workbook from the last practice, you can open the Chapter 4 workbook and then create a new worksheet named "Format."

2. Click cell A1. Type Copy formatting and then press Ctrl + Enter.

3. Widen column A so that all the text appears in the cell.

4. Type the word **Data** in cells B2 and C3. Use the AutoFill Handle on cell C3 and drag down to cell C5. Let go of your left mouse button. Drag the AutoFill Handle to cell E5. See Figure 4-52.

	A	B	C	D	E
1	Copy formatting				
2		Data			
3			Data	Data	Data
4			Data	Data	Data
5			Data	Data	Data
6					
7					
8			Data	Data	Data
9			Data	Data	Data
10					
11					
12			Data		
13			Data		

Figure 4-52. *Result of using AutoFill*

5. Type the word **Data** in C8. Use the AutoFill Handle to drag to E9. Type **Data** in C12 and use the AutoFill Handle to drag to C13.

6. Format cell A1 by selecting the following options from the Font group: Bold, Italic, and Underline; change the text color to red.

7. Click the Center button on the Alignment tab.

8. With cell A1 selected, click the Format Painter button. The Format Painter button should now be highlighted. The mouse pointer changes to a block plus sign with a paintbrush. See Figure 4-53.

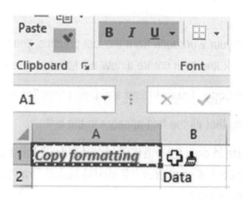

Figure 4-53. *Format Painter selected with cell A1 active*

9. Click cell B2. Cell B2 takes on all the formatting that you applied to cell A1. The Format Painter button should no longer be highlighted. Your mouse pointer should return to ⬥.

10. Select cell A1. This time double-click the Format Painter button. The mouse pointer will stay as ⬥⌗ and you can keep copying the formatting until you turn off the format painter.

11. Drag across cells C3:E9. Those cells now have the same formatting as cell A1.

12. Drag across cells C12 and C13. Those cells now have the same formatting as cell A1. See Figure 4-54.

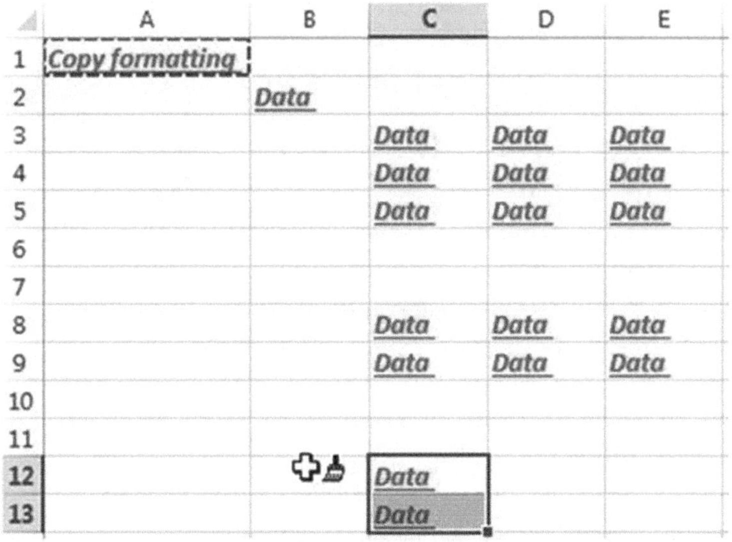

▲	A	B	C	D	E
1	Copy formatting				
2		Data			
3			Data	Data	Data
4			Data	Data	Data
5			Data	Data	Data
6					
7					
8			Data	Data	Data
9			Data	Data	Data
10					
11					
12		⊕🖌	Data		
13			Data		

Figure 4-54. *Result of applying the Format Painter*

13. Click the Format Painter button again to turn it off. Your mouse pointer should return to ⊕.

14. Since you dragged across cell C6 when you copied the formatting, if you type something in this cell, it should take on the same formatting as cell A1. Type horse in cell C6.

15. If you want to clear the formatting from a cell(s), you can clear just the formatting and leave the data in the cell. Drag across cells C6:E8. Click the down arrow for the Clear button located in the Home tab's Editing group. Click Clear Formats. See Figure 4-55.

181

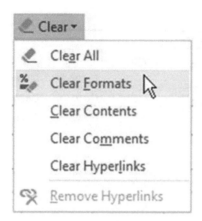

Figure 4-55. *Clear options*

16. Click the Save button 💾 on the Quick Access Toolbar.

You have been using commands from the Ribbon. Most of these commands are more quickly accessible from a mini-toolbar or the context menu which brings up commands that relate to what you are currently working on.

Using the Mini-Toolbars and the Context Menu

You have gone through Excel's tools for formatting both numeric and text data. You will no doubt be using these tools often. Microsoft gives you quick access to them via the mini-toolbars. There are two different size mini-toolbars.

If you select any text within a cell, you will see the smaller of the two mini-toolbars as shown in Figure 4-56. Moving your mouse pointer away from the mini-toolbar will make it fade away. When you bring your cursor back toward the mini-toolbar, it will fade back in. You can remove the mini-toolbar by doing one of the following:

- Move your cursor into the Ribbon area.

- Click another cell.

- Move the cursor far enough away from the toolbar until it is no longer visible.

You will need to reselect text in the cell if you want to get the mini-toolbar back. This smaller mini-toolbar only includes tools from the Font group.

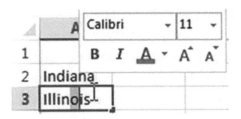

Figure 4-56. *Smaller mini-toolbar*

The larger mini-toolbar shown in Figure 4-57 includes buttons from the Font, Alignment, and Number groups along with the Format Painter from the Clipboard group. This mini-toolbar can be accessed by right-clicking any cell even if the cell is empty.

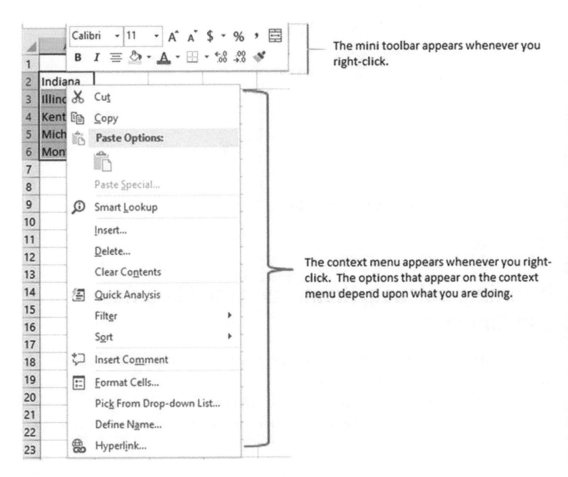

Figure 4-57. *Context menu and larger mini-toolbar*

A context menu (also known as a shortcut menu) is also displayed when you right-click a cell or an object. What items are displayed in the context menu depends upon the object you clicked and what you are working on at the time. You can make a selection from either the mini-toolbar or the shortcut menu. If you decide you don't want to use either one, just click another cell and they both will disappear.

Notice that the context menu has some options that have three dots after them. See Figure 4-58. The three dots are called an **ellipsis**. The ellipsis signifies that another dialog box will be opened when this option is selected. For example, clicking the Insert... option on the context menu will bring up the Insert window. Some of the options on the context menu have an arrow. Moving your mouse pointer over an arrow brings up a submenu of additional options.

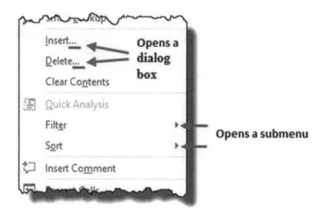

Figure 4-58. *Context menu*

Inserting, Deleting, Hiding, and Unhiding Rows and Columns

You can add new blank rows between existing spreadsheet rows. You can also delete rows to remove data or to remove blank rows.

You can hide a single row or multiple rows of data from the view of users. Rows that have been hidden can be unhidden. You don't have to unhide all the rows, only the ones you choose.

Hiding and Unhiding Columns and Rows

Columns and rows can be hidden and unhidden so that the users can concentrate on only the data that is necessary and relevant to them.

Hiding Columns and Rows

There are many reasons for hiding columns or rows in a spreadsheet:

- There is too much data to display on a single screen, so you hide the data you really don't need to see.

- You want to hide some of the data so it is not printed.

- The data is not relevant to the user.

- You want to hide the formulas.

- You don't want the user to see some of the data.

Hiding a column or row does not affect the data in that column or row, nor does it prevent it from being referenced.

A single column or row of data can be hidden as well as multiple columns or rows. If you are hiding multiple columns or rows, they do not need to be adjacent.

How to Hide a Single Column or Row

1. Right-click the column or row header.

2. Select Hide from the context menu.

How to Hide Adjacent Columns or Rows

1. Drag across the headers of the columns or rows you want to hide.

2. Right-click the selected headers.

3. Select Hide from the context menu.

How to Hide Nonadjacent Columns

1. Click a column header.

2. Hold down the Ctrl key while you click other column heads.

3. Right-click one of the selected column heads.

4. Select Hide from the context menu.

How to Hide Nonadjacent Rows

1. Hide nonadjacent rows by using the same steps that you do for nonadjacent columns.

Note   You can't select columns and rows in the same operation. You must first hide either columns or rows and then do the other.

Unhiding Columns and Rows

Columns and rows that you hide will remain hidden until you run the **Unhide** command.

Unhide All Columns and Rows

You can unhide all columns and rows by first clicking the **Select All** button located to the left of row head A and then right-clicking any header and selecting Unhide from the context menu. If you wanted to unhide columns and rows from multiple worksheets, you would select those sheet tabs before performing these steps.

Unhide Adjacent Columns or Rows

In the worksheet example shown in Figure 4-59, columns B and C are hidden.

Figure 4-59. *Columns B and C are hidden*

Columns B and C can be unhidden by dragging across column heads A and D, right-clicking, and then selecting Unhide.

Rows would be unhidden using the same steps, except of course you would use the row heads rather than the column heads.

Unhide Nonadjacent Columns or Rows

In the worksheet example shown in Figure 4-60, columns B and C are hidden, and so are columns F, G, and H.

Figure 4-60. *Columns B and C, F, G, and H are hidden*

Columns B and C and columns F, G, and H can be unhidden by dragging across column heads A and D and then holding down the Ctrl key while dragging across column heads E and I, right-clicking, and then selecting Unhide.

Another way to accomplish this would be to unhide all five columns at once. You could drag across column heads A to I. Right-click one of the selected column heads and select Unhide.

Unhide Column A or Row 1

Unhiding column A or row 1 is a unique situation. It is unique because there is no column head before Column A to drag across, nor is there a row head before row 1 to drag across.

Unhide Column A

1. Type **A1** in the Name Box and then press Enter.

2. Click the Home tab on the Ribbon.

3. In the Cells group, click the Format button.

4. Select **Hide & Unhide** from the menu and then select **Unhide Columns**. See Figure 4-61.

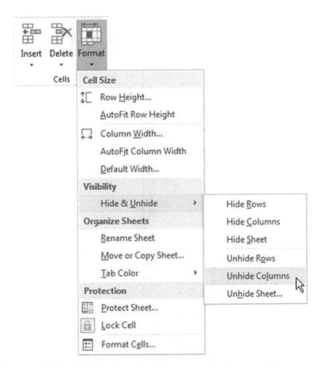

Figure 4-61. Hide & Unhide options

Unhide Row 1

1. Type **A1** in the Name Box and then press the Enter key.

2. Click the Home tab on the Ribbon.

3. In the Cells group, click the Format button.

4. Select **Hide & Unhide** from the menu and then select **Unhide Rows**.

EXERCISE 4-7: HIDING AND UNHIDING ROWS AND COLUMNS

This exercise takes you through hiding and unhiding rows and columns as well as dealing with the unique issue of unhiding column A and row 1.

1. Create a new worksheet. Name it "Rows_Columns."

2. Type **1** in cell A1 and **2** in cell A2.

3. Drag across cells A1 and A2 to select them.

4. Drag the Autofill Handle down to row 12. Let go of the mouse button. See Figure 4-62.

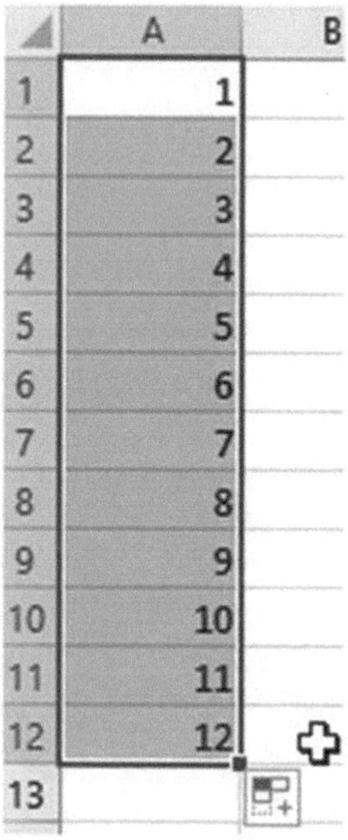

Figure 4-62. *Copied pattern of cell A1 and A2 down through cell A12*

5. Drag the Autofill Handle to the right to column M. See Figure 4-63.

▲	A	B	C	D	E	F	G	H	I	J	K	L	M
1	1	1	1	1	1	1	1	1	1	1	1	1	1
2	2	2	2	2	2	2	2	2	2	2	2	2	2
3	3	3	3	3	3	3	3	3	3	3	3	3	3
4	4	4	4	4	4	4	4	4	4	4	4	4	4
5	5	5	5	5	5	5	5	5	5	5	5	5	5
6	6	6	6	6	6	6	6	6	6	6	6	6	6
7	7	7	7	7	7	7	7	7	7	7	7	7	7
8	8	8	8	8	8	8	8	8	8	8	8	8	8
9	9	9	9	9	9	9	9	9	9	9	9	9	9
10	10	10	10	10	10	10	10	10	10	10	10	10	10
11	11	11	11	11	11	11	11	11	11	11	11	11	11
12	12	12	12	12	12	12	12	12	12	12	12	12	12

Figure 4-63. *Copied column A data by using AutoFill Handle*

Hide Columns

1. Drag across column heads B through D.

2. Hold down your Ctrl key while you click column head G and drag across column heads J and K.

3. Right-click one of the selected column heads and then select Hide.

Hide Rows

1. Drag across row heads 2 and 3.

2. Hold down your Ctrl key while you drag across row heads 6 and 7 and click row head 10.

3. Right-click one of the selected row heads and then select Hide.

Unhide Columns

1. Drag across column heads A and E. Right-click the selected area and then select Unhide.

2. Drag across column heads F through L. Right-click the selected area and then select Unhide.

Unhide Rows

1. Drag across row heads 1 through 12.

2. Right-click the selected area and select Unhide.

Hide Column A

1. Right-click column head A.

2. Select Hide.

Unhide Column A

1. Type **A1** in the name box and then press Enter.

2. Click the Home tab on the Ribbon. In the Cells group, click the Format button.

3. Select Hide & Unhide from the menu and then select Unhide Columns.

Hide Row 1

1. Right-click row head 1.

2. Select Hide.

Unhide Row 1

1. Enter A1 in the name box and then press Enter.

2. Click the Home tab on the Ribbon. In the Cells group, click the Format button.

3. Select Hide & Unhide from the menu and then select Unhide Rows.

Inserting Columns and Rows

After you have entered data into your workbook, you may discover that you need to add a column of data between your other columns or maybe you want to add a blank column to make your worksheet easier to read. The same holds true for rows; you may discover that you forgot to add rows of data or you merely want to add additional blank rows to improve the appearance of your worksheet.

You can add a new column(s) or row(s) anywhere in your worksheet.

How to Insert Columns

1. Select the column head where you want the column inserted.

2. Right-click the worksheet and select Insert. The column you selected will be shifted to the right and the column that you created will take its place. The new column will have the same width as the column to the left of it.

You can insert multiple columns by selecting the columns where you want the new ones inserted. If you selected four columns, then four new columns will be inserted when you right-click the worksheet and select Insert.

How to Insert Rows

1. Select the row head where you want the row inserted.

2. Right-click the worksheet and select Insert. The row you selected will be shifted down, and the row that you created will take its place. The new row will have the same height as the row above it.

You can insert multiple rows by selecting the rows where you want the new ones inserted. If you selected four rows, then four new rows will be inserted when you right-click the worksheet and select Insert.

Deleting Columns and Rows

You can just as easily remove a column or row as you can add them. Some of the reasons for removing a column would be that

- You have entered the wrong data into a column.

- You have determined that you no longer need the data in a column.

- You want to remove empty columns to improve the appearance or use of your spreadsheet.

If you entered invalid data in a column, it might be easier just to clear the data from the column rather than delete the entire column. This could be done by selecting all the cells containing the invalid data, then selecting Home tab ➤ Editing Group ➤ clear button , and then selecting one of the clear options.

How to Delete Columns

You can delete one column or multiple columns at the same time.

1. Select the column heads of the columns you want to delete.

2. Right-click the worksheet and select Delete. The columns to the right of the deleted columns will be shifted left to replace those columns that you deleted (to replace the vacated space).

How to Delete Rows

You can delete one row or multiple rows at the same time.

1. Select the row heads of the rows you want to delete.

2. Right-click the worksheet and select Delete. The rows below the deleted columns will be shifted up to replace those rows that you deleted.

Inserting and Deleting Cells

You can insert blank cells above or to the left of an active cell. You can bring up the Insert dialog box (Figure 4-64) to perform this operation by right-clicking any cell and then selecting Insert from the context menu.

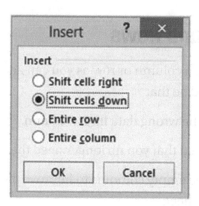

Figure 4-64. *Insert window*

When you insert blank cells, you can choose to move the other cells in the same column down (Shift cells down) or to move the other cells in the same row to the right (Shift cells right) to accommodate the new cells. Notice that the Insert dialog box also has options for inserting entire rows and columns.

You can remove cells by selecting the cells you want to remove and then right-clicking and selecting Delete from the context menu. This brings up the Delete dialog box (Figure 4-65) which provides options for deleting cells, entire rows, or entire columns.

Figure 4-65. *Delete window*

You can choose to have the deleted cells replaced with cells from the right (Shift cells left), or you can choose to have the deleted cells replaced with those from below (Shift cells up).

EXERCISE 4-8: INSERT AND DELETE CELLS, ROWS, AND COLUMNS

This exercise takes you through inserting and deleting cells and then deciding how the other cells need to be shifted over.

1. Create a new worksheet. Name the worksheet "Stocks."

2. Type **Stocks Purchased 06-15-2016** in cell A1. Select cells A1 to D1. Right-click the selected cells. The larger mini-toolbar should appear above your selected text. Click the Merge & Center button ▥.

3. Type **Stock** in cell A2.

4. Type **Closing** in cell B2. Press Alt + Enter. Pressing Alt + Enter creates an additional line in the cell and turns on text wrapping. Type **Price**.

5.   Select cells C2 and D2. Right-click in the selected area. Select Format Cells. Select Wrap text from the Alignment tab. Click the OK button.

6.   Type the data from Figure 4-66 in cells C2 and D2. You may need to widen your column to see all of the word Purchased. Then type the rest of the data in Figure 4-66.

◢	A	B	C	D
1		Stocks Purchased 06-15-2016		
2	Stock	Closing Price	Shares Purchased	Total Shares
3	A	38.85	200	300
4	B	14.12	100	500
5	C	27.72	300	300
6	D	35.5	300	500
7	E	18.85	200	600

Figure 4-66. *Type the data from this worksheet*

7.   Right-click column head A and then select Insert from the context menu.

8.   Right-click row head 1 and then select Insert from the context menu.

9.   Right-click cell B5 and then select Insert. Select Shift cells down. Click the OK button.

10.   Right-click cell B5 and then select Delete. Select Shift cells up. Click the OK button.

11.   You can select multiple cells before you select to shift the cells down or to the right. The number of cells shifted over will be the same number of cells you selected:

a.   Select cells B5:D5. Right-click a cell in the selected area and then select Insert.

b.   Select Shift cells right. Click the OK button.

12.   Right-click the three cells that are still selected and then select Delete. Select Shift cells left. Click the OK button.

13. Right-click cell B4 and then select Insert. Select Entire row. Click the OK button. A row is inserted above the selected cell.

14. Right-click cell C4 and then select Insert. Select Entire column. Click the OK button. A column is inserted to the left of the selected cell.

15. Right-click cell C4 and then select Delete. Select Entire column. Click the OK button.

16. Click row head 6. Hold down your Ctrl key and click row heads 7, 8, and 9. Right-click any of the selected row heads and then select Insert.

17. You should now have a blank row between each of your rows.

Note   When making selections, your selected rows, columns, or cells do not have to be adjacent.

You have learned how to insert and delete rows and columns as well as how to hide and unhide them. You should only let users who need to view the data have access to it; you should hide anything that they don't require. You also learned to insert and delete individual cells. You will often discover that you forgot to make an entry. Excel makes it easy to correct this oversight by letting you insert a new cell where it is needed and then moving the data below it down one row or moving the data to the right of it over one column.

Summary

Formatting a worksheet can make it more visually appealing and easier to understand. It can also be used to change how numeric and date data are displayed. Formatting can be used to create word wrapping, rotating text, and shrinking text. Formatting is also used to align data. You can use Excel's Format Painter to copy all the formatting applied to a cell to other cells. You've practiced using color effectively and learned how to create a uniform graphic style for worksheets. This is an important aspect of developing presentations that are professional in appearance and providing a signature look for your projects. Excel makes formatting easier by providing quick access to the tools you need most often by placing them on two mini-toolbars. You have seen how to hide and unhide rows and columns and how to insert and delete cells, rows, and columns. You have also seen how to apply borders and how to draw them.

In Chapter 5, you'll continue to refine how the workbook is set up by using headers and footers, setting the page breaks, dividing worksheets into multiple panes, and learning more functions. You'll also be learning about custom views and how to print select portions of the worksheet for specific data reporting purposes.

CHAPTER 5

Different Ways of Viewing and Printing Your Workbook

Excel allows you to enter a tremendous variety and amount of data into one worksheet and then extract only the data you need to share with clients or employees. Sometimes you don't need to print the entire workbook or the entire worksheet. In this chapter, you'll be learning how to view and print only the data relevant to your task. You may be dealing with a geographic area or a certain salesperson, and you only want to show what is relevant to each person. Print layout view lets you see how worksheets will look when printed, which is very important because screenshots can be quite different from the final printed product.

After reading and working through this chapter, you should be able to

- Insert, remove, reset, and change the location of page breaks

- Repeat row and column headers on all pages

- Use preset headers

- Create your own headers and footers

- Create a print area, add to it, and remove it

- Print a portion of a worksheet

- Use Paste Special to assemble a printable worksheet

- Create your own custom views

- Use the Synchronous Scrolling feature

© David Slager and Annette Slager 2020
D. Slager and A. Slager, *Essential Excel 2019*, https://doi.org/10.1007/978-1-4842-6209-2_5

- Divide a worksheet into multiple panes

- Freeze rows and columns

Views

So far, we have been looking at the worksheets in the **Normal view**, but this isn't the only view available. In addition to the Normal view, there is a **Page Break Preview** and a **Page Layout view**. The different views can be accessed from the View tab on the Ribbon. **Page Break Preview** is used for adjusting page breaks. This view lets you insert, change, remove, and reset page breaks. **Page Layout view** provides a quick way to add, edit, and delete headers and footers. It also lets you visualize how your workbook will look when printed. You can also create your own **custom views**.

The Workbook Views are located on the Ribbon's View tab. See Figure 5-1.

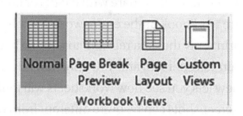

Figure 5-1. *Workbook View buttons*

The Normal view, Page Break Preview, and Page Layout view buttons also appear on the right side of Excel's task bar. These three buttons ▦ | ▣ ▥ perform the same functions as those in the Workbook Views group. If you do not see them on your task bar, right-click the task bar and select View Shortcuts.

Note   Normal view and Page Break Preview do not display headers and footers, but Page Layout view does.

There is also a Print Preview which you will use in this chapter. The Print Preview is not a part of Workbook views. Print Preview is only used for seeing exactly how your worksheet will be printed.

Page Break Preview

There are two types of page breaks: those that are created automatically by Excel and those you create manually. Automatic page breaks adjust whenever you do such things as change the row height or column width, delete or add columns or rows, change page size, change page orientation, change margins, and so on.

In Normal view, page breaks appear as dashed lines. In Normal view, you can insert, remove, and reset page breaks. You can't, however, drag page breaks to a new location like you can in Page Break Preview. Page break lines are also much easier to see in Page Break Preview. The way you insert, remove, and reset page breaks is done the same way in both Normal and Page Break Preview. Therefore, we will only look at performing these steps in the Page Break Preview mode.

Printing a worksheet that contains a lot of columns can be a challenge. Excel may not be breaking up your report (printed worksheet) the way you want it to. You can change from **Portrait** to **Landscape** orientation so that you have more width to your paper, but you still may end up with some columns printing on a separate page. **Page Break Preview** is used for adjusting page breaks. This view lets you insert, change, remove, and reset page breaks.

Page Break Preview shows a minimized view of your data. The data is separated by horizontal and vertical blue dashed lines that represent page breaks. You can change the location of a page break by dragging a blue line to where you want the break to occur. If you change the location of a page break, the line changes from a dashed line to a solid line. If you insert a new page break, it will also display as a solid line. This makes it easy to identify changes that you have made from the page breaks that Excel originally created.

EXERCISE 5-1: USING PAGE BREAKS AND PREVIEWS

In this exercise, you will practice creating vertical and horizontal page breaks. You will also practice moving and removing them.

Note   You can download the sample files included with this book from the Source Code/Download area of the Apress website (`www.apress.com/source-code`).

---

1. Open the workbook Chapter 5. Click the Months worksheet.
2. Click the File tab.
3. Select Options in the left pane.
4. Select Advanced in the left pane.

5. Look for the **Display options for this worksheet** group. If the Show page breaks isn't checked, then check it.

6. Click the OK button. This only affects Normal view. You can't turn off the Page Break lines in Page Break Preview.

Now that Show Page Break is selected, you should see page breaks that are represented by dashed lines. There is a vertical page break between columns E and F and another vertical page break between columns J and K (Figure 5-2). There is a horizontal page break between rows 47 and 48 (Figure 5-3).

E	F	G	H	I	J	K
Apr	May	Jun	Jul	Aug	Sep	Oct
579	172	385	895	244	108	619
557	398	509	795	109	363	960
958	740	310	896	281	243	879
697	663	126	966	386	250	649
707	685	128	238	239	221	816
446	851	663	662	331	290	664

Figure 5-2. *Vertical page break line*

45	Product 44	815	270	749	877	476	562
46	Product 45	323	763	797	302	800	151
47	Product 46	918	854	834	582	164	406
48	Product 47	712	425	720	722	254	276
49	Product 48	788	870	562	515	564	355
50	Product 49	670	470	635	573	359	768

Figure 5-3. *Horizontal page break line*

7. Click the View tab on the Ribbon. In the Workbook Views group, click the Page Break Preview button. The rest of the exercise pertains only to Page Break Preview. Look at the Zoom tool at the bottom right of your window. The zoom is automatically set lower than 100% to make it easier to see your page breaks.

The worksheet displays with two vertical dashed lines and one horizontal dashed line separating the worksheet into six pages. See Figure 5-4.

	Jan	Feb	Mar	Apr	May	Jun	Jul	Aug	Sep	Oct	Nov	Dec
Product 1	300	346	388	573	172	385	895	244	108	619	270	855
Product 2	390	658	750	557	398	509	795	109	363	360	556	190
Product 3	651	777	233	358	740	310	896	281	243	879	140	476
Product 4	867	382	161	697	663	126	966	386	250	649	417	225
Product 5	430	663	446	707	685	128	238	239	221	816	897	265
Product 6	752	305	577	446	851	663	662	331	230	664	323	322
Product 7	726	114	321	323	281	272	867	351	433	288	173	183
Product 8	366	281	309	776	831	831	738	151	516	148	495	133
Product 9	255	702	725	529	533	503	604	775	527	563	630	520
Product 10	312	302	725	645	553	734	677	878	379	511	129	478
Product 11	320	271	803	448	655	639	579	785	930	588	766	338
Product 12	685	415	213	603	361	278	160	134	105	466	603	255
Product 13	158	351	261	311	187	373	435	702	618	739	120	546
Product 14	262	784	421	741	526	133	193	839	783	581	116	934
Product 15	142	464	267	234	571	663	935	853	199	337	725	382
Product 16	756	543	750	586	635	606	468	128	243	736	565	422
Product 17	300	384	100	346	519	493	721	698	174	256	179	147
Product 18	368	881	336	161	597	523	824	590	817	194	241	340
Product 19	387	863	367	321	394	261	297	199	616	880	850	159
Product 20	815	270	749	877	476	562	404	198	244	366	118	813
Product 21	323	763	737	302	800	151	725	289	859	113	195	377
Product 22	318	854	834	582	164	406	173	820	856	138	339	823
Product 23	712	425	720	722	254	276	420	401	444	133	613	734
Product 24	788	870	562	515	564	355	622	736	331	726	607	723
Product 25	670	470	635	573	359	768	327	508	358	919	560	626
Product 26	633	384	539	367	341	166	338	637	143	442	217	239
Product 27	715	829	169	210	142	592	181	603	341	713	375	391
Product 28	437	283	307	645	774	794	334	217	184	336	420	242
Product 29	111	638	438	400	504	387	211	963	328	606	693	704
Product 30	533	837	688	651	516	865	938	117	841	172	234	773
Product 31	148	533	504	253	411	528	685	701	834	233	808	970
Product 32	347	357	892	458	970	537	857	718	908	539	184	356
Product 33	230	351	504	430	503	407	816	582	884	569	313	830
Product 34	803	375	804	745	932	693	115	445	773	105	600	652
Product 35	751	512	539	191	611	273	837	579	911	373	578	307
Product 36	576	630	396	174	488	309	820	477	856	519	520	435
Product 37	373	179	244	688	721	179	635	624	394	545	812	807
Product 38	527	420	394	126	390	583	313	187	603	201	330	228
Product 39	387	352	635	121	523	581	327	603	861	674	755	763
Product 40	756	543	750	586	635	606	468	128	243	736	565	422
Product 41	300	384	100	346	519	493	721	638	174	256	179	147
Product 42	368	881	336	161	597	523	824	590	817	194	241	340
Product 43	387	863	367	321	394	261	297	199	616	880	850	159
Product 44	815	270	749	877	476	562	404	198	244	366	118	813
Product 45	323	763	737	302	800	151	725	289	859	113	195	377
Product 46	318	854	834	582	164	406	173	820	856	138	339	823
Product 47	712	425	720	722	254	276	420	401	444	133	613	734
Product 48	788	870	562	515	564	355	622	736	331	726	607	723
Product 49	670	470	635	573	359	768	327	508	358	919	560	626
Product 50	633	384	539	367	341	166	338	637	143	442	217	239
Product 51	715	829	169	210	142	592	181	603	341	713	375	391
Product 52	437	283	307	645	774	794	334	217	184	336	420	242
Product 53	111	638	438	400	504	387	211	963	328	606	693	704
Product 54	533	897	688	651	516	865	938	117	841	172	234	773
Product 55	148	533	504	253	411	528	685	701	834	233	808	970
Product 56	347	357	892	458	970	537	857	718	908	539	184	356
Product 57	236	651	504	430	503	407	816	582	884	563	313	830
Product 58	803	375	804	745	932	693	115	445	773	105	600	652
Product 59	751	512	539	191	611	273	837	579	911	373	578	307
Product 60	576	630	396	174	488	309	820	477	856	519	520	435
Product 61	373	179	244	688	721	179	635	624	394	545	812	807
Product 62	255	702	725	529	533	503	604	775	527	563	630	520
Product 63	312	302	725	645	553	734	677	878	373	511	123	478
Product 64	320	271	803	448	655	639	579	785	330	588	766	338
Product 65	685	415	213	603	361	278	160	134	105	466	603	255
Product 66	158	351	261	311	187	373	435	702	618	739	120	546

Figure 5-4. *Page Break Preview*

View the Report in Print Preview

1. Click the File tab. Select Print in the left pane. The Print Preview pane on the right shows how the worksheet would look when printed. You can click the Forward and Back buttons ◄ 1 of 6 ► at the bottom of the preview area to view the different pages.

2. Click the Return button ⊜ to leave Print Preview.

Insert a Page Break in Page Break Preview

1. Click column head H so that the column is selected. When you insert a vertical Page Break, the break will occur to the left of the selected column.

2. On the Ribbon, click the Page Layout **tab**. In the Page Setup group, click the Breaks button. Click Insert Page Break. Click any cell so that you can see the Page Break. It is a solid line rather than a dashed line. A solid line is one that you created rather than the dashed lines that Excel automatically creates.

Remove a Page Break in Page Break Preview

1. Right-click column head H. Select Remove Page Break.

2. You can also remove a page break line by dragging it off the worksheet. Drag the vertical page break between columns J and K off the worksheet.

Reset Page Breaks in Page Break Preview

1. Right-click the worksheet and then select **Reset All Page Breaks**. This puts the page breaks back to the way Excel originally set them.

Change the Location of the Page Breaks in Page Break Preview

1. Drag the horizontal page break line so that it falls between rows 36 and 37. This line turns into a solid line. This makes it easy to identify page separators that Excel set (dashed lines) and those that you set (solid lines).

2. Drag the dashed break line that is between columns E and F so that it is between columns G and H. The line turns into a solid line.

View the Spreadsheet in Print Preview

1. Right-click the worksheet and then select Page Setup. On the Page Setup dialog box, click the Print Preview button.

2. The right side of the window is the Print Preview area. This shows exactly what your printout would look like. At the bottom of the window, you will see a page selector ◄ [1] of 4 ►. Click the right arrow to go to the next page. Notice that on page 2, there are no column headers. Click the right arrow of the page selector to go to page 3. Notice there are no row headers. Go to page 4. This page has neither row nor column headers. See Figure 5-5. It would be very difficult for anyone to read this report unless you taped the pages together.

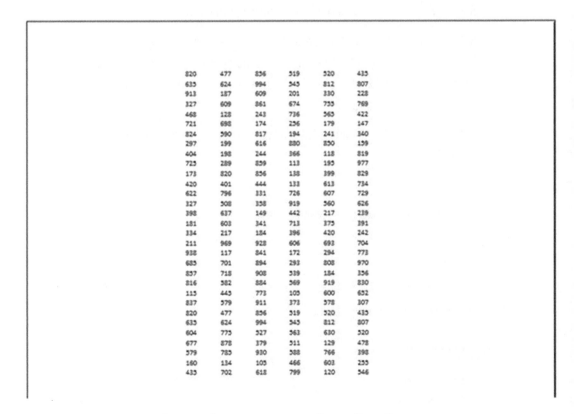

Figure 5-5. *Page 4 of report has no row or column headings*

Note   When you viewed the spreadsheet in Page Break Preview, it appeared with a grid, but in Print Preview, there is no grid. It would print without any gridlines.

3. Click the Return button ⬅️ and then click the Ribbon's Home tab.

Repeat the Row and Column Headers on All of the Pages

1. Right-click the worksheet and then select Page Setup. On the Page Setup dialog box, click the Sheet tab.

2. If we want the row headings to appear on every printed page, we need to identify the heading row in the **Rows to repeat at top** text box. Click inside the text box. You can type **$1:$1** in the text box, or you can click cell A1 and it will put the address in the text box for you. This address tells Excel to use the headings in row 1 as the column headings for all of the printed pages.

3. Tab to the **Columns to repeat at left** text box. You can type **$A:$A** in the text box, or you can click cell A1 and it will put the address in the text box for you. This address tells Excel to use the row headings in column A as the row headings for all printed pages.

4. Select the check box for Gridlines. This will print all of the data within gridlines. See Figure 5-6.

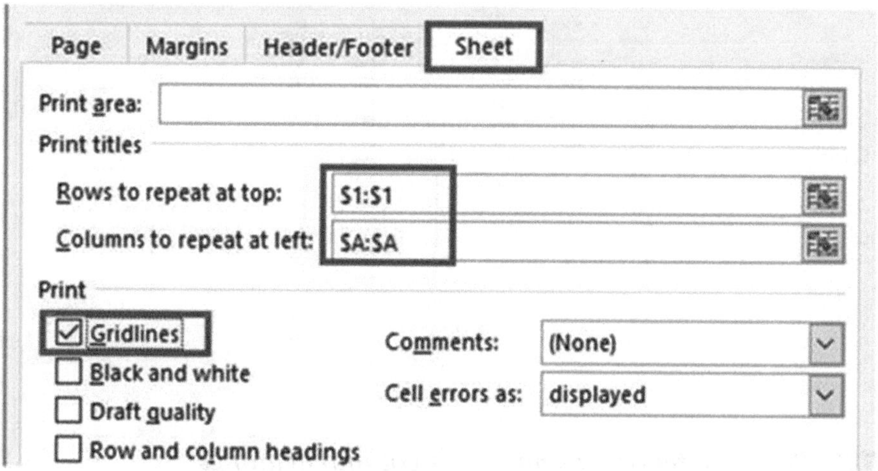

Figure 5-6. *Select Gridlines to print the data within gridlines*

5. Click the Print Preview button. The page now displays with gridlines.

6. Click the **Zoom to Page** button located at the bottom right of the Print Preview area. See Figure 5-7.

Figure 5-7. *Click the Zoom to Page button*

7. Click the left arrow of the page selector until you are at page 1 of 4 ◄ 1 of 4 ►. Page 1 shows the data for months January through June. Click the right arrow of the page selector to go to page 2. Page 2 shows the remaining records for months January through June. Page 2 now has column headers.

8. Go through the other pages; they all have column and row headers.

9. Click the Return button.

You have learned how to insert, remove, and rearrange page breaks so that you can control where page breaks occur. You also learned how to make the row and column heads appear on each page of a report. Next, you will see how to add headers and footers to your worksheets in the Page Layout view.

Page Layout View

Page Layout view provides a quick way to add, edit, and delete headers and footers. It shows what your printout will look like relative to a horizontal and vertical ruler along with the breaks between pages. You can easily change margins and then instantly see the results. Page Layout view works similar to Normal view; you can still add and edit data and apply formatting.

Page Layout view displays a horizontal ruler and vertical ruler that moves to another page when you move to another page. You can adjust your borders by dragging the ruler to the location you want. As you are dragging your mouse pointer, a line moves at the same time showing you where the margin is. If you want your margin at a certain inch, this is not the place to do it because it doesn't display the change in inches as you are dragging.

The Page Layout view displays three header sections and three footer sections. If you click one of the header or footer sections, a new contextual tab will be added to the Ribbon named **Header & Footer Tools.** See Figure 5-8. A header prints at the top of each page. A footer prints at the bottom of each page.

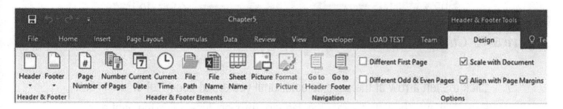

Figure 5-8. *Contextual tab for headers and footers*

EXERCISE 5-2: USE PAGE LAYOUT VIEW

In this exercise, you will use the Page Layout view to add and change headers and footers and to remove whitespace between pages.

1.   Use workbook Chapter 5. Click the Months worksheet.

2.   Click the Page Layout view button ▦ on the Workbook's status bar.

3.   You should see on the status bar that we are looking at page 1 of 4.

4.   Change the Zoom tool so that you can see all pages at the same time.

5.   Press Ctrl + Home to go to cell A1.

Change the Header and Footers

Note   The Header and Footer sections have a habit of not displaying when they are empty or first used. You may need to move your cursor to the area where they should exist to be able to see them.

---

1.   Change the Zoom tool so that you can clearly see the text **Add Header** above the worksheet. Move your mouse pointer to the **Add header** text. The section should be selected. See Figure 5-9.

Figure 5-9. *Add a header to the workbook*

2.   Move your mouse pointer to the left of that section and the left section will be selected. Click inside the left section. Whenever a Header or Footer section is active, the Ribbon will have a **Design** tab appearing under the heading **Header and Footer Tools**.

3.   Click the Design tab. In the Header & Footer group, click the Header button. Excel displays a list of preset headers. See Figure 5-10.

The data that will print in each section of the header is separated by commas:

- The preset headers that contain no commas will display in the middle header section.

- For those that contain one comma, the text to the left of the comma will appear in the middle Header section. The text to the right of the comma will appear in the right section.

- For those that contain two commas, the text to the left of the first comma will appear in the left-most Header section. The text to the right of the first comma will appear in the middle section. The text to the right of the second comma will appear in the right-most section.

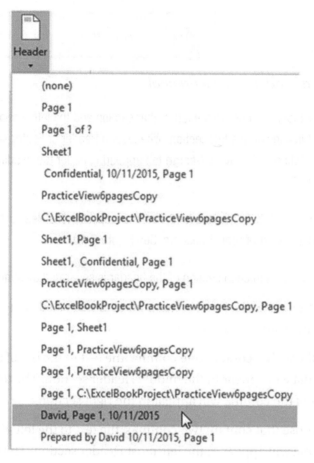

Figure 5-10. *Select a header from one of these preset headers*

4. Click the preset header that consists of the username, the page number, and the current date.

5. Click inside the first header section and change the name to your full name. Use the worksheet's vertical and horizontal scroll bars to see that the header appears at the top of each page.

6. Scroll to the bottom of page 1. You should see the text **Add footer**.

7. Click the words Add Footer so that you can see the three footer sections.

8. Click the Design tab. In the Header & Footer group, click the Footer button.

   Excel displays a list of preset footers. Just like with the headers, the commas are used to identify which section the text will go in.

9. Click the preset footer **Page 1 of ?**. The preset footer didn't contain a comma; therefore, it displays in the middle section.

10. Click inside the middle section. The code **Page &[Page] of &[Pages]** used to create the footer is displayed. We would prefer the text to appear in the right section. Press **Ctrl + X** to cut the text. Click inside the right section. Press Ctrl + V to paste it. Click inside the middle section. You should now see the text in the right section.

11. Click the Ribbon's Design tab, in the Navigation group, and then click the **Go to Header** button.

Print the Worksheet

1. Click the Ribbon's File tab. Click **Print** in the left pane.

2. You should see the worksheet with the three header sections. Scroll down to the bottom of the page if you need to, to see the footer **Page 1 of 4**. Go through the other pages. The headers and footers are the same on each page.

3. Click the Print button if you want to print the report.

4. Click the Return button ⬅.

Change the Scale and Alignment of Your Headers and Footers

1. Click any one of the three Header sections at the top of the page. Click the Design tab.

   The Options group (Figure 5-11) contains options for your headers and footers:

   - One option is for making one set of headers and footers for the first printed page, while all of the other pages will have a different set of headers and footers.

   - Another option is to have one set of headers and footers for the odd pages and another set for the even pages.

   - The **Scale with Document** option when selected scales the size of the headers and footers so that they are proportionate to any changes you made to the print scale to fit in more or less information on your printed page.

   - The **Align with Page Margins** when selected places the headers and footers within the page margins that you set up.

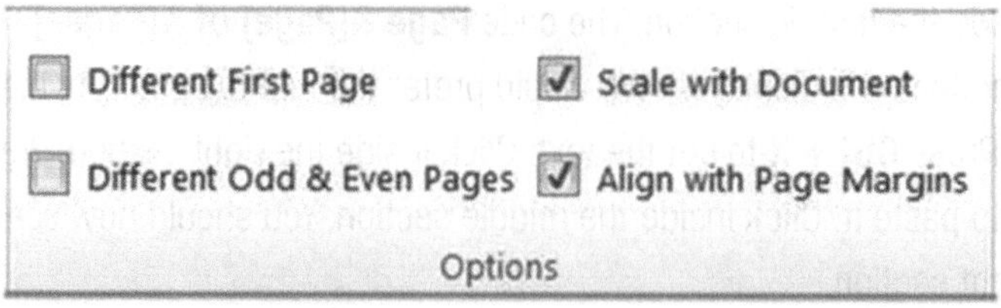

Figure 5-11. *Options for how you want your headers and footers to appear*

2.   Remove the check mark for the **Align with Page Margins** option; as you do so, notice how your header spreads out.

3.   Remove the check mark for **Scale with Document**. Notice how your header text increases in size.

4.   Turn the **Scale with Document** and **Align with Page Margins** back on.

5.   Select **Different First Page** if it isn't already selected.

6.   Click the Ribbon's Design tab. In the Header & Footer group, click the Header button. Select **Months, Confidential, Page 1**.

7.   Click inside the Page Header area to see the results.

8.   Scroll through the pages; you will see that only the first page header has changed.

9.   Click the Page Header area for page 1.

10. On the Ribbon's Design tab in the Options group, remove the check from Different First Page.

11. Click **Different Odd & Even Pages**.

12. Click the Left Header section on page 1 and change the text to Accounting Department.

13. Scroll down to page 2. Click inside the Page Header section.

14. Click the Ribbon's Design tab. In the Header & Footer group, click the Header button. Select **Page 1 of ?**.

15. Click inside the Page Header area to see the results.

16. Scroll through the pages. You should see that the headers are different for odd and even pages.

Remove Whitespace Between Pages

There is a lot of whitespace between each of our pages. Removing the whitespace would make the worksheet easier to read.

1.   Move your mouse pointer to the vertical blank space between pages 1 and 3 and click. See Figure 5-12.

May	Jun
172	385
398	509
740	310
663	126
685	128
851	663
281	272
891	891

	Jul
Product 1	895
Product 2	795
Product 3	896
Product 4	966
Product 5	238
Product 6	662
Product 7	867
Product 8	738

Figure 5-12. *Remove whitespace between pages*

2. Scroll down. You should see that all of the whitespace between the pages has been removed.

Note   The worksheet is easier to read now, but you can't see the headers or footers. Clicking between the pages will put the whitespace back. You can again view the headers and footers.

Turn the Show Options Off and On

1. Click the Ribbon's View tab. In the Show group, there are check boxes for Ruler, Formula Bar, Gridlines, and Headings. See Figure 5-13.

Figure 5-13. *The Ribbon's Show group on the View tab*

2. These options only affect how the worksheet is displayed. They have no effect on how the worksheet is printed. The ruler option is only available in Page Layout view. The **Headings** are the row numbers and column head letters. Remove the check mark for each one of these options. As you do so, notice its effect. Then turn them all back on.

3. Switch to Normal view by clicking the Normal View button ▦ on the status bar.

You've used the Page Break Preview and Page Layout views. Now, you'll return to the Normal view to see different ways of selecting what you to want to print.

Printing

Spreadsheets can be huge. Aligning the pages and getting the margins just the way you want on a large spreadsheet can be difficult. But chances are you don't want to print all of the information from a large spreadsheet. If you only want to print a portion of a worksheet, you can do this in several ways:

- Hide the rows or columns you do not want to print.

- If the cells that you want to print are not adjacent, you can select the cells from one area that you want to print, then hold down your Ctrl key, and select other areas to print.

- Copy the different ranges you want to print to another worksheet in the order that you want to see it printed. You can use paste links so that the data is updated dynamically. Then print the worksheet that contains only the data you want to print in the order that you want to print it.

- Set up print areas. Print areas can be expanded as needed or can be cleared. Only the print areas will be printed unless you select the option to Ignore Print Area.

The following sections will guide you through using Print Areas and assembling a printed report in the order you want.

Creating a Print Area

You can elect to print only the portions of the spreadsheet you need by selecting one or more ranges that you want to print and then making those ranges a **Print Area**. Set up a Print Area if you will be frequently printing only those areas of the worksheet.

A Print Area can be set up by selecting the cells you want to print and then clicking the Print Area button located in the Page Setup group on the Page Layout tab. See Figure 5-14. Selecting the **Set Print Area** creates the print area. The **Add to Print Area** option only appears if you have already set up a Print area.

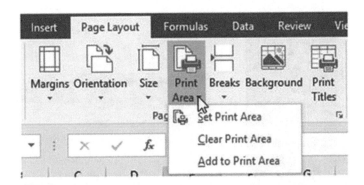

Figure 5-14. *Print Area button*

Excel saves the selected cells as a named range automatically naming it Print Area. You can only have one Print Area, but the Print Area can contain multiple ranges.

Adding Additional Cells to the Print Area

Additional cells can be added to the Print Area after you have created it by selecting those cells and then clicking the Print Area button and selecting **Add to Print Area**.

Note   There is one irritating problem when using multiple ranges for Print Areas and that is Excel places a page break between the ranges. This means that each addition to the Print Area will appear on a separate page.

Removing the Print Area

The Print Area can be removed by simply clicking the Print Area button and then selecting **Clear Print Area**.

You would only use a Print Area if you were going to use it multiple times; otherwise, you could select the areas you want to print and then click the Ribbon's File tab, select Print from the left pane, click the down arrow for Print Active Sheets, and then select Print Selection. See Figure 5-15.

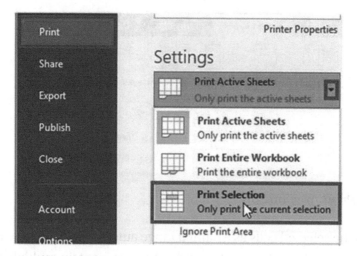

Figure 5-15. *Print only the selected area of the worksheet*

EXERCISE 5-3: PRINT A PORTION OF A WORKSHEET

The worksheet for this example contains data for a large corporation that is tracking sales for 39 products sold in 17 of its stores.

1. Open the workbook Chapter 5, which you have been using for this chapter, if it isn't currently open. Click the Stores worksheet.

2. Drag across column heads D through G. Right-click one of the selected column heads. Select Hide.

3. Drag across column heads L through R. Right-click one of the selected column heads. Select Hide.

4. Click the Ribbon's File tab. Select Print from the left pane.

5. Notice that the hidden columns do not appear in the print preview.

6. Click the Return button ⬅.

7. Add the data in Figure 5-16 to the worksheet as shown starting in cell U3.

U	V
Cost of Goods Sold	193,700
Average Store Sale	21251
Highest Store Sales	24216
Minimum Store Sales	18192

Figure 5-16. *Add this data to the worksheet*

8.   Select cells A1 through H9 (A1:H9).

9.   On the Ribbon's Page Layout tab in the Page Setup group, click the Print Area button and then select Set Print Area. The name Print_Area should appear in the Name box.

10. Select cells U3 through V6 (U3:V6). You should see a thin line around the cells A1:H9 indicating that you selected those cells for the print area.

11. In the Page Setup group, click the Print Area button and select **Add to Print Area**.

12. Click the Ribbon's File tab. Select Print from the left pane.

13. The print preview area should show that you are looking at page 1 of 2. The cells you selected when you clicked Add to Print Area are on the second page. Every time you use **Add to Print Area** to add another selection, it will print on a different page. If you want each section on a separate page, this works great; however, if you do not, you can use Paste Special which is discussed in the next section.

Note   If you change Settings to **Print Entire Workbook**, the Print Area will be ignored and the entire workbook will be printed. If you select **Ignore Print Area**, the Print Area will not be recognized by any of the Print options. See Figure 5-17.

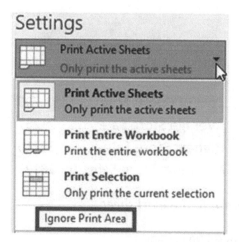

Figure 5-17. *Select Ignore Print Area to prevent any print options from using the Print Area*

14. Click the Return button ⬅.

15. On the Ribbon's Page Layout tab in the Page Setup group, click the Print Area button and then select **Clear Print Area**.

16. Unhide the hidden columns.

You've learned how to print only the cell ranges that you want to print. Next, you will see how using Paste Special gives you more control over printing and helps overcome the problem of Excel placing page breaks between ranges.

Using Paste Special for Printing

If you want more control of where the different sections of your report will print, you can copy the sections you want to see in a report and then use Paste Special to place them on a separate worksheet in the location that you want them printed. Another benefit of this is that you can copy the data from different worksheets.

EXERCISE 5-4: USING PASTE SPECIAL TO ASSEMBLE A PRINTABLE WORKSHEET

In this exercise, you will take different portions of a worksheet and assemble them on another worksheet which can be printed as a single page.

1. Use the Chapter 5 workbook. Create a new worksheet. Name the worksheet "Pasted."

2. Click the Stores worksheet. Select the cell range A1:E9. Press Ctrl + C to copy the data.

3. Click the worksheet you named Pasted. Click inside cell A1.

4. Click the Ribbon's Home tab, and in the Clipboard group, click the down arrow of the Paste button. Select **Paste Special** which is the last option.

5. The Paste Special dialog box displays. Click **Paste Link**. See Figure 5-18. This creates a link between the data on the Stores worksheet and the data on the Pasted worksheet.

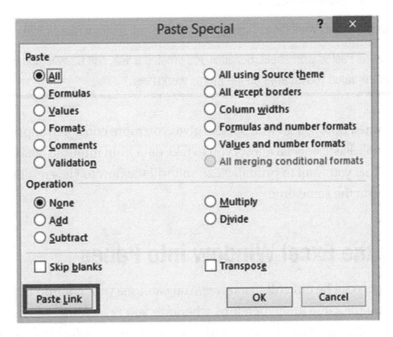

Figure 5-18. *Paste Link creates a link between the copied and pasted data*

6. Click the Stores worksheet. Select the range U3:V6. Press Ctrl + C to copy the data.

7. Click the worksheet you named Pasted. Click cell G2. On the Ribbon's Home tab in the Clipboard group, click the down arrow of the Paste button. Select **Paste Special**.

8. The Paste Special window displays. Click **Paste Link**. Widen column G to see all the data.

9. Click the Stores worksheet. Select the range A17:D21. Press Ctrl + C to copy the data.

10. Click the worksheet you named Pasted. Click inside cell A11. On the Ribbon's Home tab in the Clipboard group, click the down arrow of the Paste button. Select **Paste Special**.

11. The Paste Special window displays. Click **Paste Link**.

12. Click the Ribbon's File tab. Select Print from the left pane. The print preview shows that you have managed to get the data you want in the location you want and still keep it on the same page.

13. Click the Return button ⬅.

14. Click the Stores worksheet. Change the value in cell B2 from 900 to 820.

15. Click the Pasted worksheet. Because you created a link, cell B2 was updated with the value you entered on the Stores worksheet.

You've learned how using Paste Special gives you more control over printing than using Print Areas. Paste Special allows you to take data from multiple areas and arrange it in the order that you want to print it. Next, you will see how to view multiple copies of your worksheet at the same time.

Dividing the Excel Window into Panes

Large worksheets can be difficult to analyze. You can't see your column heads and your column totals at the same time; therefore, whenever you're looking at a total, you have to scroll back to the top of the worksheet to view its column heading in order to know what the total represents. The same thing applies to your row totals and row headings. Another potential problem is that you want to compare the values in column B with the values in column V, but you can't see both of them at the same time, so you have to scroll horizontally back and forth to do your comparisons. These types of problems can be solved by dividing your Excel screen into separate panes. A copy of your worksheet is automatically placed into each of the panes you create. This gives you the benefit of being able to view different areas of the worksheet at the same time.

The screen can be divided into two vertical panes, two horizontal panes, or a combination of the two which would give you four panes.

Here is how to create two vertical panes:

1. Click a column head.

2. Click the Ribbon's View tab. In the Window group, click the Split button. See Figure 5-19. The pane will split to the left of the selected column.

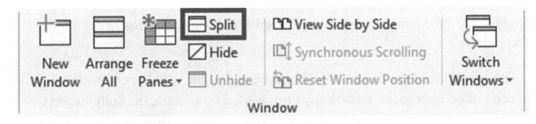

Figure 5-19. *Split Window button is highlighted*

Here is how to create two horizontal panes:

1. Click a row head.

2. Click the Ribbon's View tab. In the Window group, click the Split button. The pane will split above the selected row.

Here is how to create four panes:

1. Click a cell.

2. Click the Ribbon's View tab. In the Window group, click the Split button. Split Bars appear above and to the left of the selected cell. See Figure 5-20.

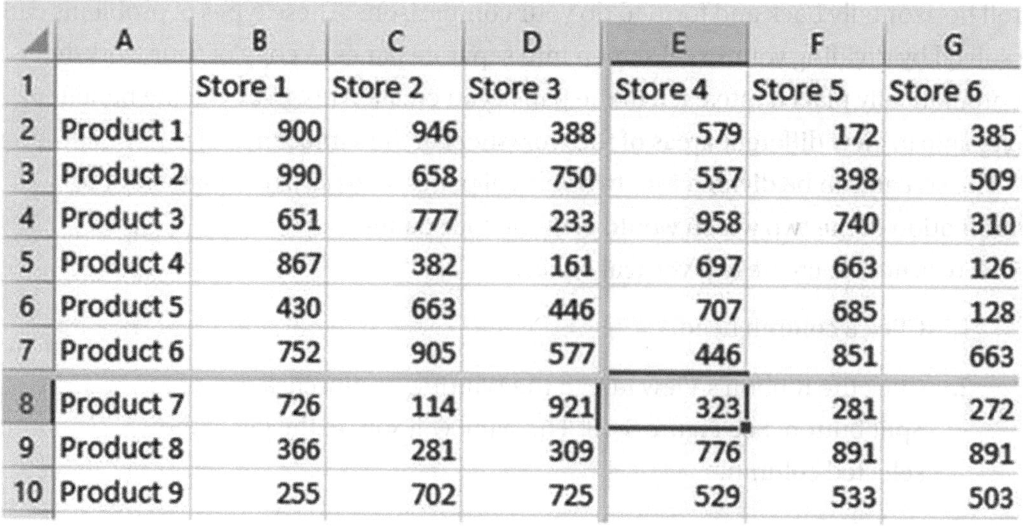

	A	B	C	D	E	F	G
1		Store 1	Store 2	Store 3	Store 4	Store 5	Store 6
2	Product 1	900	946	388	579	172	385
3	Product 2	990	658	750	557	398	509
4	Product 3	651	777	233	958	740	310
5	Product 4	867	382	161	697	663	126
6	Product 5	430	663	446	707	685	128
7	Product 6	752	905	577	446	851	663
8	Product 7	726	114	921	323	281	272
9	Product 8	366	281	309	776	891	891
10	Product 9	255	702	725	529	533	503

Figure 5-20. *Window split into 4 panes. Split is above and to the left of selected cell*

The size of the panes can be adjusted by dragging the split bars to the location you want. If no Split bars are currently displayed, then clicking the Split button will display a horizontal and vertical bar at the center of the window (not the worksheet). See Figure 5-21.

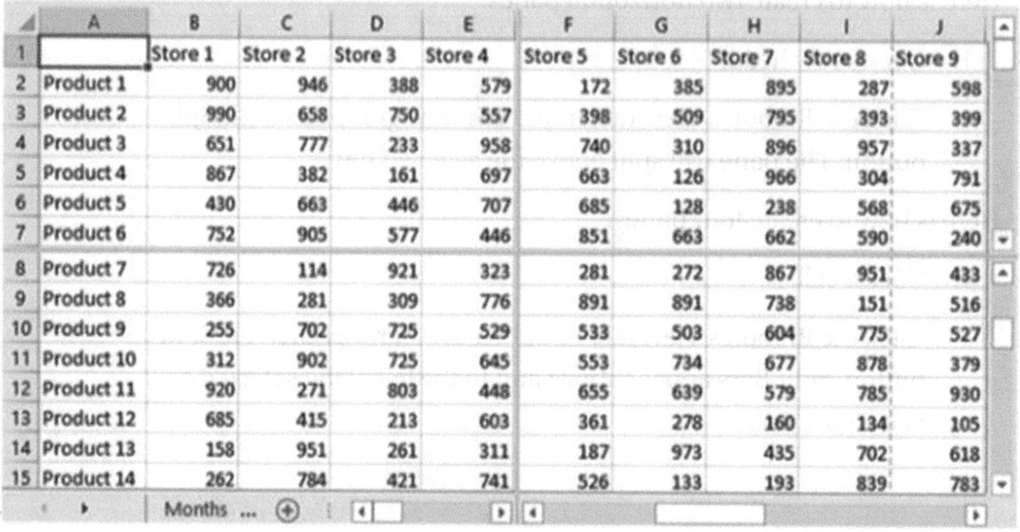

	A	B	C	D	E	F	G	H	I	J
1		Store 1	Store 2	Store 3	Store 4	Store 5	Store 6	Store 7	Store 8	Store 9
2	Product 1	900	946	388	579	172	385	895	287	598
3	Product 2	990	658	750	557	398	509	795	393	399
4	Product 3	651	777	233	958	740	310	896	957	337
5	Product 4	867	382	161	697	663	126	966	304	791
6	Product 5	430	663	446	707	685	128	238	568	675
7	Product 6	752	905	577	446	851	663	662	590	240
8	Product 7	726	114	921	323	281	272	867	951	433
9	Product 8	366	281	309	776	891	891	738	151	516
10	Product 9	255	702	725	529	533	503	604	775	527
11	Product 10	312	902	725	645	553	734	677	878	379
12	Product 11	920	271	803	448	655	639	579	785	930
13	Product 12	685	415	213	603	361	278	160	134	105
14	Product 13	158	951	261	311	187	973	435	702	618
15	Product 14	262	784	421	741	526	133	193	839	783

Months ...

Figure 5-21. *Split bars appear at the center of the window*

Placing a horizontal split bar on a worksheet adds an additional vertical scroll bar. Placing a vertical split bar on a worksheet adds an additional horizontal scroll bar. Placing a horizontal and vertical split bar on a worksheet adds an additional horizontal and vertical scroll bar. This provides the capability to scroll through each pane individually. See Figure 5-22.

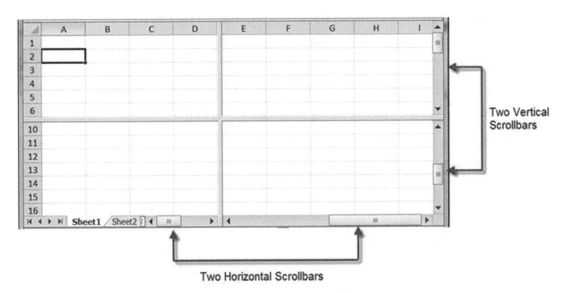

Figure 5-22. *Horizontal and vertical scroll bars for separate scrolling of the four panes*

A split bar can be removed from the worksheet by double-clicking it. You can remove both split bars at the same time by double-clicking the area where the two split bars meet. The Split button on the Ribbon works like a toggle switch. Click it to add split bars. Click it again to remove the split bars.

Freezing Rows and Columns

Freezing rows and columns is another way of resolving the problem of not being able to view your row headings and row totals at the same time or your column headings and column totals at the same time.

You can select one of three options when freezing your worksheet:

- **Freeze Panes**—You can freeze both the columns and rows that come before the current cell. If the current cell is C4, this means that columns A and B are frozen and rows 1, 2, and 3 are frozen. The rest of the columns and rows are scrollable.

- **Freeze Top Row**—You can freeze the top row of your worksheet so that it remains visible while scrolling through the rest of the worksheet.

- **Freeze First Column**—You can freeze the first column so that it remains visible while scrolling through the rest of the worksheet.

You can select one of the three options by clicking the View tab and then clicking the Freeze Panes button located in the Window group. See Figure 5-23.

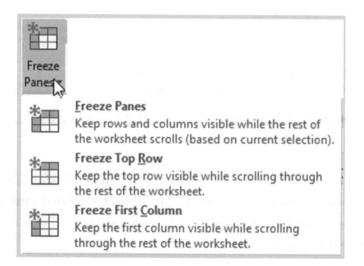

Figure 5-23. *Freeze Panes options*

Once you have frozen the worksheet using one of the three options, the first option will change from Freeze Panes to Unfreeze Panes. Clicking Unfreeze Panes will return you to normal viewing.

Synchronizing Scrolling

You can use the New Window, View Side by Side, and Synchronous Scrolling options in the Window group to open a second copy of your workbook and view the workbooks side by side with the same worksheets having synchronized scrolling.

EXERCISE 5-5: SYNCHRONIZE SCROLLING, SPLIT THE WINDOW INTO MULTIPLE PANES, AND FREEZE WINDOWS

In this exercise, you will split a worksheet into multiple worksheets. It appears as if there are multiple copies of the worksheet, but they are just separate views of the same worksheet that you can scroll together. You will also practice freezing rows and columns.

1. Open workbook Chapter 5 if it isn't opened already. Click the Stores worksheet.

2. Click the Ribbon's View tab. In the Window group, click New Window. A second window opens with the Stores worksheet active. You may not see the original window because it may be hidden behind the new window.

3. Click the Ribbon's View tab. In the Window group, click Arrange All. Select Vertical and Windows of active workbook. See Figure 5-24.

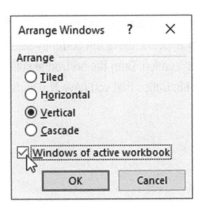

Figure 5-24. *Select Windows of active workbook*

4.   Click the Ribbon's View tab. In the Window group, click View Side by Side.

Note   If you have other workbooks open in addition to the Chapter 5 workbook, a Compare Side By Side window appears from which you can choose the workbook you want to view side by side with it. You would select Chapter 5:1.

---

5.   Click the Ribbon's View tab. In the Window group, click Arrange All. Select Vertical and Windows of active workbook.

The name at the top of the workbook shows as Chapter 5:2. The original workbook title bar shows Chapter 5:1.

6.   Synchronous Scrolling is now highlighted under the View Side by Side button. Use the vertical and horizontal scroll bars on one of the workbooks. Notice that as you drag the scroll bar on one workbook, the other workbook scrolls as well.

7.   Close one of the workbooks by clicking the X in the upper right corner of the workbook. The workbook title goes back to its original name of Chapter 5.

Create the Split Bars

1.   Click inside cell A3. On the Ribbon's View tab, in the Window group, click the Split button. A split bar appears above the currently selected cell. You can drag the split bar to another location. Drag the horizontal split bar down so that it is between rows 7 and 8. Notice that you have two vertical scroll bars. See Figure 5-25.

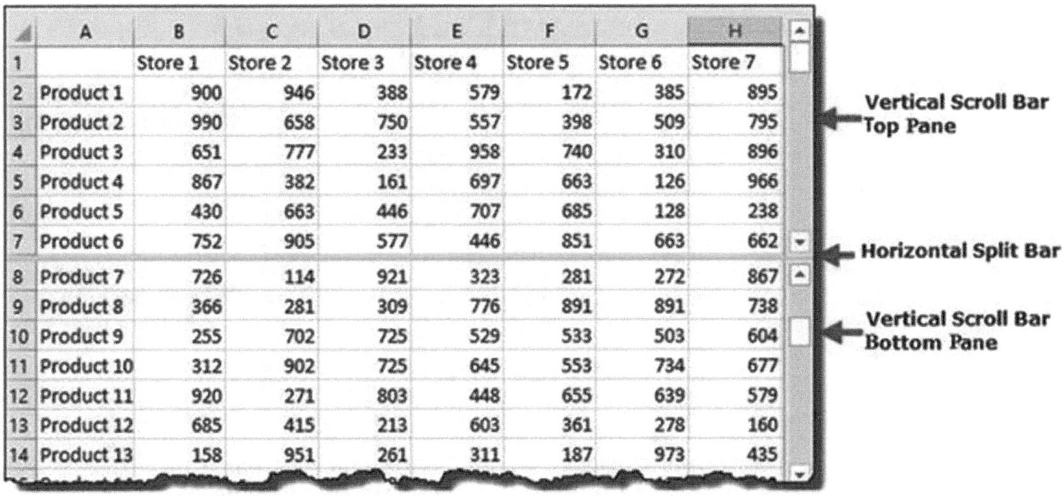

Figure 5-25. Window divided into two panes with two separate scroll bars

2. Use the scroll bar in the top pane to scroll through the data in the top pane.

3. Use the scroll bar in the bottom pane to scroll through the data in the bottom pane. As you can see, you have two complete copies of your worksheet. Adjust your worksheets so that you can see cell B2 in both worksheets.

4. Click inside cell B2 in the upper pane. When you select a cell in one pane, it also selects it in the other pane.

5. Type **500** in the cell. Press Enter. Cell B2 should now have a value of 500 in both worksheets.

6. Double-click the split bar to remove it.

7. Click cell G8. On the Ribbon's View tab, in the Window group, click the Split button. This creates intersecting split bars above and to the left of the cell. See Figure 5-26.

⊿	A	B	C	D	E	F	G	H	I	
1		Store 1	Store 2	Store 3	Store 4	Store 5	Store 6	Store 7	Store 8	
2	Product 1	900	946	388	579	172	385	895	28?	
3	Product 2	500	658	750	557	398	509	795	39:	
4	Product 3	651	777	233	958	740	310	896	95?	
5	Product 4	867	382	161	697	663	126	966	304	
6	Product 5	430	663	446	707	685	128	238	568	
7	Product 6	752	905	577	446	851	663	662	59(	
8	Product 7	726	114	921	323	281	272	867	951	
9	Product 8	366	281	309	776	891	891	738	151	
10	Product 9	255	702	725	529	533	503	604	77!	
11	Product 10	312	902	725	645	553	734	677	878	
12	Product 11	920	271	803	448	655	639	579	78!	
13	Product 12	685	415	213	603	361	278	160	134	

← **Vertical scroll bar adjusts both top panes**

← **Vertical scroll bar adjusts both bottom panes**

Months   St ...

Horizontal scroll bar adjusts both left panes

Horizontal scroll bar adjusts both right panes

Figure 5-26. *Window split into four panes with two vertical and two horizontal scroll bars*

Practice moving both horizontal and vertical scroll bars and then adjust the four worksheets so that you can see the same cell in all four panes. Click one of the cells. Notice that it highlights the same cell in all four panes. If you make a change to one of the cells, it affects the worksheet in all four panes.

8. Move your mouse pointer where both split bars meet. Your mouse pointer should turn into a double arrow. See Figure 5-27. Hold down your mouse button and drag in any direction; both split bars will move with your mouse.

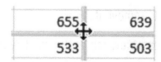

655	639
533	503

Figure 5-27. *The mouse pointer displays as a double arrow when placed where split bars connect*

9. Double-click where the two split bars meet to remove both of them.

Freeze Panes

1. In the Window group, click the Freeze Panes button. The three options for freezing appear:

   a. Click the one for freezing the top row. Scroll down through your data. The column headings will remain in their fixed location so that no matter what row you are in, you can always see the column headings.

   b. In the Window group, click the Freeze Panes button. The first option has changed to Unfreeze Panes. Click this option to remove freezing of the top row.

   c. Click the Freeze Panes button. Click the one for freezing the first column. Scroll to the right to the end of your data. The row headings will remain in their fixed location so that no matter what row you are in, you can always see the row headings.

   d. Click the Freeze Panes button. Click Unfreeze Panes.

2. Click cell B2 and complete these steps:

   a. Click the Freeze Panes button.

   b. Click Freeze Panes.

   c. Scroll through to the end of your data vertically and horizontally. Both your column headings and row headings remain stationary.

   d. Unfreeze the panes.

3. Practice with a different active cell and then freeze the panes. How does this affect the rows above and the columns to the left of the active cell?

You've learned how to freeze row and column heads so they remain stationary as you scroll. You have also learned to view multiple copies of your worksheet at the same time while keeping their scrolling synchronized. Next, you will see how you can specify the different ways you want to view a worksheet and then save those views as custom views.

Custom Views How to Create, Show, and Delete

If you or other users view your workbook in different ways, such as with different headers or footers, with different print areas or different margins, with different columns or rows showing, or with the data filtered, and so on, you can save the different views as custom views. Each custom view is assigned its own name. When a custom view is opened, it opens with all of its stored settings.

Before you create a custom view, you need to set the workbook to appear exactly how you want it to appear and how you want it to print. When you are ready to save your current view, click the View tab. In the Workbook Views group, click the Custom Views button. See Figure 5-28.

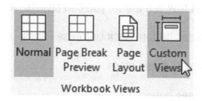

Figure 5-28. *Custom Views button*

Clicking the Custom Views button brings up the Custom Views dialog box shown in Figure 5-29 where you click the Add button.

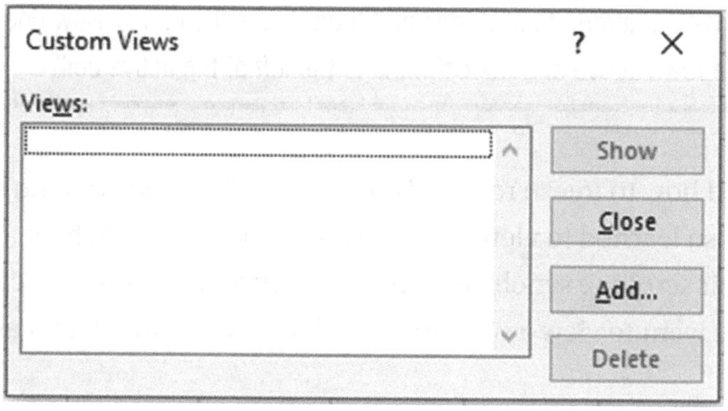

Figure 5-29. *Custom Views dialog box*

Clicking the Add button brings up the Add View dialog box shown in Figure 5-30 where you enter the name you want to assign to the view.

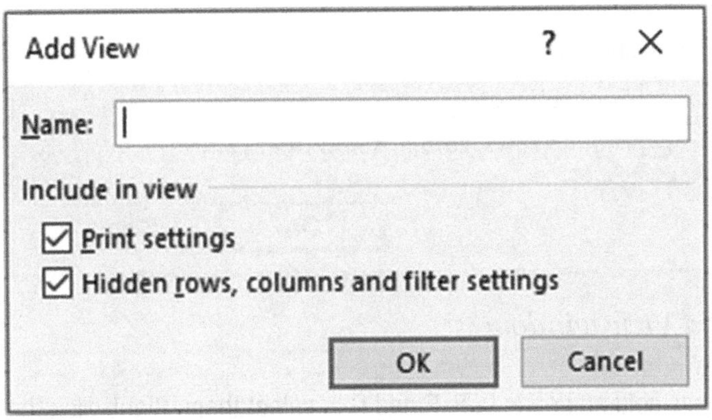

Figure 5-30. *The Add View dialog box*

After you have created your views, you can switch to another view by selecting it from the Custom Views dialog box and then clicking the Show button.

Note   You can't create a custom view that contains a table.

EXERCISE 5-6: CREATING, SHOWING, AND DELETING CUSTOM VIEWS

In this exercise, you will create three views, display the views, and then delete them.

Create Views

1. Open workbook Chapter 5 if it isn't already open. Click the Custom tab.

2. Click the Ribbon's View tab. In the Workbook Views group, click the Custom Views button.

3. Click the Add button in the Custom Views window.

   Enter **Original View** for the name of the custom view as shown in Figure 5-31. Click the OK button.

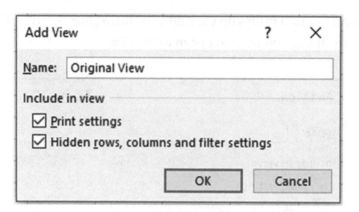

Figure 5-31. *Add View window*

4. Drag across column heads D, E, F, and G to select them. Right-click the selected text and select Hide.

5. Click the Ribbon's View tab. In the Workbook Views group, click the Custom Views button.

6. Click the Add button in the Custom Views window.

    Enter **Hidden Columns** for the name of the custom view as shown in Figure 5-31. Click the OK button.

7. Drag across column heads C and H to select them. Right-click the selected text and select Unhide.

8. Drag across row heads 15 to 20 to select them. Right-click the selected text and select Hide.

9. Click the Ribbon's View tab. In the Workbook Views group, click the Custom Views button.

10. Click the Add button in the Custom Views window.

11. Enter **Hidden Rows** for the name of the custom view. Click the OK button.

12. Drag across row heads 14 and 21. Right-click the selected text and select UnHide.

13. Click the Ribbon's Page Layout tab. In the Page Setup group, click the Orientation button and then select Landscape.

14. Click the Page Setup group's dialog box launcher 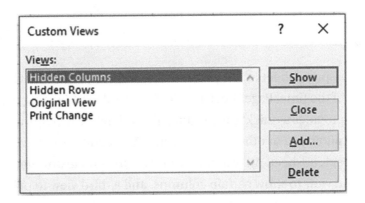 :

   a. Click the Sheet tab.

   b. Enter A1:K6 in the Print area text box.

   c. Select Gridlines.

   d. Click the OK button.

15. Click the Ribbon's View tab. In the Workbook Views group, click the Custom Views button.

16. In the Custom Views window, click the Add button. Enter **Print Change** for the Name in the Add View window. Click the OK button.

<u>Show a View</u>

1. Click the Custom Views button. Click Hidden Columns and then click the Show button. See Figure 5-32 (columns D through G are hidden).

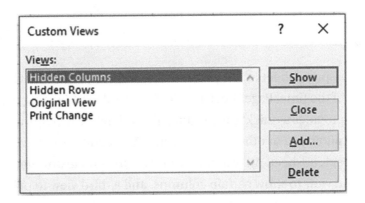

Figure 5-32. *Custom Views window*

2. Click the Custom Views button. Click Hidden Rows and then click the Show button. Rows 15 through 20 are hidden.

3. Click the Custom Views button. Click Print Change and then click the Show button. Click the Ribbon's File tab. Click Print in the left pane.

You can see that only what you selected for the print area is going to print and it is in landscape mode. Click the Return button.

4. Click the Custom Views button. Click Original View and then click the Show button.

Delete a View

1. In the Workbook Views group, click the Custom Views button.

2. Select a View and then click the Delete button. You will be asked to confirm the deletion. Click Yes.

3. Repeat steps 1 and 2 for the other three views.

This section showed you how to create your own custom views so that the same worksheet can be viewed in different ways. Each custom view is assigned its own name. All the custom views for the workbook are stored within the workbook. If you didn't use custom views, you would have to save different versions of the workbook to get the same result.

Summary

Excel can display a spreadsheet in one of three different views. Normal view is the view you will use most of the time. Page Break Preview is useful for adjusting page breaks. Page Layout view provides a quick way to add, edit, and delete headers and footers.

Custom views allow you or other users to switch back and forth between different views of the same spreadsheet. An example would be you create one view to show all your data, a second view to show certain columns, and a third view to print a portion of a worksheet.

You have split worksheets into multiple panes and then scrolled through the individual panes. You've practiced making the worksheet look the way you want it to and used many of the buttons you can see when you open Excel. Next, you will be learning about the "Backstage" which includes more options for saving the workbook in different formats and how to keep track of information about when the workbook was created and why. You'll also practice how to share workbooks via email. Other features discussed next include how to keep track of your recent work and how to use the Navigation pane.

CHAPTER 6

Understanding Backstage

The **File** tab, unlike the other tabs, doesn't bring up a series of groups with buttons on them; rather, it brings up what is called the Backstage. This area is called the Backstage because the options on the Backstage are handled behind the scenes from the workbook such as handling files, printing, sharing and exporting workbooks, account information, and selecting the options that change how Excel works. It also displays information about the workbook.

After reading and working through this chapter, you should be able to

- Save a workbook in different formats

- Add and edit information about a file and its properties

- Use various options to print the workbook the way you want it to appear

- Send a workbook as an email attachment

- Search for a template that meets your needs

- Open a workbook by browsing for it or by using the Recent list

- Specify the number of workbooks that appear in the Recent list

- Keep workbooks on the Recent list

- Place recently opened workbooks in the Navigation pane

- Set Excel Options

233

© David Slager and Annette Slager 2020
D. Slager and A. Slager, *Essential Excel 2019*, https://doi.org/10.1007/978-1-4842-6209-2_6

Backstage Overview

Figure 6-1 shows the Backstage. Clicking a group item in the Navigation pane such as Info, New, Open, and so on displays information and options for that grouping in the area to the right. We will cover most of the groups in the Navigation pane.

The list that follows shows some of the things that you can do from Backstage:

- Save files in many different formats.

- Change printer settings.

- Protect the workbook.

- Save the workbook to OneDrive.

- Give others rights to your workbook.

- Email the workbook in various formats.

- Add and remove personal information from the workbook.

- Check the accessibility of your workbook for those with disabilities.

- Check for compatibility with previous Excel versions.

- Update Excel.

- Change information on your Microsoft Account.

- Specify the number of workbooks that appear in the Recent list.

- Pin workbooks to the Backstage pane.

- Recover unsaved workbooks.

Table 6-1 lists some of the groups from the Navigation pane and describes them.

Table 6-1. *Backstage Navigation Groups and Their Descriptions*

Group	Description
Info	Displays information about the currently opened workbook.
New	Creates a new blank workbook or creates a workbook from a template.
Open	Opens an existing Excel workbook.
Save	Saves updates to the currently open workbook. Doesn't change the name of the existing workbook.
Save As	Saves a workbook that hasn't been previously saved. It can also be used to create a second copy of the currently open workbook and assign it a different name.
Print	Prints options for printing the workbook.
Share	Provides options for sharing your workbook with others.
Close	Closes the workbook.
Account	Provides your Microsoft Office Account Information.
Options	Shows options you can set to change how Excel works.

Info Group—Viewing, Adding, and Editing Information About the Workbook

The **Info** group is displayed by default in the Backstage when you click the File tab. See Figure 6-1.

The Info group displays information about the currently open workbook and handles information related to it. We will first look at the Properties Pane tab; then we will look at the options in the middle pane for protecting and inspecting the workbook.

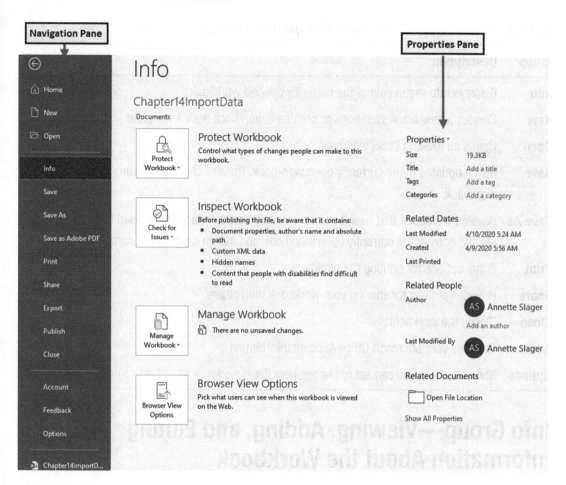

Figure 6-1. *Backstage when accessed from a workbook with data in it*

Properties Pane

When you look at a file in Windows Explorer, you can see the file's name, its date, and size, but often this isn't enough information. Additional information about the file can be entered or viewed in the Properties pane. At the bottom of the Properties pane, you can click **Show All Properties** if not all the properties are currently showing. You can click **Show Fewer Properties** if all properties are currently showing.

You can enter or edit the properties directly on this window. You can enter such things as the title, tags, categories, Company, and Manager.

Clicking the down arrow next to **Properties** in the Properties pane brings up the Advanced Properties button shown in Figure 6-2. Clicking the **Advanced Properties** button displays the Properties dialog box as shown in Figure 6-3.

Figure 6-2. *Advanced Properties button*

Oct_Week1_Income Properties	?	×

General	Summary	Statistics	Contents	Custom

Title: | The Lake Front Inn

Subject: | Room Income

Author: | David

Manager: |

Company: | Lake Lodging Associates

Category: | Income

Keywords: | Income Rooms Available Charges

Comments: | Shows gross income for room charges for the first week of October|

Hyperlink base: |

Template:

☐ Save Thumbnails for All Excel Documents

| OK | Cancel |

Figure 6-3. *Workbook Properties dialog box*

Some of the information you enter in the Properties dialog box can be viewed in the Properties pane. It is also displayed at the bottom of the Save As window when you click the Browse button as shown in Figure 6-4. The information you enter here is useful as documentation and can help you search for the workbook.

File name:	Lodging Charges.xlsx			
Save as type:	Excel Workbook (*.xlsx)			
Authors: David Slager		Tags: Income Rooms Availa...	Title: The Lake Front Inn	Subject: Room Income

☐ Save Thumbnail

ide Folders   Tools ▼ | Save |

Figure 6-4. *Some of the information you enter in the Properties dialog box appears when you save a file using Save As from the Backstage's Navigation pane*

Right-clicking an Excel file in Windows Explorer and then selecting Properties displays the Properties window. Select the Details tab to see the information you entered. See Figure 6-5.

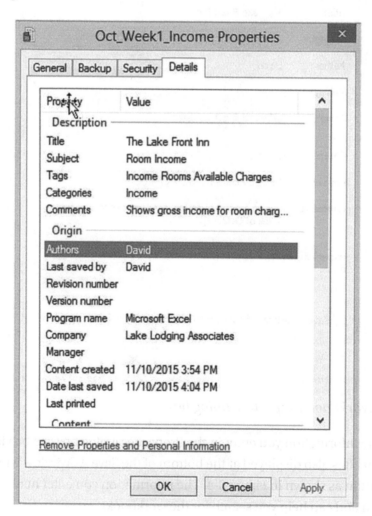

Figure 6-5. *Workbook information for the Workbook named Oct_Week1_Income*

Protect Workbook Options

In the Info pane, clicking the Protect Workbook button brings up a menu of options for different ways of protecting your workbook. See Figure 6-6. We will cover Mark as Final, Encrypt with Password, and Protect Workbook Structure. These aren't the only options for protecting your workbook. We look at other ways to protect your workbook later in this chapter.

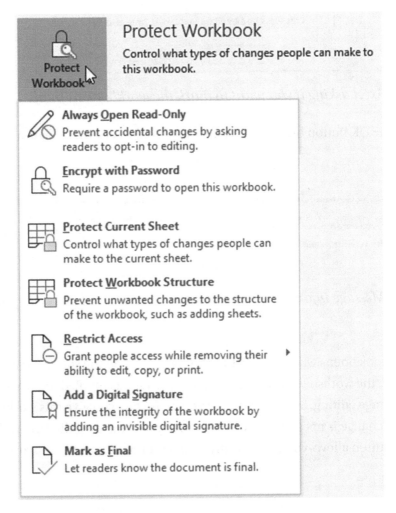

Figure 6-6. *Options for protecting a workbook*

Mark as Final

Selecting Mark as Final will make your workbook read only. You won't be able to make any changes to any worksheet, and you won't be able to create any new worksheets. Clicking the Mark as Final button will bring up the message box shown in Figure 6-7.

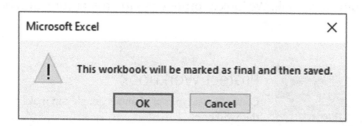

Figure 6-7. *Excel asking if you want to mark the workbook as final*

Clicking the OK button brings up the message window shown in Figure 6-8.

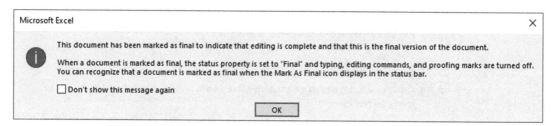

Figure 6-8. *Message informing that editing is complete and this is the final version*

The workbook opens with a **MARKED AS FINAL** icon on the status bar and a message above the worksheet indicating that the author has marked the workbook as final to discourage editing. See Figure 6-9. If you or anyone you give this workbook to decides that a change is necessary, it can be done by clicking the Edit Anyway button. Clicking this button allows you to make any changes or additions to any worksheet in the workbook.

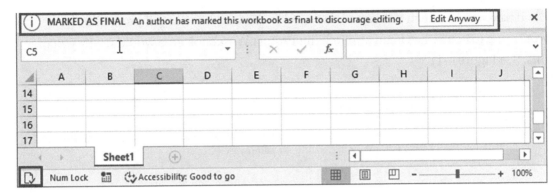

Figure 6-9. *Workbook that has been Marked as Final*

Encrypt with Password

Selecting this option requires you to enter a password and then verify it. It uses 128-bit advanced encryption to make your file more secure. You will need to use this password the next time you open the file.

Note Microsoft does not provide a way to recover passwords created in Microsoft Office.

Protect Workbook Structure

This option prevents users from adding or deleting worksheets; hiding and unhiding worksheets; and moving, copying, or renaming worksheets. You can add a password to this option if you desire. See Figure 6-10.

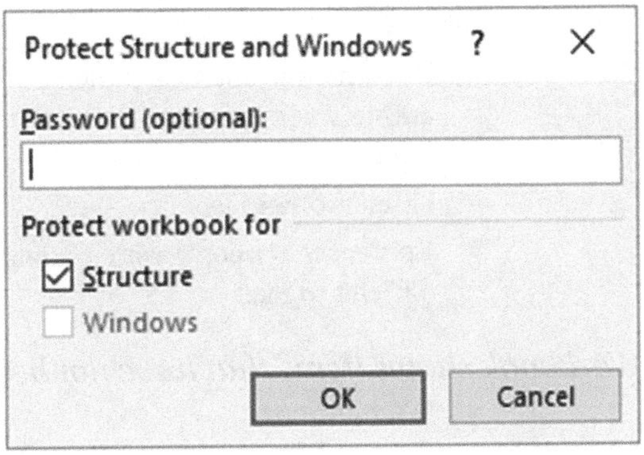

Figure 6-10. *Structure protection selected to prevent user from making changes*

If password protection has been applied to the workbook and if protection for workbook structure has been applied, the Protect Workbook area would notify you of this. See Figure 6-11.

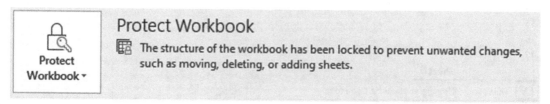

Figure 6-11. *Message notifying user that the structure has been locked and a password is required to open the workbook*

If you have protected the workbook and you click the Ribbon's Review tab**,** you will notice that in the Protect group the Protect Workbook button is highlighted. You will need to click this button if you want to unprotect it. If you entered a password when you protected it, you will need to enter the same password to unprotect it. Passwords are case sensitive.

Check for Issues

Just to the right of the Check for Issues button is the Inspect Workbook area. The Inspect Workbook shows items that have not been resolved. See Figure 6-12. These are not problems that will keep your workbook from working.

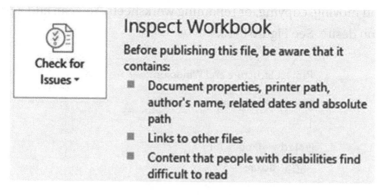

Figure 6-12. *Inspect Workbook shows items that have not been resolved*

The Check for Issues button has options for checking if the workbook contains personal information, checking if the workbook can be easily read by users with disabilities, and checking for compatibility with previous Excel versions. See Figure 6-13. We will cover how to use the Inspect Document, Check Accessibility, and Check Compatibility next.

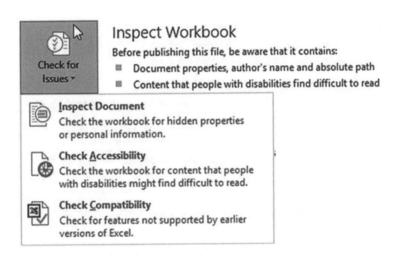

Figure 6-13. *Check for Issues options*

Inspect Document

The Document Inspector helps you find personal and hidden information that you may want to remove before sharing this document with anybody else. If you run Inspect Document before you have saved your Excel workbook, you will get the warning message in Figure 6-14 that you should click Yes to save the file now; otherwise, you might lose some of your data.

If you have already saved the file or you have responded to the message in Figure 6-14, then the Document Inspector dialog box displays. See Figure 6-15. Check each check box in the Document Inspector that you want to check.

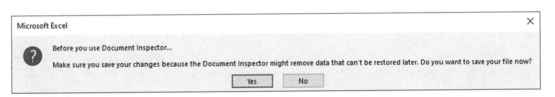

Figure 6-14. *Message informing user to click the Yes button to save the file*

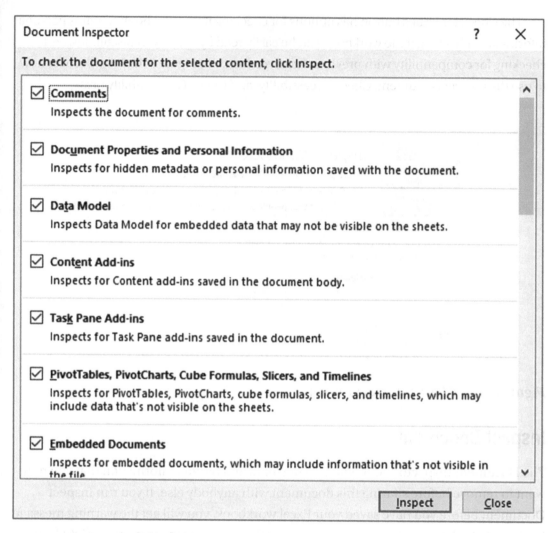

Figure 6-15. *Select items you want inspected*

Clicking the Inspect button inspects the document for those items that you have checked. If Excel finds personal or hidden data for any option you have selected, it will display a **Remove All** button for that option. See Figure 6-16.

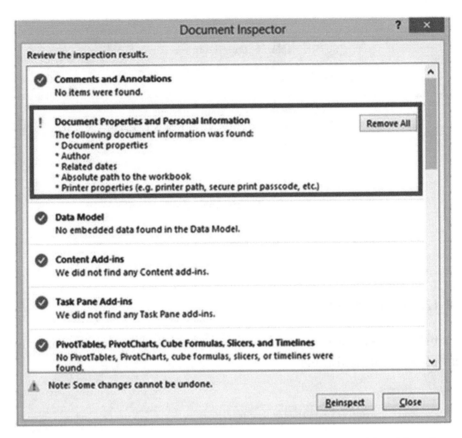

Figure 6-16. *Excel reports on personal or hidden data it has found for the options you selected*

Check Accessibility

Check Accessibility checks your workbook for issues pertaining to users who have disabilities that make it difficult for them to read the workbook. Some people use a screen reader to read the workbook out loud for them. An example of an issue would be that a screen reader can't read an image. If you want to make your workbook compatible for these users, you would need to enter alternative text for an image. When you click the Check Accessibility button, the Accessibility Checker appears to the right of your workbook. The Accessibility Checker breaks down issues into errors and warnings. Figure 6-17 shows how the Accessibility Checker would look if no issues were found.

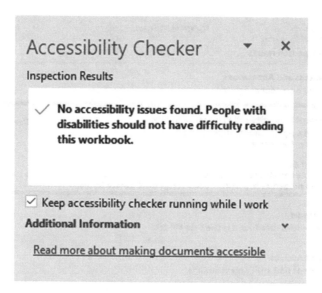

Figure 6-17. *No errors or warnings*

Figure 6-18 shows a warning error because the worksheets weren't given names so they were left as Sheet1 and Sheet2.

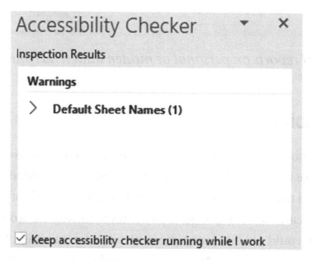

Figure 6-18. *Warning that worksheets still have default names of Sheet1 and Sheet2*

Check Compatibility

If you want to share a workbook with someone else who doesn't have the same version of Excel, you will want to run the Compatibility Checker. The Compatibility Checker gives a report on what functionality may be lost and how many times that functionality occurs in the workbook. See Figure 6-19. If you save a workbook in the Excel 97-2003 format, the Compatibility Checker will run automatically. Workbooks in the 97-2003 format have a file extension of **.xls**. Workbooks starting with version 2007 have an extension of **.xlsx**.

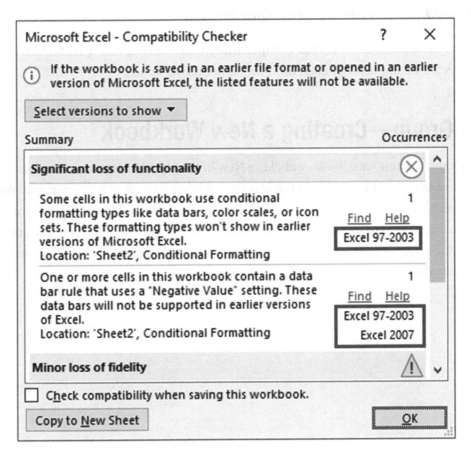

Figure 6-19. *Compatibility Checker*

Clicking the **Select versions to show** button in the Compatibility Checker brings up the Excel versions to test for. See Figure 6-20. The Compatibility Checker will test for compatibility with the Excel versions you have selected.

Figure 6-20. *Select which Excel versions to test for*

Select the check box for **Check compatibility when saving this workbook** to have Excel check for compatibility each time you save this workbook.

New Group—Creating a New Workbook

You can create a new workbook by clicking New in the Navigation pane of Backstage or by pressing Ctrl + N while not in Backstage. Selecting **New** from the Navigation pane allows you to create a new blank document from scratch, but if the project you are going to create is similar to one of the vast number of prebuilt templates, you might want to consider using one of them instead (see Figure 6-21). As the saying goes, "No sense in reinventing the wheel."

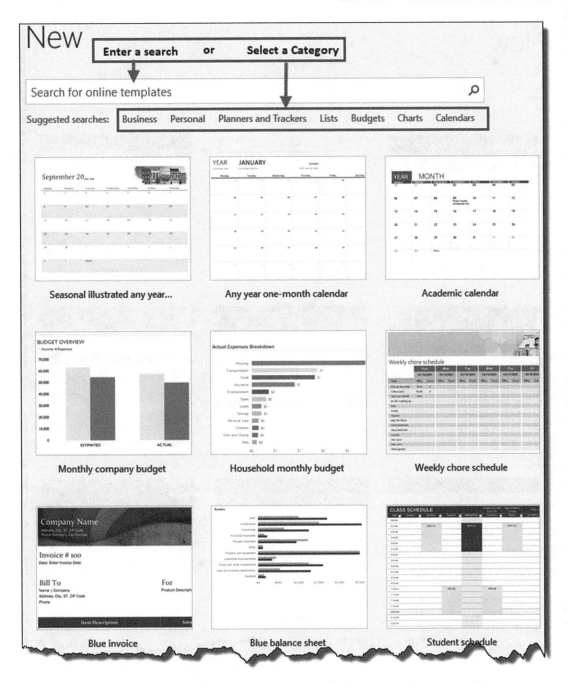

Figure 6-21. *Search for a template or select one*

If you select one of the template categories, Excel displays available templates for that category. See Figure 6-22.

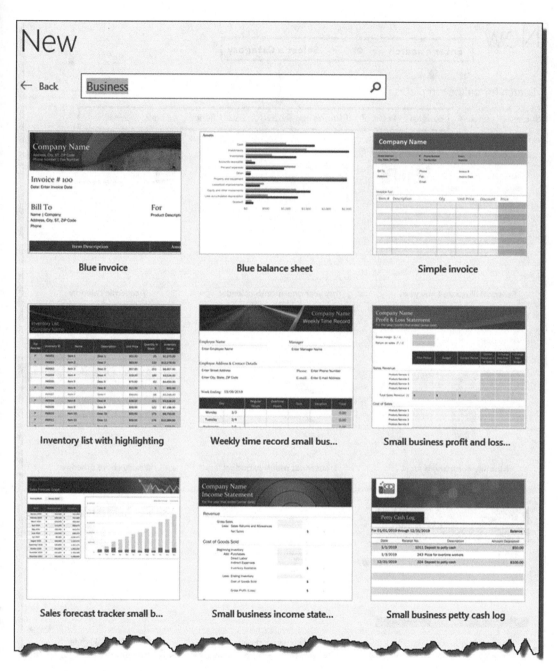

Figure 6-22. *Available templates for selected category*

Double-click the template you want to use. Excel opens a new workbook that uses that template. If you want to see a larger view of the template before you open it, you can single-click it. See Figure 6-23. You can use the Next and Previous buttons to view other templates. You can click the Create button to open the template.

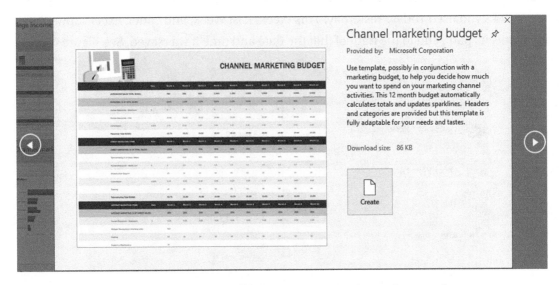

Figure 6-23. Single-clicking a template shows an enlarged view of it

Open Group—Open a Workbook

From the Open group in the Backstage pane, you can browse for a workbook or select it from a Recent list. You can pin a workbook if you want it to remain in the Recent list.

Opening an Existing Workbook

From the Open group in the Backstage view, you can open a workbook by doing one of the following:

- Select a workbook from the Recent list.

- Select a location of an Excel file such as OneDrive or a web location.

- Select PC to find an Excel file on your local computer or network. You may need to click the Browse button to get to the workbook file location.

Opening a Workbook from the Recent List

If you used the workbook recently, it would be easier to open the file by selecting Recent in the middle pane. Selecting Recent displays a list of the workbooks that you have opened recently. The most recent one would be at the top of the list. Excel breaks it down into workbooks you used Today, Yesterday, This Week, Last Week, and Older. Excel shows the file name, its path, and to the right of that the date and time it was saved. See Figure 6-24.

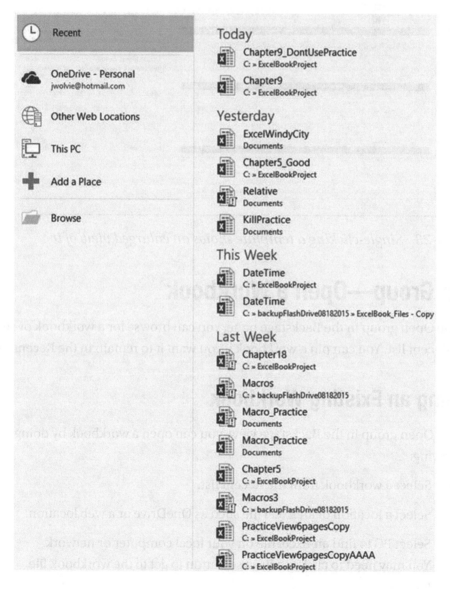

Figure 6-24. *Recently used workbooks and where they are stored*

If the workbook you need to open isn't in the Recent list, you can use one of the other options such as accessing workbooks from OneDrive or other locations on your PC.

Keeping Workbooks on the Recent List

When you move your mouse cursor over a workbook name in the Recent list in the Backstage, you see a pin to the right of the workbook name. See Figure 6-25.

ImportData.xlsx
Documents
4/12/2020 5:31 AM

Figure 6-25. *Pins*

Clicking a pin will make the workbook associated with that pin remain on the Recent list no matter how many files you open and close. Pinned items are displayed at the top of the Recent list. See Figure 6-26. You can unpin the workbook by clicking the pin again.

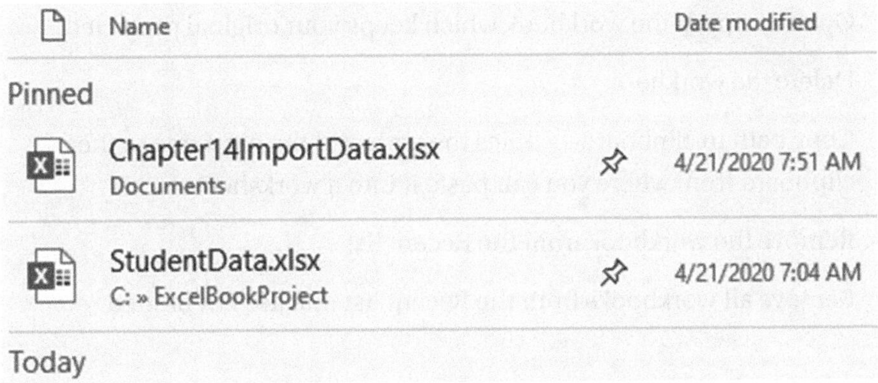

Name
Date modified
Pinned
Chapter14ImportData.xlsx
Documents
4/21/2020 7:51 AM
StudentData.xlsx
C: » ExcelBookProject
4/21/2020 7:04 AM
Today

Figure 6-26. *Pinned workbooks stay at the top of the Recent list*

Right-clicking a workbook in the Recent list brings up a context menu with six options, including Pin to list, which works the same way that clicking the pin next to the file does. See Figure 6-27.

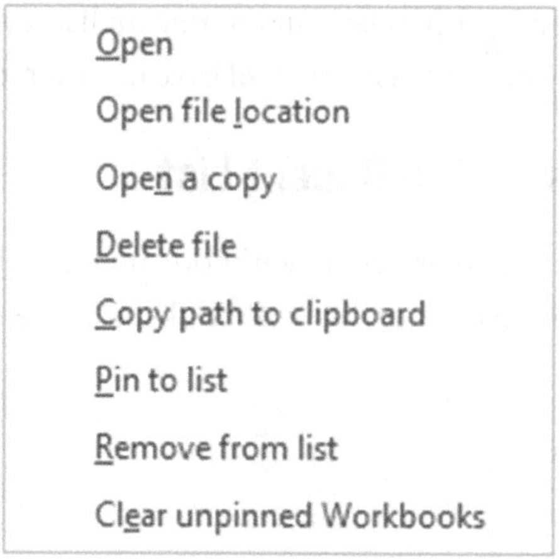

Open

Open file location

Open a copy

Delete file

Copy path to clipboard

Pin to list

Remove from list

Clear unpinned Workbooks

Figure 6-27. *Context menu with six options*

In addition to pinning or unpinning a workbook from this menu, you also can choose to

- Open the workbook

- Open a copy of the workbook which keeps your original protected

- Delete the workbook

- Copy path to clipboard—copies the path and the file name to the clipboard from where you can paste it into a worksheet

- Remove the workbook from the Recent list

- Remove all workbooks from the Recent list that are not pinned

Options Affecting the Open Group

There are two options that you can set in the Options group that affect how many workbooks appear in the Recent list and how many will appear on another available list on the Navigation pane.

Specifying the Number of Workbooks That Appear in the Recent List

How many workbooks appear in the recent list depends upon how many are specified in the Excel Options window. By default, the list shows the 25 most recent workbooks, but you can adjust that number higher or lower. To change the number that appears in the list, click the File tab on the Ribbon, click Options in the Navigation pane, and then click Advanced in the left pane. You will have to scroll down to find the Display section. Change **Show this number of Recent Workbooks** to whatever you want. See Figure 6-28. Click the OK button.

Figure 6-28. *Select the number of recent documents to display in Backstage*

Placing Recently Opened Workbooks in the Navigation Pane

If you want even quicker access to your workbooks, select the next option in the Display section which is **Quickly access the number of Recent Workbooks**. Here, you can enter the number of workbooks that you want to access from the Navigation pane. See Figure 6-29.

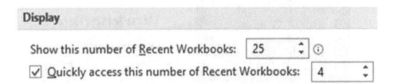

Figure 6-29. *Checking the Quickly access option displays the number you entered for the latest used workbooks on the Navigation pane*

Figure 6-30 displays an example of how the Navigation pane looks when four workbooks are selected for quick access. These four workbooks will remain on the Navigation pane. Clicking one of these workbooks will open it.

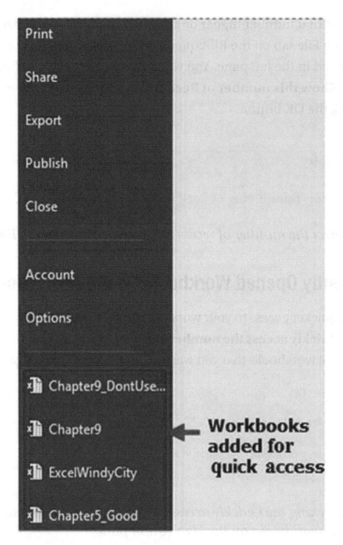

Figure 6-30. *The four latest workbooks used*

Save and Save As Groups—Saving a Workbook Using Save or Save As

While you are working in Excel, everything that you create and import such as your data, charts, graphics, and so on are all stored in the computer's random-access memory (RAM). Your RAM is only temporary storage. When you shut off your computer, everything in RAM is cleared. This is why if something happens to your computer or if the power goes off, you will lose all the additions and changes you have made since your last save. Additions and changes are not saved to your hard drive or other type of permanent storage until you perform a Save. When you save the workbook, the information is taken from your RAM and stored in a file on your permanent storage.

If something goes wrong, such as the power going out, and you haven't saved your work for quite some time, don't panic because Excel may have already saved your work to your permanent storage device using the AutoRecover feature which we will look at later in this chapter.

There are two Backstage groups dedicated to saving a workbook: the Save group and the Save As group. Actually, Save isn't much of a group because it contains no options. The Save As group is almost identical to the Open group which you have already seen earlier in this chapter. If you click the Save group and the workbook has previously been saved, then Excel updates the workbook with your most recent changes. If you select Save and this is the first time you are saving the workbook, Excel opens the Save As window (this works the same as if you had clicked Save As).

In the Save As window, you must select a folder where the workbook will be saved. Next, you need to enter the file name for the workbook file. Excel uses the file name **Book1** by default. You will want to change the name to something that matches with what type of data the workbook contains; for example, if the workbook contains your Inventory as of June 30, 2020, then name it something like "Inventory_063020." In Figure 6-31, you see that the current folder is Documents. If you click the up arrow to the left of the folder icon, you will move up to the folder level above Documents. The folders shown at the bottom right of the figure are folders that are inside of the Documents folder. If the folder you want to save to is not showing, you will need to click Browse or click More options.... Clicking either one will bring up the Save As dialog window shown in Figure 6-32.

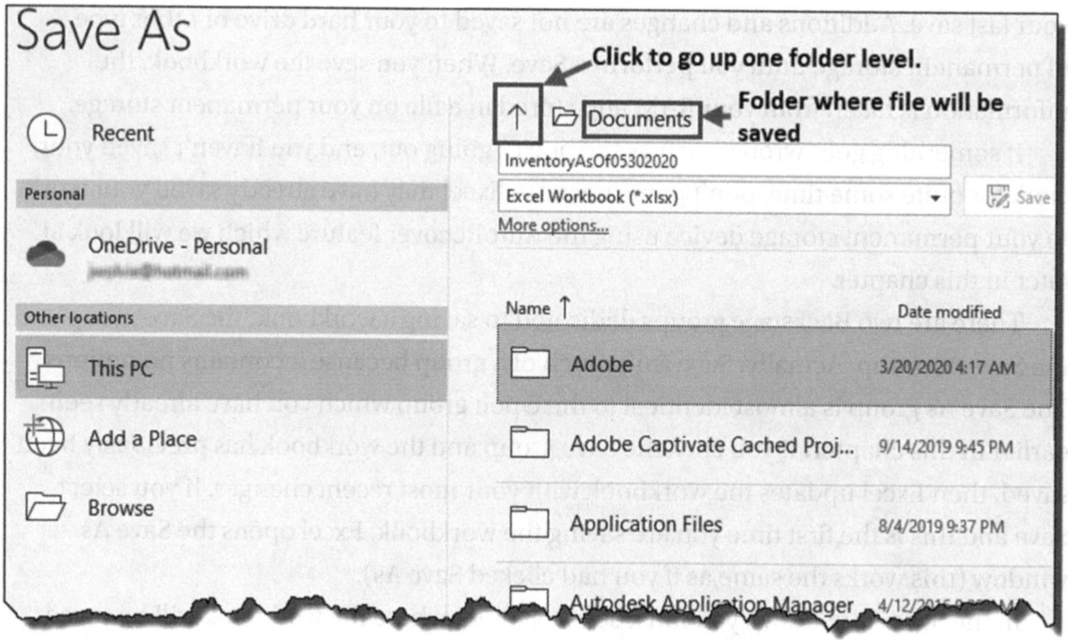

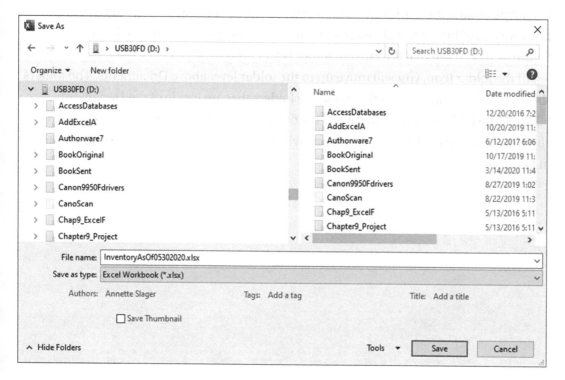

Figure 6-31. Save As—select folder and enter File name

Figure 6-32. Save As dialog box

Next, you will need to select the **Save as type** for the file you are saving. The default file type for Workbooks is **Excel Workbook**. This is the format you will probably use most often. If you want to use a different format, you can click the down arrow for the Save As type. As you can see in Figure 6-33, the workbook can be saved in many different formats.

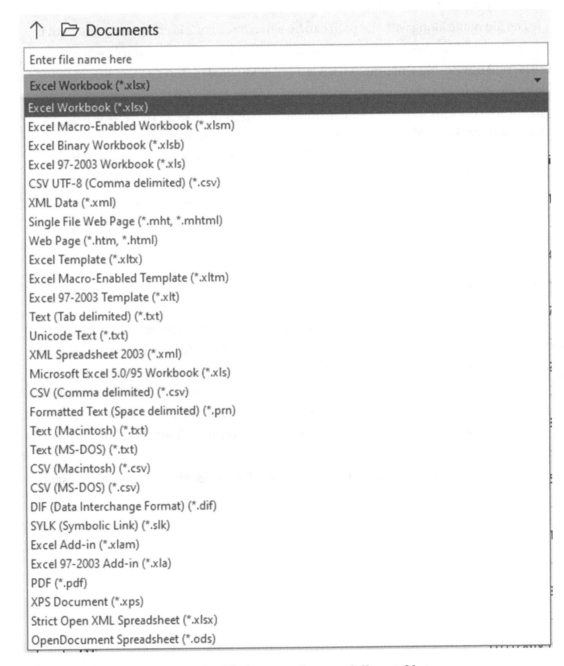

↑ 📂 Documents

Enter file name here

Excel Workbook (*.xlsx)   ▼

Excel Workbook (*.xlsx)
Excel Macro-Enabled Workbook (*.xlsm)
Excel Binary Workbook (*.xlsb)
Excel 97-2003 Workbook (*.xls)
CSV UTF-8 (Comma delimited) (*.csv)
XML Data (*.xml)
Single File Web Page (*.mht, *.mhtml)
Web Page (*.htm, *.html)
Excel Template (*.xltx)
Excel Macro-Enabled Template (*.xltm)
Excel 97-2003 Template (*.xlt)
Text (Tab delimited) (*.txt)
Unicode Text (*.txt)
XML Spreadsheet 2003 (*.xml)
Microsoft Excel 5.0/95 Workbook (*.xls)
CSV (Comma delimited) (*.csv)
Formatted Text (Space delimited) (*.prn)
Text (Macintosh) (*.txt)
Text (MS-DOS) (*.txt)
CSV (Macintosh) (*.csv)
CSV (MS-DOS) (*.csv)
DIF (Data Interchange Format) (*.dif)
SYLK (Symbolic Link) (*.slk)
Excel Add-in (*.xlam)
Excel 97-2003 Add-in (*.xla)
PDF (*.pdf)
XPS Document (*.xps)
Strict Open XML Spreadsheet (*.xlsx)
OpenDocument Spreadsheet (*.ods)

Figure 6-33. You can save the file in one of many different file types

Excel automatically adds an extension to your file based on the file type that you have selected.

If you had entered **Inventory_063020** for the file name, your complete file name would be **Inventory_063020.xlsx**. Windows Explorer uses these types to determine what icon to display next to the file name.

The file name along with the path can be a maximum of 255 characters. The path is the drive and folders you need to go through to get to your file. An example would be **C:\users\Annette\My Documents\Inventory_063020.xlsx**. You can use spaces in the file name, but it is better to use an underscore to separate words. The following characters can't be used in a file name: < > (angle brackets), | (pipe symbol), \ (backslash), ? (question mark), : (colon), * (asterisk), and "" (quotation marks).

An alternative to saving the workbook to your computer is to store the workbook online using OneDrive. OneDrive is an online storage area provided for free from Microsoft. Clicking OneDrive brings up a Save As window just like when you save files to your personal computer or network. You can create folders in OneDrive just like you can on your hard drive.

Document Recovery

You should save your workbook often, but if you haven't done a save and your system crashes or you suddenly lose power, don't panic. Excel has a feature called AutoRecover which automatically saves your file at the interval you specify. The interval can be set between 1 and 120 minutes. The default time interval is 10 minutes. I like to set mine at 5 minutes or less because I can make a lot of changes within that time that I wouldn't want to lose. Shortening the time might slow down your computer depending upon the size of the file and number of changes made.

Clicking Options in the Navigation pane of the Backstage view brings up the Excel Options window. See Figure 6-34.

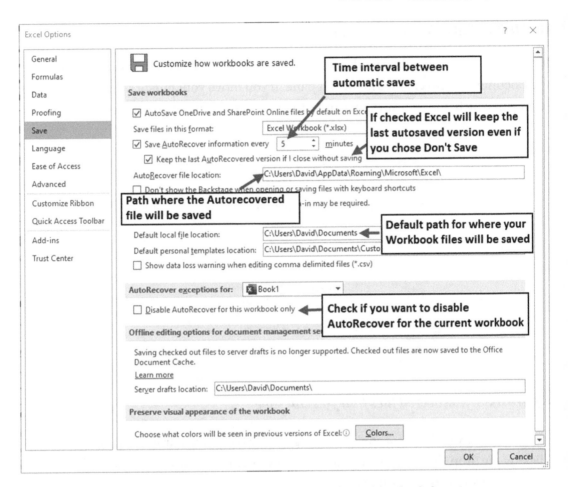

Figure 6-34. *Excel Options window with Save selected in the left pane*

Clicking Save in the left pane of the Excel Options window displays the options for saving workbooks and using AutoRecover. Here, you can set the time interval between saves, the path where the AutoRecover files will be stored, if you want to keep the last autosaved version if you closed without saving, whether or not you want to disable AutoRecover for the current workbook, and so on.

Using the default time interval of 10 minutes, if you save your Excel workbook and then 29 minutes later the power goes out, you will have lost only the changes that you made during the last 9 minutes because Excel would have backed up your file at 10 and 20 minutes. When you reopen the file later, Excel will display the Document Recovery pane containing the Autosaved version of the file and the original version of the file.

Recover from a System Crash

If your system crashes while working on a workbook when you open Excel, you will see the Document Recovery files in the left pane. If you move your cursor over the files, it will show you which one is the original and which is the recovered version. See Figure 6-35.

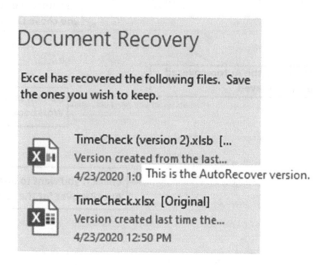

Figure 6-35. *Recovered files that you can save*

Clicking any of the recovered files in Figure 6-36 displays a drop-down arrow. Clicking the drop-down arrow displays a context menu from which you can select Open, Save As, or **Delete,** or Show Repairs.

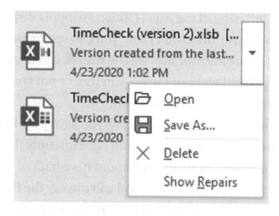

Figure 6-36. *Options for recovered files*

Selecting **Show Repairs** restores your workbook with the changes you made before the crash. It will display a message if the file contains any errors and if any repairs were made to the file.

You can click Save As to give the recovered workbook a new name.

If you click the Close button at the bottom of the Document Recovery pane while there are still recoverable files showing, Excel will ask if you want to view the files later or remove them. See Figure 6-37. If you select "Yes, I want to view these files later," Excel will keep the files and display them in the Document Recovery pane the next time you start Excel.

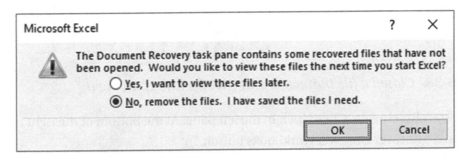

Figure 6-37. *Message asking if you want to view the recovered files later or if you want to remove them*

Closed a Workbook Without Saving

Closed a Workbook That Has Never Been Saved Without Saving

If you close a workbook without saving it and later decide that you really did need to save it, you may still be able to get back some or all of the changes depending upon what you set the AutoRecover time interval for and how much time elapsed since your last save.

If you are working on a New workbook that has never been saved and you close it without saving it, you will see the message in Figure 6-38.

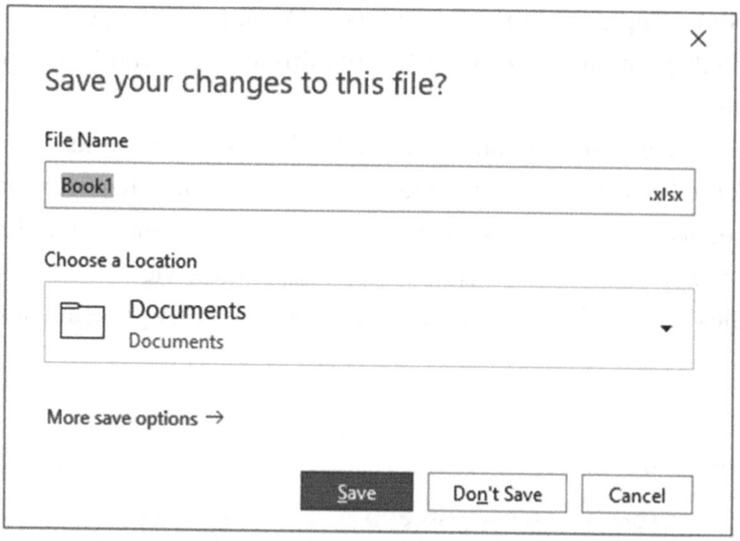
Save your changes to this file?
File Name
Book1
.xlsx
Choose a Location
Documents
Documents
More save options →
Save
Don't Save
Cancel

***Figure 6-38.** Closed a file that never was saved without saving it*

Now start Excel again. Click Open in the left pane. At the bottom of the right pane, you will see a Recover Unsaved Workbooks button.

Click Recover Unsaved Workbooks. The Open dialog box displays showing any unsaved files. The file displays the workbook's name followed by a long number. See Figure 6-39.

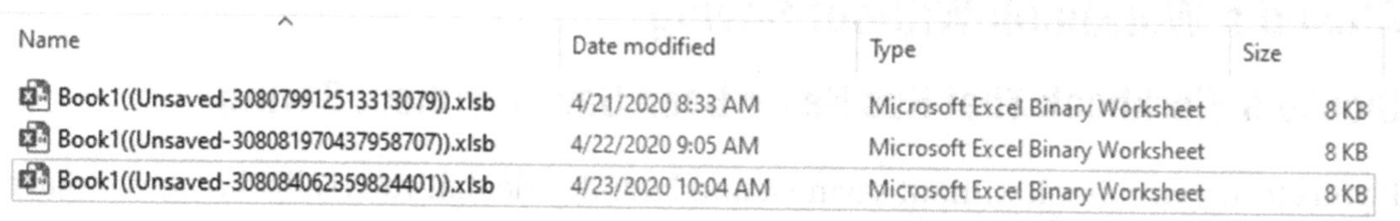
Name
Date modified
Type
Size
Book1((Unsaved-308079912513313079)).xlsb
4/21/2020 8:33 AM
Microsoft Excel Binary Worksheet
8 KB
Book1((Unsaved-308081970437958707)).xlsb
4/22/2020 9:05 AM
Microsoft Excel Binary Worksheet
8 KB
Book1((Unsaved-308084062359824401)).xlsb
4/23/2020 10:04 AM
Microsoft Excel Binary Worksheet
8 KB

***Figure 6-39.** Unsaved files have a long number added to their name*

Double-clicking the Unsaved file opens the file in Excel. You will see the message shown in Figure 6-40. Click Save As and assign the workbook a new name.

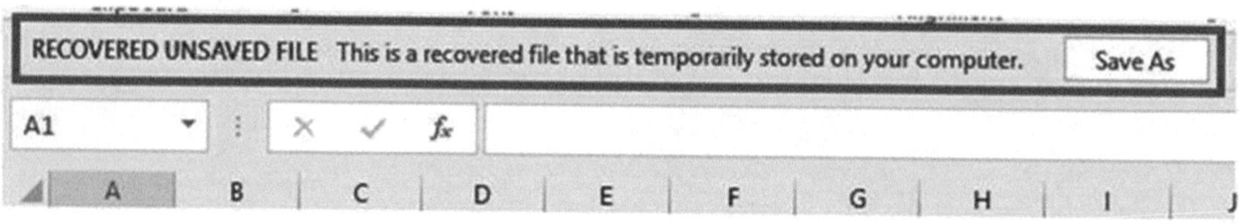
RECOVERED UNSAVED FILE This is a recovered file that is temporarily stored on your computer.
Save As
A1
fx
A
B
C
D
E
F
G
H
I
J

***Figure 6-40.** RECOVERED UNSAVED FILE message*

You can also check if there are unsaved workbooks by clicking the Ribbon's File tab and then clicking Info in the left pane. Click the Manage Workbook button to see if there are any other recoverable unsaved workbooks. See Figure 6-41.

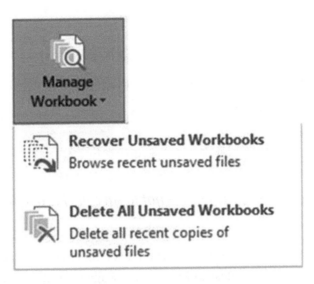

Figure 6-41. *Click Manage Workbook button to see if there are any Unsaved Workbooks*

Closed a Saved Workbook Without Saving

Use the information in this section if your workbook has already been saved at least once.

If you close your Excel workbook and you haven't saved it and the changes you have made to it have been done during the AutoRecover time interval, Excel will display the message in Figure 6-42. If you click Don't Save, you won't be able to recover and your changes will be lost. For example, if the time interval is 5 minutes and you enter some data and close it without saving it before 5 minutes has passed, you will lose your data.

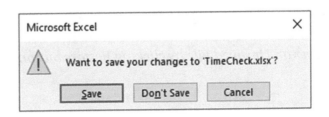

Figure 6-42. *Warning that you are closing the workbook before saving it*

If you have made changes to your workbook but haven't saved it and you then close the workbook after the time interval you set for AutoRecover, Excel will display the message in Figure 6-43. If you see this message, you will be able to recover at least some of your data.

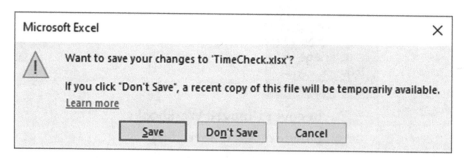

Figure 6-43. *Message that if you choose not to save the file a temporary one will still be available*

You should click the Save button, but if you click Don't Save, you can recover at least some of the changes you made. Restart Excel. A message should appear to the right of the Manage Workbook button. See Figure 6-44.

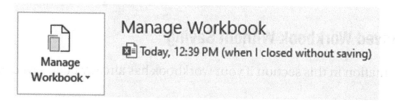

Figure 6-44. *Message when opening a workbook that was closed without being saved*

Click the message. The Workbook will open with the message in Figure 6-45 appearing above the worksheet. Click the Restore button to restore your workbook.

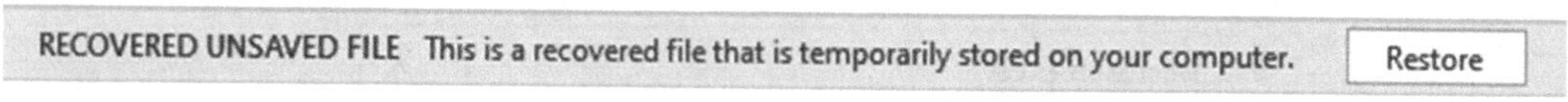

RECOVERED UNSAVED FILE This is a recovered file that is temporarily stored on your computer.   Restore

Figure 6-45. *Recover an Unsaved File by clicking Restore button*

Saving Workbooks with Protections: Backups and Limiting Changes

You can set a password at the file level to prevent an unauthorized user from opening a workbook. You can also allow a user to open the workbook but not save any changes, except as a new file. In this section, we'll see how to set up regular backups for your workbooks, protect the original file using Read Only, and password-protect your workbooks.

With the workbook you want protected open, click the Ribbon's File tab. Click Save As in the Navigation pane. Click Browse. In the Save As dialog box, click the Tools drop-down menu and select General Options. See Figure 6-46.

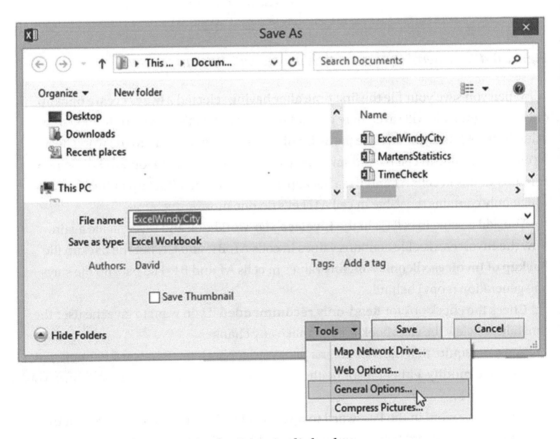

Figure 6-46. *Tools options in the Save As dialog box*

The General Options dialog box opens. Click the check box for **Always create backup**. This will create a backup file of the file you are saving in the same location. The file will be named **Backup of** followed by the name you give the workbook. It will be given an extension of **.xlk**. For example, if you name your file **Invoices.xlsx**, the backup file will be named **Backup of Invoices.xlk**. See Figure 6-47.

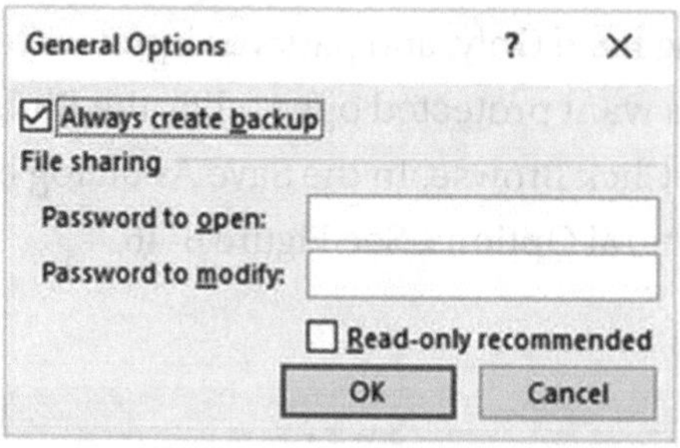

Figure 6-47.   *General Options with Always create backup selected*

When you save your file the first time after having selected **Always create backup**, only the original file will save; there will not be a backup. Let's say you only entered a value in cell A1 and you save the file as **Invoices.xlsx**. Now you add to the worksheet by entering a value in cell B1 and save the workbook again. The **Invoices.xlsx** file now contains values in cells A1 and B1; a backup is created named **Backup of Invoices.xlk** which only contains the value in cell A1. Let's do one more.

You add a value in cell C3 in the **Invoices.xlsx** workbook and save the file again. Now the **Invoices.xlsx** file contains values in cells A1, B1, and C1, but the backup file **Backup of Invoices.xlk** only contains values in cells A1 and B1. The backup file stays one generation (copy) behind.

Check the check box for **Read-only recommended** if you want to give the user the capability to view the workbook but not make any changes.

You can require that the user enter a password for both the **Password to open** and **Password to modify**, either one or neither. Remember, if you forget your password, you can't get it back.

Enter a password in the **Password to open** text box if you want to require that the user enter this password before opening the workbook. The user can view or change the workbook.

Enter a second password in the **Password to modify** text box if you want to require that the user enter this password if he or she wants to modify the workbook. The user will be able to view the workbook as read only if he doesn't enter the password.

You should enter both passwords if you want to require that the user enter a password to open the workbook and then require them to enter a second one to modify the workbook. The passwords should not be the same.

If you entered a password for Password to open, you will need to confirm that password. See Figure 6-48.

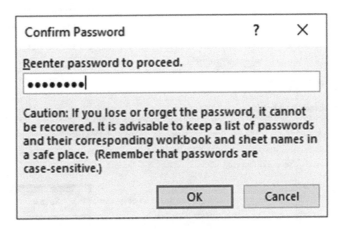

Figure 6-48. *Confirm password if you entered a password for Password to open*

If you entered a password for Password to modify, you will need to confirm that password. See Figure 6-49. If you entered both passwords, you will be required to confirm both of them.

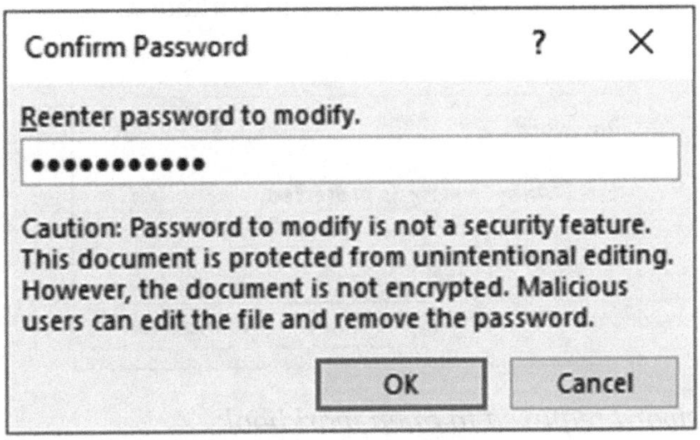

Figure 6-49. *Confirm password if you entered a password for Password to modify*

Clicking the Save button saves the workbook with the options you entered for General Options. See Figure 6-50.

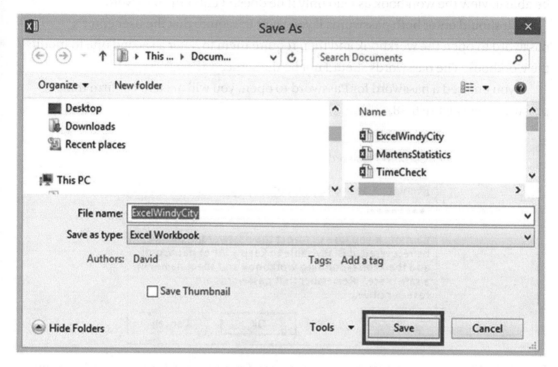

Figure 6-50. *Save the file using the selections you made*

If you entered passwords, you will be required to enter them once you have closed the workbook and reopened it.

- If you required a password to open the workbook, you will see the window in Figure 6-51 when opening the workbook.

Figure 6-51. *Password required to open workbook*

- If you required a password for modifying the workbook, another Password window opens. The user must enter a password in order to modify the workbook. The workbook can be opened in Read Only without entering a password by clicking the Read Only button. See Figure 6-52.

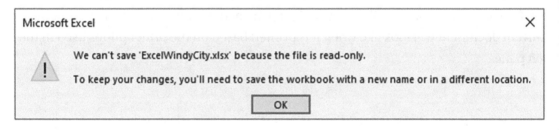

Figure 6-52. *Enter password to modify workbook or click Read Only button*

If the user clicks the Read Only button, they can make changes to the workbook but will not be able to save it, thus preventing any changes from being made. The user will see the message in Figure 6-53.

Microsoft Excel	✕
⚠	We can't save 'ExcelWindyCity.xlsx' because the file is read-only.
	To keep your changes, you'll need to save the workbook with a new name or in a different location.
	OK

Figure 6-53. *Message saying the file is read only. If you want to save changes, you will need to save the file with a new name*

If a password is required to open the workbook and the user elected to use Read Only instead of entering the password for modifying, the Backstage will show as in Figure 6-54.

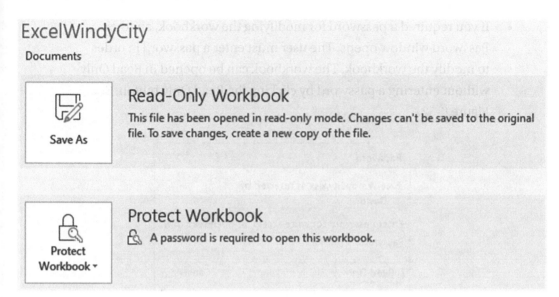

Figure 6-54. *Backstage messages*

Print Group—Printing a Workbook

The left pane of the Print group (Figure 6-55) contains Settings for controlling how the worksheet or workbook is to be printed. Many of these same options are available on the Page Layout tab on the Ribbon. The right pane of the Print group displays what the current document will look like when it is printed based on the print options used in the left pane.

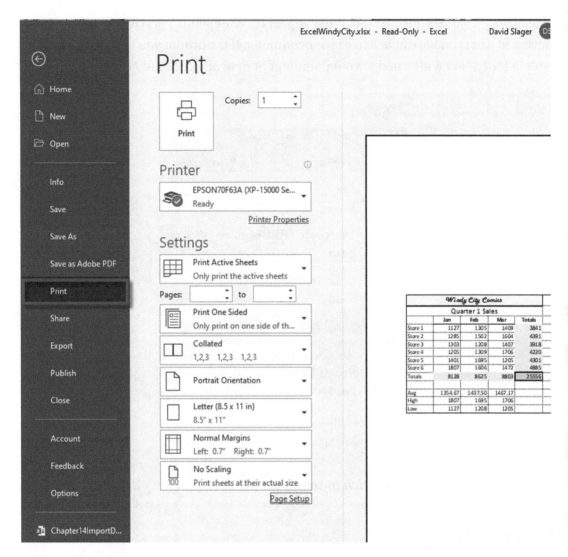

Figure 6-55. *Printing options*

You should first make any necessary changes to the **Printer** and **Settings** properties and then click the Print button to do the printing.

Selecting a Printer

One of the decisions you will need to make is where you want your report printed to. You don't have to print to a printer. You can print to programs, to other devices such as a fax machine, or to a file. Clicking the down arrow for the Printer lists all your currently available options. See Figure 6-56. What appears here depends upon what programs you have and what you have attached to your computer. If the printer you wish to print to isn't in the list but is attached to your computer or network, click the Add Printer option.

Figure 6-56. *Locations to which to send your printed data*

Printer Settings

The Printer Properties options depend entirely upon what you have selected for a printer. If you have selected an actual printer, you may see loads of options across three or more tabs. Printer companies have different options; you may see options for presets, quality of print, type of paper, orientation, document size, duplex printing, borderless, maintenance, and so on. If you select Fax, you will only see a few options.

Selecting to Print Active Worksheets, an Entire Workbook, or a Selected Portion

The first option in the Settings area is for selecting whether you want to print only the current worksheet, the Entire Workbook, or just the area that you have selected. See Figure 6-57. What you select here is reflected in the right pane. The **Print Active Sheets** only prints those worksheets that you currently have selected. The **Print Selection** prints the cells that you currently have selected.

Figure 6-57. *Print options*

Selecting Which Pages to Print

The next Settings option lets you select what pages you want to print. The entries in Figure 6-58 will print only pages 2, 3, and 4. This is true even if you have selected **Print Entire Workbook**.

Figure 6-58. *Enter the page range you want to print*

Collating

Collating in Excel works just like collating on a copy machine. If you are printing a worksheet that is five pages and you select to print four copies, then selecting **Collated** will give you four sets of pages 1 through 5. Selecting **Uncollated** will print four copies of page 1, then four copies of page 2, then four copies of page 3, and so on. See Figure 6-59.

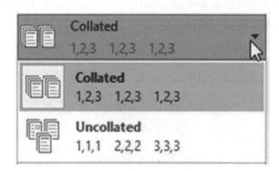

Figure 6-59. *Collating options*

Orientation

You can print in a vertical format (Portrait) or a horizontal format (Landscape). Selecting Landscape is one way of printing more columns on a page.

Selecting Paper Size

You will probably be printing on letter size (8.5" x 11") paper, but if not, you can select another paper size. See Figure 6-60. If you have a spreadsheet with a lot of columns, you might try using legal size paper. Excel will automatically adjust your spreadsheet to reflect the selected paper size.

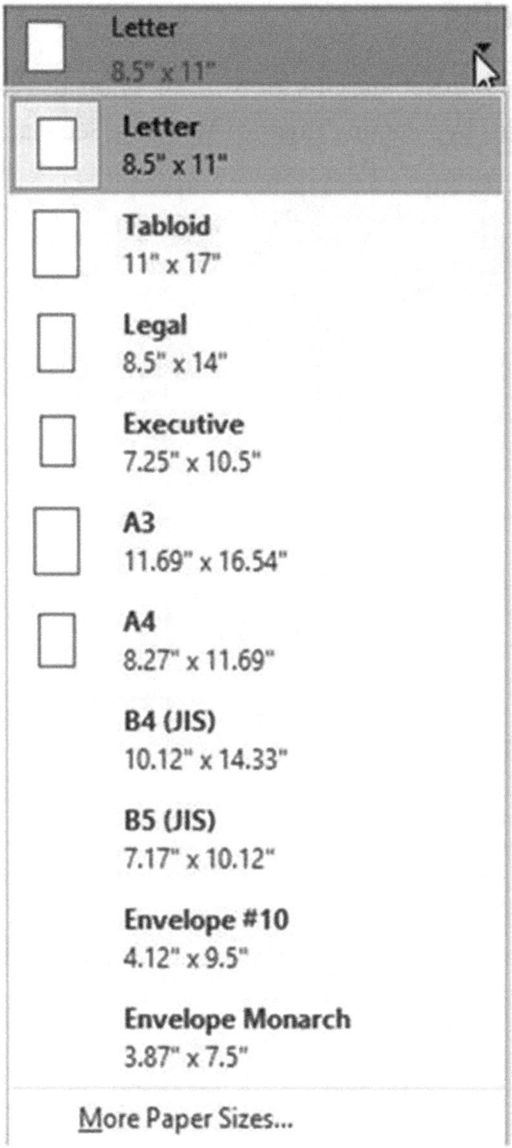

Letter
8.5" x 11"

Letter
8.5" x 11"

Tabloid
11" x 17"

Legal
8.5" x 14"

Executive
7.25" x 10.5"

A3
11.69" x 16.54"

A4
8.27" x 11.69"

B4 (JIS)
10.12" x 14.33"

B5 (JIS)
7.17" x 10.12"

Envelope #10
4.12" x 9.5"

Envelope Monarch
3.87" x 7.5"

More Paper Sizes...

Figure 6-60. *Select output size*

Setting Page Margins

You can use one of Excel's preset margins for Normal, Wide, or Narrow, or you can create your own custom margins by selecting **Custom Margins**. See Figure 6-61.

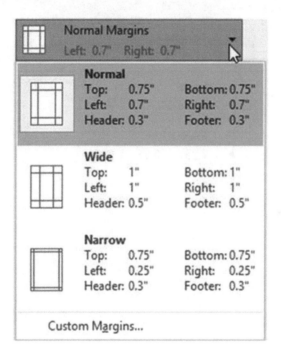

Figure 6-61. *Select margins*

Clicking Custom Margins brings up the Page Setup window with the Margins tab selected. See Figure 6-62.

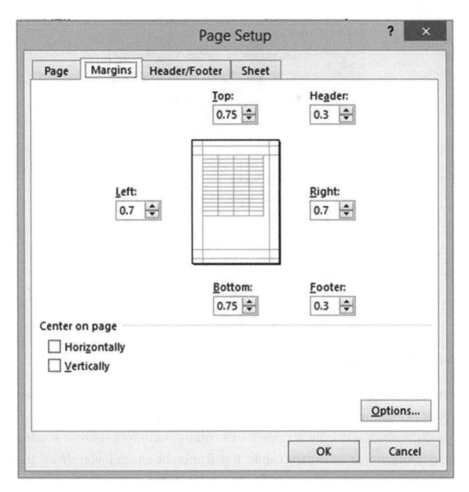

Figure 6-62. *Enter margins and how you want the data centered on the page*

The Margins tab has options for setting the top, left, right, and bottom margins of your page as well as setting margins for the header and footers. You can use the check boxes to specify how you want the data centered on the page. When you select ☑ Horizontally, ☑ Vertically, or both, the image in the middle of the tab will reflect your choices by showing how your data will be displayed on the report.

Selecting Scaling Options

You can print the report at its actual size, or you can select one of Excel's preset Scaling options that shrink the text and objects on the report. See Figure 6-63.

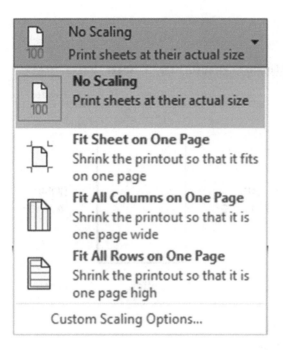

Figure 6-63. *Select a Scaling option*

When you are printing a report, you probably don't want to have an extra page for a couple of extra columns or a few extra rows. This can make reading your data very difficult. As discussed previously, if there are too many columns to fit on a single page, you could try switching to landscape mode. It will often be easier if you shrink the text and objects on a worksheet so that they fit on a single paper.

- Selecting **Fit Sheet on One Page** will print everything that is on one worksheet on a single page.

- Selecting **Fit All Columns on One Page** shrinks the printout so that it is one page wide, but your report may print vertically on more than one page.

- Selecting **Fit All Rows on One Page** shrinks the printout so that it is one page high, but your report may print horizontally on more than one page.

- If none of these options meet your needs, you can select **Custom Scaling Options**. Selecting Custom Scaling Options brings up the Page Setup dialog box with the Page tab selected. See Figure 6-64.

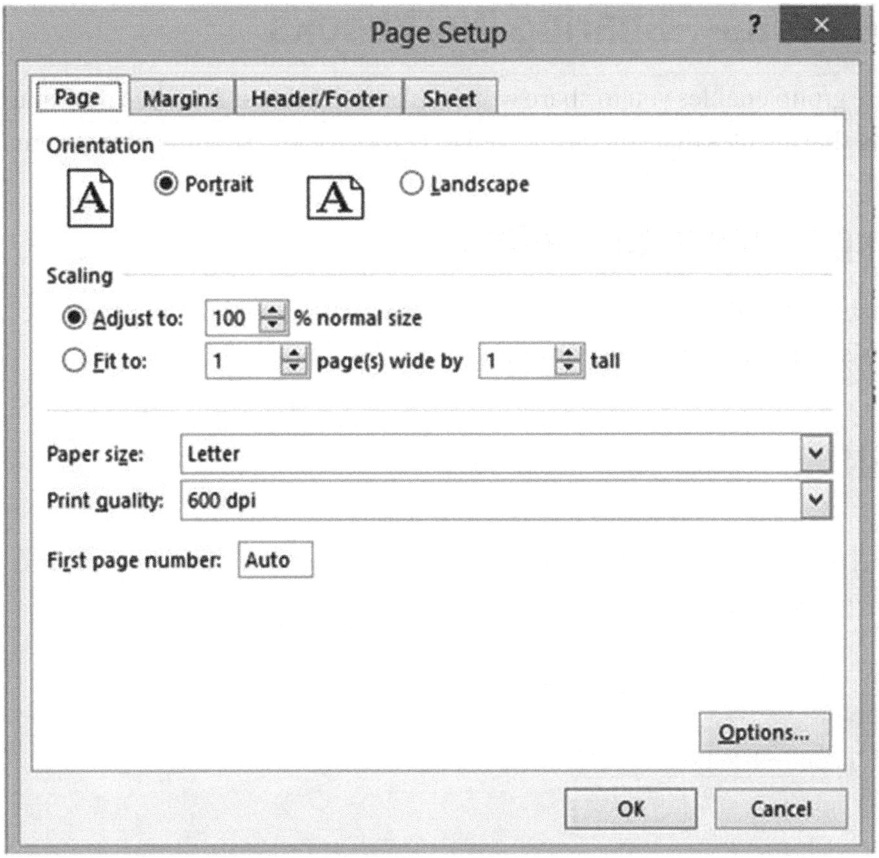

Figure 6-64. *Custom Scaling options*

The scaling options here allow you to set the percentage of the actual size you want your report to be. The value you enter must be between 10 and 400 inclusive.

You can also specify how many pages wide by how many pages high you want the report.

Note   The Page Setup option is located at the bottom of the Settings area in the Backstage. Clicking Page Setup brings up the Page Setup dialog box. This is the same Page Setup window you see when you click Custom Margins or Custom Scaling options. The Page Setup window includes some of the same options available in other areas of the Print panel plus some additional options. The Page Setup window can also be accessed from the Print Layout tab on the Ribbon.

Share Group—Sharing Workbooks

The Share group enables you to share workbooks with others. You have multiple options, including sharing files that you have saved to OneDrive and sending files via email.

Sharing Online with OneDrive

After you have saved a workbook to your OneDrive, click Share in the Backstage left panel. This brings up the Share panel in Figure 6-65.

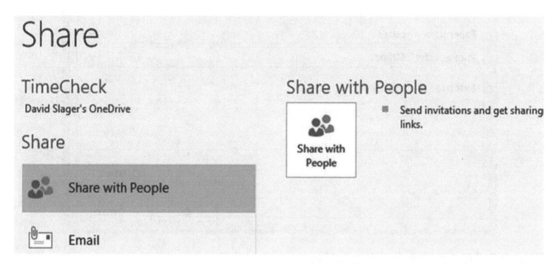

Figure 6-65. *Sharing options*

Now that you have saved your workbook on OneDrive, you can invite others to just view your workbook on OneDrive or you can let them make changes to it. Right-click the **Share with People** button in the right pane shown in Figure 6-65. This brings up the Share window shown in Figure 6-66. Enter the email address of those people you want to invite to use your workbook. You can select **Can Edit** or **Can View**. You can add a message. Click the Share button to send the emails.

Figure 6-66. *Inviting others to share your workbook*

You can also share a workbook by providing a link to your workbook. At the bottom of the Share window, you will see two options: Send as attachment and **Get a sharing link**. See Figure 6-67. Click the Get a sharing link.

Figure 6-67. *Click the text Get a sharing link*

The window in Figure 6-68 displays. You can click the **Create an edit link** button if you want to provide a user with the ability to edit your workbook online, or you can click the **Create a view-only link** button to provide a user with the ability to only view your workbook online.

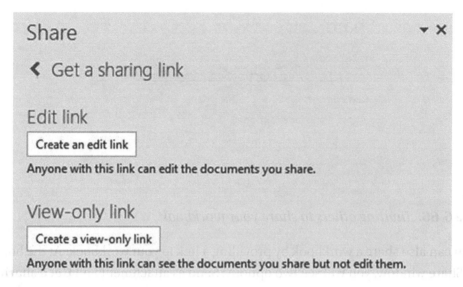

Figure 6-68. *Select if you want the user only to be able to view the workbook or if you want to allow him to edit it*

Figure 6-69 shows the link that was provided when **Create an edit link** was clicked.

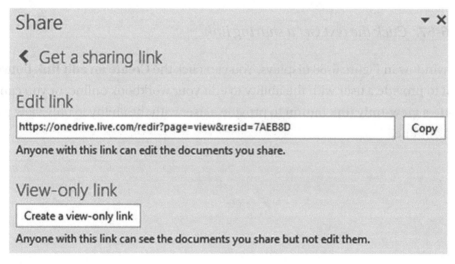

Figure 6-69. *Provide user with link to edit the workbook*

You can email this link to others, you can tell others what it is, you can post it on a website, and so on.

Click the arrow to the left of **Get a sharing link** in Figure 6-68. The Share window now displays those people to whom you have sent an email and what rights you gave them. It also shows that you now have an edit link that you can share with anyone. See Figure 6-70.

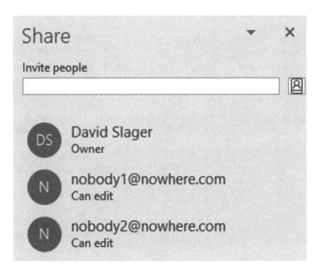

Figure 6-70. *Share window displays with users to whom you have sent an email*

Sharing Files Using Email

You can send your Excel workbook to others in any of the available formats as an email attachment. See Figure 6-71. Those people receiving your email can then open it in their copy of Excel.

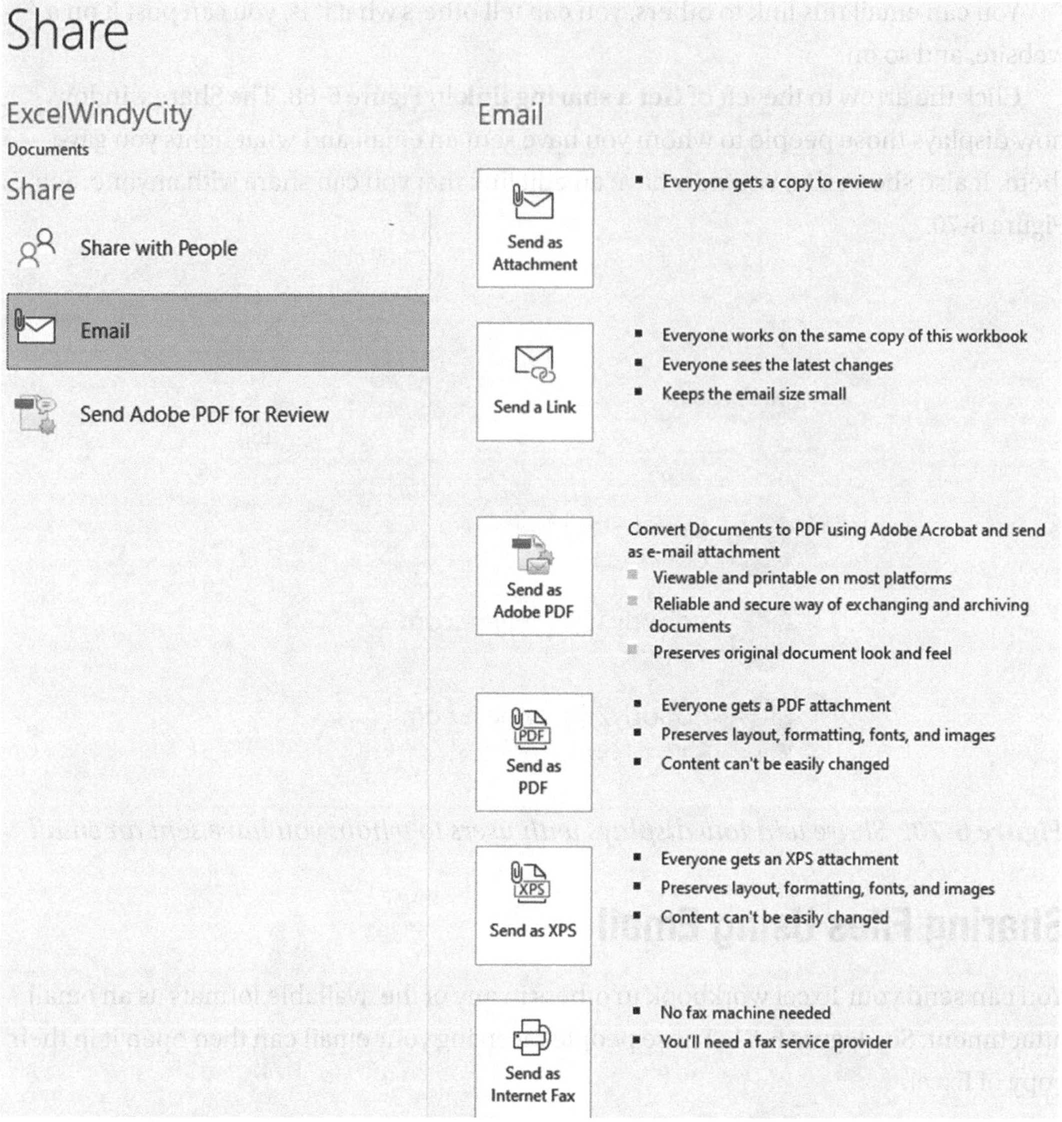

Share

ExcelWindyCity
Documents

Email

Share

- 🧑 Share with People

- 📧 Email

- 📝 Send Adobe PDF for Review

📎📧 Send as Attachment

- Everyone gets a copy to review

📧 Send a Link

- Everyone works on the same copy of this workbook
- Everyone sees the latest changes
- Keeps the email size small

📄📧 Send as Adobe PDF

Convert Documents to PDF using Adobe Acrobat and send as e-mail attachment

- Viewable and printable on most platforms
- Reliable and secure way of exchanging and archiving documents
- Preserves original document look and feel

📎 PDF Send as PDF

- Everyone gets a PDF attachment
- Preserves layout, formatting, fonts, and images
- Content can't be easily changed

📎 XPS Send as XPS

- Everyone gets an XPS attachment
- Preserves layout, formatting, fonts, and images
- Content can't be easily changed

🖨 Send as Internet Fax

- No fax machine needed
- You'll need a fax service provider

Figure 6-71. *Select the format you want the workbook emailed in*

Clicking the **Send as Attachment** button brings up Outlook (provided you have a copy of Outlook) with the Excel workbook file already set as the attachment. See Figure 6-72. The **Subject** is filled with the name of the Excel workbook. You can then fill in the email address of the person you are sending it to and perhaps a text message. Then click the Send button to send the attachment to that email address.

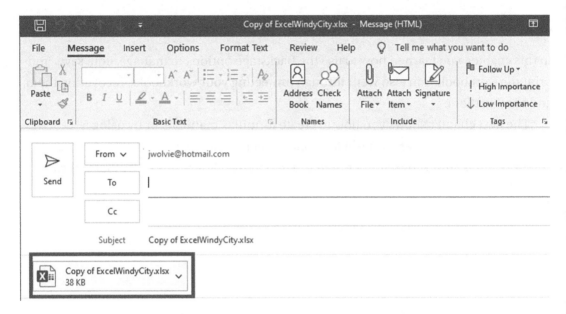

Figure 6-72. *The workbook is set as an attached file in Outlook*

If the recipient of your email doesn't need to edit the workbook or doesn't have a copy of Excel, then you can send your workbook in Adobe's PDF format or Microsoft's XPS format. The advantage of sending a file in either one of these formats is that they have many viewing options and they don't require the recipient to have the Excel program.

Account Group

Changes in your Account pane affect all your Microsoft Office products. Here, you can manage your Microsoft Office Account, check for Microsoft updates, and change your Office theme. You can also add an Office background pattern that appears on your title bar.

Summary

In this chapter, you learned how to use the Backstage options to manage, edit, save, and print the workbook. You also went through protecting your workbooks, recovering them, and sharing them.

The next chapter shows you how to create formulas manually or by using a wizard. Excel functions are built-in formulas. Functions are what turns Excel into a computational powerhouse rather than just a tool that allows data entry that you can nicely format and print.

CHAPTER 7

Creating and Using Formulas

Formulas are created to give us the results from such things as adding the items we buy at the store, computing our taxes, computing our car's miles per gallon, and so on. A formula in Excel is an expression that is computed, and the results are placed in the cell that contains the formula. You can enter your own formulas, or you can use one of Excel's many built-in formulas. Excel's built-in formulas are called **functions**.

After reading and working through this chapter, you should be able to

- Enter formulas into a cell or into the formula bar

- Enter formulas using the correct order of precedence

- Copy formulas

- Use the Auto functions

- View all of the formulas on a worksheet

- Select named ranges to use in formulas

- Use absolute and mixed cell references

- Create and edit named ranges and constants

© David Slager and Annette Slager 2020
D. Slager and A. Slager, *Essential Excel 2019*, https://doi.org/10.1007/978-1-4842-6209-2_7

Formulas

Without the capability to enter formulas, Excel would merely serve as a sheet for data entry where you could align and format data and create graphs. Excel has formulas for totaling rows and columns, but Excel allows you to go way beyond that. You can use one of Excel's built-in formulas, or you can enter just about any formula you can think of. You can use very basic formulas or those that are extremely complex. If you want to solve the area of a triangle, you can enter the formula. Do you need to solve something using the quadratic equation? You can enter the formula for it.

Introducing Formulas

An example of a formula would be = 5 * 6. The expression returns a value of 30. Formulas always start with an equal sign (=). Table 7-1 lists math symbols used by Excel and their meanings.

Table 7-1. *Math Symbols Used by Excel*

Symbols for Math Operations	
Symbol	**Meaning**
+	Addition
–	Subtraction
*	Multiplication
/	Division
%	Percent (converts the whole number to a percent)
^	Exponentiation Ex.(3^2)
=	Equal to
<	Less than
<=	Less than or equal to
>	Greater than
>=	Greater than or equal to
<>	Not equal to

Formulas can include specific values and cell addresses. When a cell address is included in a formula, it is not the address that is used in the formula but, rather, the value that is in that cell. The result of the expression is placed in the cell that contains the formula.

If I wanted to add the values in cells A3 and B3 and I wanted the result to be placed in cell C3, I would enter the formula as shown in Figure 7-1.

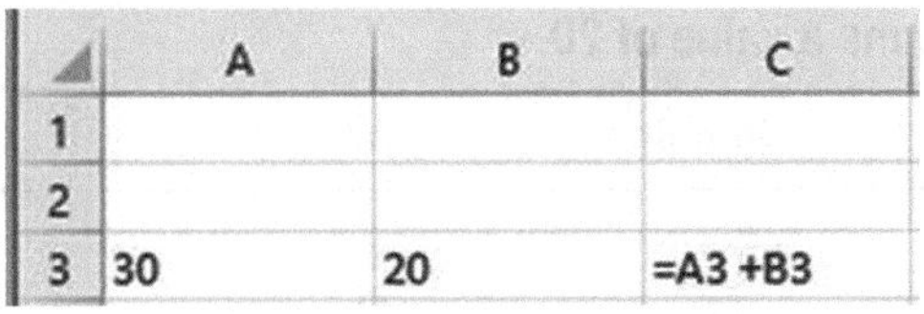

◢	A	B	C
1			
2			
3	30	20	=A3 +B3

Figure 7-1. *Cell C3 contains the formula for adding the values in cells A3 and B3*

Figure 7-2 shows the result of entering the formula in Figure 7-1.

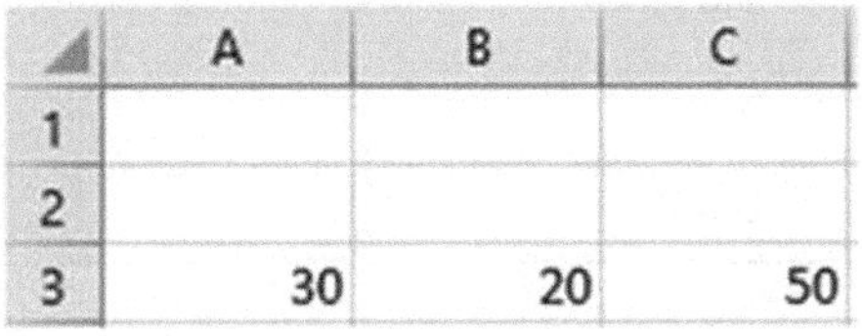

◢	A	B	C
1			
2			
3	30	20	50

Figure 7-2. *Cell C3 contains the result of adding cells A3 and B3*

If there is a formula in the current cell, the results of the formula are displayed in the cell, but the formula is displayed in the formula bar. See Figure 7-3.

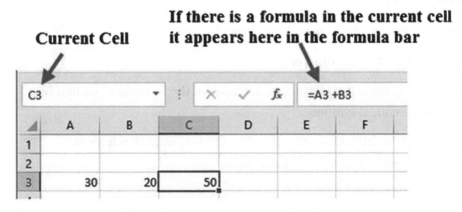

Figure 7-3. *The formula for cell C3 appears in formula bar*

Table 7-2 shows examples of formulas and the results they would produce if cell A3 contains a value of 30 and cell B3 contains a value of 20.

Table 7-2. *Excel Formulas and Their Results (the Results Would Be Placed in the Cell That Contains the Formula)*

Cell A3 contains a value of 30.

Cell B3 contains a value of 20.

Formula	Result
= A3 + 25	55
= A3 – B3	10
= B3 – A3	−10
= A3 * 10	300
= A3 / B3	1.5
= A3 * 10%	3
= B3 ^ 3	8000

Entering an = sign as the first character in a cell tells Excel that what follows is a formula.

Entering Formulas

Formulas can be entered in one of three ways:

- Typing it directly into a cell
- Clicking the cells you want to reference in the formula separated by mathematical symbols
- Using your arrow keys to select the cells you want to reference separated by mathematical symbols

Typing the Formula into a Cell

When typing the complete formula, you start by typing an equal sign and then entering any values or cell references that would be separated by mathematical symbols. If you enter any cell references in the formula, Excel will highlight the cell address in the formula and it will place a border around the cell being referenced. It will highlight both in the same color.

The example in Figure 7-4 shows the formula you would enter in cell C1 to add the values in cells A1 and B1. When entering A1 in the formula, the background color of cell A1 and the text A1 will be the same color. Cell B2 will have a different matching color for its cell background and the text B1.

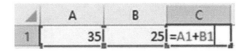

Figure 7-4. *Colors are used to show which cell is used for which operand*

Creating the Formula by Clicking the Cells You Want to Reference

Creating formulas by clicking the cells you want to reference is done by first entering an equal sign and then clicking any cells you want to reference separated by mathematical symbols.

Using Figure 7-4 as an example, you would do the following:

1. Click inside cell C1.

2. Type the equal sign.

3. Click cell A1 (Excel enters A1 into the formula).

4. Type the plus sign.

5. Click cell B1 (Excel enters B1 into the formula).

6. Press Enter.

Your formula would be the same as if you had entered it manually =A1+B1.

Creating the Formula Using Your Arrow Keys

The benefit of using your arrow keys to enter a formula is that your hands never have to leave the keyboard. Using the same example, you would create the formula by doing the following:

1. Click inside cell C1.

2. Type an equal sign.

3. Press the left arrow key twice to select cell A1.

4. Type a plus sign.

5. Press the left arrow key once to select cell B1.

6. Press Enter.

EXERCISE 7-1: ENTERING FORMULAS

In this exercise, you will enter four formulas using different math operations.

1. Create a new workbook named Chapter 7. Rename the worksheet Sheet1 tab to "MarkUp."

2. Make the worksheet look like Figure 7-5.

▲	A	B	C	D	E	F	G	H
1			Southwest Sports World					
2	Product	Items Sold	Cost Per Item	Price Per Item	Mark up Amount	Mark Up %	COGS	Profit
3	Basketballs	120	18	28.5				

Figure 7-5. *Enter this data in your own workbook*

Table 7-3 shows the formulas needed for each unfilled column.

Table 7-3. *Formulas Needed for Spreadsheet*

Spreadsheet Formulas

Markup Amount	Price Per Item – Cost Per Item
Markup %	Cost Per Item / Price Per Item
COGS	Items Sold * Cost Per Item
Profit	(Items Sold * Price Per Item) – COGS

Computing the Markup Amount

1. Enter an **equal sign** (=) in cell E3. Enter **d**. Excel knows you are entering formulas and so it tries to be helpful by suggesting one of Excel's built-in functions that starts with a D. See Figure 7-6.

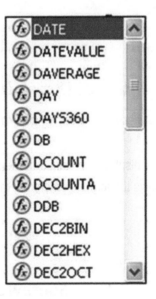

Figure 7-6. *Built-in functions*

2. We will look at built-in functions later, but for now ignore the function choices and just enter **d3**. Excel provides a visual cue for your selection by putting a box around the cell that you entered in your formula. It makes the cell address and the box both blue. See Figure 7-7.

◢	A	B	C	D	E	F
1				**Southwest Sports World**		
2	**Product**	**Items Sold**	**Cost Per Item**	**Price Per Item**	**Mark up Amount**	**Mark Up %**
3	Basketballs	120	18	28.5	=d3	

Figure 7-7. Cell address and its corresponding cell are the same color

3. Enter a **minus sign** (–) and then **c3**. Excel makes the cell address and the box surrounding the referenced cell both red. See Figure 7-8.

◢	A	B	C	D	E	F
1				**Southwest Sports World**		
2	**Product**	**Items Sold**	**Cost Per Item**	**Price Per Item**	**Mark up Amount**	**Mark Up %**
3	Basketballs	120	18	28.5	=d3-c3	

Figure 7-8. Cell address and its corresponding cell are both red

4. Press the **Tab** key. The cell displays the value returned from the formula. The formula bar, however, shows the formula entered in the cell. See Figure 7-9.

E3	▼	⋮	✕	✓	*fx*	=D3-C3

◢	A	B	C	D	E
1					**Southwest Sports World**
2	**Product**	**Items Sold**	**Cost Per Item**	**Price Per Item**	**Mark up Amount**
3	Basketballs	120	18	28.5	10.5

Figure 7-9. Formula bar displays the formula entered in the active cell

Computing the Markup %

1. Enter an **equal sign** (=) in cell F3. This time instead of entering cell addresses in the formula, you will click the cell you want to use and Excel will automatically put the address of that cell in your formula. Click **cell C3**. Excel places an animated box around cell C3 and places its address in the formula.

2. Enter a **divide sign** (/).

3. Click cell **D3**. See Figure 7-10.

Cost Per Item	Price Per Item	Mark up Amount	Mark Up %
18	28.5	10.5	=C3/D3

Figure 7-10. Formula entered for Mark Up %

4. Press **Ctrl + Enter**. The result is 0.631578947. This number might be rounded depending upon the width of the column.

5. We would like to show the value as a percentage. Select the Home tab if it isn't already selected and then click the % icon in the Number group. The result is rounded up to 63%.

6. We would like to have the number displayed with two decimal positions. Click the increase decimal button twice. The value changes to 63.16%.

Computing the Cost of Goods Sold

1. Click cell G3 and enter **=b3***.

2. Click cell C3 and press the Tab key.

Compute the Profit

1. You can use any combination of typing the cell reference or clicking the cell to get its reference for the profit formula. Enter the formula **=(b3*d3)–g3** in cell H3.

2. Press Enter. See Figure 7-11.

Southwest Sports World							
Product	Items Sold	Cost Per Item	Price Per Item	Mark up Amount	Mark Up %	COGS	Profit
Basketballs	120	18	28.5	10.5	63.16%	2160	1260

Figure 7-11. *Result of entering formula for profit*

Copying Formulas

You used the Fill Handle previously to AutoFill a series of values. The Fill Handle can also be used to copy formulas across rows or columns while it automatically adjusts the formula for the proper column and row.

If you copy formulas down through a column, the row number in the formula will be automatically adjusted to the current row where the formula is being copied to. If you copy formulas across rows, the column letter in the formula will be automatically adjusted to the current column of where the formula is being copied to. This automatic adjustment of row and column addressing is called **relative cell referencing** because the cell addresses change relative to where the formula is placed.

EXERCISE 7-2: COPYING FORMULAS USING AUTOFILL

This exercise continues with the same worksheet from the previous exercise. You will add additional records and then copy the formulas you made previously into the new rows.

1. Add the additional records (rows 4–7) shown in Figure 7-12.

	A	B	C	D	E	F	G	H
1				Southwest Sports World				
2	Product	Items Sold	Cost Per Item	Price Per Item	Mark up Amount	Mark Up %	COGS	Profit
3	Basketballs	120	18	28.5	10.5	63.16%	2160	1260
4	Wood Bats	140	16	24.5				
5	Steel Bats	90	18.5	26				
6	Skis	50	120	160.75				
7	Tennis Rackets	70	20.25	38				

Figure 7-12. *Add these additional records*

2. Click cell E3. Drag the AutoFill Handle down from cell E3 through cell E7. See Figure 7-13.

| 10.5 |
| 8.5 |
| 7.5 |
| 40.75 |
| 17.75 |

Figure 7-13. *Result of dragging the AutoFill Handle down from cell E3 to cell E7*

AutoFill copies the formulas to the other cells automatically. When AutoFill is used to copy formulas down through a column, it automatically adjusts the cell references in the formula to the current row.

3. Click cell E4. The formula in cell E4 is =D4-C4. The row was adjusted from 3 to 4.

4. Click other cells in the column to see how their formulas were adjusted.

5. Drag the AutoFill Handle down from cell F3 through F7.

6. Drag the AutoFill Handle down from G3 to G7.

7. Drag the AutoFill Handle down from H3 to H7.

Your result should look like Figure 7-14.

◢	A	B	C	D	E	F	G	H
1				**Southwest Sports World**				
2	**Product**	**Items Sold**	**Cost Per Item**	**Price Per Item**	**Mark up Amount**	**Mark Up %**	**COGS**	**Profit**
3	Basketballs	120	18	28.5	10.5	63.16%	2160	1260
4	Wood Bats	140	16	24.5	8.5	65.31%	2240	1190
5	Steel Bats	90	18.5	26	7.5	71.15%	1665	675
6	Skis	50	120	160.75	40.75	74.65%	6000	2037.5
7	Tennis Rackets	70	20.25	38	17.75	53.29%	1417.5	1242.5

Figure 7-14. *Result of copying formulas*

EXERCISE 7-3: COPY FORMULAS ACROSS COLUMNS

You have seen how the AutoFill Handle automatically adjust the **row reference** when copying formulas down in a column. Now we will look at how the AutoFill Handle can be used to automatically adjust the column reference when copying formulas across a row.

1. Create a new worksheet named "ColumnRef." Enter the values as shown in cells A1:D2. A colon between two cell addresses means the range between the cells. The first cell address, in this case A1, is the upper left-hand cell of the range. The second cell address, in this case D2, is the bottom right cell of the range. See Figure 7-15.

	A	B	C	D
1	135	205	50	49
2	179	150	80	180

Figure 7-15. *Enter these values*

2. Enter = **a1-a2** in cell A3. See Figure 7-16.

	A	B	C	D
1	135	205	50	49
2	179	150	80	180
3	=a1-a2			

Figure 7-16. *Enter formula in cell A3*

3. Press Ctrl + Enter.

4. Drag the AutoFill Handle in cell A3 across through cell D3.

Figure 7-17 shows the result of using the AutoFill Handle to copy the formula from cell A3 to cells B3, C3, and D3.

	A	B	C	D
1	135	205	50	49
2	179	150	80	180
3	-44	55	-30	-131

Figure 7-17. *Result of using copying formula*

You can see in Figure 7-18 that Excel adjusted the column in the cell reference when it copied the formula.

◢	A	B	C	D
1	135	205	50	49
2	179	150	80	180
3	=A1-A2	=B1-B2	=C1-C2	=D1-D2

Figure 7-18. *Copied formula*

You've learned how to enter formulas. Next, you will use Excel's AutoCalculate Tools. These tools allow you to select a range of data and then click the Calculate tool you need. The formula is automatically applied to that data without you having to enter it.

AutoCalculate Tools

The AutoCalculate Tools feature includes tools for summing, averaging, counting the number of cells that contain numeric data, and finding the largest or smallest value in a cell range.

AutoSum

The AutoSum feature provides a quick way to sum a range of numbers. The range could consist of a single cell or thousands of cells. The AutoSum feature automatically adds values that are in a *contiguous* range. A contiguous range is one in which there are no empty cells.

The AutoSum feature is located on the Ribbon's Home tab in the Editing group. The icon for the AutoSum feature is the Greek capital letter Sigma Σ. Math and Science use Σ as a summation operator. See Figure 7-19.

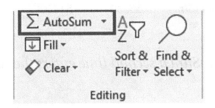

Figure 7-19. *AutoSum feature*

When you click the AutoSum button, Excel first checks to see if there are any values above the current cell. If there are, Excel will sum the values above it; otherwise, it will check to see if there are values to the left of it, and if there are, it will sum those values.

In Figure 7-20, cell F2 was selected when the AutoSum button was pressed. Excel highlighted the numeric values to its left. If only one cell is selected when you click the AutoSum button, the Sum formula will display with the range of values that is being summed. If this isn't the range you want to sum, then you can change either the starting or ending cell addresses.

◢	A	B	C	D	E	F	G	ꓞ
1		Qtr1 Sales	Qtr2 Sales	Qtr3 Sales	Qtr4 Sales	Annual Sales		
2	Store1	35540	49250	55175	39980	=SUM(B2:E2)		
3	Store2	55583	65490	62225	99985	SUM(number1, [number2], ...)		
4	Store3	47395	55380	55422	59120			
5	Store4	44180	35280	51350	54175			
6	Total Sales							

Figure 7-20. *Excel selects numeric values to the left when using AutoSum and displays the range that it is going to use*

Clicking the AutoSum button a second time or pressing the Enter key will display the results as shown in Figure 7-21. The formula bar is displaying the formula created by AutoSum.

| F2 | ▼ | ⋮ | × | ✓ | *fx* | =SUM(B2:E2) | |

◢	A	B	C	D	E	F
1		Qtr1 Sales	Qtr2 Sales	Qtr3 Sales	Qtr4 Sales	Annual Sales
2	Store1	35540	49250	55175	39980	179945
3	Store2	55583	65490	62225	99985	
4	Store3	47395	55380	55422	59120	
5	Store4	44180	35280	51350	54175	
6	Total Sales					

Figure 7-21. *Clicking AutoSum a second time enters the resulting value in the cell*

Figure 7-22 shows the results of having used the AutoFill Handle to copy the formula down through cell F5.

▲	A	B	C	D	E	F
1		Qtr1 Sales	Qtr2 Sales	Qtr3 Sales	Qtr4 Sales	Annual Sales
2	Store1	35540	49250	55175	39980	179945
3	Store2	55583	65490	62225	99985	283283
4	Store3	47395	55380	55422	59120	217317
5	Store4	44180	35280	51350	54175	184985
6	Total Sales					

Figure 7-22. *Results of copying formula*

Since we are summing rows, Excel automatically adjusted the row address for each of the copied cells:

- Cell F3 contains the formula =SUM(B3:E3).

- Cell F4 contains the formula =SUM(B4:E4).

- Cell F5 contains the formula =SUM(B5:E5).

Next, we will look at adding both the columns and the rows at the same time, but before we do that, let's look at how you would clear the totals from column F.

The cell contents can be cleared by first selecting the cells to clear and then doing one of the following:

- Clicking the Clear button ✐ in the Editing Group and then selecting Clear Contents

- Right-clicking the selected range and then selecting Clear Contents from the context menu

Totaling the columns and rows all at once is done by first selecting the row and column cells where the totaled results will be placed. In the spreadsheet in Figure 7-23, you would do it by selecting cells F2:F5 and then holding down the Ctrl key and selecting cells B6:F6.

◢	A	B	C	D	E	F
1		Qtr1 Sales	Qtr2 Sales	Qtr3 Sales	Qtr4 Sales	Annual Sales
2	Store1	35540	49250	55175	39980	
3	Store2	55583	65490	62225	99985	
4	Store3	47395	55380	55422	59120	
5	Store4	44180	35280	51350	54175	
6	Total Sales					

Figure 7-23. *Holding down the Ctrl key lets you select multiple cell ranges*

Clicking the AutoSum button now will sum both the rows and columns. See Figure 7-24.

◢	A	B	C	D	E	F
1		Qtr1 Sales	Qtr2 Sales	Qtr3 Sales	Qtr4 Sales	Annual Sales
2	Store1	35540	49250	55175	39980	179945
3	Store2	55583	65490	62225	99985	283283
4	Store3	47395	55380	55422	59120	217317
5	Store4	44180	35280	51350	54175	184985
6	Total Sales	182698	205400	224172	253260	865530

Figure 7-24. *Result of summing rows and columns*

A faster way of summing the rows and columns for a range of values is to select the values themselves along with one column to the right and one row below as shown in Figure 7-25.

◢	A	B	C	D	E	F
1		Qtr1 Sales	Qtr2 Sales	Qtr3 Sales	Qtr4 Sales	Annual Sales
2	Store1	35540	49250	55175	39980	
3	Store2	55583	65490	62225	99985	
4	Store3	13.25	55380	55422	59120	
5	Store4	44180	35280	51350	54175	
6	Total Sales					

Figure 7-25. *Select data plus the column to right and one row below before clicking AutoSum*

Then press the AutoSum button. The result will be the same as Figure 7-24.

Handling Blank Cells in a Range

AutoSum does not include any cells in its range after reaching a blank cell. See
Figure 7-26. If you wanted to include cell A2 in the total, you would need to change A4 in
the formula to cell A2.

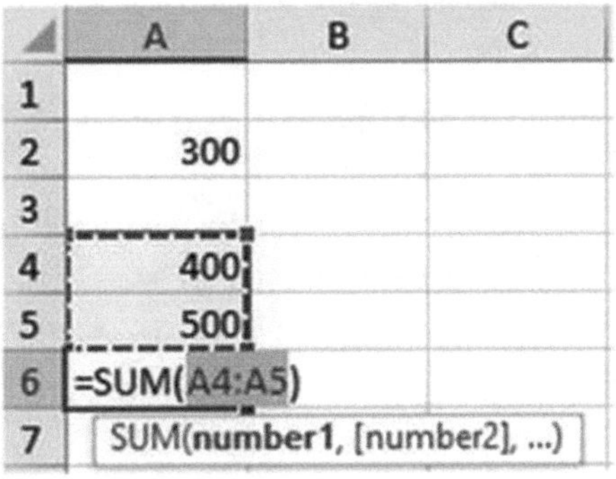

	A	B	C
1			
2	300		
3			
4	400		
5	500		
6	=SUM(A4:A5)		
7	SUM(**number1**, [number2], ...)		

Figure 7-26. *Change A4 in the SUM function to A2*

Another way to change the range is to take one of the handles of the selected range
and drag it so that it includes all the cells you want, as shown in Figure 7-27.

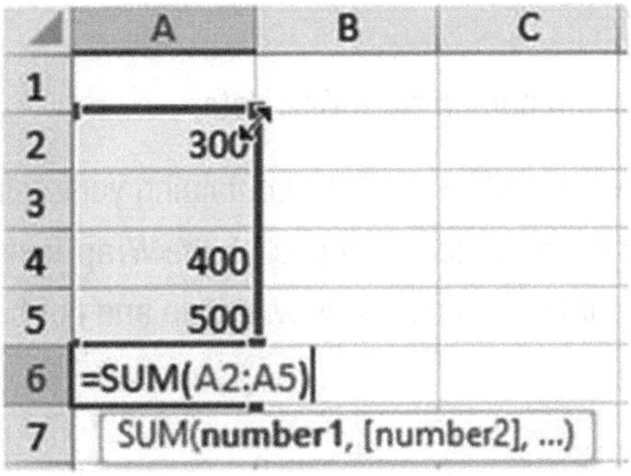

	A	B	C
1			
2	300		
3			
4	400		
5	500		
6	=SUM(A2:A5)		
7	SUM(**number1**, [number2], ...)		

Figure 7-27. *Drag selection handle to include cells you want included*

Average, Count Numbers, Max, Min

If you click the down arrow besides the AutoSum button (Figure 7-28), you can see that Excel includes several other "Auto" functions. The functions are Average, Count Numbers, Max(imum), Min(imum), and More Functions. These "Auto" functions work the same way as AutoSum.

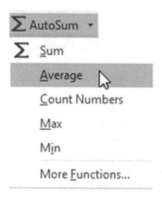

Figure 7-28. *Auto functions*

EXERCISE 7-4: AUTOCALCULATE TOOLS

In this exercise, you will apply the AutoCalculate Tools AutoSum, Average, Max, and Min across rows and columns.

1. Create a new worksheet named "AutoCalculate."

2. Select cells B1:I1, which will be the cells containing your column headings. On the Home tab in the Alignment group, click the Wrap Text button. See Figure 7-29. Now your column headings will wrap and not be partially hidden.

Figure 7-29. *Select Wrap Text*

3. Enter the data for the workbook as shown in Figure 7-30.

◢	A	B	C	D	E	F	G	H	I
1		Qtr1 Sales	Qtr2 Sales	Qtr3 Sales	Qtr4 Sales	Store Totals	Store Average	Store Max	Store Min
2	Store1	35540	49250	55175	39980				
3	Store2	55583	65490	62225	99985				
4	Store3	47395	55380	55422	59120				
5	Store4	44180	35280	51350	54175				
6	Total Sales								
7	Average Qtr Sales								
8	Highest Qtr Sale								
9	Lowest Qtr Sale								

Figure 7-30. *Enter this data*

4. Select the cell range B6:E6. Click the AutoSum button.

5. Select the cell range B2:E5. Click the down arrow next to the AutoSum button. Select Average.

Note   When you use an "Auto" function, Excel will place the results in the first blank row below the selected range when the range is in a column or, as in our case, include both rows and columns. If the range is just in a row, Excel places the result to the right of the range.

---

6. Cell range B2:E6 is currently selected. Change the selection back to B2:E5. Click the down arrow next to the AutoSum button. Select Max. The Max function returns the highest value in a selected column or row. Store2 sold the most in each quarter.

7. Select the cell range B2:E5 again. Click the down arrow next to the AutoSum button. Select Min. See Figure 7-31. The Min function returns the lowest value in a selected column or row. Store1 sold the least in quarters 1 and 4. Store4 sold the least in quarters 2 and 3.

◢	A	B	C	D	E	F	G	H	I
1		Qtr1 Sales	Qtr2 Sales	Qtr3 Sales	Qtr4 Sales	Store Totals	Store Average	Store Max	Store Min
2	Store1	35540	49250	55175	39980				
3	Store2	55583	65490	62225	99985				
4	Store3	47395	55380	55422	59120				
5	Store4	44180	35280	51350	54175				
6	Total Sales	182698	205400	224172	253260				
7	Average Qtr Sales	45674.5	51350	56043	63315				
8	Highest Qtr Sale	55583	65490	62225	99985				
9	Lowest Qtr Sale	35540	35280	51350	39980				

Figure 7-31. *Result of using Auto functions*

8. Select cells F2:F5. Click the AutoSum button.

9. Select cells B2:E2. Click the down arrow of the AutoSum button. Select Average. Click cell G2. Use the AutoFill Handle to drag the formula down through cell G5.

10. Select cells B2:E2. Click the down arrow of the AutoSum button. Select Max. Click cell H2. Use the AutoFill Handle to drag the formula down through cell H5.

Let's look at another way of doing this.

1. Click cell I2. Click the down arrow of the AutoSum button. Select Min. See Figure 7-32. The function displays and a marquee appears around cells B2:H2. This is not the range we want.

◢	A	B	C	D	E	F	G	H	I	J	K
1		Qtr1 Sales	Qtr2 Sales	Qtr3 Sales	Qtr4 Sales	Store Totals	Store Average	Store Max	Store Min		
2	Store1	35540	49250	55175	39980	179945	44986.25	55175	=MIN(B2:H2)		
3	Store2	55583	65490	62225	99985	283283	70820.75	99985	MIN(number1, [number2], ...)		

Figure 7-32. *Wrong range selected*

2.   Select cells B2:E2. The formula will adjust accordingly. See Figure 7-33.

B	C	D	E	F	G	H	I
Qtr1 Sales	**Qtr2 Sales**	**Qtr3 Sales**	**Qtr4 Sales**	**Store Totals**	**Store Average**	**Store Max**	**Store Min**
35540	49250	55175	39980	179945	44986.25	55175	=MIN(B2:E2)

Figure 7-33.   *Use the range B2:E2 for the MIN function*

3.   Press Ctrl + Enter. Use the AutoFill Handle to copy the formula down through cell I5. See Figure 7-34.

	A	B	C	D	E	F	G	H	I
1		**Qtr1 Sales**	**Qtr2 Sales**	**Qtr3 Sales**	**Qtr4 Sales**	**Store Totals**	**Store Average**	**Store Max**	**Store Min**
2	Store1	35540	49250	55175	39980	179945	44986.25	55175	35540
3	Store2	55583	65490	62225	99985	283283	70820.75	99985	55583
4	Store3	47395	55380	55422	59120	217317	54329.25	59120	47395
5	Store4	44180	35280	51350	54175	184985	46246.25	54175	35280
6	Total Sales	182698	205400	224172	253260				
7	Average Qtr Sales	45674.5	51350	56043	63315				
8	Highest Qtr Sale	55583	65490	62225	99985				
9	Lowest Qtr Sale	35540	35280	51350	39980				

Figure 7-34.   *Use AutoFill Handle to copy the formula down from I2 to I5*

4.   Let's improve the look of the spreadsheet. Let's add commas to make it easier to read and color to distinguish the original data from the computed data.

5.   Type A1:I9 in the Name Box and then press Enter.

6.   On the Home tab in the Styles group, click the Cell Styles More button. See Figure 7-35. Figure 7-36 shows what appears after clicking the Cell Styles More button.

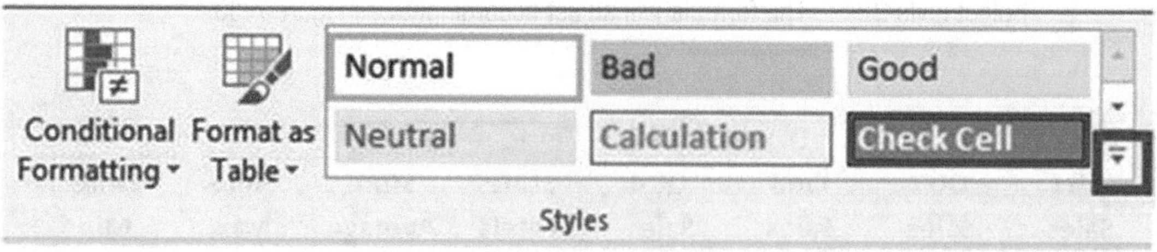

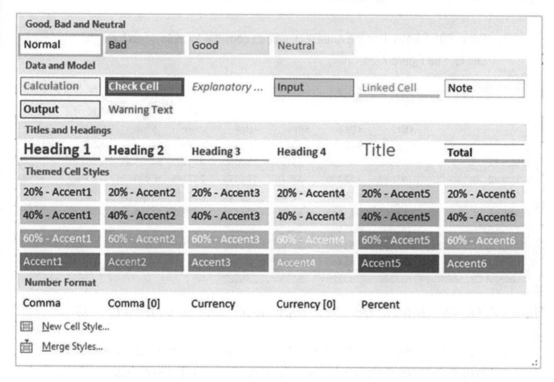

Figure 7-35. *Click the More button to view more styles*

Figure 7-36. *All of the Cell Styles*

7. Move your cursor over Comma[0] in the Number Format area. You can see how selecting this option will affect your worksheet by changing the data to having commas with no decimal position. Click Comma[0].

8. Enter A1:E5 in the Name Box and then press Enter.

9. On the Home tab in the Styles group, click the Cell Styles More button.

10. In the Data and Model area, click Input.

11. Use the border button in the Font group to create a thick outside border around the same cells. See Figure 7-37.

◢	A	B	C	D	E	F	G	H	I
1		Qtr1 Sales	Qtr2 Sales	Qtr3 Sales	Qtr4 Sales	Store Totals	Store Average	Store Max	Store Min
2	Store1	35,540	49,250	55,175	39,980	179,945	44,986	55,175	35,540
3	Store2	55,583	65,490	62,225	99,985	283,283	70,821	99,985	55,583
4	Store3	47,395	55,380	55,422	59,120	217,317	54,329	59,120	47,395
5	Store4	44,180	35,280	51,350	54,175	184,985	46,246	54,175	35,280
6	Total Sales	182,698	205,400	224,172	253,260				
7	Average Qtr Sales	45,675	51,350	56,043	63,315				
8	Highest Qtr Sale	55,583	65,490	62,225	99,985				
9	Lowest Qtr Sale	35,540	35,280	51,350	39,980				

Figure 7-37. *Thick outside border*

You've learned how to enter formulas manually and by using Excel's AutoCalculate tools. When viewing the worksheet, you see the results of formulas but not the formulas themselves. In order to see the formula, you need to click the cell that contains the formula and then view it in the formula bar.

Viewing Formulas

Clicking a cell that contains a formula will display the formula in the formula bar, but wouldn't it be nice to see all of your formulas at once? You can do this by holding down the **Ctrl** key and pressing the tilde key (~).

You can print your spreadsheet while all of the formulas are visible. This gives you a record of your formulas and provides a convenient way of overviewing your work to make sure all your formulas have been entered correctly. Figure 7-38 shows the results of using **Ctrl + ~**.

⊿	A	B	C	D	E
1		Qtr 1 Sales	Qtr 2 Sales	Qtr 3 Sales	Qtr 4 Sales
2	Store 1	35540	49250	55175	39980
3	Store 2	55583	65490	62225	99985
4	Store 3	47395	55380	55422	59120
5	Store 4	44180	35280	51350	54175
6	Total Sales	=SUM(B2:B5)	=SUM(C2:C5)	=SUM(D2:D5)	=SUM(E2:E5)
7	Average Qtr Sales	=AVERAGE(B2:B5)	=AVERAGE(C2:C5)	=AVERAGE(D2:D5)	=AVERAGE(E2:E5)
8	Highest Qtr Sale	=MAX(B2:B5)	=MAX(C2:C5)	=MAX(D2:D5)	=MAX(E2:E5)
9	Lowest Qtr Sale	=MIN(B2:B5)	=MIN(C2:C5)	=MIN(D2:D5)	=MIN(E2:E5)

Figure 7-38. *Formula view*

You can return to normal view by clicking **Ctrl + ~** again.

Creating Named Ranges and Constants

Up to this point, we have been using cell addresses in our formulas, but when we are thinking about computations, we don't normally think of cell addresses; we usually think in descriptive words. For example, when computing the company payroll, I am thinking **Hours * Rate** not **B2 * C2**. We can also assign names to values that we know will not change. For example, I can assign the name PI a value of 3.14 and then use the word **PI** in my formulas rather than the value 3.14.

Naming Ranges

We can assign a descriptive name to a range of cells and then use that name in our formulas. A name is assigned to a range of cells by selecting those cells and then entering a name for that range in the **Name** Box. See Figure 7-39.

Enter the name for the range in the
Name Box

| Rate | ▼ | ⋮ | ✕ | ✓ | *fx* | 2(|

◢	A	B	C
1		**Weekly Payrc**	
2		**Rate**	**Hours**
3	Carson, Russell	20	40
4	Charles, Ron	13.25	38
5	Darcy, Marcy	15	39
6	Harold, Lee	18	40
7	Lee, Jack	18	38
8	Tam, Tom	14.75	35
9	Totals		

Figure 7-39. *Enter range name in the Name Box*

Names can't be longer than 255 characters, and they must start with a letter, underscore, or backslash. Names are not case sensitive. Clicking the down arrow next to the Name Box displays a list of all the named ranges in the workbook. Clicking one of them will highlight the cells belonging to that named range.

Note   A common mistake people make is that they do not press the Enter key after they have entered a name in the Name Box. If you do not press the Enter key, the name will not be saved.

Another way of creating named ranges is to use the options available in the Defined Names group which is located on the Ribbon's Formulas tab. See Figure 7-40.

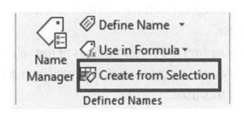

Figure 7-40. *Defined Names group*

Clicking the Define Name button brings up the **New Name** dialog box. See Figure 7-41. You can either select the cells you want to become a named range before you open the New Name dialog box or wait until it is opened and then select a range. You can also bring up the New Name dialog box by right-clicking the spreadsheet and selecting **Define Name** from the context menu. Creating a named range by using the New Name dialog box gives you a couple of options that would not be available if you just entered a name in the Name Box.

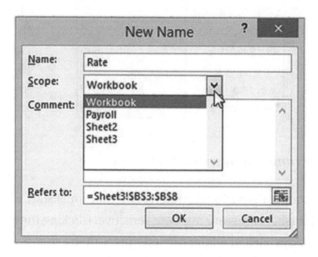

Figure 7-41. *New Name dialog box*

The **Name** text box in the New Name dialog box uses the column or row heading by default for the cells you selected before opening the New Name dialog box. You can change the name to whatever you want.

The purpose of the **Scope** drop-down box is to let you select whether you want to make the named range available for formulas in the entire workbook or for just a particular worksheet.

You can enter a **Comment**. The comment does not appear on the worksheet. It will appear only in the Name Manager which we will look at shortly. You can also name ranges in the Name Manager.

The **Refers to** text box displays the range address of the selected cells. If you haven't already selected the cells you want to name before you bring up the New Name dialog box, you can select them at this time.

Naming Noncontiguous Ranges

To create a named range for a group of noncontiguous cells, select the cells while holding down the Ctrl key and then enter a name for the group. In Figure 7-42, you could assign an appropriate name for all of the Quarter 1 values such as QTR1. You could add all of the Quarter 1 values by entering =SUM(QTR1) in a cell or find the average value by entering =AVERAGE(QTR1).

◢	A	B	C	D	E	F	G
1		Quarter 1	Quarter 2	Quarter 1	Quarter 2	Quarter 1	Quarter 2
2		349	421	370	186	308	414
3		241	188	124	267	133	117
4		268	202	286	422	204	263
5		411	366	193	200	166	284
6		203	279	328	139	237	268
7		242	446	306	168	256	263
8		343	170	313	126	261	197
9		345	347	248	430	245	359
10		165	343	313	123	425	376
11		260	179	386	160	244	141

Figure 7-42. Create a named range for noncontiguous cells

Naming Constants

You can also use the New Name dialog box to assign names to constants. A constant is a value that doesn't change. For example, you can assign names to interest or tax rates. Enter the name you wish to assign to the constant in the **Name** text box. Next, replace the current cell address in the **Refers to** text box with the constant value. If you are entering a constant that is a percent, you can enter the value with a percent sign such as 4.5%, or you could enter the decimal equivalent of .045. See Figure 7-43. Assigning a name to a constant lessens the chance of errors. If I used the tax rate of 4.5% in ten cells and now the tax rate changes to 4.7%, how many cells am I going to have to change? That's right; I would have to find the ten cells and then change all of them. But, if I used a constant, I would just change the constant value once to 4.7%.

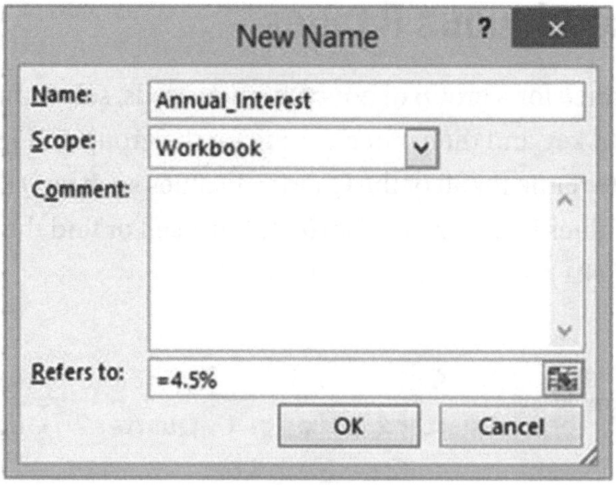

Figure 7-43. *Assign a name to a constant*

Name Manager

Clicking the Name Manager button in the Defined Names group displays the Name Manager which lists the named ranges and constants. The Name Manager has buttons for Adding, Editing, and Deleting named ranges. It also has a Filter option for limiting which named ranges to display.

You can see in Figure 7-44 that the named range **Rate** had values in its range when it was created. The Gross Pay and Hours ranges were created before any values were placed in their cells.

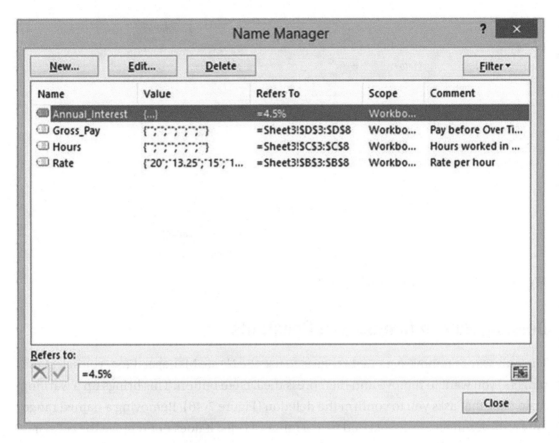

Figure 7-44. *Range name Rate has values in it because they were there when the range was selected*

Editing Named Ranges and Constants

Clicking the **New** button in the Name Manager window brings up the same **New Name** dialog box we saw when we clicked the Define Name button in the Defined Names group. To edit a named range, select the range you wish to edit from the list and then click the **Edit** button. The Edit Name dialog box (Figure 7-45) is similar to the New Name dialog box. From here, you can change the Name, the Comment, and the location of a range or change the Constant value.

317

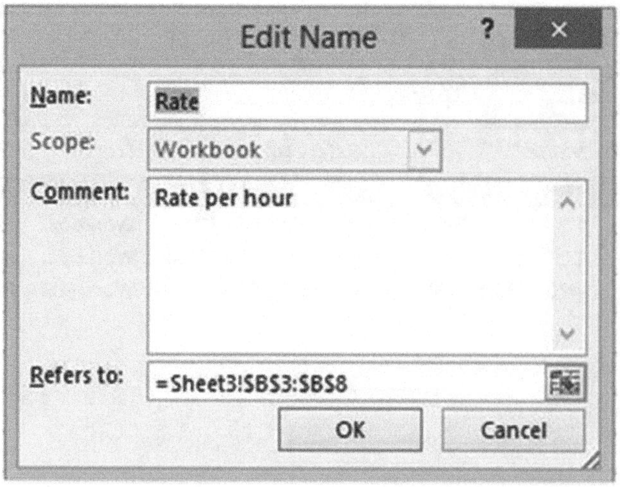

Figure 7-45. *Edit Name dialog box*

Deleting Named Ranges and Constants

The only way to remove a named range is to use the Name Manager. First, you select the range(s) you want to remove and then press the Delete button. This brings up a warning dialog box that asks you to confirm the deletion (Figure 7-46). Removing a named range only removes the name associated with a range of cells. It does not remove the cells or the data in them.

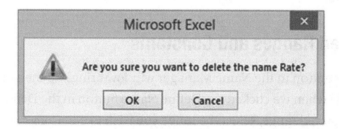

Figure 7-46. *Message confirming that you want to delete the named range*

Using Column or Row Headings for Range Names

If you are going to create named ranges for multiple rows or columns, you may find it easier to let Excel use your row or column headings as names for the named ranges. To do this, you would first select the cells that your named ranges would include along with their column or row headings. See Figure 7-47.

	A	B	C	D	E	F	G
1		**Weekly Payroll Report**					
2		Rate	Hours	Gross Pay	Fed Tax	State Tax	Net Pay
3	Carson, Russell	20	40				
4	Charles, Ron	13.25	38				
5	Darcy, Marcy	15	39				
6	Harold, Lee	18	40				
7	Lee, Jack	18	38				
8	Tam, Tom	14.75	35				

Figure 7-47. *Select headings along with data when creating range names for multiple rows or columns*

You would then click the **Create from Selection** button in the Defined Names group. See Figure 7-48. This would bring up the **Create Names from Selection** dialog box (Figure 7-49).

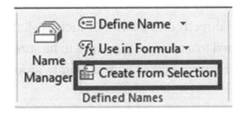

Figure 7-48. *Create from Selection button*

Create Names from Selection   ?   ✕

Create names from values in the:

☑ Top row
☐ Left column
☐ Bottom row
☐ Right column

OK   Cancel

Figure 7-49. *Create Names from Selection dialog box*

You would need to select Top row to get the column headings. Excel would assign the named ranges as follows:

- **Rate** to cells B3:B8

- **Hours** to cells C3:C8

- **Gross_Pay** to cells D3:D8

- **Fed_Tax** to cells E3:E8

- **State_Tax** to cells F3:F8

Note   Named ranges can't contain spaces. Excel puts an underscore in place of spaces when it uses row or column headings.

Selecting Named Ranges Rather Than Typing Them into Formulas

You can type named ranges directly into a formula such as =Rate*Hours. If you don't remember the name assigned to a range or you want to be sure you don't misspell it, you can use the **Use in Formula** button in the Defined Names group. See Figure 7-50.

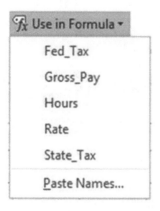

Figure 7-50. *Use in Formula button*

If you are just starting a formula, clicking one of the named ranges will place the equal sign along with the named range in the formula. For example, using our payroll example, we want to compute the gross pay in cell D3 by multiplying Rate times Hours. You would click inside cell D3 and then click the Rate in the **Use in Formula** list box. This would create =Rate. You would then type the * for multiplication. You would finish the formula by clicking Hours and then pressing the Enter key.

A quicker way to select named ranges or constants is to press F3. Pressing F3 brings up all the available names for you to select from.

EXERCISE 7-6: NAMED RANGES AND CONSTANTS

In this exercise, you will create range names using different methods. You will also create constants that will hold the tax values.

1. Type the data into a new worksheet as shown in Figure 7-51.

◢	A	B	C	D	E	F	G
1	Weekly Payroll Report						
2		Rate	Hours	Gross Pay	Fed Tax	State Tax	Net Pay
3	Carson, Russell	20	40				
4	Charles, Ron	13.25	38				
5	Darcy, Marcy	15	39				
6	Harold, Lee	18	40				
7	Lee, Jack	18	38				
8	Tam, Tom	14.75	35				
9	Totals						

Figure 7-51. *Enter data*

2. Format the title:

    a. Select cells A1:G1.

    b. Click the Merge & Center button located on the Home tab in the Alignment group.

    c. Bold the text.

    d. On the Home tab, in the Font Group, click the **Increase Font Size** button twice.

3. Center and bold the column headings. Your spreadsheet should now look like Figure 7-52.

	A	B	C	D	E	F	G
1			**Weekly Payroll Report**				
2		Rate	Hours	Gross Pay	Fed Tax	State Tax	Net Pay
3	Carson, Russell	20	40				
4	Charles, Ron	13.25	38				
5	Darcy, Marcy	15	39				
6	Harold, Lee	18	40				
7	Lee, Jack	18	38				
8	Tam, Tom	14.75	35				
9	Totals						

Figure 7-52. *Formatted worksheet*

Creating Named Ranges Using the Name Box

1. Select cells B3:B8. Type **Rate** into the Name Box. Press Enter.

2. Select cells C3:C8. Type **Hours** into the Name Box. Press Enter.

3. Select cells D3:D8. Type **GrossPay** into the Name Box. Press Enter.

4. Select cells E3:E8. Type **FedTax** into the Name Box. Press Enter.

Creating Named Ranges Using Define Name

1. Select cells F3:F5.

2. Right-click any one of the selected cells. Select **Define Name** from the context menu. Excel assigns a name of State_Tax to the range. Excel got the name from the column heading. Names can't contain spaces; consequently, Excel filled in the space with an underscore. The **Refers to** text box contains the selected range.

3. Click the OK button.

Changing a Name or Its Range

You can't change a name or its range in the Name Box. These changes can only be done through the Name Manager. The State Tax range is currently F3:F5, but it needs to be changed to F3:F8. Let's also change the name by removing the underscore.

1. Click the Formulas tab; in the Defined Names group, click the Name Manager button.

2. Click State_Tax. Click the Edit button. The Edit Name dialog box appears.

3. Remove the underscore from the name so that the name is StateTax.

4. Tab to the **Refers to** text box. See Figure 7-53. An animated marquee appears around cells F3:F5.

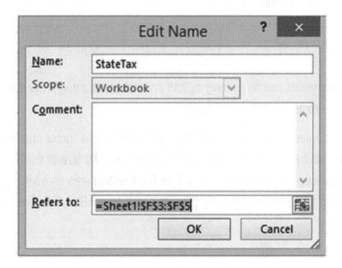

Figure 7-53. *Change the range name and location*

5. If you can't see cells F3:F8, then move the Edit Name window over until you can.

6. Drag across the cell range F3:F8. Click the OK button. Click the Close button.

7. Click the down arrow of the Name Box. It should display all of your named ranges. See Figure 7-54.

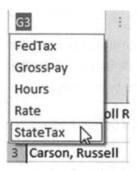

Figure 7-54. *Range names displayed in Name Box*

8. Click StateTax. This highlights the StateTax range.

9. Select the cell range B3:B8 on the worksheet. The name box should now display the name Rate.

Changing the Size of the Name Box

1. Since your names can be as long as 255 characters, you might need to widen the Name Box to see more of the name.

2. Move your cursor to the right of the Name Box. When the cursor changes to a horizontal double arrow, drag to the right to increase the size of the Name Box. As you increase the size of the Name Box, the function area shrinks and vice versa. Drag to the left to decrease the size of the Name Box. See Figure 7-55.

Figure 7-55. *Change the size of the Name box*

Removing a Named Range

1. Click the **Formulas** tab on the Ribbon.

2. In the **Defined Names** group, click the **Name Manager** button. The Name Manager displays the existing named ranges. They are displayed in alphabetic order.

3. To select all of the named ranges, click FedTax, then hold down your Shift key, and click StateTax.

4. Click the Delete button in the Name Manager window.

5. A dialog box appears asking "Are you sure you want to delete the selected names?"

6. Click the OK button.

7. Click the Close button.

Creating Named Ranges Using Create from Selection

1. Select cells B2:F8.

2. In the Defined Names group, click the **Create from Selection** button.

3. Select **Top Row** from the **Create Names from Selection** dialog box. Click the **OK** button.

4. Click the Name Manager button. The named ranges were created from the column heads. Spaces in the column names were replaced with an underscore. See Figure 7-56.

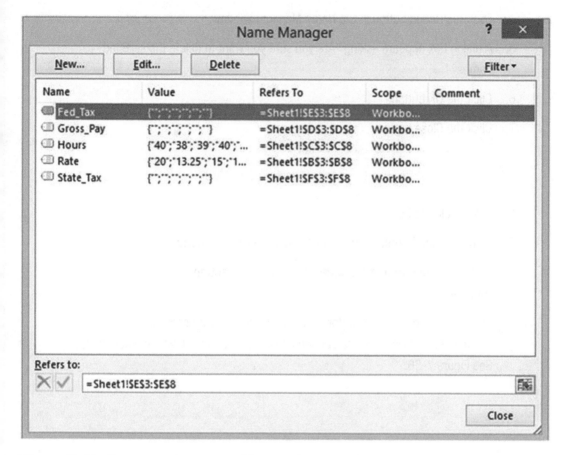

Figure 7-56. *Range names created from column heads*

5.   Click the Close button.

Creating Named Constants

For the purposes of this example, let's just say the federal tax is 32%. We will assign a named constant to this value.

1.   In the Defined Names group, click the **Define Name** button.

2.   Type **FederalTaxRate** in the Name text box.

3.   Type **Fictitious Federal Tax rate for practice example** in the Comment box.

4.   Type **=32%** in the Refers to text box over the existing text. See Figure 7-57. If you prefer, you can use the decimal equivalent and type **=.32**.

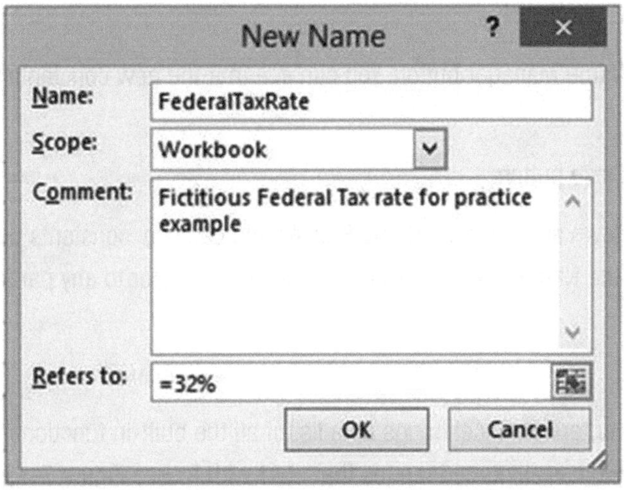

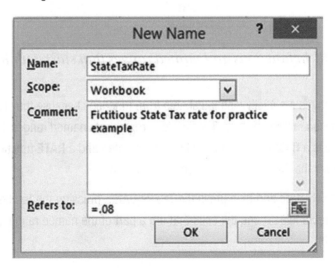

Figure 7-57. *Name of FederalTaxRate assigned to the constant 32%*

5. Click the OK button.

Let's say the state tax is 8%.

1. Click the Define Name button.

2. Type StateTaxRate in the Name text box.

3. Type Fictitious State Tax rate for practice example in the Comment box.

4. This time, let's use the decimal equivalent of 8%. Type **=.08** in the Refers to text box. See Figure 7-58.

Figure 7-58. *Name of StateTaxRate assigned to the constant .08*

5. Click the OK button.

6. Click the Name Manager button. You can see that the new constants have been added.

7. Click the Close button.

8. Click the down arrow for the Name Box. As you can see, constants do not appear in the Name Box. This is because they don't refer to any particular cell.

Using Named Ranges in Formulas

1. In cell D3, enter **=R**. Excel brings up a list of all the built-in functions that start with R as well as all named ranges that start with R. See Figure 7-59.

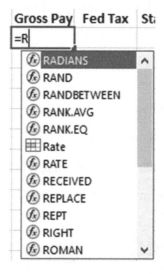

Figure 7-59. *Built-in functions and named ranges that start with R*

Named ranges are easily distinguishable from functions because they have a different icon in front of them. The ⊞ icon represents named ranges. Notice that there is a **Rate** named range which you created and a **RATE** function. We want to use the Rate with the ⊞ icon in front of it.

2. Click the ⊞ **Rate** and then press the Tab key. After pressing the Tab key, Excel places a rectangle around the cells that are a part of the named range. See Figure 7-60.

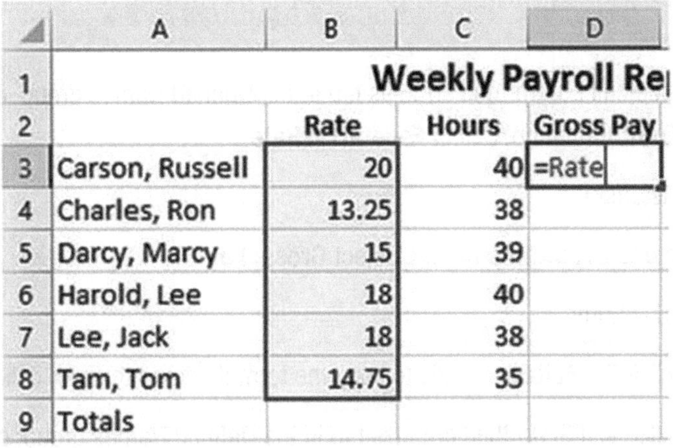

◢	A	B	C	D
1		**Weekly Payroll Re**		
2		Rate	Hours	Gross Pay
3	Carson, Russell	20	40	=Rate
4	Charles, Ron	13.25	38	
5	Darcy, Marcy	15	39	
6	Harold, Lee	18	40	
7	Lee, Jack	18	38	
8	Tam, Tom	14.75	35	
9	Totals			

Figure 7-60. *When entering a range name, Excel highlights the cells in that range*

3. Type ***Ho**. See Figure 7-61.

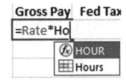

Figure 7-61. *Typing Ho brings up the IntelliSense choices for the function Hour and the named range Hours*

4. Use the down arrow to select ▦ **Hours** and then press the Tab key. Excel places a different color rectangle around the cells that encompass the Hours range.

5. Press Ctrl + Enter.

6. Use AutoFill to copy the formula down through cell D8.

7. Click cell D4. Look in the formula bar. The formula for every cell in the column will show **=Rate*Hours**. You won't see any row adjusting as you do when you are using cell references in your formulas.

Selecting Named Ranges from Use in Formula

1. Click inside cell E3. On the Formulas tab in the Defined Names group, click the **Use in Formula** button. Select **FederalTaxRate**.

2. Type * (asterisk).

3. Click the **Use in Formula** button. Select **Gross_Pay**.

4. Press Ctrl + Enter.

5. Double-click the AutoFill Handle to copy the formula down through cell E8.

6. Click inside cell F3. On the Formulas tab in the Defined Names group, click the **Use in Formula** button. Select **StateTaxRate**.

7. Type * (asterisk).

8. Click the **Use in Formula** button. Select **Gross_Pay**.

9. Press Ctrl + Enter.

10. Double-click the AutoFill Handle to copy the formula down through cell F8.

Setting the Scope for a Name

The scope of a name is where that name can be used. The scope can pertain to an entire workbook or to an individual worksheet. When you create a named range using the Name Box, the named range has a scope of the entire workbook. In other words, you can use that named range on any worksheet in the workbook that you declared it in. It doesn't matter how many worksheets are in the workbook; it will work on any of them.

Let's see what happens when we declare a scope of an individual worksheet for a named range or named constant.

1. Add another worksheet to the workbook. Change the name on the worksheet tab to "TestScope."

2. Select the cell range A1:A5.

3. On the Formulas tab on the Ribbon in the Defined Names group, click the Define Name button.

4. Type **SheetScope** for the Name.

5. Click the down arrow for Scope and select TestScope. The Refers to text box should reflect that you have selected the range A1:A5. See Figure 7-62.

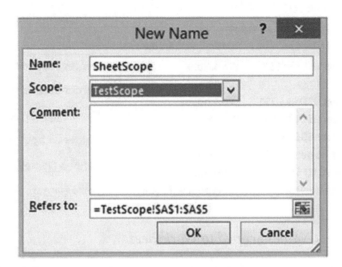

Figure 7-62. *Range A1:A5 is selected*

6. Click the OK button.

7. Press F3. The SheetScope range appears in the list along with the Names you created previously. Since the other names were created with a scope of Workbook, they are accessible from any worksheet.

8. Click the Cancel button.

9. Click another worksheet. Press F3. SheetScope will not appear in this list because its scope is limited to the TestScope worksheet.

10. Click the Cancel button.

11. Click the Name Manager button. Click the Filter button. Select Names **Scoped to Worksheet**. See Figure 7-63. SheetScope is the only Name displayed because it is the only one whose scope is limited to a worksheet.

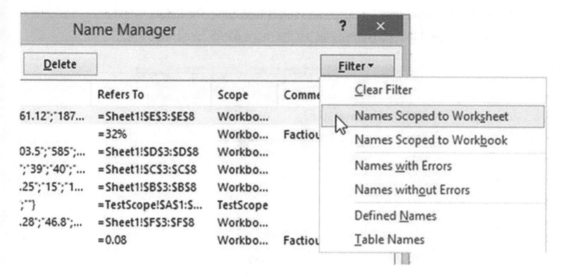

Figure 7-63. *Select Names Scoped to Worksheet*

12.   Click the Filter button again. Select Names Scoped to Workbook. All the Names
      are displayed except for SheetScope.

Note   If you assign a name with a scope of worksheet, you can create a range on
another worksheet with that same name.

13.   Click the Close button.

Now you've seen how to simplify formulas by using named ranges and constants. Up
to this point, when you copied a formula down a column or across a row, the formula
automatically adjusted for the row or column change, but there are times when you
don't want that to happen. You learn how to prevent these changes in the next section.

Absolute Cell References

As you have seen when using relative cell addressing with the AutoFill feature, the cell
addresses are adjusted automatically. As you will see, **relative cell referencing** doesn't
work for every situation.

Binder Corporation is thinking about giving each of its employees a 5% raise. Each employee's salary is listed in column B under the heading Current Salary. The amount of the increase for each employee is to be displayed in column C. The formula in column C4 is entered as =B4*B1. See Figure 7-64. The result is 2087.5. This is the result of taking 41750 * 5%.

	A	B	C
1	Proposed Raise	5%	
2			
3	Employees	Current Salary	Increase
4	Carson, Florence	41,750	=B4*B1
5	Carson, Annette	39,950	
6	Burns, Robert	38,895	
7	Edwards, Denise	52,105	
8	Huges, Harold	60,300	
9	Mathews, Luann	48,850	
10	Perkins, Paula	29,985	
11	Stork, Larry	30,000	

Figure 7-64. *The formula for 5% raise is entered in cell C4*

Say you use the AutoFill Handle to copy the formula for the rest of the employees. Figure 7-65 shows the resulting copied formulas.

	A	B	C
1	Proposed Raise	0.05	
2			
3	Employees	Current Salary	Increase
4	Carson, Florence	41750	=B4*B1
5	Carson, Annette	39950	=B5*B2
6	Burns, Robert	38895	=B6*B3
7	Edwards, Denise	52105	=B7*B4
8	Huges, Harold	60300	=B8*B5
9	Mathews, Luann	48850	=B9*B6
10	Perkins, Paula	29985	=B10*B7
11	Stork, Larry	30000	=B11*B8

Figure 7-65. *Formulas after being copied with AutoFill Handle*

Looking at our formulas, we can see that we are headed for trouble starting with cell C5 and continuing down through the other formulas. The row portion of the relative cell address B1 was automatically adjusted from 1 to 2. Since there is no value in B2, the value 39950 will be multiplied by 0 giving the result of 0.

Figure 7-66 shows the results of the formulas.

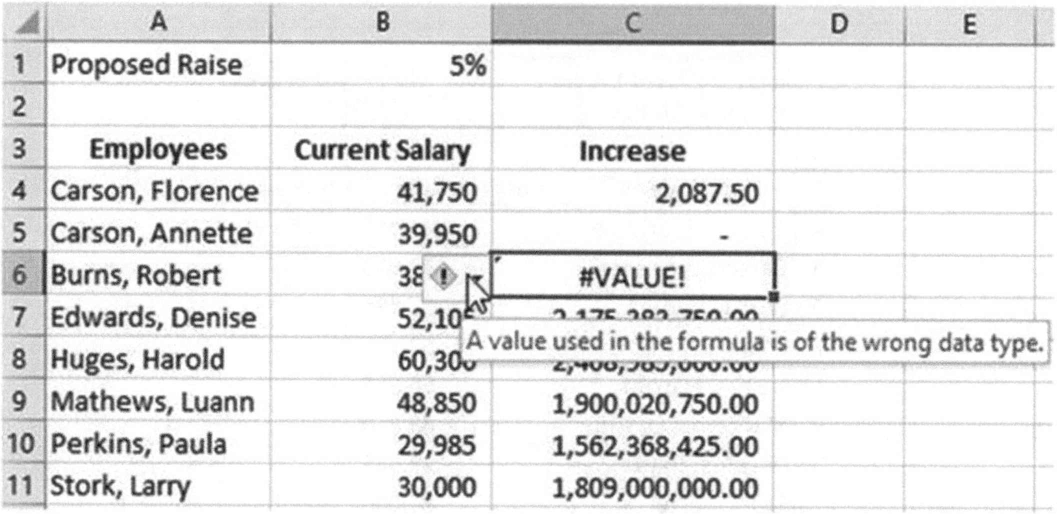

	A	B	C	D	E
1	Proposed Raise	5%			
2					
3	**Employees**	**Current Salary**	**Increase**		
4	Carson, Florence	41,750	2,087.50		
5	Carson, Annette	39,950	-		
6	Burns, Robert	38	#VALUE!		
7	Edwards, Denise	52,10	2,175,282,750.00		
8	Huges, Harold	60,30	2,408,985,000.00		
9	Mathews, Luann	48,850	1,900,020,750.00		
10	Perkins, Paula	29,985	1,562,368,425.00		
11	Stork, Larry	30,000	1,809,000,000.00		

A value used in the formula is of the wrong data type.

Figure 7-66. *Results of formula not what we wanted*

Figure 7-65 shows that the formula in cell C6 is =B6*B3. Cell B3 contains the alphabetic data "Current Salary." You can't use alphabetic data in a formula, which is why it shows the error #**VALUE!**. If you click the smart tag next to the error, Excel displays a tool tip which states that the value used in the formula is of the wrong data type.

Cell C7 multiplies 52105 * 41750. This is clearly not what we want. We do not want Excel to perform any automatic adjusting for cell B1.

If you want formulas to refer to a single cell, you need to use **absolute cell referencing**. Absolute cell referencing locks the cell address in a formula so that when you copy it, it doesn't automatically adjust for the row or column change. Absolute cell addresses are made by placing a dollar sign in front of both the column and the row. This can be done by manually typing them in or you can type the address as you normally do and then press F4. Pressing F4 places a $ sign in front of both the column and the row.

In our example, if we entered the formula =B4*$B$1 in cell C4 and then used the AutoFill handle to copy the formula down through cell C1, the formulas would look like those in Figure 7-67.

◢	A	B	C
1	Proposed Raise	0.05	
2			
3	Employees	Current Salary	Increase
4	Carson, Florence	41750	=B4*$B$1
5	Carson, Annette	39950	=B5*$B$1
6	Burns, Robert	38895	=B6*$B$1
7	Edwards, Denise	52105	=B7*$B$1
8	Huges, Harold	60300	=B8*$B$1
9	Mathews, Luann	48850	=B9*$B$1
10	Perkins, Paula	29985	=B10*$B$1
11	Stork, Larry	30000	=B11*$B$1

Figure 7-67. *Making cell B1 an absolute address in the formulas in column C*

The result would be as shown in Figure 7-68. This is what we wanted.

◢	A	B	C
1	Proposed Raise	5%	
2			
3	Employees	Current Salary	Increase
4	Carson, Florence	41,750	2,087.50
5	Carson, Annette	39,950	1,997.50
6	Burns, Robert	38,895	1,944.75
7	Edwards, Denise	52,105	2,605.25
8	Huges, Harold	60,300	3,015.00
9	Mathews, Luann	48,850	2,442.50
10	Perkins, Paula	29,985	1,499.25
11	Stork, Larry	30,000	1,500.00

Figure 7-68. *Result of using absolute cell referencing*

EXERCISE 7-7: ABSOLUTE CELL REFERENCING

In this example, you will create a travel expense report for the second quarter of 2020. You will add computations for computing the mileage expense and the total amount of travel expenses for the quarter. You will then compare the quarter totals to the same quarter in 2019 and determine the dollar amount and percent difference between the two.

1. Create a new worksheet named "Absolute." Make your worksheet look exactly like the one in Figure 7-69.

	A	B	C	D	E	F	G	H	I	J
1	Second Quarter (April-June) 2020 Travel Expense Report									
2										
3	Reimbursement Per Mile	0.6								
4										
5		Miles	Mileage Expense	Hotels	Meals	Entertainment	Qtr 2 2020	Qtr 2 2019	$ diff	% diff
6	George Traveller	3575		3775.5	185.76	300		10385		
7	Johnny Sales	4000		3209.85	205	175		9375.25		
8	Marten Moonlighter	5900		4205.85	421.3	1275		15929.53		

Figure 7-69. Enter the data here in a new worksheet named Absolute

2. Merge and center the title in row 1 across cells A1:J1. Click the Increase Font Size button twice.

3. Center the column headings both vertically and horizontally. Bold the column headings.

4. Compute the values for the Mileage Expense column by taking the miles traveled times the reimbursement rate per mile. The reimbursement rate stored in cell B3 is 60 cents for every mile traveled:

    a. In cell C6, enter the formula =B6*B3.

    b. Use the AutoFill handle and drag the formula down through C8.

Problem:

Cell C8 displays #Value!. Click cell C8. Look at the formula in the formula bar. The formula is =B8*B5. The formula adjusted for the row when you used the AutoFill handle. Cell B5 contains the text "Miles." If you use text in a formula, Excel displays a #Value error.

Fix the Problem:

Since you want to use cell B3 in each formula, you will need to make it absolute.

5. Double-click cell C6 to place the cell in Edit mode. Move your cursor in front of B3 in the formula, and then press F4 to make the cell address absolute. Your formula should now be =B6*$B$3.

6. Press Ctrl + Enter. The address needs to be absolute because cells C7 and C8 will also use the same rate from cell B3 in their formulas.

7. Use the AutoFill handle to copy the formula in cell C6 down through cell C8.

8. Select cell G6. Click Σ AutoSum. The formula is displayed as =Sum(B6:F6). Cell B6 contains the miles traveled which are not an expense. The formula should start with column C.

9. In the formula, change B6 to C6 and then press Ctrl + Enter.

10. Use the AutoFill Handle to copy the formula in cell G6 down through cell G8.

11. We need to know the dollar difference between the second quarter in 2020 and the same quarter in 2019. In cell I6, enter the formula =G6–H6 and then press Ctrl + Enter.

12. Use the AutoFill Handle to copy the formula in cell I6 down through cell I8.

13. Divide the 2020 totals by the 2019 totals by entering the formula =(G6/H6)-1 in cell J6. Press Ctrl + Enter.

14. Use the AutoFill handle to copy the formula in cell J6 down through cell J8.

15. We want to convert the decimal values in cells J6:J8 to a percent. Select cells J6:J8, and then select the Home tab and the % button in the Number group. Cells J6:J8 are displayed without any decimal positions.

16. Change the cells so that they appear with two decimal positions by clicking two times on the increase decimal button ⁺⁰⁸₀₀. Your worksheet should appear as shown in Figure 7-70.

◢	A	B	C	D	E	F	G	H	I	J
1	Second Quarter (April-June) 2020 Travel Expense Report									
2										
3	Reimbursement Per Mile	0.6								
4										
5		Miles	Mileage Expense	Hotels	Meals	Entertainment	Qtr 2 2020	Qtr 2 2019	$ diff	% diff
6	George Traveller	3575	2145	3775.5	185.76	300	6406.26	10385	-3978.74	-38.31%
7	Johnny Sales	4000	2400	3209.85	205	175	5989.85	9375.25	-3385.4	-36.11%
8	Marten Moonlighter	5900	3540	4205.85	421.3	1275	9442.15	15929.53	-6487.38	-40.73%

Figure 7-70. *Add two decimal positions to the cells in the range J6:J8*

17.   Press Ctrl + ~ to view all the formulas on the worksheet. Press Ctrl + ~ again to return to normal view.

Mixed Cell References

Mixed cell referencing is a mixture of relative cell referencing and absolute cell referencing. Mixed cell referencing is when the row is fixed but the column is relative (the formula will be automatically adjusted to the column where the formula is being copied to) or vice versa.

Remember, in absolute cell referencing, a dollar sign is placed in front of both the column and row to lock them both. Mixed cell references look like those in Table 7-4.

Table 7-4. *Mixed Cell References*

Cell Address	Description
$A4	The column remains the same, but the row can change when the formula is copied to another location.
A$4	The column can change when the formula is copied to another location, but the row remains the same.

The dollar signs can be typed directly into the cell address, or you can place your cursor anywhere in the cell address or one position to the left or right of it and then press F4. This will create an absolute address by placing dollar signs in front of both the column and row such as $A$4. Pressing F4 a second time places the dollar sign in front of just the row, such as A$4. Pressing F4 a third time places the dollar sign in front of the column, such as $A4. Pressing F4 a fourth time removes the dollar signs, thus returning the address to a relative address.

Mixed cell referencing can best be illustrated by creating a table.

EXERCISE 7-8: MIXED CELL REFERENCING

In this exercise, we will create a multiplication table that looks like Figure 7-71. The table could represent the multiplication tables for numbers 1 through 7, or it could represent square feet, the result of multiplying length * width.

◢	A	B	C	D	E	F	G	H
1		1	2	3	4	5	6	7
2	1	1	2	3	4	5	6	7
3	2	2	4	6	8	10	12	14
4	3	3	6	9	12	15	18	21
5	4	4	8	12	16	20	24	28
6	5	5	10	15	20	25	30	35
7	6	6	12	18	24	30	36	42
8	7	7	14	21	28	35	42	49

Figure 7-71. *Multiplication table*

1. Create a new worksheet. Name the worksheet "Mixed."

2. Enter a 1 in cell A2. Enter a 2 in cell A3 and then press Ctrl + Enter.

3. Drag across cells A2 and A3. Use the AutoFill Handle to copy the pattern down through row 8.

4. Enter a 1 in cell B1. Enter a 2 in cell C1 and then press Ctrl + Enter.

5. Drag across cells B1 and C1. Use the AutoFill Handle to copy the pattern across through column H.

6. Bold the values you entered and give them a background color to distinguish them.

7. Enter =A2*B1 in cell B2. Press Ctrl + Enter.

8. Drag the AutoFill Handle down through cell B8. Now drag the AutoFill Handle to the right to cell H8. Expand the width of the columns so that you can see all the cell data. See Figure 7-72.

◢	A	B	C	D	E	F	G	H
1		1	2	3	4	5	6	7
2	1	1	2	6	24	120	720	5040
3	2	2	4	24	576	69120	49766400	2.51E+11
4	3	6	24	576	331776	2.29E+10	1.14E+18	2.86E+29
5	4	24	576	331776	1.1E+11	2.52E+21	2.88E+39	8.25E+68
6	5	120	69120	2.29E+10	2.52E+21	6.37E+42	1.84E+82	1.5E+151
7	6	720	49766400	1.14E+18	2.88E+39	1.84E+82	3.4E+164	#NUM!
8	7	5040	2.51E+11	2.86E+29	8.25E+68	1.5E+151	#NUM!	#NUM!

Figure 7-72. *Results of multiplication using relative cell addressing*

This is definitely not the results we wanted. We wanted the spreadsheet to look like Figure 7-71.

9. Press Ctrl + ~ to see the formulas. Your formulas should look those in Figure 7-73. Looking at the formulas, we can see that the columns have changed from the original formula to adjust for the current column and the row address has changed to adjust for the current row.

◢	A	B	C	D	E	F	G	H
1		1	2	3	4	5	6	7
2	1	=A2*B1	=B2*C1	=C2*D1	=D2*E1	=E2*F1	=F2*G1	=G2*H1
3	2	=A3*B2	=B3*C2	=C3*D2	=D3*E2	=E3*F2	=F3*G2	=G3*H2
4	3	=A4*B3	=B4*C3	=C4*D3	=D4*E3	=E4*F3	=F4*G3	=G4*H3
5	4	=A5*B4	=B5*C4	=C5*D4	=D5*E4	=E5*F4	=F5*G4	=G5*H4
6	5	=A6*B5	=B6*C5	=C6*D5	=D6*E5	=E6*F5	=F6*G5	=G6*H5
7	6	=A7*B6	=B7*C6	=C7*D6	=D7*E6	=E7*F6	=F7*G6	=G7*H6
8	7	=A8*B7	=B8*C7	=C8*D7	=D8*E7	=E8*F7	=F8*G7	=G8*H7

Figure 7-73. *Formulas used in multiplication table*

When creating a mixed cell reference, you need to determine if it is the rows or columns in the formulas that should remain constant. In our example, we want all of the formulas to use column A. Therefore, we want the formula to lock in column A. We need to make the first address in the formula in cell B2 **$A2**. Placing a $ in front of just the A will lock in the column but still allow the row in the formula to change to the current row.

We want all the formulas to use only row 1, but we want the column portion of the formula to change to the current column. We need to lock in row 1 in the second address in cell B2. Thus, the formula we need to use in cell B2 is =$A2*B$1.

10. Double-click the address A2 in cell B2. Press F4. The address should now be $A$2.

11. Press F4 again. The address is now A$2. Press F4 again. Your address should now be $A2.

12. Double-click the address B1 in the formula. Press F4 twice. Your address should be B$1. Press Ctrl + Enter.

13. Drag the AutoFill Handle down through cell B8. Now drag the AutoFill Handle to the right to cell H8. Your formulas should look those in Figure 7-74. You can see that only column A was used in the formulas for the first cell address and only row 1 was used in the second cell address.

◢	A	B	C	D	E	F	G	H
1		1	2	3	4	5	6	7
2	1	=$A2*B$1	=$A2*C$1	=$A2*D$1	=$A2*E$1	=$A2*F$1	=$A2*G$1	=$A2*H$1
3	2	=$A3*B$1	=$A3*C$1	=$A3*D$1	=$A3*E$1	=$A3*F$1	=$A3*G$1	=$A3*H$1
4	3	=$A4*B$1	=$A4*C$1	=$A4*D$1	=$A4*E$1	=$A4*F$1	=$A4*G$1	=$A4*H$1
5	4	=$A5*B$1	=$A5*C$1	=$A5*D$1	=$A5*E$1	=$A5*F$1	=$A5*G$1	=$A5*H$1
6	5	=$A6*B$1	=$A6*C$1	=$A6*D$1	=$A6*E$1	=$A6*F$1	=$A6*G$1	=$A6*H$1
7	6	=$A7*B$1	=$A7*C$1	=$A7*D$1	=$A7*E$1	=$A7*F$1	=$A7*G$1	=$A7*H$1
8	7	=$A8*B$1	=$A8*C$1	=$A8*D$1	=$A8*E$1	=$A8*F$1	=$A8*G$1	=$A8*H$1

Figure 7-74. Formulas using mixed cell referencing

14. Press Ctrl + ~ to return to the results of the formulas. Your data should look like the table in Figure 7-75.

◢	A	B	C	D	E	F	G	H
1		1	2	3	4	5	6	7
2	1	1	2	3	4	5	6	7
3	2	2	4	6	8	10	12	14
4	3	3	6	9	12	15	18	21
5	4	4	8	12	16	20	24	28
6	5	5	10	15	20	25	30	35
7	6	6	12	18	24	30	36	42
8	7	7	14	21	28	35	42	49

Figure 7-75. *Results of multiplication table*

You have created formulas using absolute and mixed cell references. Knowing how and when to use these are very important if you want to become a skilled Excel user. Improper use of cell references is one of the mistakes new Excel students make most often. You should always review your formulas and their results before relying on those results.

Order of Precedence

The order of precedence is a set of predefined rules that determines the order in which operations in a formula will be performed. Looking at the given formula, it is difficult to know which operation to perform first:

$$= 3+2*18 /-3^2 - 2$$

Excel follows the normal mathematical **order of precedence** also known as the **order of operations**. If an expression contains operators with the same precedence, they are evaluated from left to right. Placing portions of the expression within parentheses allows you to control the order in which the operators are evaluated. Excel performs mathematical operations in the order shown in Table 7-5.

Table 7-5. *Order of Precedence*

Operator	Description	Precedence
( )	Within parentheses	1
-	Unary operator (negative sign)	2
%	Percent	3
^	Exponentiation	4
*	Multiplication	5
/	Division	5
+	Addition	6
–	Subtraction	6

Now that you know the order in which operations take place, you can use this knowledge to determine the outcome of this formula:

$$=3+2*18/-3\wedge2-2$$

Negative numbers are evaluated first and then Exponents. The expression is now equivalent to the following:

$$=3+2*18/9-2$$

Multiplication and division have the same level. Since the multiplication is to the left of the division, it is done first. The expression is now equivalent to the following:

$$=3+36/9-2$$

The division is performed next:

$$=3+\mathbf{4}-2$$

Addition and subtraction have the same precedence. Since the addition is to the left of the subtraction, it is done first:

$$=\mathbf{7}-2$$

The subtraction is done next giving a result of **5**.

The normal operator precedence can be overridden by putting parentheses around the operations you want performed first. If you have parentheses within parentheses, the operation in the innermost parentheses is performed first.

The formula could have been made clearer by the use of parentheses as in the formula that follows:

$$=3+((2*18)/(-3\wedge2))-2$$

Summary

All formulas in Excel must start with an = sign.

Formulas can be entered in one of three ways:

- Typing directly into a cell

- Clicking the cells you want to reference in the formula separated by mathematical symbols

- Using your arrow keys to select the cells you want to reference separated by mathematical symbols

Excel follows the normal mathematical order of precedence. Be sure to use parentheses to make sure your mathematical operations are performed in the order you want.

Excel has an Auto button that has the aggregate functions Sum, Average, Count, Min, and Max. You don't need to enter the function to use one of these often-used functions. Just select the cells above or to the left of where you want to place the result and then select the function you want to use by clicking the down arrow to the right of the AutoSum button.

You can assign a name to a range of cells. This can make your formulas easier to read and understand.

There are three places where you can name a range:

- Name Box

- Define Names dialog box

- Name Manager

Press Ctrl + ~ to display all of the functions on your worksheet. You can print your worksheet with the formulas displayed. Press Ctrl + ~ again to display the results of your functions rather than the functions themselves.

You can copy your formulas using the Autofill Handle. Any addresses used in your formulas will automatically adjust in the copied cells. Addresses that adjust are called relative addresses. You can lock an address in a formula so that it will not adjust when being copied. A locked address is called an absolute address. The dollar sign is used to lock a row or a column. Press F4 to place a dollar sign in front of the row and column. If you lock only a row or only a column in an address, it is called a mixed cell address.

You can select your named ranges and constants from a list by pressing F3.

You can place the names you assigned to ranges along with the range addresses in your worksheet by clicking the down arrow to the right of **Use in Formula** and then selecting Paste Names. This creates good documentation for your worksheet.

CHAPTER 8

Excel's Pre-existing Functions

This chapter covers some of the most used functions that allow you to do such things as sum values only if they meet certain conditions, something based on whether a condition is true or not, determine the number of days between dates, get the current date and time, and so on. You will learn how to enter functions manually or by using the Function wizard.

After reading and working through this chapter, you should be able to

- Create a nested function

- Use the Function wizard

- Find the number of days between two days

- Sum a range of values based on no conditions, one condition, or multiple conditions

- Get the month, day, or year from a date

- Display different values based on whether a condition is true or false

- Get the highest or lowest value based on criteria you set

- Get the current date and time and use them in formulas

© David Slager and Annette Slager 2020
D. Slager and A. Slager, *Essential Excel 2019*, https://doi.org/10.1007/978-1-4842-6209-2_8

Excel's Built-In Functions

Excel has more than 450 predefined built-in formulas which are called **functions**. You are not limited to these functions. You can create your own. Each function has a unique name. The functions are grouped into categories. The function categories are located on the Formula tab's Function Library group. Additional categories are available by clicking the More Functions button. See Figure 8-1.

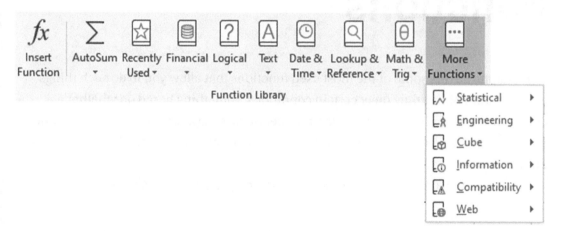

Figure 8-1. *Excel functions are categorized*

Table 8-1 describes the main built-in functions.

Table 8-1. *Function Categories and Their Descriptions*

Function Category	Function Description
AutoSum	The AutoSum drop-down arrow displays the "Auto" functions. These are the exact same functions we looked at previously, and they are available on the Home tab in the editing group. These functions can return the sum, average, count, highest value, and lowest value for a range of cells.
Compatibility	These are functions that have been kept for compatibility with previous versions of Excel. These functions have all been replaced with newer functions, so unless you need backward compatibility, you should not use them.

(continued)

Table 8-1. (*continued*)

Function Category	Function Description
Cube	These functions return values from analysis of multidimensional data sets used in databases.
Date & Time	Functions for dealing with days, dates, and times.
Engineering	Functions for engineering calculations.
Financial	Functions dealing with loans, investments, interest, depreciation, future values, etc.
Information	Functions that help you determine the type of data that exists in a cell.
Logical	Logical functions are used primarily in comparisons in which the result would be True or False.
Lookup & Reference	Excel's Lookup & Reference functions can be used to simplify finding values in a list or data table or finding the reference of a cell.
Math & Trig	Functions used for performing simple and complex mathematical calculations.
Statistical	Functions that perform statistical analysis on ranges of data.
Text	Functions that manipulate text strings.
Web	Gets data from a web service.

The Recently Used button is not really a category. It displays the last ten functions that you have used. This creates a quick way to jump to some of the functions you probably use most often.

Function Construction

Each category consists of separate functions that relate to their category. Figure 8-2 shows some of the Logical functions.

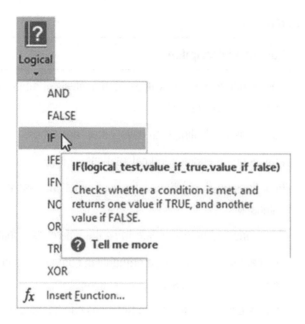

Figure 8-2. *Logical functions*

Moving your cursor over a function provides a tool tip. The tool tip provides a description of the function and its syntax. Each function has its own particular syntax. **Syntax** is the rules for how a function can appear in a cell. Functions usually contain **arguments**. Arguments are the values that are passed to the function to work on. There are three arguments for the IF function (logical_test, value_if_true, and value_if_false). Arguments are always separated by commas. You would enter values or cell addresses for the arguments. Clicking **Tell me more** will provide more help.

An example of an IF function is =IF(C3>25000,C3*6%,C3*5%). The IF function will be fully explained later in this chapter. Now that we have looked at how a function is created, we will look at the functions that you are likely to use the most often.

Functions That Sum Values

There are three functions for summing cell values. Table 8-2 describes them.

Table 8-2. *Descriptions for Summing Functions*

Function	Description
SUM	Sums a series of values without testing for any conditions.
SUMIF	Used to sum values, provided they meet a condition that you specify.
SUMIFS	Used to sum values only if multiple conditions are met.

SUM Function

We will look how a SUM function is constructed. The SUM function is one of the Math & Trig functions. The SUM function, used for adding a group of values, is the most used function. The SUM function should be familiar to you, because when you use AutoSum, you are running the SUM function.

The syntax for the Sum function is

```
=SUM(number1, [number2], [number3]. . .)
```

Function names are written in all capital letters; however, they are not case sensitive. You could enter the function name in small letters. If you enter a function in small letters, Excel will change it to all caps.

Function names are followed by a set of parentheses. If the function requires arguments, they are placed within the parenthesis. The SUM function has the arguments number1, number2, and number3. The arguments number2 and number3 are placed within square brackets. Arguments placed within square brackets are optional. The three dots (ellipsis) after the last argument mean that this is not the end of the list. You could place any number of arguments within the parenthesis of the SUM function.

Excel is very helpful when entering a function. As we saw in Chapter 7, entering a function in a cell starts by entering an = sign followed by the name of the function. As you type the letters of the function, Excel narrows down the function choices based on the letters you have entered so far. See Figure 8-3.

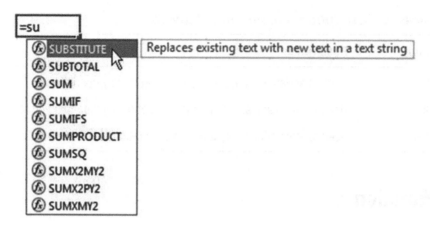

Figure 8-3. *As you type the letters of a function, Excel helps find a function that starts with those letters*

Clicking a function name brings up a description of what that function does.

After you have entered enough characters to narrow down the function to the one you want, press the Tab key. Excel will place a left parenthesis after the function name, and it will bring up a tool tip to help you with entering the function. The first argument is highlighted in the tool tip. See Figure 8-4.

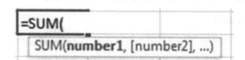

Figure 8-4. *First argument is highlighted*

After entering the first argument and a comma (**23,** in this example), the second argument becomes highlighted in the tool tip and a third argument option is displayed. See Figure 8-5.

=SUM(23,
SUM(number1, **[number2]**, [number3], ...)

Figure 8-5. *Second argument is highlighted*

Every time you enter another argument, an additional argument appears in the tool tip. See Figure 8-6.

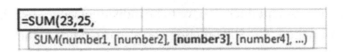

Figure 8-6. *Additional argument is added to tool tip*

If you need to change one of the arguments you entered, click the argument in the tool tip. This will select the argument in the function and then you can make your changes. See Figure 8-7.

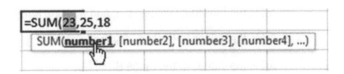

Figure 8-7. *Click an argument in the tool tip, to select that argument in the function*

You do not need to enter a comma after the last argument. It also is not necessary to enter the closing parenthesis. When you accept the entry, Excel will automatically enter the closing parenthesis for you. The result returned by the function will display in the cell where you entered the function. The complete function will display in the formula bar. See Figure 8-8.

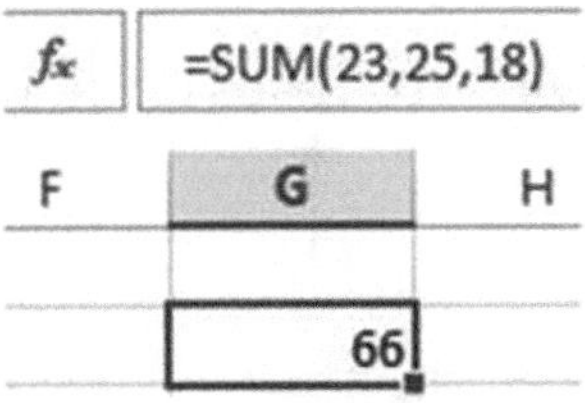

f_x	=SUM(23,25,18)

F	G	H
	66	

Figure 8-8. *Result of function appears in the cell. The formula for that cell displays in the Formula bar*

Single values entered in a formula are called constants. A **constant** is a value that does not change.

The entry for an argument doesn't have to be individual values like those we have entered in the SUM function. The SUM function can also use range names, individual cell references, and cell ranges for its arguments. What can be entered for each argument depends upon the function used. Table 8-3 shows several examples of the SUM function with different arguments.

Table 8-3. *The SUM Function with Different Arguments*

Function	Description
=SUM(3,5,8)	Adds the individual values 3, 5, and 8 giving a result of 16.
=SUM(Sales)	Adds the values in the range named Sales.
=SUM(B3:B9)	Adds the values in the range B3 to B9.

If you wanted to sum the cells in the range C5 to C10 on a different worksheet named "Purchases," you would enter the formula as shown in Figure 8-9.

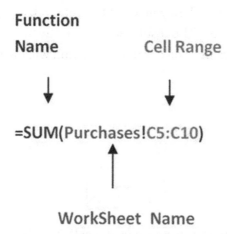

Figure 8-9. *Using the SUM function to sum a range on a different worksheet*

You could type **Purchases!** or you could click the worksheet named Purchases and Excel would enter the name along with the exclamation point in the formula for you.

If you only wanted to add the values in column C, then you could use the shortcut method C:C. The formula could be written as =SUM(Purchases!C:C).

Using the Insert Function Option

As we discussed in Chapter 7, clicking the **Insert Function** button in the Function Library group brings up the Insert Function dialog box. See Figure 8-10.

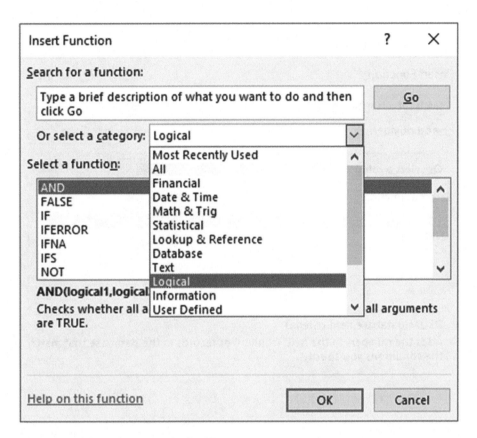

Figure 8-10. *Insert Function dialog box*

Clicking the down arrow for the **Or select a category** list box brings up the same categories that are available in the Function Library group. The only additional category here is the All category. If you don't know the category for the function you want to use, clicking the All category will bring up every function from every category in alphabetical order. The syntax and description for each function appears as you click the different functions.

A primary reason for using the Insert Function option is for its search capability. You can type a description of what you want a function to do into the search text box and then click the Go button. Excel will provide a list of functions that may meet your needs. (The search capability for the Insert Function in Excel 2019 doesn't work nearly as well as it did in Excel 2016. I hope that Microsoft will get this fixed.) In Figure 8-11, "add numbers" was entered for the search. Excel returned with a list of functions that would meet that criterion. It would be up to you then to decide which one in the list best meets your needs.

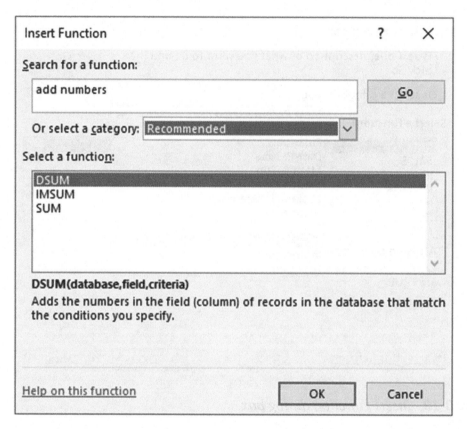

Figure 8-11. *Enter what you want to do in the Search for a function text box and then Excel attempts to find a function that meets your needs*

EXERCISE 8-1: USING THE SUM FUNCTION

We will add three columns of numbers using the SUM function.

1. Create a new workbook. Enter the data in Figure 8-12 into a worksheet. Name the worksheet "Sum."

▲	A	B	C	D	E	F
1	Store A		Store B		Store C	
2	Qtr 1	Qtr 2	Qtr 1	Qtr 2	Qtr 1	Qtr 2
3	125	200	900	776	500	400
4	500	400	300	400	400	385
5	700	600	200	425	300	275
6	800	750	100	180	200	192
7						
8	Total Qtr 1					
9	Total Qtr 2					

Figure 8-12. *Enter your data in a worksheet in this format*

2. Click inside cell B8. On the Formulas tab in the Function Library group, click the Insert Function button.

3. Enter **add numbers** in the Search for a function box. Press Enter or click the Go button. Click **SUM** in the Select a function list box. Click the OK button.

4. The Function Arguments dialog box opens. Move the dialog box if it is covering the numbers in your worksheet. An entry box appears for each required argument. Click inside the Number1 box. Required arguments are bolded. Number1 is the only required argument for the Sum function. See Figure 8-13.

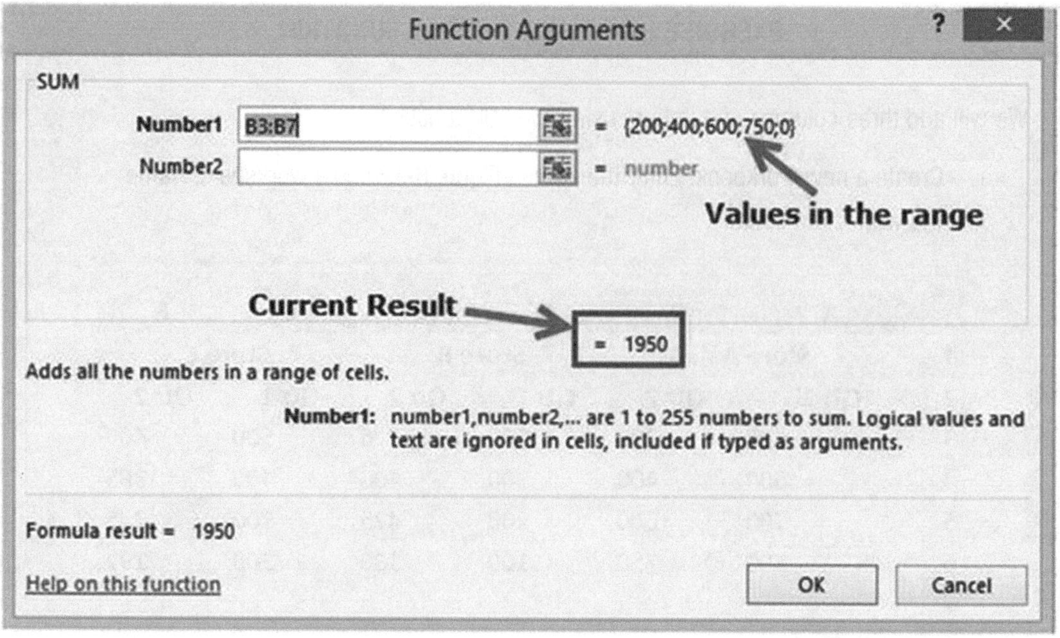

Figure 8-13. *Excel assumes you want to add the cell range B3:B7 and enters the range for the first argument*

Argument Number1 contains the range B3:B7 because it is attempting to add the values above it. The values being summed are displayed to the right of the Number1 box. It includes the 0 for cell B7. This doesn't affect the current total, which as you can see is 1950.

5. Drag across cells A3:A6. The values from this range are displayed as well as the current sum. See Figure 8-14.

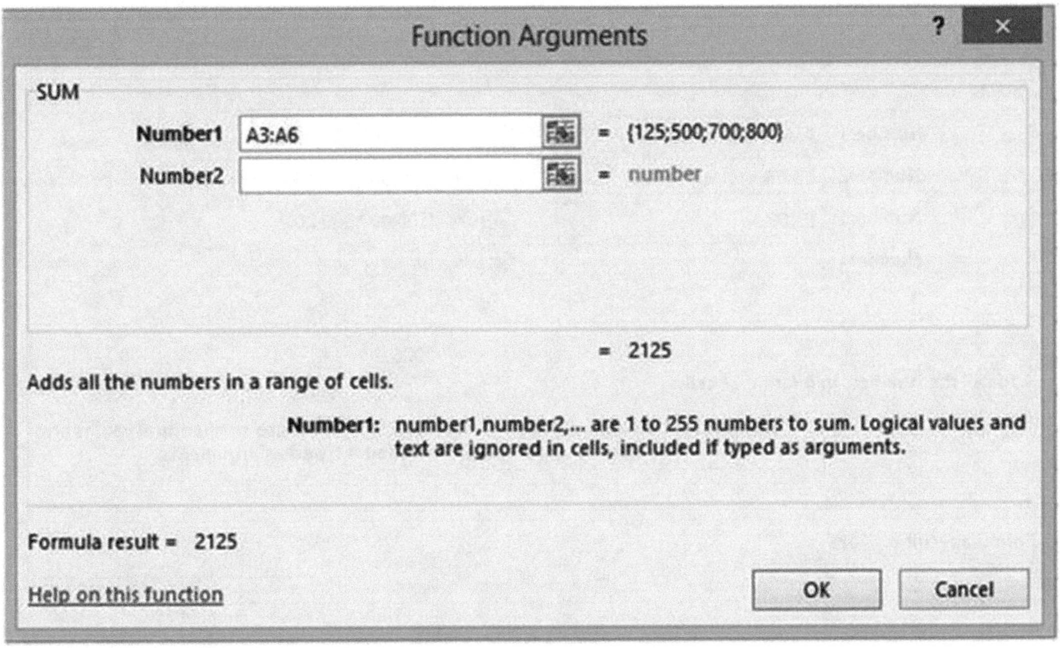

Figure 8-14. *Excel sums the values in the range A3:A6*

6. The Collapse Dialog 🔲 buttons can be used to collapse the window when it is blocking the data you need to use. Click inside the Number2 box. Click the 🔲 button to the right of the Number2 box. The window collapses to a single box for the current argument. Drag across cells C3:C6. See Figure 8-15. When you have finished your selection, click the Expand Dialog 🔲 button to expand the window back.

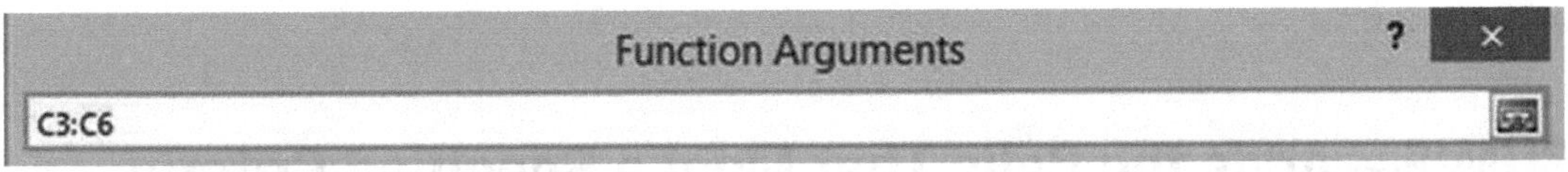

Figure 8-15. *Collapsed window*

7. The Number3 argument has been added to the window. Click inside the Number3 box. The Number4 argument has been added. Drag across cells E3:E6. Figure 8-16 displays the result. Click the OK button.

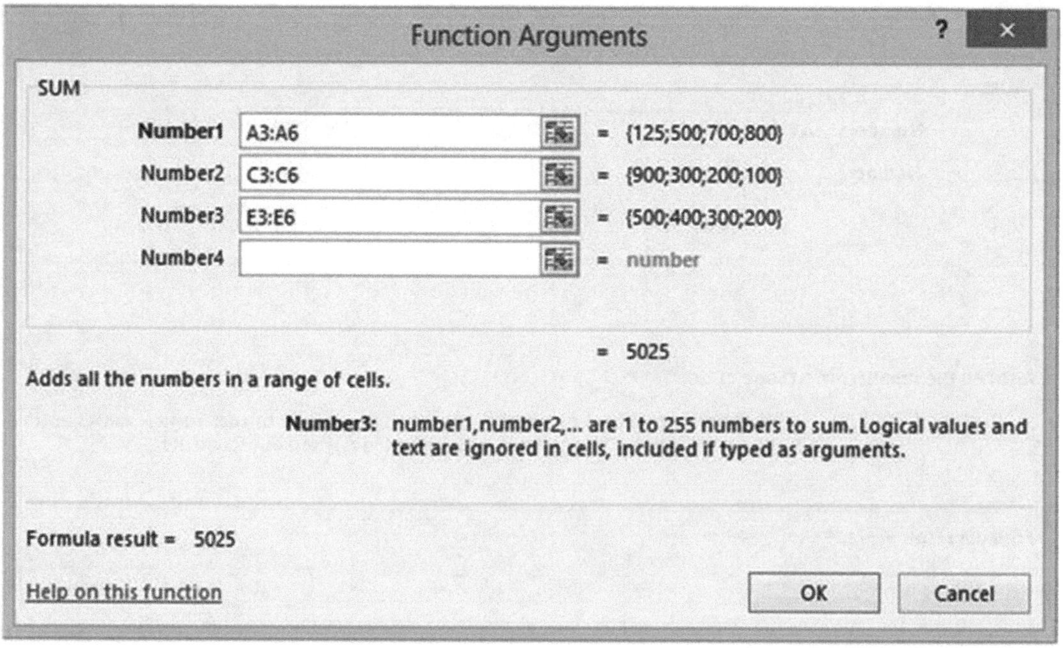

Figure 8-16. *Additional arguments were added. The first three arguments have been added and the result is 5025*

The result is placed in cell B8.

Using the same steps, see if you can get the answer for cell B9.

You have learned how to add all the values in a single range or in multiple ranges at the same time. What if you only want to add only some of the values in a range such as expenses, but only if they are entertainment expenses? You'll see how to do this in the next section.

SUMIF—Adds the Cells That Meet a Specified Criteria

The SUMIF function is used for summing a range of values, provided that they meet a condition that you specify in the criteria argument. The SUMIF function only works if you are basing your summing on a single criterion. If you need to base it on more than one criterion, you will need to use the SUMIFS function, which is covered in the next section.

The syntax for the SUMIF function is

```
SUMIF(range, criteria, [sum_range])
```

Argument descriptions are

> range is the cells that you are basing your criterion on. The range contains the values that will be checked by the criteria to determine if they are to be included in the sum.

> criteria is the condition that must be met if a value is to be included in the sum. If the condition contains any text, logical, or mathematical symbols, it must be enclosed in double quotation marks. Quotation marks are not necessary if the criterion is just a number.

> [sum_range] is optional. If this optional range is used, then it will be this range that contains the cells to be summed rather than the range in the first argument. The criteria will still be based on the first range argument.

Using SUMIF with the Required Arguments

Figure 8-17 contains a range of values that need to be summed, but we only want to sum those items whose value is greater than 900. There are only two values greater than 900 in the range B1:B6: 1100 and 1500.

	A	B
1		1100
2		800
3		400
4		900
5		500
6		1500
7	Total of items with values greater than 900	

Figure 8-17. *The values that are greater than 900 will be summed*

The arguments used in the formula for summing the values greater than 900 are

> B1:B6 is the range of values that the criteria will be based on.

> ">900" is the criteria. Since the criterion contains something other than just a number, it must be enclosed in double quotes.

The complete formula is =SUMIF(B1:B6,">900").

We will place this formula in cell B7. The result of adding the two numbers over 900 (1100 + 1500) is 2600. See Figure 8-18.

	A	B
1		1100
2		800
3		400
4		900
5		500
6		1500
7	Total of items with values greater than 900	2600

Figure 8-18. *Result of SUMIF function*

Figure 8-19 shows a list of items and their prices. We only want to sum those items whose price is exactly 99 cents. The formula entered into cell B10 is =SUMIF(B2:B8,0.99). Since we are only summing those values which are exactly .99, there is no need for any additional logical symbols because the equal sign is optional. Because the criterion is only a number, it does not need to be placed within double quotes.

◢	A	B
1	**Item**	**Price**
2	pencils	0.99
3	tablet	0.99
4	stapler	4.95
5	printpaper	5.95
6	paper clips	0.99
7	staples	1.95
8	black marker	0.99
9		
10	Total 99 cent items	3.96

Figure 8-19. *Sum of the four .99 prices*

Using the Optional sum_range Argument

When using the optional sum_range argument, you base your criteria on a different range of cells than those that you are summing.

We want to find the total number of males and females in a high school. Figure 8-20 shows the worksheet for this. The worksheet uses separate formulas for determining the number of males and females. The arguments used in the formula for determining the number of males are

> B2:B9 is the **range** that the criteria will be based on.
>
> "M" is the **criterion**.
>
> C2:C9 is the **sum_range**. It is the range of the cells that will be summed, if the criterion is met.

▲	A	B	C
1	**Class**	**Gender**	**Count**
2	Freshman	M	31
3	Freshman	F	28
4	Sophmore	M	25
5	Sophmore	F	29
6	Junior	M	33
7	Junior	F	35
8	Senior	M	28
9	Senior	F	27
10			
11	Total Males		117
12	Total Females		119

◄────── =SUMIF(B2:B9,"M",C2:C9)
◄────── =SUMIF(B2:B9,"F",C2:C9)

Figure 8-20. *Result of SUMIF functions*

The formula entered in cell C11 is

=SUMIF(B2:B9,"M",C2:C9)

The only difference in the formula for computing females is the criterion which is "F". The formula entered in cell C12 is

=SUMIF(B2:B9,"F",C2:C9)

SUMIFS—Adds the Cells That Meet Multiple Criteria

The SUMIFS function is similar to the SUMIF function, except that the SUMIFS function is used to sum values only if multiple conditions are met rather than a single condition.

The syntax for the SUMIFS function is

SUMIFS(sum_range, criteria_range1, criteria1, [criteria_range2, criteria2], . . .)

Argument descriptions are

sum_range is the cells that you are basing your criteria on. The range contains the values that will be checked by the criteria to determine if they are to be included in the sum.

criteria_range1 is the first range in which to evaluate the associated criteria.

criteria1 is the condition that defines which cells in the criteria_range1 argument will be added.

Criteria_range2, Criteria2,... You can add as many as 126 additional ranges and their associated criteria.

Figure 8-21 shows a company's bank deposits along with the date of each deposit for its different branches. We want to sum the bank deposits in column B but only if the following conditions are met:

- It is not a Hobart office.

- The amount of the deposit has to be at least 20,000.

- The date of the deposit has to be 03/12/2010 or later.

	A	B	C
1	**Office**	**Deposit**	**Date of Deposits**
2	Fort Wayne	25538.25	3/10/2010
3	Indianapolis	35694.24	3/11/2010
4	Hobart	8950	3/11/2010
5	Merrillvile	29875.14	3/25/2010
6	Fort Wayne	19808.75	4/1/2010
7	Merrillvile	29305.4	4/9/2010
8	Hobart	22815.08	4/27/2010
9	Fort Wayne	17640.72	5/1/2010
10	Indianapolis	17489.95	5/28/2010
11	Fort Wayne	24382.25	6/4/2010
12			
13		83562.79	

Figure 8-21. *Result of SUMIFS function*

The formula that will add the deposits that meet our criteria is

=SUMIFS(B2:B11,A2:A11,"<>Hobart",B2:B11,">=20000",C2:C11,">=03/01/2010")

The arguments used in the formula for summing the bank deposits are as follows:

B2:B11 is the sum_range. This range contains the values we want to sum if they meet our criteria.

We only want the branches that are not Hobart.

- A2:A11 is the criteria_range1.

- "<>Hobart" is criteria1. (<> means not equal to.)

The amount of the deposit has to be at least 20,000.

- B2:B11 is the criteria_range2.

- ">=20000" is criteria2. (>= means greater than or equal to.)

The date of the deposit has to be 3/12/2010 or later.

- C2:C11 is the criteria_range3.

- ">=03/12/2010" is criteria3.

Figure 8-22 highlights the rows that meet all the criteria. They are the only deposits that meet all three criteria of (1) not being a Hobart branch, (2) having a deposit greater than or equal to 20000, and (3) having been deposited on or after 3/12/2010. The result of adding the deposit for these three rows is 83562.79.

	A	B	C
1	**Office**	**Deposits**	**Date of Deposits**
2	Fort Wayne	25538.25	3/10/2010
3	Indianapolis	35694.24	3/11/2010
4	Hobart	8950	3/11/2010
5	Merrillville	29875.14	3/25/2010
6	Fort Wayne	19808.74	4/1/2010
7	Merrillville	29305.4	4/9/2010
8	Hobart	22815.08	4/27/2010
9	Fort Wayne	17640.72	5/1/2000
10	Indianapolis	17489.95	5/28/2010
11	Fort Wayne	24382.25	6/4/2010
12			
13		83562.79	

Figure 8-22. *Rows that meet the SUMIFS criteria are highlighted*

Using a Cell Address Rather Than a Cell Value for a Criteria

If you are checking if an item is equal to a value in the criteria range, you can specify a cell address that contains the value of that item for the criteria. In Figure 8-23, we want to sum the attendees if they took an Excel class in Fort Wayne. Cells E2 and F2 can be used for the criteria because they contain the values to be checked against the criteria range.

	A	B	C	D	E	F	G	H	I
1	**Office**	**Class**	**Attendees**		**Office**	**Class**	**Total Attendees**		
2	Fort Wayne	Excel	25		Fort Wayne	Excel	=SUMIFS(C2:C11,A2:A11,E2,B2:B11,F2)		
3	Fort Wayne	Excel	30				SUMIFS(sum_range, criteria_range1, criter		
4	Fort Wayne	Word	28						
5	Merrillville	Word	18						
6	Fort Wayne	PowerPoint	14						
7	Merrillville	Excel	19						
8	Hobart	PowerPoint	22						
9	Fort Wayne	Excel	21						
10	Indianapolis	Excel	20						
11	Fort Wayne	Word	17						

Figure 8-23. *The criteria for SUMIFS is in cells E2 and F2*

Figure 8-24 shows the formula entered in cell G2.

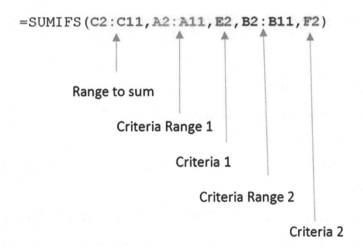

Figure 8-24. *Breaking down the SUMIFS formula*

Handling Empty Cells in SUMIFS Functions

The worksheet in Figure 8-25 totals the values for items that have been completed but haven't yet shipped. The two criteria that will be used for the SUMIFS arguments will be those items whose **Date Completed** is not blank and whose **Date Shipped** is blank. The figure highlights those items that meet both criteria.

	A	B	C	D	E
1	**Prod ID**	**Date Ordered**	**Date Completed**	**Date Shipped**	**Value**
2	SN1234	5/27/2015	6/8/2015	6/15/2015	575
3	AB2397	6/4/2015	6/14/2015		800
4	SN1234	6/8/2015			670
5	AB4823	6/21/2015	10/26/2015	11/1/2015	1025
6	DC3214	6/30/2015			400
7	DC3214	7/1/2015	9/3/2015	10/7/2015	385
8	KR1285	7/11/2015	11/17/2015		377
9	DC3872	7/14/2015			432
10	KR1285	7/16/2015	11/4/2015		175
11					1352

Figure 8-25. *Rows are highlighted where the items have been completed but not shipped*

The criteria are as follows:

The criteria for the Date Completed cells is "<>"
(<> means not blank).

The criteria for the Date Shipped is "" (two quotes together test for a blank).

The formula entered in cell E11 is =SUMIFS(E2:E10,C2:C10,"<>",D2:D10,"").

Note   When testing a criteria range for blanks, the cells must be real blanks and not the result of a formula that had no result.

IF—Returns Different Values Depending upon If a Condition Is True or False

The IF function returns different values depending upon whether a condition is true or false.

The syntax for the IF function is

IF(logical_test , [value_if_true], [value_if_false])

Argument descriptions are

> logical_test is the condition being tested.
>
> [value_if_true] if the condition being tested is true, then perform this step.
>
> [value_if_false] if the condition being tested is false, then perform this step.

If the condition being tested is true, what is in the value_if_true argument will be used; otherwise, what is in the value_if_false argument will be used.

Let's say that the value in cell A2 is 20. If you place the following formula in cell A1

=IF(A2=20,"Equal 20","Not equal 20")

then cell A1 will display **Equal 20**. If you change the value in cell A2 to 19, then the value in cell A1 will change to **Not equal 20**.

EXERCISE 8-2: USING THE IF FUNCTION

We need to determine the Gross Pay for each of our employees. If an employee works over 40 hours, he will receive overtime pay at 1.5 times the normal rate. The Gross Pay will be computed by adding an employee's regular pay and overtime pay.

An IF function will be used to compute the employee's regular pay; another IF function will be used to compute his or her overtime pay.

If an employee works 40 hours or less, his regular pay and his gross pay would be the hours he worked times his pay rate.

If an employee works 40 hours or more, then his regular pay will be 40 hours times the pay rate. The hours the employee worked in addition to the 40 hours will be used for determining overtime pay. If we assign names to our ranges, the formula needed to compute regular pay is =IF(hours>40, 40*rate, hours*rate).

We take **40 * rate** if the condition (hours >40) is true and **hours * rate** if the condition is false.

To compute the OT (overtime pay), we will again need to test if the hours worked is greater than 40. If the hours are greater than 40, we will need to take the number of hours worked minus 40 to get the overtime hours. The overtime hours will be multiplied by the pay rate times 1.5. If the hours worked are 40 or less, then the overtime will be 0.

Figure 8-26 shows the formula to compute the overtime pay.

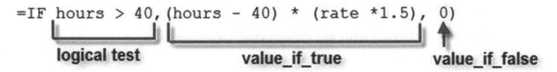

```
=IF hours > 40, (hours - 40) * (rate *1.5), 0)
```

logical test   value_if_true   value_if_false

Figure 8-26. *The overtime pay formula*

1. Enter the data in Figure 8-27 in a new worksheet. Name the worksheet "Payroll."

	A	B	C	D	E	F
1	Emp ID	Hours	Rate	Regular Pay	OT	Gross Pay
2	G125	38	9.75			
3	G232	40	10.25			
4	G238	40	10.75			
5	R114	43	11.15			
6	R116	48	10.95			
7	R119	56	12.18			
8	S202	40	15.05			

Figure 8-27. *Enter this data in a new worksheet named Payroll*

2. Select the range B1:E8, as in Figure 8-28.

	A	B	C	D	E	F
1	Emp ID	Hours	Rate	Regular Pay	OT	Gross Pay
2	G125	38	9.75			
3	G232	40	10.25			
4	G238	40	10.75			
5	R114	43	11.15			
6	R116	48	10.95			
7	R119	56	12.18			
8	S202	40	15.05			

Figure 8-28. *Select range B1:E8*

3. On the Ribbon, click the Formula tab. In the Defined Names group, click the **Create from Selection** button. Select Top row from the Create Names from Selection dialog box. See Figure 8-29.

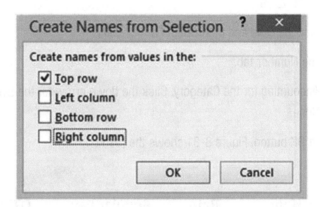

Figure 8-29. *Select Top row in the Create Names from Selection dialog box*

4. Type this formula in cell D2: =IF(hours >=40,40*rate,hours*rate). Press the Tab key.

5. Type this formula in cell E2: =IF(hours > 40,(hours - 40)*rate* 1.5,0). Press the Tab key.

6. Type this formula in cell F2: =Regular_Pay+OT. Press the Enter key.

7. Select the range D2:F2. Drag the AutoFill Handle down through row 8. Figure 8-30 shows the result.

	A	B	C	D	E	F
1	Emp ID	Hours	Rate	Regular Pay	OT	Gross Pay
2	G125	38	9.75	370.5	0	370.5
3	G232	40	10.25	410	0	410
4	G238	40	10.75	430	0	430
5	R114	43	11.15	446	50.175	496.175
6	R116	48	10.95	438	131.4	569.4
7	R119	56	12.18	487.2	292.32	779.52
8	S202	40	15.05	602	0	602

Figure 8-30. *Result of copying formulas with AutoFill Handle*

8. We need to change the Regular Pay, OT, and Gross Pay columns so that they display with two decimal positions. We also want to show dashes instead of zeroes for the overtime pay. Formatting the cells as the Accounting type will do this for us:

    a. Select cells D2:F8.

    b. Right-click the cells and then select Format Cells.

    c. Select the Number tab.

    d. Select Accounting for the Category. Click the down arrow for the Symbol and then select None.

    e. Click the OK button. Figure 8-31 shows the results.

	A	B	C	D	E	F
1	Emp ID	Hours	Rate	Regular Pay	OT	Gross Pay
2	G125	38	9.75	370.50	-	370.50
3	G232	40	10.25	410.00	-	410.00
4	G238	40	10.75	430.00	-	430.00
5	R114	43	11.15	446.00	50.18	496.18
6	R116	48	10.95	438.00	131.40	569.40
7	R119	56	12.18	487.20	292.32	779.52
8	S202	40	15.05	602.00	-	602.00

Figure 8-31. *Result of formatting the range D2:F8 as Accounting*

You have used the IF statement to test a condition; the function does one thing if the condition is true and something else if the condition is false. Next, you will look at the AND function and later the OR function. These functions don't do much on their own. They only return either a TRUE or FALSE. Their strength is shown when they are combined with other functions. The exercise following the OR function demonstrates how to combine these functions within an IF function.

AND—Returns TRUE if All of Its Arguments Are TRUE

The AND function returns either a TRUE or a FALSE. It returns a TRUE if **all** of its arguments are true, and it returns a FALSE if any of its arguments are not true.

The syntax for the AND function is

AND(logical1, [logical2], . . .)

The AND function can be used to evaluate up to 255 logical conditions.

Figure 8-32 shows the AND function with three logical conditions. All three of these conditions are true; therefore, the function returns a TRUE.

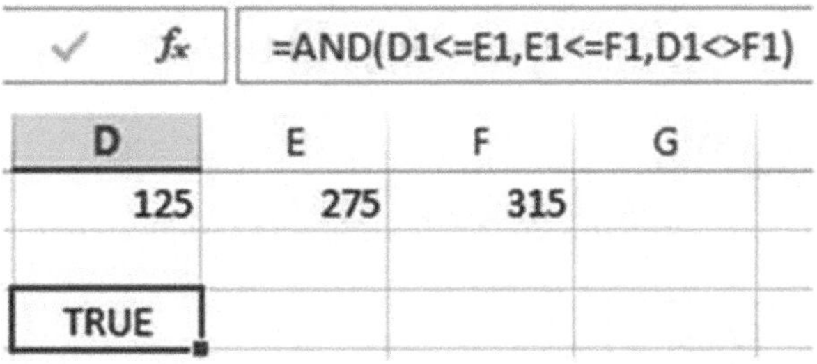

Figure 8-32. *AND function that tests three conditions*

The last condition in Figure 8-33 is false. The value in cell D1 is not equal to the value in cell F1. The AND function therefore returns a FALSE because even if only one of the logical conditions is false, it will return a FALSE.

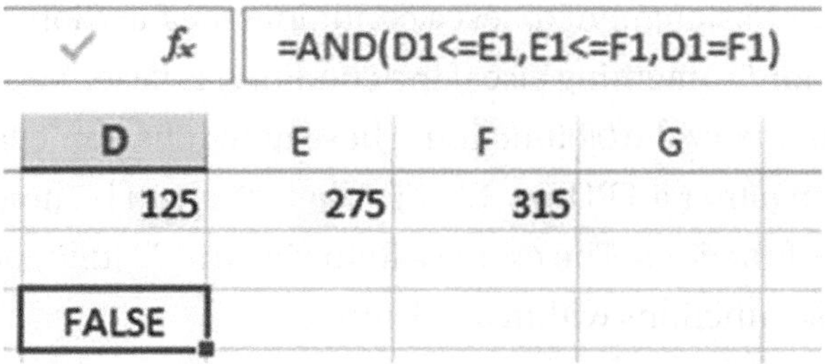

✓	f_x	=AND(D1<=E1,E1<=F1,D1=F1)

D	E	F	G
125	275	315	
FALSE			

Figure 8-33. *The third condition is false, which returns a false*

OR—Returns TRUE If Any Argument Is TRUE

The OR function returns a TRUE if any of its arguments are true, and it returns a FALSE if **all** of its arguments are false.

The syntax for the OR function is

OR(logical1,[logical2],...)

Figure 8-34 shows the OR function with three logical conditions. All three of these conditions are false. If and only if **all** of the arguments are false will the OR function return a FALSE.

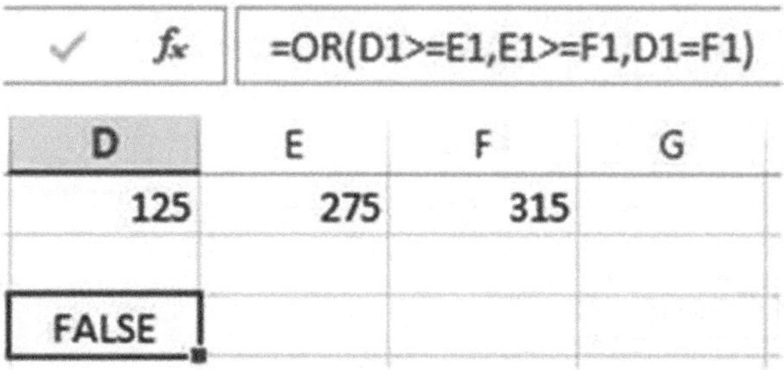

✓	f_x	=OR(D1>=E1,E1>=F1,D1=F1)

D	E	F	G
125	275	315	
FALSE			

Figure 8-34. *All conditions are false so the OR condition returns a FALSE*

The first logical condition in Figure 8-35 is true. The value in cell D1 is less than the value in cell E1. The other conditions are false. The OR function returns a TRUE because even if only one logical condition is true, the function returns a TRUE.

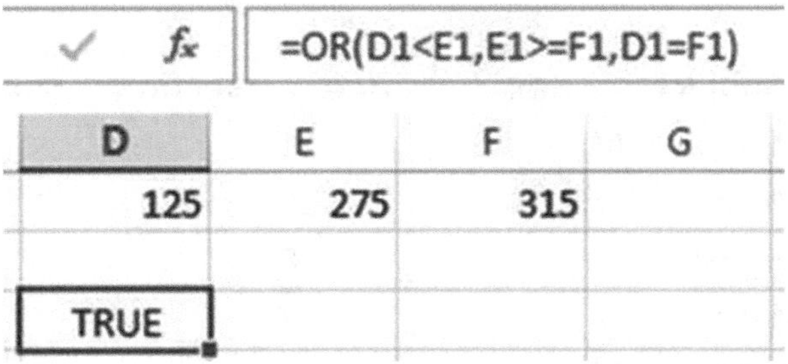

		=OR(D1<E1,E1>=F1,D1=F1)		
D	E	F	G
125	275	315	
TRUE			

Figure 8-35. *Only the first condition is true; therefore, the OR function returns a TRUE*

Nested Functions

Sometimes a single function will not give you the results you need. In these cases, you will need to create a nested function. Nested functions combine the operations of two or more functions into a single function. A nested function is used as an argument in the other function. Excel allows a maximum of 64 functions nested within each other. That should be more than you will ever need.

EXERCISE 8-3: CREATING A NESTED FUNCTION

In this exercise, you create functions nesting different operators.

Nesting an AND Function Within an IF Function

1. Enter the data in Figure 8-36 into a worksheet. Name the worksheet "Nested."

	A	B	C	D	E
1	Subject	Sex	Age	Section	Section
2	Math	M	21		
3	Math	M	19		
4	English	F	18		
5	Math	M	17		
6	History	M	18		
7	Math	M	19		
8	History	F	21		
9	Math	M	18		
10	Science	M	19		
11	English	M	23		

Figure 8-36. *Enter this data in a worksheet named Nested*

Students who are in Math, are males and are less than 21 years old will be placed in Section A. This problem will require an AND function nested within an IF function. The AND function will be the condition to be tested. If the condition is true, we will display an A in the cell. If the condition is false, we will display a blank.

2. Select the cell range A1:C11. On the Ribbon's Formula tab in the Defined Names group, click **Create from Selection**. In the **Create Names from Selection** window, only **Top row** is to be selected. Click the OK button.

3. Click the down arrow of the name box to see your named ranges.

4. Click inside cell D2. Start the formula by entering =IF(AND(. The Excel tool tip shows the needed arguments for the AND function. See Figure 8-37.

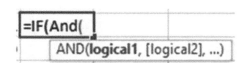

Figure 8-37. *Tool tip for the AND function*

5. We will now enter the first of the three conditions we will be testing for:

    a. Press F3 to bring up the range names. Select Subject and then click the OK button.

    b. Enter **="Math"** followed by a comma. Entering the comma highlights the second argument [logical2]. See Figure 8-38.

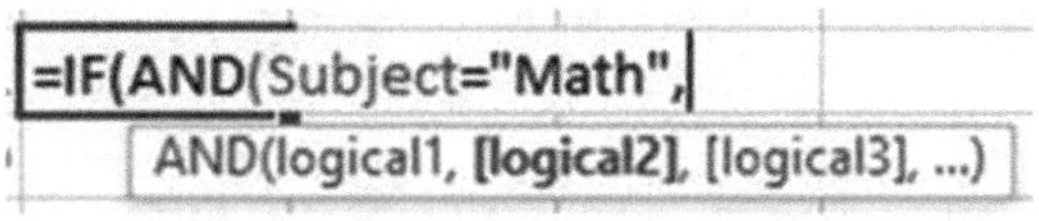

> =IF(AND(Subject="Math",|
>
> AND(logical1, **[logical2]**, [logical3], ...)

Figure 8-38. *Entering a comma in the function highlights the next argument in the tool tip*

6. Press F3 to bring up the range names. Select Sex and then click the OK button.

7. Type ="M",. The formula should now be =IF(AND(Subject="Math", Sex="M",.

8. Press F3 to bring up the range names. Select Age and then click the OK button.

9. Enter <21),. The formula should now be =IF(AND (Subject="Math",Sex="M",Age<21),.

    The AND function is finished and now you are back to the IF function and ready to enter the data for the argument [value_if_true]. See Figure 8-39.

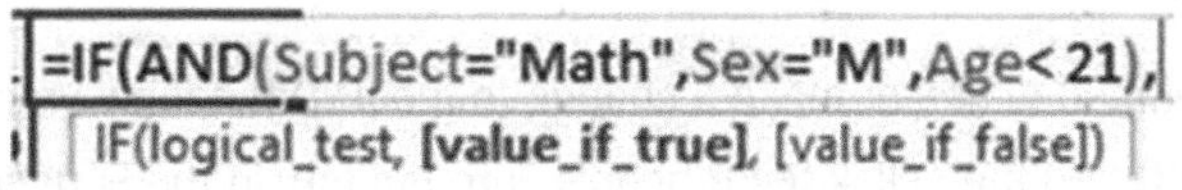

> =IF(AND(Subject="Math",Sex="M",Age< 21),|
>
> IF(logical_test, **[value_if_true]**, [value_if_false])

Figure 8-39. *The AND function is completed; [value_if_true], is highlighted to let you know that you should now enter this argument for the IF function*

10. Enter "A", for the true argument. Enter "") for the false argument. This will display a blank. The complete formula is =IF(AND(Subject="Math", Sex="M",Age<21),"A","").

    If you didn't enter "" for the false part, Excel would by default display the word **False**.

11. If you wanted to see or make a change to your formula, you could click the word **IF** and then click the Insert Function button (Formula tab ➤ Function Library).

   You can see that the AND condition is false; therefore, our result is a blank. See Figure 8-40.

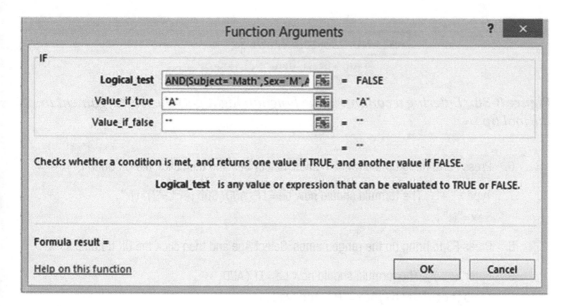

Figure 8-40. *The Function Arguments dialog box shows the results of the Logical_test*

12. Click the Cancel button.

13. If you just wanted to see or make a change to the AND function, you could click the word **AND** and then click the Ribbon's Formula tab. In the Functions Library group, click the Insert Function button. See Figure 8-41. You can see that the Age is not less than 21 for the first cell in the range.

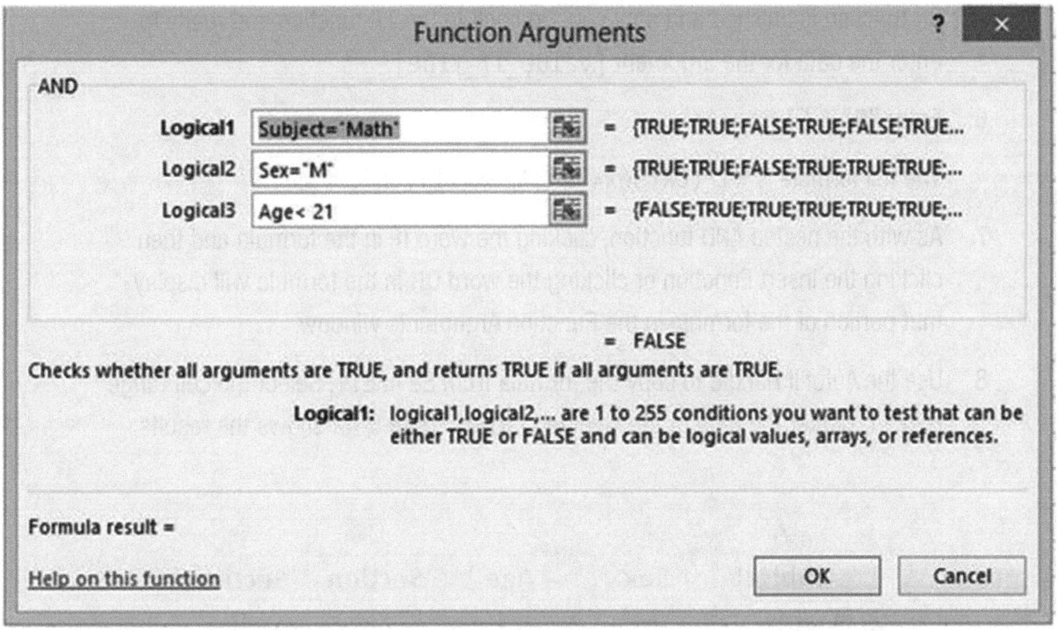

Figure 8-41. *The result of the logical argument is shown for each cell*

14. Click the Cancel button.

15. Press Ctrl + Enter to accept the formula. Cell D2 is blank.

16. Use the Autofill Handle to copy the formula from D2 to D11.

Nesting an OR Function Within an IF Function

Let's create another IF function, but this time let's nest an OR function within it that will test if the student is a female or if the student is 21 or more. If a student meets either one of these conditions, he or she will be placed in Section B.

1. In cell E2, enter the formula =IF(OR(.

2. Press F3 to bring up the range names. Select Sex and then click the OK button.

3. Type Sex = "F",.

4. Press F3 to bring up the range names. Select Age and then click the OK button.

5. Type Age >= 21),. The function is now =IF(OR(Sex="F",Age>=21),.

The OR function is finished and now you are back to the IF function and ready to enter the data for the argument [value_if_true].

6. Enter "B","").

The full formula is =IF(OR(Sex="F",Age>=21),"B","").

7. As with the nested AND function, clicking the word **IF** in the formula and then clicking the Insert Function or clicking the word **OR** in the formula will display that portion of the formula in the Function Arguments window.

8. Use the Autofill Handle to copy the formula from E2 to E11. Select the cell range B2:E11. Center the data of the selected range. Figure 8-42 shows the results.

	A	B	C	D	E
1	Subject	Sex	Age	Section	Section
2	Math	M	21		B
3	Math	M	19	A	
4	English	F	18		B
5	Math	M	17	A	
6	History	M	18		
7	Math	M	19	A	
8	History	F	21		B
9	Math	M	18	A	
10	Science	M	19		
11	English	M	23		B

Figure 8-42. *Range B2:E11 centered in cells*

Let's make a change to the application by placing a check mark in column D if the student meets the conditions of being a Male, a History major, and under 21. The check mark is part of the Wingding font Set. A check mark is created when you hold your Alt key and then on the numeric keypad type 0252 (it only works if you use the numeric keypad). The Wingding font must then be selected for the cell.

9. Delete column E.

10. Change the text in cell D1 to Section A.

11. Click cell D2. In the Formula bar, select **A** in the true part of the IF formula.

12. Hold down the Alt key and type 0252 on the numeric keypad. The formula should be =IF(AND(Subject="Math",Sex="M",Age<21),"ü","").

13. Select the Autofill handle on cell D2 and drag it down through cell D11.

14. With the range D2:D11 selected, change the Font typeface to Wingdings. Figure 8-43 shows the result.

	A	B	C	D
1	**Subject**	**Sex**	**Age**	**Section A**
2	Math	M	21	
3	Math	M	19	✓
4	English	F	18	
5	Math	M	17	✓
6	History	M	18	
7	Math	M	19	✓
8	History	F	21	
9	Math	M	18	✓
10	Science	M	19	
11	English	M	23	

Figure 8-43. *Use Wingdings to create check marks*

You created nested AND and OR functions. You can nest about any function within another. This gives your functions a lot more flexibility.

New Excel 2019 Functions

Excel 2019 introduces IFS, MAXIFS, and MINIFS functions.

We have looked at IF functions and nested functions. Prior to Excel 2019, nesting IF statements within IF statements created a long function that could get very complicated. Excel's new IFS function for 2019 allows you to perform the same function without having to use nesting.

IFS—Returns Value Based on Criteria

The syntax for the IFS functions is

```
IFS(logical test 1, value if true,[logical test 2, value if true],...)
```

We have test scores we want to assign grades to. A score of 90 or higher gets an A, 80 or higher gets a B, 70 or higher gets a C, 60 or higher gets a D, and below 60 gets an F. See Figure 8-44.

	A	B
1	Score %	Grade
2	85	
3	75	
4	40	
5	95	
6	62	
7	89	
8	77	

Figure 8-44. *Need to find Grade for Test Scores*

Let's enter the function using the Insert Function command found on the Ribbon's Formula tab in the Function Library group.

Start the function by entering =IFS(in cell B2. On the Ribbon's Formula tab, click the **Insert Function** command. The closing) is added to our function, and the Function Arguments window opens for the IFS function. See Figure 8-45.

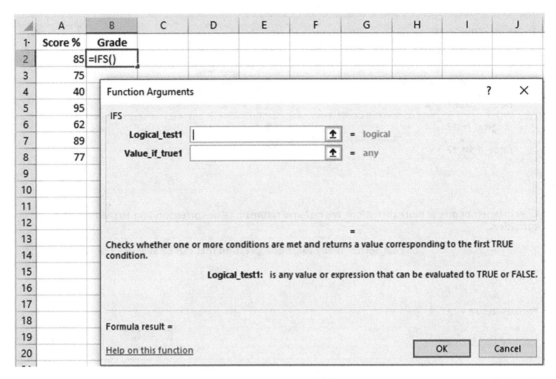

Figure 8-45. *Open IFS function in Function Arguments window*

The first logical test will be A2 >= 90. The value we want to give for that logical test is an A. The A will go in quotation marks. Anything other than a numerical value should be entered in quotes. After you have entered the data for the first test, Excel displays the text boxes for the next entry. You should press the Tab key to go to the next entry. If you press the Enter key, the window closes and what you have entered so far becomes your function. Notice the False to the right of the logical test. See Figure 8-46. Excel is telling you that cell B2 does meet the logical test condition.

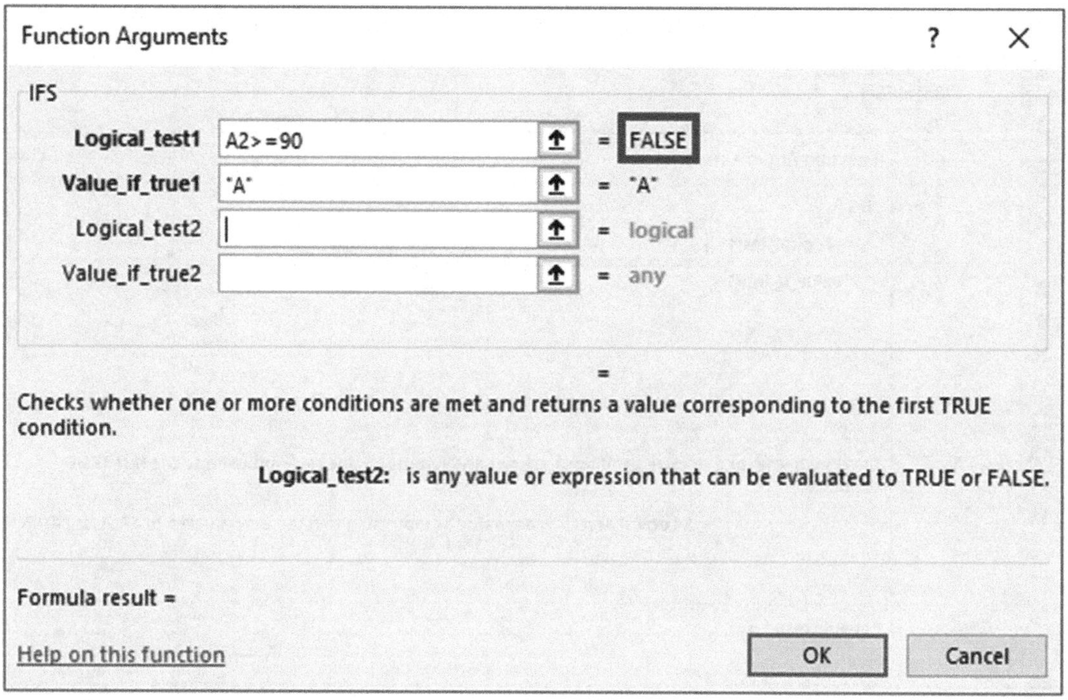

Function Arguments   ?   ✕

IFS

Logical_test1	A2>=90	⬆	=	FALSE
Value_if_true1	"A"	⬆	=	"A"
Logical_test2		⬆	=	logical
Value_if_true2		⬆	=	any

=

Checks whether one or more conditions are met and returns a value corresponding to the first TRUE condition.

Logical_test2: is any value or expression that can be evaluated to TRUE or FALSE.

Formula result =

Help on this function   OK   Cancel

Figure 8-46. *Enter arguments for first logical test*

The second logical test with its associated value has been entered in Figure 8-47. Notice the TRUE to the right of the second logical test. Excel is telling us that the value in cell B2 meets the requirements for logical test2. The other logical tests we will enter for this function will also be True, but the IFS function inserts the value of the first True logical test into the cell. At the bottom of the window, you can see that the result of the function is going to be B. Scroll bars have been added so that we can keep adding new tests.

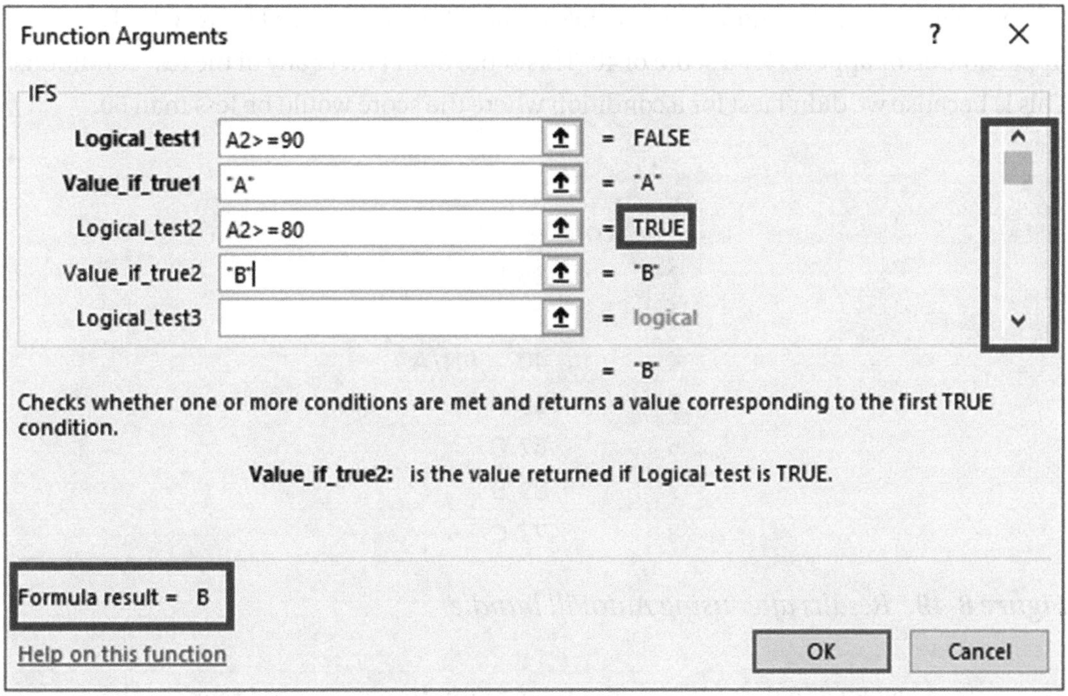

Figure 8-47. *Enter arguments for second logical test*

The result of entering the rest of the tests and values and pressing the Enter key is shown in Figure 8-48.

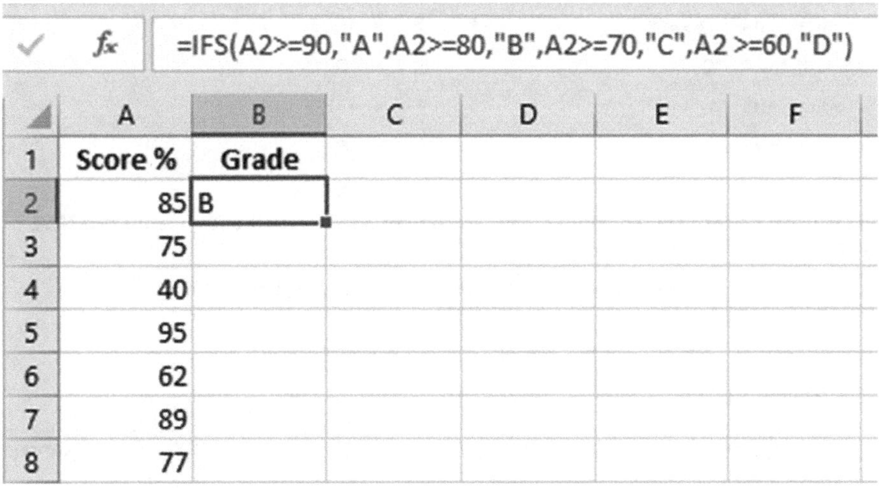

Figure 8-48. *Results for first score*

Double-clicking the AutoFill handle fills in the other grades. See Figure 8-49. The Not Applicable error appears for a score of 40. This score didn't meet any of the test conditions. This is because we didn't test for a condition where the score would be less than 60.

	A	B
1	Score %	Grade
2	85	B
3	75	C
4	40	#N/A
5	95	A
6	62	D
7	89	B
8	77	C

Figure 8-49. *Results after using AutoFill handle*

Clicking cell B2 and then clicking the **Insert Function** command reopens the function in the Function Arguments window. We can now enter the test for a score that is less than 60 and assign it a value of F. See Figure 8-50.

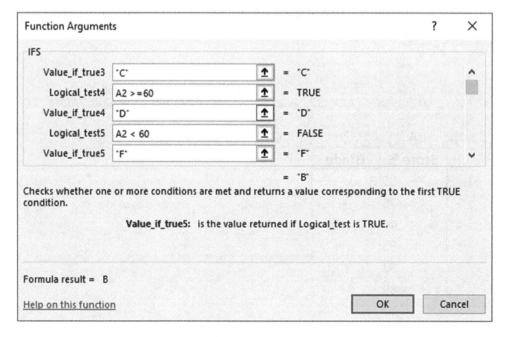

Figure 8-50. *Create Logical_test5 to account for a score of less than 60*

Now when we drag the AutoFill handle down from cell B2, we can see that the Score of 40 has been assigned a grade of F. See Figure 8-51.

	A	B
1	Score %	Grade
2	85	B
3	75	C
4	40	F
5	95	A
6	62	D
7	89	B
8	77	C

Figure 8-51. *Computed Grades using IFS function*

We could have achieved the same result by entering TRUE for the last test.

`=IFS(A2>=90,"A",A2>=80,"B",A2>=70,"C",A2 >=60,"D",`**`TRUE`**`,"F")`

If none of the test conditions have been met making the final test, True will make the value with this test the default value.

Let's do another IFS function. Our factory pays employees based on the amount of parts they produce. We want to reward or penalize employees based on the amount of scrap they produce. Figure 8-52 shows the number of items produced, the number scrapped, and the percent scrapped. We will add to or subtract from the number of items produced and place the result in column E.

	A	B	C	D	E
1	Employee #	Produced	Scrapped	Scrapped %	Production Credit
2	147	2500	25	1.0%	
3	229	2700	50	1.9%	
4	318	3800	15	0.4%	
5	432	3600	21	0.6%	
6	117	2700	30	1.1%	
7	503	3200	10	0.3%	

Figure 8-52. *Spreadsheet to be used for computing Production Credit*

Table 8-4 shows how much will be added to or subtracted from the employees production based on the amount of scrap they produced.

Table 8-4. *Amount to be added or subtracted from employees' production*

Scrap %	Change to Production
< = .5%	+ 300
< = 1%	+ 150
< = 1.5%	0
Above 1.5%	– 300

Because the IFS function uses the first True logical test, we don't have to worry that <= 1% would also include <= .5%. Once the condition for a logical test is met, the IFS function ignores everything after that.

Let's create a named range for our Scrapped percent to use in our function. Remember one way of doing this is to click cell D2, then double-click the AutoFill handle to select the other cells in the column, then enter Scrapped in the Name box, and press Enter.

Now we can start the formula in cell E2 by typing =**IFS(** and then clicking the Insert Function command on the Ribbons' Formula tab.

Figure 8-53 shows the first entries. To the right of Logical_test1, you can see that the third and sixth rows met the condition for the first test. At the bottom of the window, you can see that the result for cell E2 is going to be 2650.

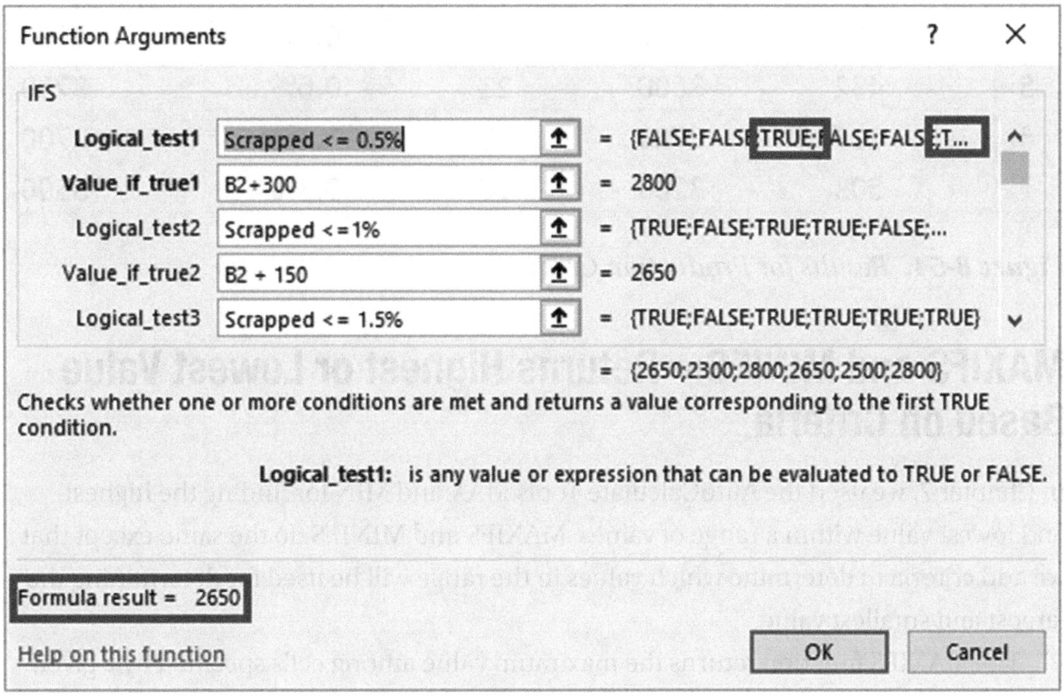

Function Arguments				?	×
IFS					
Logical_test1	Scrapped <= 0.5%	⬆	=	{FALSE;FALSE;TRUE;FALSE;FALSE;T...	∧
Value_if_true1	B2+300	⬆	=	2800	
Logical_test2	Scrapped <=1%	⬆	=	{TRUE;FALSE;TRUE;TRUE;FALSE;...	
Value_if_true2	B2 + 150	⬆	=	2650	
Logical_test3	Scrapped <= 1.5%	⬆	=	{TRUE;FALSE;TRUE;TRUE;TRUE;TRUE}	∨

= {2650;2300;2800;2650;2500;2800}

Checks whether one or more conditions are met and returns a value corresponding to the first TRUE condition.

Logical_test1: is any value or expression that can be evaluated to TRUE or FALSE.

Formula result = 2650

Help on this function   OK   Cancel

Figure 8-53. *Arguments for IFS function*

The complete function is

```
=IFS(Scrapped <= 0.5%,B2+300,Scrapped <=1%,B2 + 150,Scrapped <=
1.5%,B2,TRUE,B2 - 200)
```

The last test uses TRUE, making the default value B2 –200 for those values that didn't evaluate to true for any of the other logical tests. In this example, 1.9% would use the default value because it didn't evaluate to true for any of the logical tests.

The result is shown in Figure 8-54.

	A	B	C	D	E
1	Employee #	Produced	Scrapped	Scrapped %	Production Credit
2	147	2500	25	1.0%	2650
3	229	2700	50	1.9%	2500
4	318	3800	15	0.4%	4100
5	432	3600	21	0.6%	3750
6	117	2700	30	1.1%	2700
7	503	3200	10	0.3%	3500

Figure 8-54. *Results for Production Credit*

MAXIFS and MINIFS—Returns Highest or Lowest Value Based on Criteria

In Chapter 7, we used the AutoCalculate Tools MAX and MIN for finding the highest and lowest value within a range of values. MAXIFS and MINIFS do the same except that we add criteria to determine which values in the range will be used for determining the largest and smallest value.

The MAXIFS function returns the maximum value among cells specified by a given set of conditions or criteria.

The syntax for the MAXIFS function is

```
=MAXIFS(max_range, criteria_range1, criteria1, [criteria_range2, criteria2], ...)
```

Argument descriptions are

max_range is the cells that you are basing your criteria on. The range contains the values that will be checked by the criteria to determine if they will be used in the maximum value.

criteria_range1 is the first range in which to evaluate the associated criteria.

criteria1 is the condition defined by a number, expression, or text that will determine which cells will be evaluated as maximum.

Criteria_range2, Criteria2,... You can add as many as 126 additional ranges and their associated criteria.

We will create a MAXIFS and MINIFS function that only requires a single criterion. Exercise 8-4 gives you practice for using multiple criteria.

We want to find the highest and lowest sales amounts for three product categories: televisions Stereos and Appliances. See Figure 8-55.

	A	B	C	D	E	F	G
1	Product Category	Sales Rep	Sales Amount		Product Category	Highest Sale	Lowest Sale
2	Televisions	Sterk, Carrie	712.00		Televisions		
3	Stereos	Carson, Russell	535.00		Stereos		
4	Appliances	Barton, Rick	685.00		Appliances		
5	Appliances	Sterk, Carrie	684.00				
6	Stereos	Carson, Russell	478.00				
7	Televisions	Barton, Rick	741.00				
8	Stereos	Carson, Russell	427.00				
9	Televisions	Sterk, Carrie	430.00				
10	Televisions	Carson, Russell	430.00				
11	Appliances	Sterk, Carrie	610.00				
12	Stereos	Barton, Rick	643.00				
13	Televisions	Sterk, Carrie	759.00				
14	Televisions	Carson, Russell	546.00				
15	Appliances	Barton, Rick	701.00				
16	Stereos	Sterk, Carrie	722.00				
17	Appliances	Barton, Rick	570.00				
18	Televisions	Carson, Russell	513.00				
19	Appliances	Sterk, Carrie	544.00				

Figure 8-55. *Worksheet to be used to determine highest and lowest sales amounts*

We will start our function as shown in Figure 8-56 by entering =**MAXIFS(** in cell F2.

	E	F	G	H	I
	Product Category	**Highest Sale**	**Lowest Sale**		
	Televisions	=MAXIFS(			
	Stereos	MAXIFS(**max range**, criteria_range1, criteria1, ...)			
	Appliances				

Figure 8-56. *Starting the MAXIFS function*

The tooltip for the function displays. The first required entry is the max_range. This is the range of values from which we will get the highest value. Clicking cell C2 and then pressing Shift + Enter selects this range. We will be using the same range for the other two products as well as for the MINIFS function so we want to lock the range in. Press F4. Now we will add a comma to bring us to our next option criteria_range1. This is the range from which we will specify our criteria. Clicking cell A2 and then pressing Shift + Enter selects this range. Press F4 to make the ranges Absolute. See Figure 8-57.

	A	B	C	D	E	F	G	H	I
1	**Product Category**	**Sales Rep**	**Sales Amount**		**Product Category**	**Highest Sale**	**Lowest Sale**		
2	Televisions	Sterk, Carrie	712.00		Televisions	=MAXIFS($C$2:$C$19,$A$2:$A$19			
3	Stereos	Carson, Russell	535.00		Stereos	MAXIFS(max_range, **criteria range1**, crit			
4	Appliances	Barton, Rick	685.00		Appliances				
5	Appliances	Sterk, Carrie	684.00						
6	Stereos	Carson, Russell	478.00						
7	Televisions	Barton, Rick	741.00						
8	Stereos	Carson, Russell	427.00						
9	Televisions	Sterk, Carrie	430.00						
10	Televisions	Carson, Russell	430.00						
11	Appliances	Sterk, Carrie	610.00						
12	Stereos	Barton, Rick	643.00						
13	Televisions	Sterk, Carrie	759.00						
14	Televisions	Carson, Russell	546.00						
15	Appliances	Barton, Rick	701.00						
16	Stereos	Sterk, Carrie	722.00						
17	Appliances	Barton, Rick	570.00						
18	Televisions	Carson, Russell	513.00						
19	Appliances	Sterk, Carrie	544.00						

Figure 8-57. *Make the max-range and criteria-range1 Absolute Cell References*

Now we can enter the comma and then click cell E2 because this is our criteria. We place a dollar sign in front of the E but not in front of the 2 (mixed cell reference), so that when we copy the formula down, the column will remain the same but the row will change. See Figure 8-58.

E	F	G	H	I	J	K	L
Product Category	**Highest Sale**	**Lowest Sale**					
Televisions	=MAXIFS($C$2:$C$19,$A$2:$A$19,$E2						
Stereos	MAXIFS(max_range, criteria_range1, **criteria1**, [criteria_range2, criteria2], ...)						
Appliances							

Figure 8-58. *Make criteria1 a Mixed Cell Reference*

Press Ctrl + Enter and then double-click the AutoFill handle to get the results for the three categories. See Figure 8-59.

E	F	G
Product Category	**Highest Sale**	**Lowest Sale**
Televisions	712	
Stereos	722	
Appliances	701	

Figure 8-59. *Results for MAXIFS*

The formula in cell F3 is

`=MAXIFS($C$2:$C$19,$A$2:$A$19,$E3)`

Column E stayed the same, but the row adjusted to 3 so for that row our criteria is Stereos.

Now we can use a copy of our MAXIFS function to create our MINIFS function. We will copy the function in cell F2 and then paste it into cell G2. In the formula bar will change MAXIFS to MINIFS

=MINIFS($C$2:$C$19,$A$2:$A$19,$E2)

Press Ctrl + Enter and then double-click the AutoFill handle to get the results for the three categories. As you can see in Figure 8-60, row 13 has the highest sale for televisions and row 9 has the lowest sale for televisions.

◢	A	B	C	D	E	F	G
1	Product Category	Sales Rep	Sales Amount		Product Category	Highest Sale	Lowest Sale
2	Televisions	Sterk, Carrie	712.00		Televisions	759	430
3	Stereos	Carson, Russell	535.00		Stereos	722	427
4	Appliances	Barton, Rick	685.00		Appliances	701	544
5	Appliances	Sterk, Carrie	684.00				
6	Stereos	Carson, Russell	478.00				
7	Televisions	Barton, Rick	741.00				
8	Stereos	Carson, Russell	427.00				
9	Televisions	Sterk, Carrie	430.00				
10	Televisions	Carson, Russell	430.00				
11	Appliances	Sterk, Carrie	610.00				
12	Stereos	Barton, Rick	643.00				
13	Televisions	Sterk, Carrie	759.00				
14	Televisions	Carson, Russell	546.00				
15	Appliances	Barton, Rick	701.00				
16	Stereos	Sterk, Carrie	722.00				
17	Appliances	Barton, Rick	570.00				
18	Televisions	Carson, Russell	513.00				
19	Appliances	Sterk, Carrie	544.00				

Figure 8-60. *Results for MAXIFS and MINIFS*

EXERCISE 8-4: USING THE MAXIFS AND MINIFS FUNCTIONS

In this exercise, you will use two criteria for the MAXIFS and MINIFS functions. The first criteria will be a Sales Rep, and the second criteria will be a Product Category. Open workbook Chapter8.xlsx. See Figure 8-61.

	A	B	C	D	E	F	G	H
1	Product Category	Sales Rep	Sales Amount		Sales Rep	Product Category	Highest Sale	Lowest Sale
2	Televisions	Sterk, Carrie	712.00		Carson, Russell	Televisions		
3	Stereos	Carson, Russell	535.00			Stereos		
4	Appliances	Barton, Rick	685.00			Appliances		
5	Appliances	Sterk, Carrie	684.00					
6	Stereos	Carson, Russell	478.00					
7	Televisions	Barton, Rick	741.00					
8	Stereos	Carson, Russell	427.00					
9	Televisions	Sterk, Carrie	430.00					
10	Televisions	Carson, Russell	430.00					
11	Appliances	Sterk, Carrie	610.00					
12	Stereos	Barton, Rick	643.00					
13	Televisions	Sterk, Carrie	759.00					
14	Televisions	Carson, Russell	546.00					
15	Appliances	Barton, Rick	701.00					
16	Stereos	Sterk, Carrie	722.00					
17	Appliances	Barton, Rick	570.00					
18	Televisions	Carson, Russell	513.00					
19	Appliances	Sterk, Carrie	544.00					

Figure 8-61. *Spreadsheet for using MAXIFS and MINIFS functions*

1. Enter **=MAXIFS(** in cell G2.

2. Click cell C2. Hold down the Ctrl + Shift key and click the down arrow key. Lock in this range by clicking the F4 key.

3. Enter a comma. The First Criteria range is the Sales Rep values. Click cell B2. Hold down the Ctrl + Shift key and click the down arrow key. Lock in this range by clicking the F4 key.

4. Enter a comma. The criteria is the Sales Rep Russell Carson. Click cell E2. Press F4 to lock in this cell. See Figure 8-62.

	A	B	C	D	E	F	G	H	I	J
1	Product Category	Sales Rep	Sales Amount		Sales Rep	Product Category	Highest Sale	Lowest Sale		
2	Televisions	Sterk, Carrie	712.00		Carson, Russell	Televisions	=MAXIFS($C$2:$C$19,$B$2:$B$19,$E$2			
3	Stereos	Carson, Russell	535.00			Stereos	MAXIFS(max_range, criteria_range1, **criteria1**, [cri			
4	Appliances	Barton, Rick	685.00			Appliances				
5	Appliances	Sterk, Carrie	684.00							
6	Stereos	Carson, Russell	478.00							
7	Televisions	Barton, Rick	741.00							
8	Stereos	Carson, Russell	427.00							
9	Televisions	Sterk, Carrie	430.00							
10	Televisions	Carson, Russell	430.00							
11	Appliances	Sterk, Carrie	610.00							
12	Stereos	Barton, Rick	643.00							
13	Televisions	Sterk, Carrie	759.00							
14	Televisions	Carson, Russell	546.00							
15	Appliances	Barton, Rick	701.00							
16	Stereos	Sterk, Carrie	722.00							
17	Appliances	Barton, Rick	570.00							
18	Televisions	Carson, Russell	513.00							
19	Appliances	Sterk, Carrie	544.00							

Figure 8-62. *Criteria1 for Russell Carson using Absolute Cell Reference*

5.   Enter a comma. You are now ready to enter the second criteria. Click cell A2. Hold down the Ctrl + Shift key and click the down arrow key. Lock in this range by clicking the F4 key.

6.   Enter a comma. You want to find the highest Televisions sale for Russell Carson. Click cell F2. You need to lock in the column, not the row, so place a dollar sign in front of the F.

The formula is now

```
=MAXIFS($C$2:$C$19,$B$2:$B$19,$E$2,$A$2:$A$19,$F2
```

	A	B	C	D	E	F	G	H	I	J	K	L
1	Product Category	Sales Rep	Sales Amount		Sales Rep	Product Category	Highest Sale	Lowest Sale				
2	Televisions	Sterk, Carrie	712.00		Carson, Russell	Televisions	=MAXIFS($C$2:$C$19,$B$2:$B$19,$E$2,$A$2:$A$19,$F2					
3	Stereos	Carson, Russell	535.00			Stereos	MAXIFS(max_range, criteria_range1, criteria1, [criteria_range2, **criteria2]** [					
4	Appliances	Barton, Rick	685.00			Appliances						
5	Appliances	Sterk, Carrie	684.00									
6	Stereos	Carson, Russell	478.00									
7	Televisions	Barton, Rick	741.00									
8	Stereos	Carson, Russell	427.00									
9	Televisions	Sterk, Carrie	430.00									
10	Televisions	Carson, Russell	430.00									
11	Appliances	Sterk, Carrie	610.00									
12	Stereos	Barton, Rick	643.00									
13	Televisions	Sterk, Carrie	759.00									
14	Televisions	Carson, Russell	546.00									
15	Appliances	Barton, Rick	701.00									
16	Stereos	Sterk, Carrie	722.00									
17	Appliances	Barton, Rick	570.00									
18	Televisions	Carson, Russell	513.00									
19	Appliances	Sterk, Carrie	544.00									

Figure 8-63. *Criteria2 for Product Category using a Mixed Cell Reference*

7. Press Ctrl + Enter. Double-click the AutoFill handle to copy the formula down to Stereos and Appliances. The Appliances highest sale is zero because Russell Carson didn't sell any Appliances. See Figure 8-64.

E	F	G	H
Sales Rep	Product Category	Highest Sale	Lowest Sale
Carson, Russell	Televisions	546	
	Stereos	535	
	Appliances	0	

Figure 8-64. *Results for MAXIFS function*

8. Click cell G2. Press Ctrl + C to copy the cell. Click cell H2. Press Ctrl + V to paste.

9. Double-click cell H2 to display the formula. Change MAXIFS to MINIFS. Press Ctrl + Enter. Double-click the AutoFill handle to copy down the function. See Figure 8-65.

	A	B	C	D	E	F	G	H
1	Product Category	Sales Rep	Sales Amount		Sales Rep	Product Category	Highest Sale	Lowest Sale
2	Televisions	Sterk, Carrie	712.00		Carson, Russell	Televisions	546	430
3	Stereos	Carson, Russell	535.00			Stereos	535	427
4	Appliances	Barton, Rick	685.00			Appliances	0	0
5	Appliances	Sterk, Carrie	684.00					
6	Stereos	Carson, Russell	478.00					
7	Televisions	Barton, Rick	741.00					
8	Stereos	Carson, Russell	427.00					
9	Televisions	Sterk, Carrie	430.00					
10	Televisions	Carson, Russell	430.00					
11	Appliances	Sterk, Carrie	610.00					
12	Stereos	Barton, Rick	643.00					
13	Televisions	Sterk, Carrie	759.00					
14	Televisions	Carson, Russell	546.00					
15	Appliances	Barton, Rick	701.00					
16	Stereos	Sterk, Carrie	722.00					
17	Appliances	Barton, Rick	570.00					
18	Televisions	Carson, Russell	513.00					
19	Appliances	Sterk, Carrie	544.00					

Figure 8-65. *Results for MAXIFS and MINIFS functions*

You have learned how to use the IFS function to select which values are to be entered in a cell based on the criteria you specify. You have also learned how to find the highest and lowest value within a range of values based on criteria you specify.

Next, you will look at functions that deal with dates. You'll see how you can get the current date and time, find the number of days between two dates, and manipulate dates by adding or deleting from them.

Date Functions

Excel allows you to use dates and times in formulas. Excel can do this because it actually stores dates and times as numbers. Excel stores dates as serial numbers. A serial number treats January 1, 1900 as the first day of the calendar. It is considered as day 1. January 2, 1900 is considered day 2, January 3, 1900 is considered day 3, and so on. Because Excel treats January 1, 1900 as the first day, you can't use any dates earlier than this. If you do, Excel will treat the date as text.

You can see how Excel treats dates and times by formatting them as General.

In the example in Figure 8-66, 1/1/1900 was entered in both cells A1 and B1. The date 3/2/1010 was entered in both cells A2 and B2. Cells B1 and B2 were formatted as General.

◢	A	B
1	1/1/1900	1
2	3/2/2010	40239

Figure 8-66. *Column B shows what the date in column A looks like when formatted as General*

Cell B2 displays 40239, which is the number of days between 1/1/1900 and 3/2/2010.

Excel stores times as a value between 0 and 1. The time 12:00 AM would be stored as a value of 0. The time 11:59:59 PM would be stored as a value of .9999.

TODAY Function—Returns the Current Date

The TODAY() function returns the current date, which it gets from your computer's internal clock. The TODAY() function doesn't use any arguments. Even though the TODAY function doesn't use any arguments, you must still enter parentheses after the function name so that Excel recognizes it as a function.

The syntax for the TODAY function is

=TODAY()

See Table 8-5 for some examples.

Table 8-5. *Description and Results for Various TODAY Function Formulas*

Formula	Description	Result If Today Was 10/11/2015
=TODAY()	Returns the current date.	10/11/2015
=TODAY() + 10	Returns the date ten days into the future.	10/21/2015
=TODAY() – 5	Returns the date five days ago.	10/06/2015

The TODAY() function is recalculated every time you open the worksheet or refresh it. In other words, if you enter the formula =TODAY() in a worksheet on May 01, 2016 and then reopen that same worksheet on June 30, 2016, the date will be updated to 06/30/2016. You can prevent this dynamic updating by making the date static.

To make the date returned from the TODAY() function static

1. Double-click the cell that contains the Today() function. This places the cell in Edit mode.

2. Press the refresh key (F9). This changes the date to a serial date. See Figure 8-67.

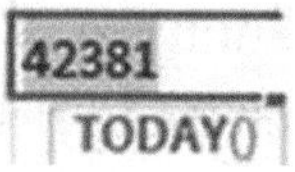

Figure 8-67.   *The date displayed as a serial date*

3. Press Enter to commit the entry. The serial number then reverts back to a date format. The date will then always show as the date you made the entry.

EXERCISE 8-5: USING TODAY FUNCTION WITH INT FUNCTION

In this exercise, you will use the TODAY function and a person's birthdate to determine his or her age.

1. Enter the data in Figure 8-68 in a new worksheet. Name the worksheet "Age." We want to determine the age for those whose birthdate appears in column B. Add your name and birthdate to the list.

	A	B	C
1	Name	Birthdate	Age
2	Carrie	9/2/1927	
3	Annette	9/7/1956	
4	Rhianna	8/18/1996	
5	Marten	9/7/2000	

Figure 8-68.   *Enter this data in a worksheet named Age*

2. Click inside cell C2. The formula for determining the age is today's date minus the birthdate divided by 365.25. We divide by 365.25 to accommodate leap year which happens every four years.

3. Type =(TODAY()-B2)/365.25.

    We place a pair of parentheses around TODAY()-B2 because we need to perform this calculation before doing the division. If we didn't place it in parentheses, Excel would first divide B2 by 365.25.

4. Press Enter. The result is a decimal amount. We usually display someone's age as an integer.

    We will use the INT function to remove the decimal portion of the age. The INT function rounds a value down to its nearest whole number.

5. Double-click cell C2.

6. Place the cursor after the equal sign in the Formula bar. Type **INT(**.

7. Press the End key on the keyboard to place the cursor at the end of the function.

8. Type the closing right parenthesis. Your formula should now be =INT((TODAY()-B2)/365.25).

9. Press Ctrl + Enter.

10. Drag the AutoFill Handle down through cell C5. If this was run in February 2016, the results would be as shown in Figure 8-69.

	A	B	C
1	Name	Birthdate	Age
2	Carrie	9/2/1927	88
3	Annette	9/7/1956	59
4	Rhianna	8/18/1996	19
5	Marten	9/7/2000	15

Figure 8-69. *Result of using TODAY function with INT function*

NOW Function—Returns the Current Date and Time

The NOW function is similar to the TODAY function, except that the NOW function includes the time along with the current date. Another difference is that the NOW function is not dynamic. It does not update to the current date when you reopen the worksheet.

The syntax for the NOW function is

=NOW()

Table 8-6 shows an example of the NOW function applied on October 20, 2015 at 3:18 PM.

Table 8-6. *NOW Applied*

Formula	Displays Current Date and Time
=NOW()	10/20/2015 15:18

Notice that the NOW function displays time in a 24-hour format. You can change the Time format so that it will display in a 12-hour format using AM or PM. Changing the time to a 12-hour format requires a custom format. This date and time 6/8/2016 16:16 is the result of entering =NOW() in a cell. To change to a 12-hour format that displays AM or PM, you would right-click the cell and select Format. This would bring up the Format Cells dialog box. See Figure 8-70. Make sure that Custom is selected for the Category and m/d/yyyy h:mm is selected for the Type.

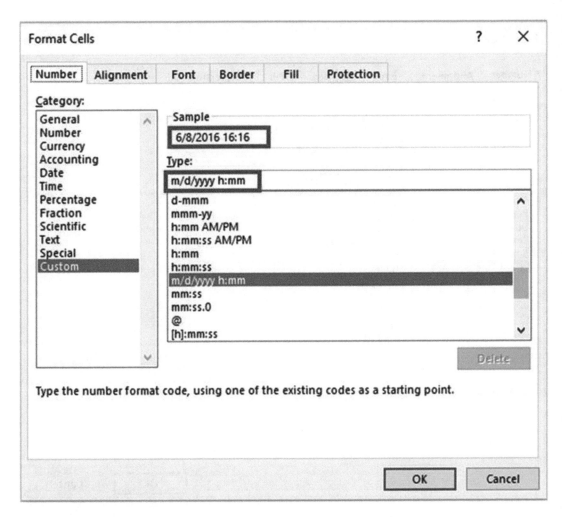

Figure 8-70. *Select the Custom category and the date and time type*

You would then add AM/PM to the end of the type. See Figure 8-71. The result is shown in the Sample area. You would then click the OK button and the cell would display the same value as what appears in the Sample area.

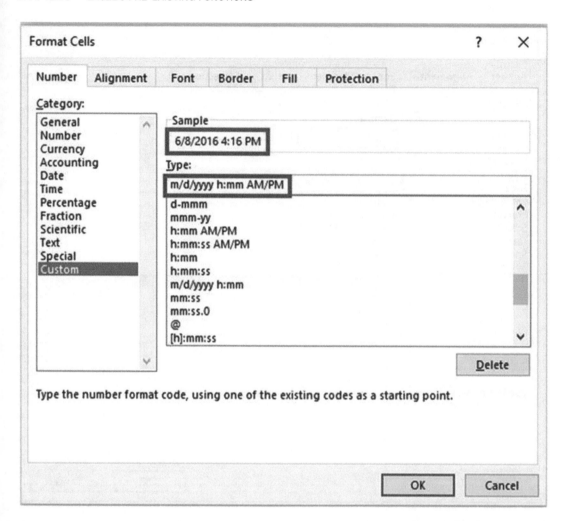

Figure 8-71. *Adding AM/PM changes the time from a 24-hour format to a 12-hour format that shows AM or PM*

If you just wanted to display the time without the date, you could use the following formula:

```
=NOW() - TODAY()
```

DATE Function—Returns the Serial Number of the DATE

The DATE function returns the serial number of the date provided in the DATE arguments.

The syntax for the DATE function is

```
=DATE(year, month, day)
```

404

All of the arguments in the DATE function are required.

When you create a new workbook, all of the cells are formatted as General by default. If you enter the formula =DATE(2015, 10, 20), Excel will automatically change the formatting of the cell to the date type and display 10/20/2015. The date is stored as a serial number so that it can be used in formulas. If you format the cell as something other than a date or time, Excel will display the date as a serial number. Remember, the serial number is the number of days that have passed since January 1, 1900.

Note   You should always enter a four-digit year when using the DATE function. If you enter a one- or two-digit year, Excel will treat these as from the 1900s. For example, if you enter =DATE(09,05,14) or =DATE(9,05,14), Excel will assume the year is 1909 not 2009.

The DATE function is most useful in situations in which you are getting the year, month, and day from other cells or when you want to use the date as part of a formula. You can't enter a date directly in a formula. If you enter a formula as =09/07/2015– 09/05/2015, Excel treats the forward slashes as divide signs. If you want to use a date in a formula, you should use the DATE function. If you wanted to know the number of days between two dates, you could enter a formula such as =DATE(2015,4,25) - DATE(2015,2,20), or let's say that you have the formula =DATE(2015, 4, 25) in cell A2 and the formula =DATE(2015,2,20) in cell B2. Entering the formula =A2-B2 will return a result of 64. The result is the number of days between the two dates.

Note   Excel returns a number rather than a date when you enter the DATE function as a part of a formula such as =Date(2015,5,13) – Date(2015,2,15) because it formats it as the General type. If you just enter =Date(2011,5,13) in a cell, Excel will display it as a date because Excel formats it as a Date type.

MONTH, DAY, and YEAR Functions

You can retrieve just the month, day, or year portion from a date by using the MONTH, DAY, and YEAR functions.

The syntax for the MONTH, DAY, and YEAR functions are

```
=MONTH(Serial Date)
=DAY(Serial Date)
=YEAR(Serial Date)
```

Retrieving the month, day, or year from a date is done by using the corresponding function along with a cell address of a date or a date itself for the argument.

Table 8-7 shows the results of using the DAY, YEAR, and MONTH functions.

Table 8-7. *Retrieving Date Values with the DAY, YEAR, and MONTH Functions*

Value in Cell A1	Formula	Result
05/15/2010	=MONTH(A1)	5
	=DAY(A1)	15
	=YEAR(A1)	2010

You can grab the current month, day, or year from the TODAY or NOW functions as shown in Table 8-8.

Table 8-8. *Returning Current Date Values with the TODAY Function*

Formula	Description	Result If Today Was 11/16/2015
=MONTH(TODAY())	Returns the current month.	11
=DAY(TODAY())	Returns the current day.	16
=YEAR(TODAY())	Returns the current year.	2015

Note   You can't use a date directly in the formula such as =MONTH(03/15/2010) because the date hasn't been stored as a serial date and the MONTH, DAY, and YEAR functions can only use dates stored as serial dates.

Values can be added or subtracted from the MONTH, DAY, and YEAR functions. Let's say that our business tries to fill its orders within three months of the order date. We can do this by adding 3 to the month in the order date. The day and year for the required date will be the same as those from the order date. We will use the data in Figure 8-72.

	A	B	C
1	Order ID	Order Date	Required Date
2	125	9/10/2012	
3	126	9/15/2012	
4	127	10/12/2012	
5	128	11/1/2012	

Figure 8-72. Data used for this example

The function that follows will be entered in cell C2 and then copied down through cell C5:

```
=DATE(YEAR(B2),MONTH(B2)+3,DAY(B2))
```

Figure 8-73 shows the result. The month in cell B4 is 10. Adding 3 to 10 should give us a month of 13, but Excel is smart enough to know there isn't any month 13 and automatically adjusts the date so that the month becomes January and the year becomes 2013. Adding 3 to the month in cell B5 automatically adjusted the month to February.

	A	B	C
1	Order ID	Order Date	Required Date
2	125	9/10/2012	12/10/2012
3	126	9/15/2012	12/15/2012
4	127	10/12/2012	1/12/2013
5	128	11/1/2012	2/1/2013

Figure 8-73. Excel automatically adjusted dates that went past the year

Excel will also automatically adjust the date if there ends up being more days than there are in a month.

DAYS—Returns the Number of Days Between Two Dates

The DAYS function is used to find the number of days between two dates.

The syntax for the DAYS function is

DAYS(end_date,start_date)

Argument descriptions are

end_date and start_date are the two dates between which you
want to know the number of days.

If the date 1/5/1994 was in cell A1 and you wanted to know how many days were
between that date and today's date, you would use today's date as the end date and
1/5/1994 as the start date. You would enter the following formula in a cell:

=DAYS(TODAY(),A1)

EXERCISE 8-6: USE THE DAYS FUNCTION

In this exercise, we will use the DAYS function to determine how many days fall between a
series of start and end dates.

1. Create a new worksheet named "DAYS."

2. Enter the data as shown in Figure 8-74:

    a. Enter and format the column heads.

    b. Enter the date in cell A2 and then hold down the Ctrl key while you use the
       AutoFill Handle to copy the date down through cell A12. Remember holding
       down the Ctrl Key copies the cell rather than creating a series.

    c. Enter the date in cell B2 and drag the AutoFill Handle from cell B2 to B5.

    d. Enter the date in cell B6 and drag the AutoFill Handle from cell B6 to B10.
       See Figure 8-74.

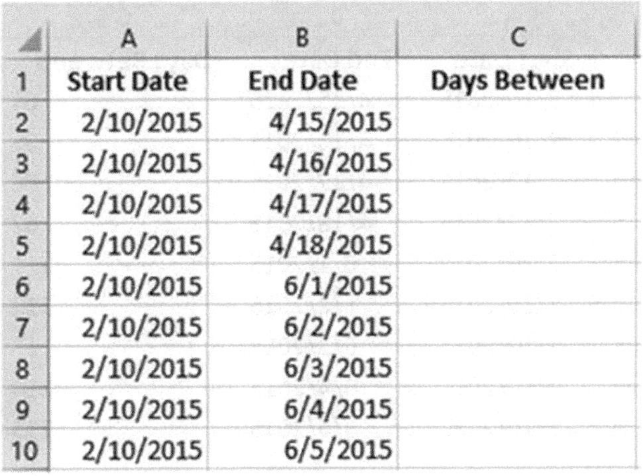

	A	B	C
1	**Start Date**	**End Date**	**Days Between**
2	2/10/2015	4/15/2015	
3	2/10/2015	4/16/2015	
4	2/10/2015	4/17/2015	
5	2/10/2015	4/18/2015	
6	2/10/2015	6/1/2015	
7	2/10/2015	6/2/2015	
8	2/10/2015	6/3/2015	
9	2/10/2015	6/4/2015	
10	2/10/2015	6/5/2015	

Figure 8-74. *Date entered using AutoFill Handle*

3. In cell C2, enter =DAYS(.

4. Select cell B2 for the end_date argument. Enter a comma.

5. Select cell A2 for the start_date argument. See Figure 8-75.

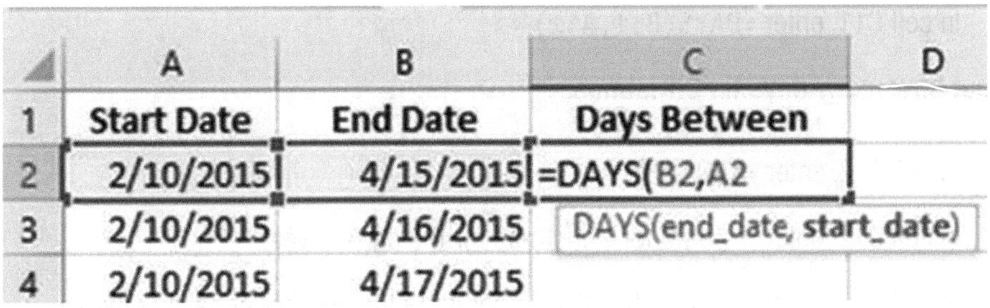

	A	B	C	D
1	**Start Date**	**End Date**	**Days Between**	
2	2/10/2015	4/15/2015	=DAYS(B2,A2	
3	2/10/2015	4/16/2015	DAYS(end_date, **start_date**)	
4	2/10/2015	4/17/2015		

Figure 8-75. *Cell A2 selected*

6. Press Ctrl + Enter. Use the AutoFill Handle to copy the formula down through cell C10. Figure 8-76 shows the result.

	A	B	C
1	Start Date	End Date	Days Between
2	2/10/2015	4/15/2015	64
3	2/10/2015	4/16/2015	65
4	2/10/2015	4/17/2015	66
5	2/10/2015	4/18/2015	67
6	2/10/2015	6/1/2015	111
7	2/10/2015	6/2/2015	112
8	2/10/2015	6/3/2015	113
9	2/10/2015	6/4/2015	114
10	2/10/2015	6/5/2015	115

Figure 8-76. *Result of using DAYS function in column C*

Now that you have experience using the DAYS function, try the following exercises:

Find out how many days you have been alive.

1. In cell A11, enter your birthday.

2. In cell B11, enter =TODAY().

3. In cell C11, enter =DAYS(B11,A11).

Find out how many days till Christmas.

1. In cell A12, enter =TODAY().

2. In cell B12, enter the date for Christmas for the current year.

3. In cell C12, enter the DAYS function to compute the number of days between the current date and Christmas.

You have learned how to handle dates and extract just the portion of a date that you want. Don't forget that you can't use a date directly in a math formula; rather, you must use the cell address of the date in the formula.

Summary

This chapter dealt with functions. You have learned the two ways of entering a formula: either directly into a cell or by using the Insert Function option. The Insert Function dialog box helps you find the function you want to use. Another benefit of the Insert Function dialog box is that it shows the result for each argument that you enter. You might want to use Insert Function until you become very familiar with the function and then switch to the quicker method of entering functions directly in the cell. You have seen how functions can determine what needs to be done based on if a condition exists or not.

The next chapter deals with handling errors and protecting the workbook from someone inadvertently or purposefully entering incorrect data.

CHAPTER 9

Auditing, Validating, and Protecting Your Data

As the creator of an Excel workbook, it is your primary responsibility to ensure that the data in the worksheets have been entered correctly. You may not have control of the source of the data, but it is your responsibility to ensure that you enter the data you are given without error. Poor data and incorrect entries can ruin a business. Errors made in one location may proliferate to other areas in the workbook. Users may be incorporating the data from your workbooks into their own projects, or they may be using your workbooks to base their decisions on. Excel provides several tools to help you prevent errors from getting into your workbooks, but it is up to you to incorporate them.

After reading and working through this chapter, you should be able to

- Identify errors

- Use data validation to prevent errors

- Circle invalid data and clear validation circles

- Check a formula for an error using the `IFERROR` function

- Correct circular references

- Audit formulas by

    - Tracing dependents

    - Tracing precedents

    - Using the Watch Window

    - Using the **Evaluate Formula** feature

© David Slager and Annette Slager 2020
D. Slager and A. Slager, *Essential Excel 2019*, https://doi.org/10.1007/978-1-4842-6209-2_9

- Have Excel read back your cell entries

- Use Spell Checker and Thesaurus

- Protect worksheets and cells

Validating Your Data and Preventing Errors

Excel has many tools to help you prevent errors from being entered into your applications, but it is still up to you to review all your entries and to check that the results of your calculations and functions look credible.

Data Validation

One of the most important functions an Excel developer can perform is to make sure that errors do not get into the spreadsheets. You need to make it as easy as possible for the user to understand what is to be entered and as easy as possible to enter the data. You don't want the user to have to guess what it is he is supposed to enter and then make a wrong assumption. One way to make sure data is entered correctly is by using the **Data Validation** feature. Data validation consists of various validation rules that specify the criteria for how the data is to be entered. If the data is still entered incorrectly, you can stop the data from being accepted. If the user has entered something questionable, you can provide her with a warning or information message and then let her decide if the data should be accepted or not.

Data being entered can be restricted to the following:

- Values in a drop-down list

- A whole number within a specified range

- A decimal number within a specified range

- A date within a specified time frame

- A time within a specified time frame

- A specified length

- A value based on the content of another cell

- A value based on a formula

The Data Validation button is located on the Ribbon's Data tab in the Data Tools group (Figure 9-1).

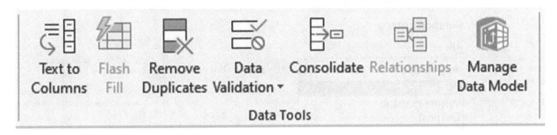

Figure 9-1. *The Ribbon's Data Tools group*

The Data Validation button is divided into two sections. Clicking the upper half of the Data Validation button brings up the Data Validation dialog box.

The Data Validation dialog box contains three tabs (see Figure 9-2):

- Settings tab—Used for setting the rules for what can be entered in a cell.

- Input Message tab—Creates a screen tip that tells the user what he should enter in the cell.

- Error Alert tab—Can be used for warning that the entry does not follow the rules you have set up, or it can be used to stop the entry if rules have been broken.

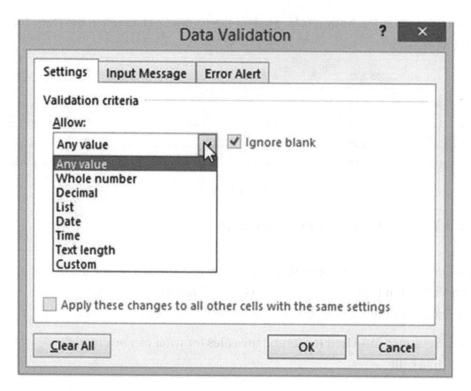

Figure 9-2. *Data Validation dialog box*

Clicking the lower half of the Data Validation button brings up the three options shown in Figure 9-3.

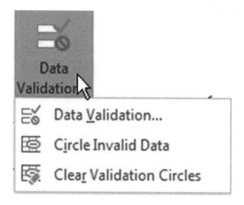

Figure 9-3. *Data Validation options*

Clicking the first option (Data Validation) gives the same result as clicking the upper half of the button.

EXERCISE 9-1: CREATING DATA VALIDATION

For this exercise, you will create data validation for six fields to warn users about potential problems or to prevent them from making invalid entries.

All of the entries you will need to make for this practice are in workbook Chapter 9 on the Validation tab. The worksheet shows what you will need to enter for each option on the three tabs of the Data Validation dialog box. You can print the worksheet and then use it as you go through this exercise, or you can follow along with the instruction given for each column.

1. Open workbook Chapter 9. Click the Val_Entry worksheet tab. Figure 9-4 shows the column headings from worksheet Val_Entry. The entries that you will need to make for this exercise are to the right of the column headings. There is a second copy on the Validation worksheet.

	A	B	C	D	E	F
1	Order ID	Credit Limit	Order Time	Shipping Method	Shipping Weight	Ship Zip

Figure 9-4. *Column headings for Chapter 9 workbook*

The Order ID must start with either an SR or a GR. We will prevent the user from entering any other text for the first two characters. We will perform a Validation check for the Order ID whenever the user enters a value in any cell in the range A2:A12.

2. Select the cell range A2:A12.

3. Click the Data tab. In the Data Tools group, click the Data Validation button.

    The Data Validation window appears. See Figure 9-5. The worksheet starting in column H contains all the entries you will be making on the Data Validation window.

4. Click the Settings tab if it isn't already selected.

5. Click the down arrow for Allow and select **Custom**. The Custom option is for entering formulas.

6. Because we want to check the first two characters in the Order ID, we will use the LEFT function.

The syntax for the LEFT function is

= LEFT(text, [num_chars])

Arguments are

text is either text or a cell address containing the text.

num_chars is the number of characters you want from the beginning of the string.

Enter the formula =OR(LEFT(A2,2) = "SR",LEFT(A2,2) = "GR") in the Formula box.

The formula uses the OR function. The formula tests if the first two characters entered in cell A2 are SR or GR, then the result is True; otherwise, it returns a False. If either argument is true, then the result is true and Excel will accept it. We are using cell A2, which is the first cell in the range. Excel will automatically adjust the formula for each cell in the range.

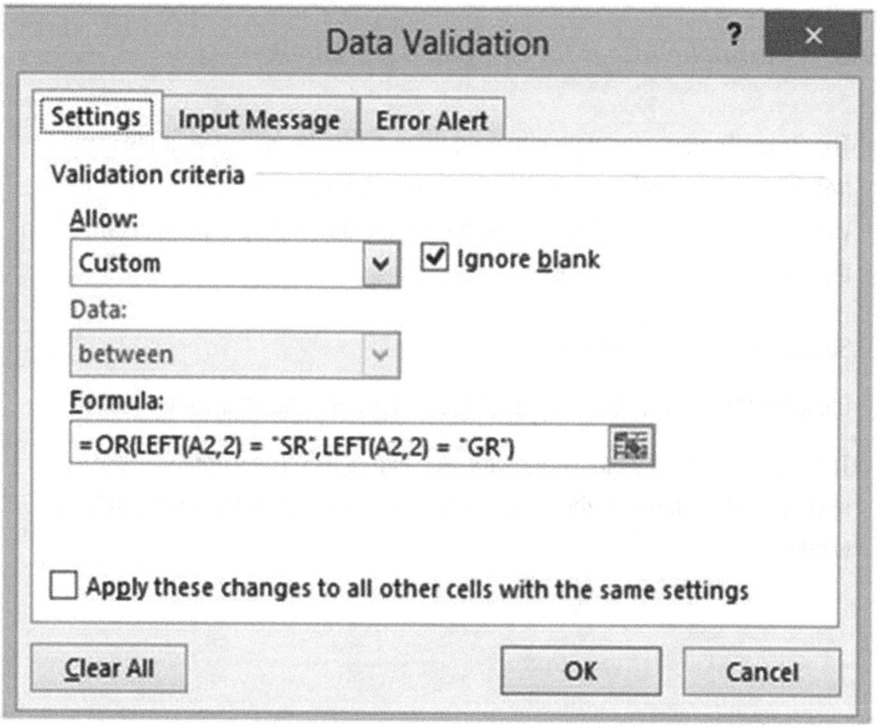

Figure 9-5. *Formula ensures that Order ID starts with an SR or GR*

7. Click the Input Message tab. If you look at column H under Order ID, you can see that on the Input Message tab you need to enter a Title of "Order ID Restrictions" and an Input message of "Order ID's must start with the letters SR or GR." Make those entries now. Your Input Message tab should look like Figure 9-6.

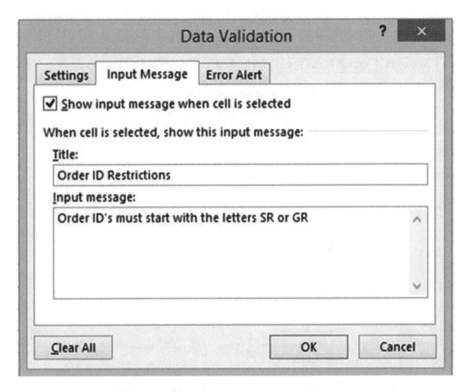

Figure 9-6. *Message appears when an Order ID cell is selected*

The input message on the Input Message tab informs the user about any entry restrictions, but if the user breaks the rule entered on the Settings tab, we want to display an error message.

8. Click the Error Alert tab.

9. Select Stop for the style. The style of **Stop** prevents an invalid entry from being accepted.

10. Enter the title "Order ID Error."

11.   Enter the error message "Order IDs can only start with the letters SR or GR."
      See Figure 9-7.

12.   Click the OK button.

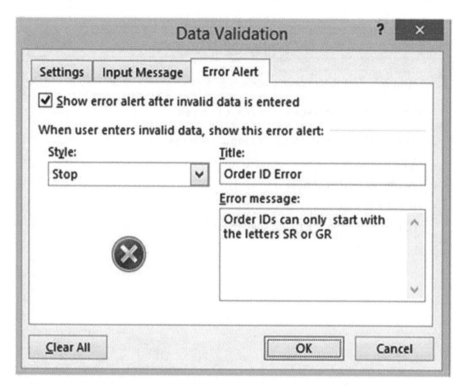

Figure 9-7.   *Create error message for making an invalid entry*

13.   Click inside cell A2. You should see the message you created on the Input
      Message tab (Figure 9-8).

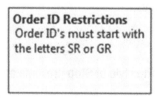

Figure 9-8.   *Message informing you what needs to be entered in the cell*

14.   Enter **KY14** in the cell and then press the Tab key. You should see the error
      message you created on the Error Alert tab. See Figure 9-9.

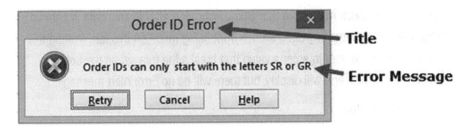

Figure 9-9. *Error message if user makes invalid entry*

15. Click the Retry button. Enter SR14 in the cell. Press the Tab key.

Creating an Error Alert Using No Style

For this Validation, we will create a message informing the user of what the entry should normally be, but we will not show an Error Alert message or stop the user from making an entry that doesn't meet the entry suggestion of entering a value of less than 30000.

1. Select the cell range B2:B12. Click the Data tab. In the Data Tools group, click the Data Validation button. Click the Settings tab.

2. Enter the data in Table 9-1 for the three tabs.

Table 9-1. *Enter the Data in This Table for the Three Tabs*

Tab	Option	Value Selected or Entered
Settings		
	Allow	Whole number
	Data	less than
	Maximum	30000
Input Message		
	Title	Credit Limit Restrictions
	Message	Credit limits are usually less than 30000
Error Alert		
		Remove the check mark from the "Show error alert after invalid data is entered" check box.

Removing the check mark from the "Show error alert after invalid data is entered" (Figure 9-10) prevents any other entries from being made on this tab. Making this change allows the user to enter a value that breaks the rule set on the Settings tab. The Input Message will display, but there will be no Error Alert message.

3. Click the OK button.

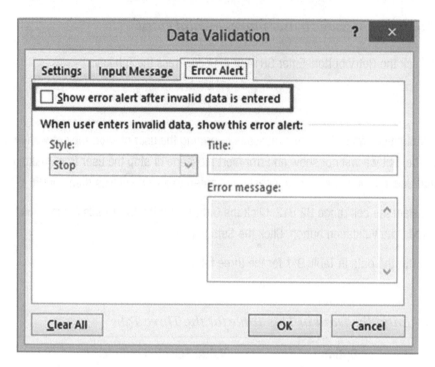

Figure 9-10. *Remove the check mark from "Show error alert after invalid data is entered"*

4. Double-click cell B2. The input message displays.

5. Enter 90000 and press Tab. The entry is accepted and no error message appears.

Creating a Warning Style and Using a Time Restriction

For this Validation, you will create a Warning for the Order Time column for entries that are not between 8:00 AM and 5:00 PM.

1. Select the cell range C2:C12. Click the Data Validation button. Click the Settings tab.

2. Enter the data in Table 9-2.

Table 9-2. *Enter the Data in This Table*

Tab	Option	Value Selected or Entered
Settings		
	Allow	Time
	Data	Between
	Start time	8:00 AM
	End time	5:00 PM
Input Message		
	Title	Order Time Restrictions
	Message	The time entered should be between 8 AM and 5:00 PM
Error Alert		
	Style	Warning
	Title	Possible Error
	Message	The time should be between 8:00 AM and 5:00 PM unless there is a specific reason for this.

3. The **Warning** style warns you that an error may have been made and lets you decide if you want to accept the entry or not. Click the OK button.

4. Enter 6:00 PM in cell C2 and then press the Tab key. The error message in Figure 9-11 appears.

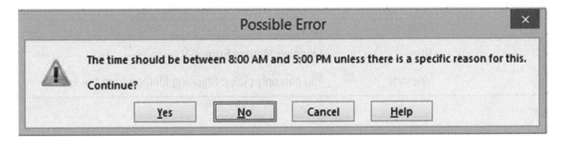

Figure 9-11. *Warning message that appears if an invalid time was entered*

If you click the No button, Excel selects the cell for editing and you can make a change. If you click the Yes button, the entry you made will be accepted.

5. Click the No button, then change the entry to 1:00 PM, and then press the Tab key.

Creating a List Restriction

This Validation limits the entries for the Shipping Method column to those of a list that we create.

1. Select the cell range D2:D12. Click the Data Validation button. Click the Settings tab.

2. Enter the data in Table 9-3.

Table 9-3. *Enter the Data in This Table*

Tab	Option	Value Selected or Entered
Settings		
	Allow	List
	Source	Freightways, FED EX, Parcel Post, UPS
Input Message		
	Title	Shipping Method Restrictions
	Message	You can only select a Shipping Method from the list.
Error Alert		
	Style	Stop
	Title	Shipping Method Error
	Message	You can only pick a Shipping Method from the list.

Figure 9-12 shows how the Settings tab should look.

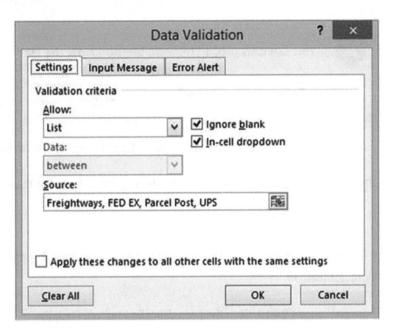

Figure 9-12. *The Settings after entering the data in Table 9-3*

3.   Click the OK button.

4.   Double-click inside cell D2. Click the down arrow to the right of the cell. A list box appears from which you can select one of the entries you created (Figure 9-13).

Figure 9-13. *List box from which to select an entry*

If you type something else into the cell that is not in the list, you will get the message shown in Figure 9-14.

Figure 9-14. *Error message if you enter something that is not in the shipping list*

5.   Click the down arrow in cell D2 and select UPS from the list. Press the Tab key.

Creating an Information Style Error Alert

This Validation for the Shipping Weight column displays a warning message if someone tries to make an entry greater than 35.

1.   Select the cell range E2:E12. Click the Data Validation button. Click the Settings tab.

2.   Enter the data in Table 9-4.

Table 9-4. *Enter This Data*

Tab	Option	Value Selected or Entered
Settings		
	Allow	Whole number
	Data	less than or equal to
	Maximum	35
Input Message		
	Title	Shipping Weight Information
	Message	Orders Weighing more than 35 lbs. are usually shipped using Freightways.
Error Alert	Style	Information
	Title	Shipping Weight Information
	Message	The weight on this Order is more than 35 lbs. Did you ship it using Freightways????

The Information style will not prevent someone from entering a value that does not meet the criteria set on the Settings tab.

3. Click the OK button.

4. Double-click cell E2. Enter 45 and then press tab. A Warning message appears (Figure 9-15).

Figure 9-15. *Information-style message*

5. Click the OK button. The cursor moves to cell F3.

Creating a Custom Restriction That Restricts the Number of Characters That Can Be Entered

This Validation restricts the Zip Code column to either 5 or 9 characters.

1. Select the cell range F2:F12. Click the Data Validation button. Click the Settings tab.

2. Enter the data in Table 9-5.

Table 9-5. *Enter This Data*

Tab	Option	Value Selected or Entered
Settings		
	Allow	Custom
	Formula	=OR(LEN(F2)=5, LEN(F2) =9)
Input Message		
	Title	Zip Code Restrictions
	Message	Please enter a 5 or 9 digit zip code.
Error Alert		
	Style	Stop
	Title	Zip Code Error
	Message	You can only enter 5 or 9 digits for the zip code.

An OR function tests if any of the arguments are true; if one is true, then the result is true and Excel will accept it. Excel will automatically adjust the formula for each cell in the range.

3. Click the OK button.

4. Double-click inside cell F2. Enter 3845 and then press the Tab key. The message in Figure 9-16 displays.

Figure 9-16. *Error message*

5. Click the Retry button. Change the zip code to 38451 and then press the
    Tab key.

6. Click inside cell A3. Type the rest of the data as shown in Figure 9-17. Cells C3
    and C4 will give a time warning. Click the Yes button for both. Cells E2 and E4
    will give a weight warning. Click the OK button for both.

	A	B	C	D	E	F
1	Order ID	Credit Limit	Order Time	Shipping Method	Shipping Weight	Ship Zip
2	SR14	90000	1:00 PM	UPS	45	38451
3	GR15	27000	6:00 PM	FED EX	25	46804
4	SR27	52000	8:00 PM	Parcel Post	53	46341
5	SR29	55000	3:00 PM	UPS	32	46342

Figure 9-17. Enter this data

Circling Invalid Data

We set the Credit Limit, Order Time, and Shipping Weight to use the Warning and Information
styles rather than the Stop Style on the Error Alert tab. This allowed us to break the rule that
we had set up on the Settings tab. We can have Excel place a circle around all those cells that
broke a rule.

1. Click the Data Validation button drop-down arrow and then select Circle Invalid
    Data. See Figure 9-18.

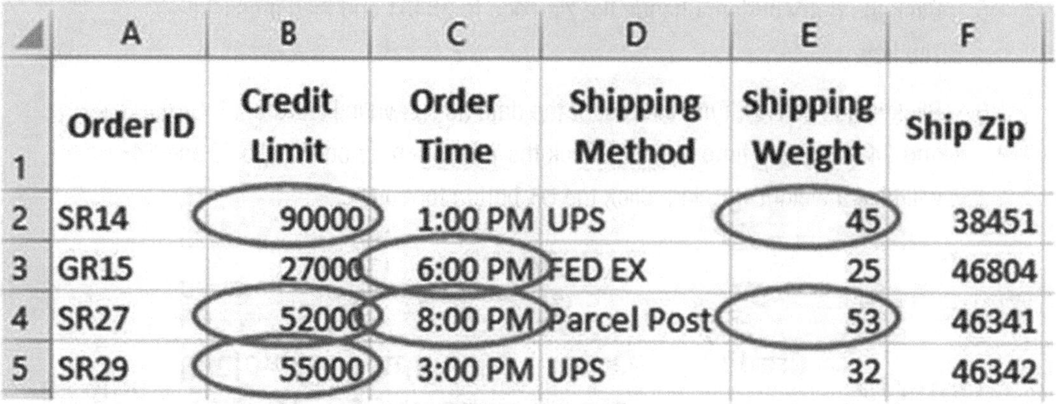

	A	B	C	D	E	F
1	Order ID	Credit Limit	Order Time	Shipping Method	Shipping Weight	Ship Zip
2	SR14	90000	1:00 PM	UPS	45	38451
3	GR15	27000	6:00 PM	FED EX	25	46804
4	SR27	52000	8:00 PM	Parcel Post	53	46341
5	SR29	55000	3:00 PM	UPS	32	46342

Figure 9-18. *Cells that broke a rule are circled*

2. Click the Data Validation button drop-down arrow and then select Clear Validation Circles.

Expanding the Size of the Validation area

We have set the Validation to work on rows 2 through 12. Now, let's extend it through row 20.

1. Click inside cell D12. There is a drop-down arrow to the right of it.

2. Click inside cell D13. There is no drop-down arrow to the right of it.

   What if we later need to add data beyond row 12? We can extend the Validation area by using Paste Special.

3. Select the cell range A2:F2. Press Ctrl + C to copy the cells.

Note   You can actually use any of the cells that contain the Validation you want to use when performing the copy.

---

4. Select the cell range A13 to F20.

5. Click the Home tab. In the Clipboard group, click the Paste button down arrow. Select Paste Special. Select Validation. See Figure 9-19. Click the OK button.

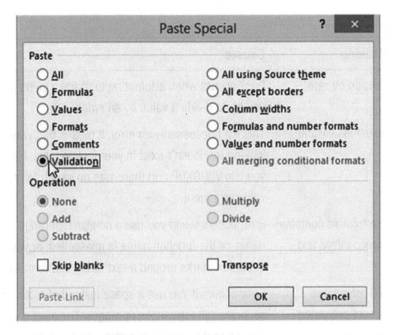

Figure 9-19. *Select Validation from the Paste Special window*

The Validation Range has now been extended to row 20.

6. Enter the following data in row 13. See Figure 9-20.

| 13 | GR15 | 28000 | 2:00 PM | FED EX | | 30 | 60601 |

Figure 9-20. *Enter this data in row 13*

You have learned how to use data validation to prevent users from making data entries that you didn't intend for them to make. Next, you will see that when Excel comes across an entry it can't understand, it displays an error code.

Evaluating Formulas

Excel can often recognize when there is a problem with a function, such as informing you when there are circular references. Table 9-6 shows some of the errors that Excel displays when it finds what it thinks is an error.

Table 9-6. *Error Value Messages*

Error Value	Meaning	Causes
#DIV/0	Division by zero.	Error occurs when attempting to divide a value by zero or if you try to divide a value by an empty cell.
#N/A	No value available.	This isn't necessarily an error. It means that what you are looking for doesn't exist in your lookup table. For example, if you use VLOOKUP and there was no match for your lookup_ value argument.
#NAME?	The formula contains unrecognized text.	Error occurs when you use a nonexistent range name or sheet name, or the function name is misspelled, or you forget to put quotation marks around a text string in a formula.
#NULL!	Cell references are not separated correctly.	Error occurs if you use a space instead of a comma to separate cell references or ranges. The error can also occur if you forget to put an operator between cell references, such as forgetting to put one of the plus signs in a formula such as = B3 + C3 D3.
#NUM!	Formula uses an invalid number.	The formula is using an invalid number such as entering an invalid serial date number.
#REF!	Invalid cell reference.	Error occurs when the formula uses a cell reference that for some reason no longer exists, such as if you deleted a value that was referenced by a formula.
#VALUE!	Formula used the wrong type of data for an argument.	Used text for an argument that was used for a mathematical operation.

Using IFERROR

The IFERROR function checks a formula (or expression) and returns the results if there is no error; otherwise, it returns a value you specify.

The syntax for the IFERROR function is

```
=IFERROR(value, value_if_error)
```

Arguments:

> value is the expression being tested.
>
> value_if_error is the text that will be returned if there is an error in the expression.

The IFERROR can't distinguish the type of error. The error could be DIV/0, #NAME, #N/A, #REF, and so on. What is used as the value_if error argument will be displayed regardless of the error type.

The formula =IF(Sales > 25000,B3 + B2, B4 + B2) would return a #NAME? error if the Sales range didn't exist. If you wanted to test for such a condition, you would use the formula

=IFERROR(IF(Sales > 25000,B3 + B2, B4 + B2),"Range name doesn't exist")

If the Sales range didn't exist, Excel would display **Range name doesn't exist** in the cell where you entered the formula. If the range did exist, the IF function would perform as normal.

In Figure 9-21, cell C1 contains the formula =a1/b1 and cell C2 contains the formula =a2/b2. Cell C1 has a divide by zero error. Cell C2 has a #VALUE! error because it attempted to divide by an alpha character.

◢	A	B	C
1	8	0	#DIV/0!
2	2	b	#VALUE!

Figure 9-21. *Invalid values for formulas*

You could make these messages more meaningful by using the IFERROR function. Entering the formula =IFERROR(A1/B1,"Divided by Zero") into cell C1 and the formula =IFERROR(A2/B2,"Divided by text") in cell C2 would produce the result shown in Figure 9-22.

	A	B	C
1	8	0	Divided by Zero
2	2	b	Divided by text

Figure 9-22. *Meaningful messages created from IFERROR function*

Note   Be careful how you use the IFERROR function. If we used the formula =IFERROR(A1/B1,"Divided by Zero") in both cells C1 and C2, the result would be the same; both would display "Divided by Zero."

Correcting Circular References

A circular reference occurs when a formula in a cell refers to itself either directly or indirectly. This is usually something you do not want to occur; however, there are rare occasions when you do (e.g., when doing iterative calculations).

Figure 9-23 shows a circular reference. The cell address A4 is used in a formula in cell A4.

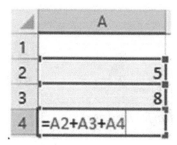

Figure 9-23. *The formula in cell A4 references its own cell address*

The first time you create a circular reference on a worksheet, you will see the message in Figure 9-24.

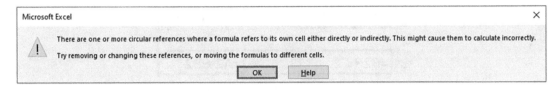

Figure 9-24. *Message for first circular reference*

When you click the OK button, Excel places a zero in the cell that contains the circular reference. See Figure 9-25. You will have to evaluate the formula to determine what it should be.

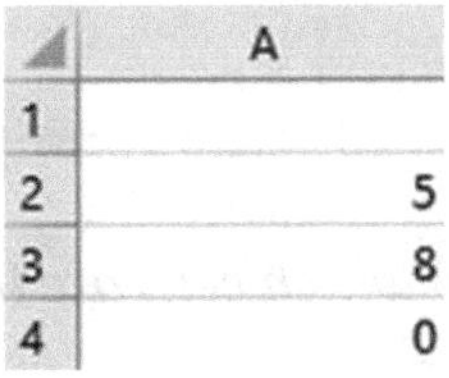

	A
1	
2	5
3	8
4	0

Figure 9-25. *Excel places a zero in a cell that contains the circular reference*

It is easy to accidentally create a circular reference when you are using the SUM function and you drag across the cells that are to be summed and you include the formula cell. See Figure 9-26.

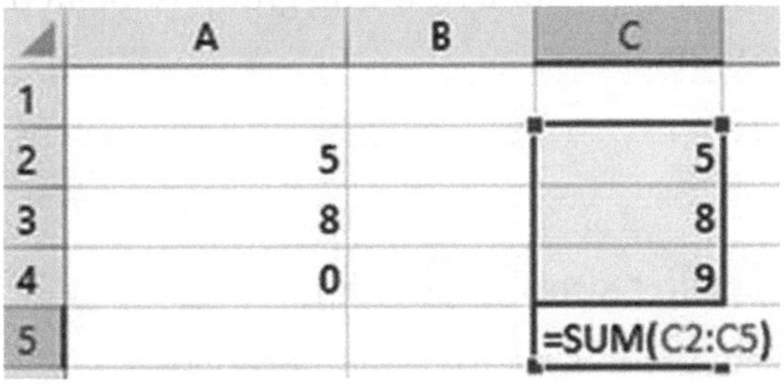

	A	B	C
1			
2	5		5
3	8		8
4	0		9
5			=SUM(C2:C5)

Figure 9-26. *The SUM function includes its own address in its argument*

Figure 9-27 shows an *indirect circular reference*. The formula in cell A2 includes the address C2, which in turn uses cell A2 in its address.

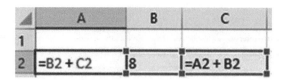

Figure 9-27. *Indirect circular reference*

After you have entered an indirect circular reference, Excel displays tracer arrows to show you which cells are involved in the circular reference. See Figure 9-28.

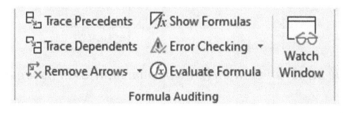

Figure 9-28. *Tracer arrows show which cells are involved in the circular reference*

Double-clicking the arrow in cell A2 makes C2 the active cell. Double-click it again and cell A2 becomes the active cell again.

Formula Auditing

The Excel Formula Auditing group has tools for tracking the relationship between cells and the formulas they use. The Formula Auditing group is located on the Ribbon's Formulas tab. See Figure 9-29.

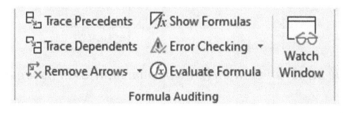

Figure 9-29. *Formula Auditing tools*

Tracing Precedents and Dependents

Precedents are values from other cells that are used in a formula. Dependents are values that are affected by a formula. Tracing precedents and dependents gives you a visual view of the relationship between cell values and formulas. A value can be both a precedent and a dependent. Select the cell that contains the formula you want to check and then click the Trace Precedents button. Clicking a formula and then clicking the precedents button will show all of the cells whose values are used in the formula.

Figure 9-30 shows the precedents for cells A5 and C5. Since individual cells were used in the cell A5 formula, it shows a dot for each precedent. A cell range was used in the C5 formula, so it shows a dot for the first precedent and a line representing the others.

	A	B	C
1	35		107
2	48		102
3	27		135
4	65		140
5	=A1+A2+A3+A4		=SUM(C1:C4)

Figure 9-30. *Precedents for cells A5 and C5*

EXERCISE 9-2: TRACING PRECEDENTS AND DEPENDENTS

This exercise will take you through tracking what formulas the selected cell was used in (dependents) and finding which cells feed into it (precedents). You will also remove the precedent and dependent arrows.

1. Open workbook Chapter 9.

2. Click the tab for the Precedents worksheet.

Figure 9-31 shows the formulas used in the worksheet.

	A	B	C	D	E	F
1	Emp ID	Hours	Rate	Regular Pay	OT	Gross Pay
2	G125	38	9.75	=IF(Hours>40, 40*rate, Hours*rate)	=IF(Hours > 40,(Hours - 40)*rate* 1.5,0)	=D2+E2
3	G232	40	10.25	=IF(Hours>40, 40*rate, Hours*rate)	=IF(Hours > 40,(Hours - 40)*rate* 1.5,0)	=D3+E3
4	G238	40	10.75	=IF(Hours>40, 40*rate, Hours*rate)	=IF(Hours > 40,(Hours - 40)*rate* 1.5,0)	=D4+E4
5	R114	43	11.15	=IF(Hours>40, 40*rate, Hours*rate)	=IF(Hours > 40,(Hours - 40)*rate* 1.5,0)	=D5+E5
6	R116	48	10.95	=IF(Hours>40, 40*rate, Hours*rate)	=IF(Hours > 40,(Hours - 40)*rate* 1.5,0)	=D6+E6
7	R119	56	12.18	=IF(Hours>40, 40*rate, Hours*rate)	=IF(Hours > 40,(Hours - 40)*rate* 1.5,0)	=D7+E7
8	S202	40	15.05	=IF(Hours>40, 40*rate, Hours*rate)	=IF(Hours > 40,(Hours - 40)*rate* 1.5,0)	=D8+E8
9				=SUM(D2:D8)	=SUM(E2:E8)	=SUM(F2:F8)

Figure 9-31.   Formulas used in worksheet

Creating Dependent Tracer Lines

1. Click cell B2. Click Trace Dependents. The Tracer lines show that cell B2 is used in the formulas in cells D2 and E2.

2. Click cell C3. Click Trace Dependents. The Tracer lines show that cell C3 is used in the formulas in cells D3 and E3.

3. Click inside cell D4. Click Trace Dependents. The Tracer lines show that cell D4 is used in the formulas in cells D9 and F4.

4. Click inside cell E5. Click Trace Dependents. The Tracer lines show that cell E5 is used in the formulas in cells E9 and F5.

5. Click inside cell F6. Click Trace Dependents. The Tracer lines show that cell F6 is used in the formula in cell F9. Figure 9-32 shows all the Dependent lines you have created.

	A	B	C	D	E	F
1	Emp ID	Hours	Rate	Regular Pay	OT	Gross Pay
2	G125	38	9.75	370.5	0	370.5
3	G232	40	10.25	410	0	410
4	G238	40	10.75	430	0	430
5	R114	43	11.15	446	50.175	496.175
6	R116	48	10.95	438	131.4	569.4
7	R119	56	12.18	487.2	292.32	779.52
8	S202	40	15.05	602	0	602
9				3183.7	473.895	3657.595

Figure 9-32.   Dependent lines

Removing Dependent Tracer Lines

1. Click the Remove Arrows down arrow. Click Remove Dependent Arrows. See Figure 9-33.

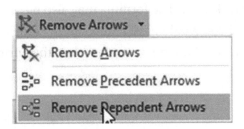

Figure 9-33. *Select Remove Dependent Arrows*

This removes the Dependent Tracer line from the selected cell.

2. Click the Remove Arrows down arrow. Click Remove Arrows. This removes all the Tracer lines.

Creating Trace Precedent Lines

1. Click cell D2. Click Trace Precedents. The Precedent lines show the cells that feed into the formula in cell D2.

2. Click cell D9. Click Trace Precedents. The Precedent lines show the cells that feed into the formula in cell D9.

3. Click cell F9. Click Trace Precedents. The Precedent lines show the cells that feed into the formula in cell F9.

4. Click the Remove Arrows down arrow. Click Remove Arrows. All arrows should now be removed.

You have learned how to create tracer lines that show where the values used in your formulas are coming from and what cells are using the results of those formulas. Next, you will learn to add the cell addresses of those cells you want to closely monitor to a Watch Window. The Watch Window will reflect any changes in the values of these cells.

Using the Watch Window

The Watch Window is useful when you want to see how the values you are entering are affecting the results of formulas that you can't see. The formula may not be visible because it is on another worksheet. The formula may also be on the same worksheet, but it is in a location where you would have to repeatedly scroll back and forth. The Watch Window button is located on the Formulas tab in the Formula Auditing group. See Figure 9-34.

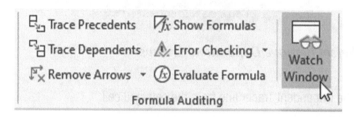

Figure 9-34. *The Ribbon's Formula Auditing group*

Cells are added to the Watch Window by clicking the Watch Window button, selecting the cells to be added, and then clicking the Add Watch button. Cells are removed from the Watch Window by selecting the items you want to remove and then clicking the Delete Watch button.

EXERCISE 9-3: USING THE WATCH WINDOW

In this exercise, you will add cells to the Watch Window and then remove them.

1. Open Workbook Chapter 9. The workbook contains worksheets for January, February, and March Sales and a worksheet that Totals those sales as well as providing the January Sales total amounts for salesmen 127 and 235.

2. Click the Sales-Totals worksheet.

Adding Cells to the Watch Window

1. Click the Ribbon's Formulas tab. In the Formula Auditing group, click the Watch Window button.

2. Hold down the Ctrl key while selecting cells B1, B4, B6, and B8.

3. Click the Add Watch button. See Figure 9-35.

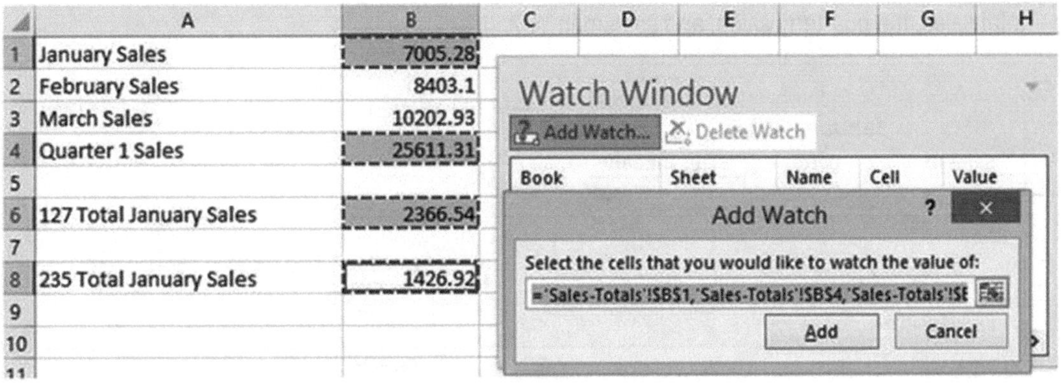

Figure 9-35. *Click the Add Watch window*

The Watch Window displays values for the cells you selected as well as their formulas. Click the Add button on the Add Watch dialog box.

4. Click the Jan-Sales worksheet tab. See Figure 9-36. Change the value in cell C3 to 600.00.

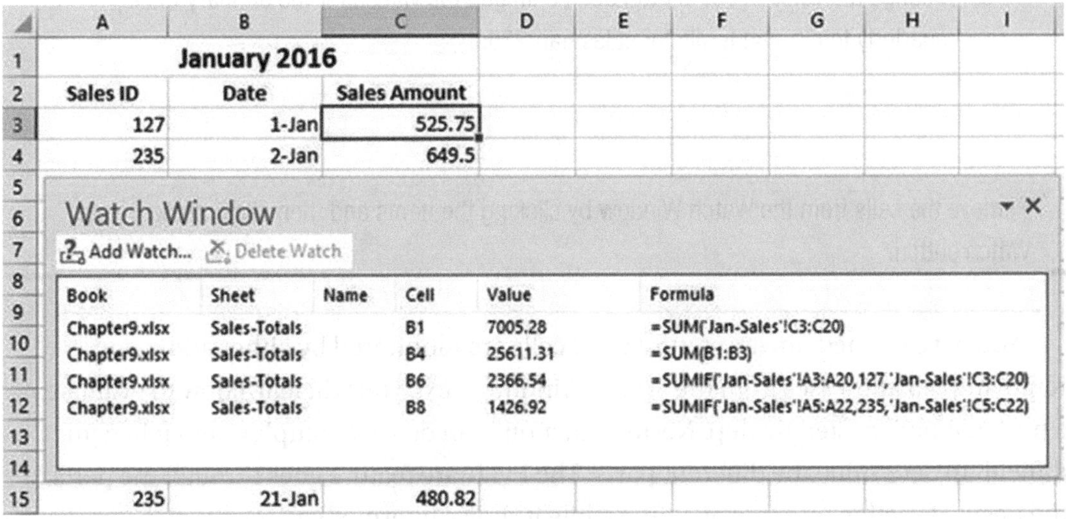

Figure 9-36. *Change value in cell C3 to 600.00*

You can see how changing the value in cell C3 to 600 has changed the values in cells B1, B4, and B6 in the Watch Window. See Figure 9-37. These cells contained the formulas for January total sales, the quarterly sales, and salesman 127.

	A	B	C	D	E	F	G	H	I
1		**January 2016**							
2	**Sales ID**	**Date**	**Sales Amount**						
3	127	1-Jan	600						
4	235	2-Jan	649.5						

Watch Window ▾ ✕

🔲 Add Watch...   ✕ Delete Watch

Book	Sheet	Name	Cell	Value	Formula
Chapter9.xlsx	Sales-Totals		B1	7079.53	=SUM('Jan-Sales'!C3:C20)
Chapter9.xlsx	Sales-Totals		B4	25685.56	=SUM(B1:B3)
Chapter9.xlsx	Sales-Totals		B6	2440.79	=SUMIF('Jan-Sales'!A3:A20,127,'Jan-Sales'!C3:C20)
Chapter9.xlsx	Sales-Totals		B8	1426.92	=SUMIF('Jan-Sales'!A5:A22,235,'Jan-Sales'!C5:C22)

| 15 | 235 | 21-Jan | 480.82 | | | | | | |

Figure 9-37. *Watch Window*

5. Change the value in cell C4 to 700. It changed the values for the January totals, quarterly totals, and totals for salesman 235.

Removing Cells from the Watch Window

Remove the cells from the Watch Window by clicking the items and then clicking the Delete Watch button.

You have learned how to control what cells are monitored by either adding or removing cell addresses from the Watch Window. Next, you will learn how to evaluate a nested function, step by step. Nested functions can be very complex, and it becomes difficult to determine the different parts. The **Evaluate** feature goes through the parts of the nested function one step at a time while it shows intermediate results.

Using the Evaluate Formula Feature to Evaluate a Nested Function One Step at a Time

The Evaluate Formula feature helps you see step by step how a nested function works.

EXERCISE 9-4: EVALUATING A NESTED FORMULA

In this exercise, you evaluate a nested formula.

1. Open workbook Chapter 9.

2. Click the Evaluate worksheet tab.

Cell B2 has a range name of Sex. Cell C2 has a range name of Age. The formula in cell D2 tests if Sex has a value of F or Age has a value of 21 or greater. If either condition is True, it will display a B; otherwise, it will display a blank. See Figure 9-38.

	A	B	C	D
1	Subject	Sex	Age	Section
2	Math	M	21	=IF(OR(Sex="F",Age>=21),"B","")

Figure 9-38. *Nested formula*

3. Click inside cell D2.

4. Click the Ribbon's Formulas tab. In the Formula Auditing group, click Evaluate Formula to open the Evaluate Formula dialog box shown in Figure 9-39. The inner function **OR** will be evaluated first and then the outer function **IF**. Sex is underlined. The underlined portion is what is to be evaluated.

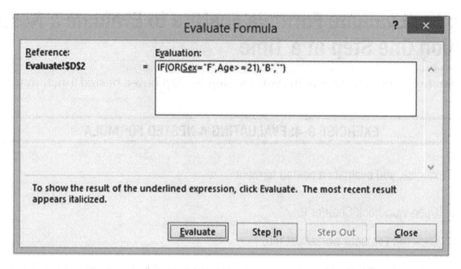

Figure 9-39. *Evaluate Formula window*

5.   If you want to know where the underlined value is coming from, you can click
      the Step In button. Click the Step In button. The Step In button shows that cell
      B2 is associated with the range name Sex. See Figure 9-40.

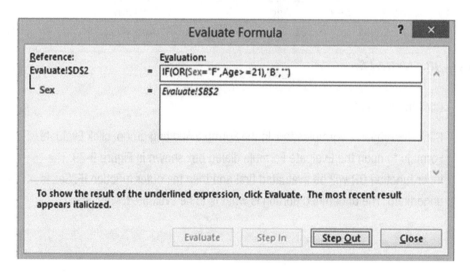

Figure 9-40. *Using Step In*

6.   Click the Step Out button. The Evaluate Formula dialog box shows that there is
      an M in cell B2. It compares to see if it is equal to F. See Figure 9-41.

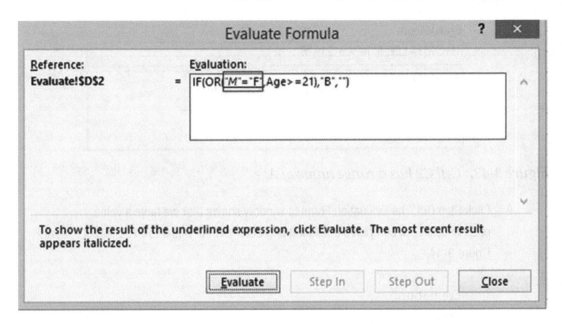

Figure 9-41. *Using Step Out*

7. Click the Evaluate button. M is not equal to F so the expression displays a
    FALSE as shown in Figure 9-42.

Figure 9-42. *M = F is evaluated to FALSE*

8. Age is underlined so it is to be evaluated next. Click the Step In button. You can
    see in Figure 9-43 that cell C2 has a range name of Age.

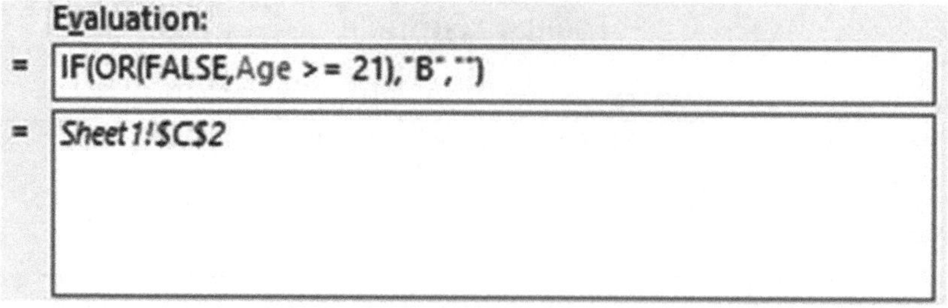

> **Evaluation:**
>
> = | IF(OR(FALSE,Age >= 21),"B","")
>
> = | Sheet1!$C$2

Figure 9-43. *Cell C2 has a range name of Age*

9. Click Step Out. The Evaluation Formula window shows that we have a value of 21 in cell C2. It compares to see if it is greater than or equal to 21. See Figure 9-44.

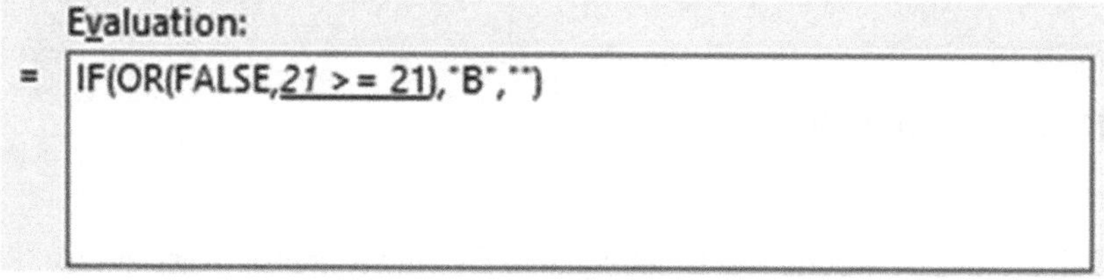

> **Evaluation:**
>
> = | IF(OR(FALSE,*21 >= 21*),"B","")

Figure 9-44. *The Age range has a value of 21*

10. Click the Evaluate button. The condition 21>=21 evaluates to TRUE. See Figure 9-45.

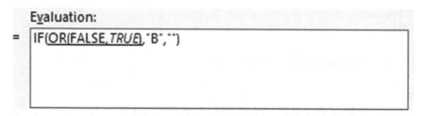

> **Evaluation:**
>
> = | IF(OR(FALSE,*TRUE*),"B","")

Figure 9-45. *The second condition 21>=21 evaluates to True*

11. Click the Evaluate button. The OR function returns a TRUE if either condition is true. The second condition was True so the function returns a TRUE. See Figure 9-46.

Evaluation:

= IF(*TRUE,*B*,*)

Figure 9-46. *The result of the OR function is TRUE*

12. The evaluation for the OR function is complete, so now it will evaluate the IF function. Click the Evaluate button. The formula evaluates to TRUE so it displays the true value which in this case is B. Close the dialog box.

Proofreading Cell Values—Have Excel Read Back Your Entries

Excel can help you proofread your spreadsheet by speaking the values entered in the cells. You can verify that you have made the correct entries by listening to the Excel speaker while you look at the paper from which you made the entries.

This option is not included on the Ribbon; it needs to be added there or to the Quick Access Toolbar.

EXERCISE 9-5: PROOFREADING CELL VALUES

Instead of adding the Proofreading options to the Ribbon, we will add them to the Quick Access Toolbar.

1. Right-click the Quick Access Toolbar and select **Show Quick Access Toolbar Below the Ribbon**.

2. Right-click the Quick Access Toolbar and select **Customize Quick Access Toolbar**.

3. In the Choose commands from drop-down box, select **Commands Not in the Ribbon**.

4. Scroll down using the scroll bar until you come to Speak Cells. Click Speak Cells and then click the Add button. Do the same thing for the next four options. See Figure 9-47.

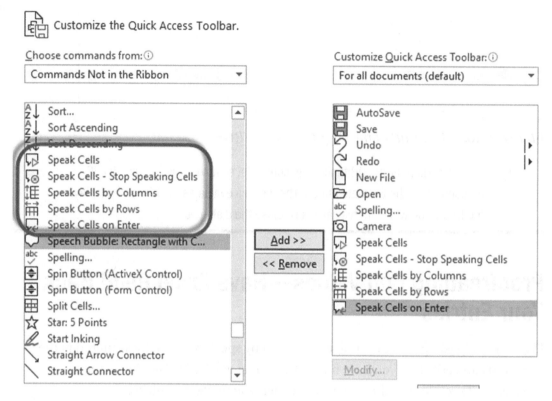

Figure 9-47. *Add Speak commands to the Quick Access Toolbar*

5. Click the OK button. You should now have five new buttons on your Quick Access Toolbar. See Figure 9-48.

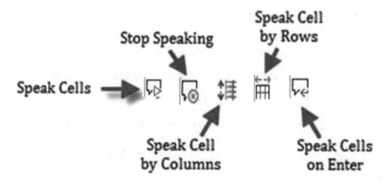

Figure 9-48. *Speak commands placed on the Quick Access Toolbar*

6. Click the **Speak Cells on Enter** button on the Quick Access Toolbar. You will get a voice confirmation that anything that you enter into the cells will be read back to you.

7. Enter the data in Figure 9-49 into a new spreadsheet. As you enter the data, the information will be read back to you.

	A	B	C
1	**Shirts**	**Shoes**	**Socks**
2	117	105	305
3	129	89	205
4	96	93	207

Figure 9-49. *Enter this data*

8. Click **Speak Cells on Enter** again. This turns off the **Speak as you Enter** option.

9. Select cells A1:C4. Click the **Speak Cells by Rows** button on the Quick Access Toolbar. Click the **Speak Cells** button on the Quick Access Toolbar. Excel speaks the data in row order.

10. Click the **Speak Cells by Columns** button. Click the **Speak Cells** button. Excel speaks the data in column order. Click the **Speak Cells—Stop Speaking Cells** button before Excel has finished reading all of the data. Excel provides the capability to stop speaking at any point and then continue from that point. Now if you click **Speak Cells**, Excel will start reading from the point where you clicked the **Speak Cells—Stop Speaking Cells** button.

Spell Checking

The Spelling and Thesaurus buttons are located on the Review tab, in the Proofing group. See Figure 9-50.

Figure 9-50. *The Proofing group*

To avoid the embarrassment of turning in a workbook with misspelled words, you should always check it with Excel's Spell Checker. You could test a range of cells or the entire worksheet. If you select a range of cells before clicking the Spell Checker, only that range of cells will be checked. If you have only one cell selected or if you click the Select All button to the left of column A, then it will check the entire worksheet.

If Excel can't find a word in its dictionary, it will display it in the Not in Dictionary text box. It provides one or more suggestions of how it thinks it should be spelled in the Suggestions: list box. See Figure 9-51.

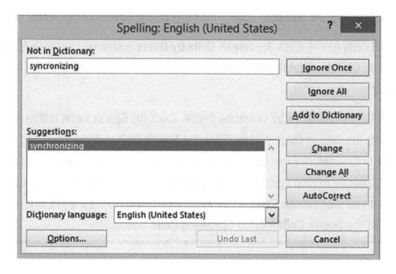

Figure 9-51. *Spell Checker*

The purpose of the buttons are as follows:

- **Ignore Once**—Tells Excel that you don't want to change the spelling at the current cell, but you still want to test the word in other locations.

- **Ignore All**—Tells Excel you don't want to change the spelling of the word in the **Not in Dictionary** text box in this location or in any other location on the worksheet.

- **Add to Dictionary**—Adds the word to Excel's dictionary. Excel will never tell you again that the word is misspelled.

- **Change**—Replaces the word in your worksheet cell with the word you have selected in the Suggestions: list box.

- **Change All**—Changes all occurrences in your worksheet of the word in the Not in Dictionary text box with the one you have selected in the Suggestions: list box.

- **AutoCorrect**—Adds the word selected in the Suggestions: list box as the replacement for the word in the Not in Dictionary for Excel's AutoCorrect feature.

- **Dictionary Language**—Allows you to select from one of many different dictionaries.

Thesaurus

If you are looking for a word that has a similar meaning to a word in your spreadsheet, you can use Excel's Thesaurus. Just select the word and then click the Thesaurus button. Figure 9-52 shows that the word **Synchronizing** was selected when the Thesaurus button was clicked.

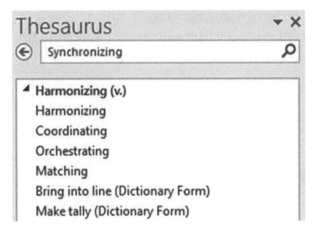

Figure 9-52. Thesaurus

The (v.) next to Harmonizing means that it is a verb. Click a word in the list, and it displays similar meaning words for it.

Figure 9-53 shows that the word **Harmonizing** was selected from the list in Figure 9-52 and now words with similar meaning to Harmonizing are displayed. The (adj.) means the word is an adjective.

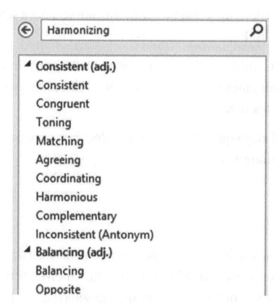

Figure 9-53. *Words with similar meaning to Harmonizing*

To move back in the list of words, click the ⊖ button.

To replace the word in your cell with one from the Thesaurus list, right-click the word in the Thesaurus list and then select Insert. See Figure 9-54.

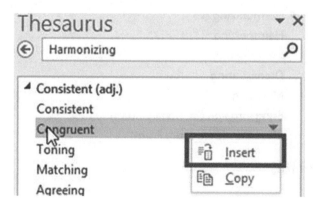

Figure 9-54. *Replace a word in your spreadsheet with one from the Thesaurus list*

Protect Worksheets and Cells from Accidental or Intentional Changes

You can protect your workbook data by preventing users from accidentally or intentionally changing, adding, or deleting your data.

You can protect your data at the workbook level, the worksheet level, or the cell level. You have already seen how to protect your data on the Workbook level in Chapter 6. You will now see how to use worksheet and cell level protection.

Protect Your Data at the Worksheet Level

The commands needed for Worksheet-level and cell-level protection can be found on the Review tab in the Protect group. See Figure 9-55.

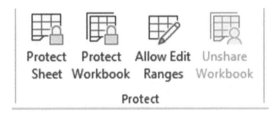

Figure 9-55. *The Ribbon's Changes group*

All cells in a workbook are locked by default. Locked Cells don't provide any protection until you click the Protect Sheet button. Right-clicking a cell and then selecting Format Cells brings up the Format Cells window. If you click the Protection tab, you will see that locked is checked. See Figure 9-56.

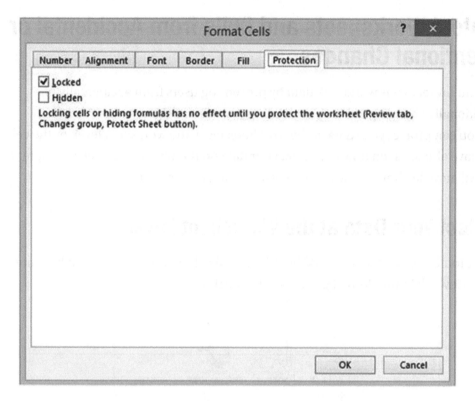

Figure 9-56. *Cells are locked by default*

EXERCISE 9-6: PROTECTING A WORKSHEET

In this exercise, you protect a worksheet with a password.

1. Open workbook Chapter 9. Click the Protect worksheet tab.

2. Click the Ribbon's Review tab. In the Changes group, click the Protect Sheet
    button. The Protect Sheet dialog box displays. See Figure 9-57.

Figure 9-57. *Protect Sheet dialog box*

The Protect Sheet dialog box lets you assign capabilities to users by checking those options you want to give them, such as the ability to format cells or insert rows and columns.

3. Leave **Select locked cells** and **Select unlocked cells** checked.

4. Enter a password.

Caution   Make sure that you remember the password. You should save it somewhere safe. If you forget it, you will not be able to open the worksheet again.

---

5. Click the OK button.

6. Reenter your password to confirm the password you entered in the Protect Sheet window. See Figure 9-58.

Figure 9-58. *Confirm Password dialog box*

7. Click the OK button.

8. Right-click a cell. Notice that Format Cells is dimmed.

9. Right-Click a row head. Notice that Insert and Delete are dimmed.

   These as well as the other options you didn't select in the Protect Sheet window are unavailable.

10. Try changing a value. You will get the message in Figure 9-59. Click the OK button.

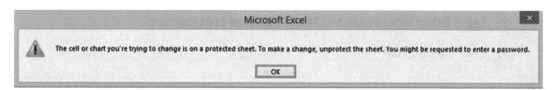

Figure 9-59. *Message informing you that what you are trying to change is protected*

11. Try entering a value in a blank cell. You will get the same message as in Figure 9-59.

12. In the P group, click the Unprotect Sheet.

13. Enter the same password you entered previously into the Unprotect Sheet dialog box. See Figure 9-60. Click the OK button.

Figure 9-60. *Enter password to unprotect the worksheet*

The worksheet is now unprotected. You can make any additions or changes to the data.

Protect Your Data at the Cell Level

You saw previously where all cells are locked by default. When you click the Project Workbook button in the Changes group, all those locked cells become protected and you will not be able to enter any data. Therefore, when allowing users to make changes to cell data, you don't protect cells because they are already protected. Rather, you unprotect those cells that you are willing to let users make changes to. You need to specify the cells that you are unprotecting before clicking the Protect Workbook button.

EXERCISE 9-7: PROTECTING CELLS

In this exercise, you will prevent users from unhiding column F and from making any changes to the formulas or headings.

1. Open workbook Chapter 9. Click the Protection worksheet tab.

2. Right-click column head F and then select Hide.

3. Select cell range A2:C8. Right-click the selected cells and then select Format Cells.

4. Click the Protection tab and do the following:

   a. Uncheck Locked. This will unlock the cell range A2:C8.

   b. Check Hidden. This will prevent users from unhiding column F.

   c. Click the OK button.

   On the Ribbon's Review tab, in the Protect group, click the Protect Sheet button. Enter a password. Confirm the password by reentering it.

5. Try to change a value in the cell range A2:C8. You can do so.

6. Try changing one of the formulas in column D or E. You will get the message displayed in Figure 9-61.

Figure 9-61. *Message informing you that what you are trying to change is protected*

7. Drag across column heads E and G. Right-click the selection; you will see that unhide is dimmed.

8. In the Protect group, click the Unprotect Sheet button.

9. Enter your password. Click the OK button.

   You can now make changes to all cells.

10. Drag across column heads E and G. Right-click the selection and then select unhide.

   Column F should now be visible.

Summary

Excel has many helpful features to make sure that data entries are made correctly and, if they aren't, to notify you immediately so that you can correct the error. You have seen how Excel can correct your misspellings and suggest other words to use from their Thesaurus. You have seen how you can get visual cues to track the movement in and out of Formulas. Excel also shows tracer arrows to show which cells are involved in a Circular Reference.

The next chapter covers how to create, edit, and remove hyperlinks. A hyperlink is often just called a link. You have most likely used them a lot on the Internet, linking from one location on the website to another or to an entirely different website. Hyperlinks are text or objects that you can click to take you to somewhere within the current application or applications such as to another workbook, a website, an email application, and so on. We will also look how you can combine text from multiple columns along with any other text you wish to add.

CHAPTER 10

Using Hyperlinks, Combining Text, and Working with the Status Bar

A hyperlink can be either text or an object such as a picture or a chart element that, when clicked, takes you to another location (e.g., a specific location in your document) or brings up information from another file or web page. You have probably used hyperlinks on the Internet. Clicking a hyperlink in a web page can either take you to a different location on that page or to a completely different web page.

The **Concatenation** and **Flash Fill** features allow you to combine text from multiple cells and add additional text.

In this chapter, you will also learn how to display information related to your spreadsheet on the status bar.

After reading and working through this chapter, you should be able to

- Hyperlink to a web page

- Hyperlink to a file

- Hyperlink to an email

- Hyperlink to a location in the current workbook

- Create a hyperlink that will create and open a new workbook

- Remove hyperlinks

© David Slager and Annette Slager 2020
D. Slager and A. Slager, *Essential Excel 2019*, https://doi.org/10.1007/978-1-4842-6209-2_10

- Create Concatenation

- Use Flash Fill

Working with Hyperlinks

Hyperlinks are usually displayed in a different color (usually blue) and often underlined so that they are easily identifiable. To make them even more identifiable, the mouse cursor changes to a pointing hand indicator when moved over a hyperlink. You can use hyperlinks in Excel to link to the following:

- A web page or a file (the file could be another spreadsheet file, or it could be another type of file such as a word document)

- A specific cell in the current workbook

- A new document

- An email address

Hyperlinking to another file opens the file in the program associated with that file type. For example, if you create a hyperlink to a file with the extension of .docx, Excel will open that file in Word when that hyperlink is clicked. If you create a hyperlink to a file with the extension of .psd, Excel will open that file in Photoshop.

To bring up the Insert Hyperlink window, perform the following steps:

1. Click the cell in which you want to place the hyperlink.

2. On the Ribbon, click the Insert tab. In the Links group, click the upper half of the Links button.

3. From the Insert Hyperlink dialog box, do the following:

   a. Select the type of link you wish to create from the **Link to pane**.

   b. Enter the text the hyperlink should display in the **Text to display** text box.

   c. Select or enter the address you want to hyperlink to and enter any screen tips you want to display. See Figure 10-1.

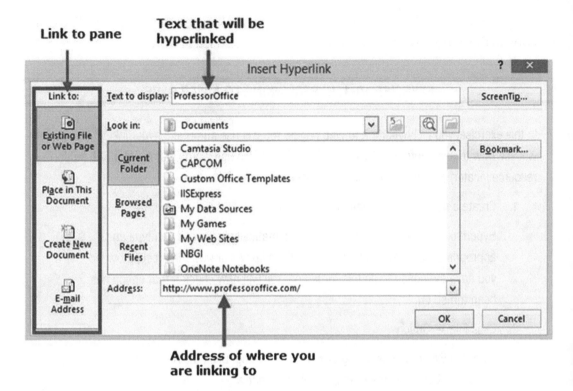

Figure 10-1. *Insert Hyperlink dialog box*

The first thing you need to do when creating a hyperlink is to decide what type of link you want to create. You select the type of link you want to create by clicking the appropriate button in the **Link to pane** located on the left-hand side of the window. See Figure 10-1. The options that appear to the right of the **Link to pane** change depending on the button you selected in the **Link to pane**. For example, clicking the **Existing File or Web Page** button will display different options to the right then if you had clicked the **Place in This Document** button.

If there was any text in the spreadsheet cell when you inserted the hyperlink, it appears in the **Text to display** text box. Whatever you enter in this text box will display in your cell; it will override any text currently in the cell. To the right of the text box is a ScreenTip... button. Clicking the ScreenTip button brings up a window in which you can enter a description of the hyperlink. This description will display for a short time when the user moves the mouse over the hyperlink. The location of where you are linking to would be entered in the **Address** box.

A hyperlink can be removed by right-clicking inside the cell of the hyperlink and then selecting **Remove Hyperlink** from the context menu.

The next four exercises provide practice in hyperlinking to a web page, to a file, to a location in the current workbook, and to a newly created Excel workbook.

EXERCISE 10-1: HYPERLINKING TO A WEB PAGE

In this exercise, you will hyperlink to web pages. You will use automatic hyperlinking, and you will see how to undo automatic hyperlinking. You will also use words instead of a URL (uniform resource locator) address to access a web page.

1. Create a new workbook named "Hyperlinks."

    Hyperlinks to web pages are created automatically in Excel if you type an appropriate web address that starts with `http://` or `www.` in a cell, provided you have the option to do this turned on. You will now check if that option has been turned on.

2. Click File ➤ Options.

3. From the Excel Options dialog box, select Proofing. Click the AutoCorrect Options… button on the right pane of the window.

4. Click the **AutoFormat As You Type** tab. Select **Internet and network paths with hyperlinks** if it isn't already selected. See Figure 10-2. Click OK.

Figure 10-2. *AutoFormat As You Type tab in the AutoCorrect dialog box*

5. Click the OK button on the Excel Options dialog box.

6. Type www.apress.com in cell A1 and then press Enter. Because you set the option to automatically create hyperlinks whenever you type an appropriate web address (starts with http:// or www.), a hyperlink is automatically created.

7. Move your cursor over the text you just entered in cell A1. The cursor should display as a pointing finger. A tooltip displays the hyperlink address. Click the hyperlink.

8. Close the web page.

    The hyperlink works, but what if you wanted the words **Apress Corporation** to appear in the cell instead of www.apress.com? You will need to edit the hyperlink.

9. Right-click cell A1 and then select **Edit Hyperlink**.

10. Change the text in the Text to display text box to **Apress Corporation**. See Figure 10-3. The address should be correct. Click OK.

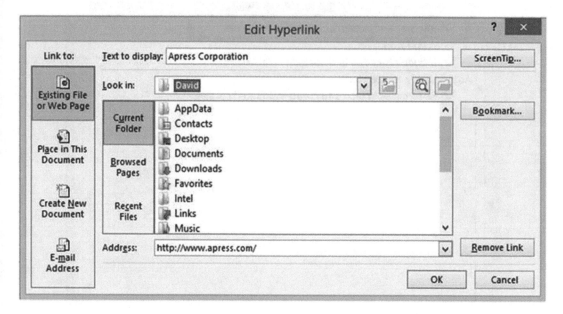

Figure 10-3. *Edit Hyperlink dialog box*

11. Cell A1 now displays <u>Apress Corporation</u>. Click the text. The Apress website should display. Close your Internet browser.

    Next, let's link to an email address.

12. Type jnobody@hotmail.com in cell A3 and then press Enter.

13. Click the email address you just created.

    When you click a hyperlink to an email address, your email program automatically starts and creates an email message with the correct address in the **To** box, provided that you have an email program installed, such as Outlook.

    What if you want to enter a URL that is not hyperlinked?

14. Type www.MadeinUsa.com in cell A5 and then press Enter.

15. Widen the column so that all the text fits within the cell. Move your cursor over the text you just entered in cell A5. The first w in the hyperlink is underlined. Move your mouse cursor over the underline and a menu Tag should appear.

16. Click the menu tag to bring up a context menu. See Figure 10-4.

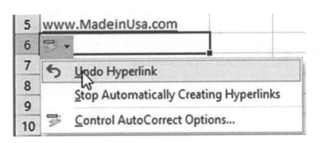

Figure 10-4. *Context menu for hyperlink*

17. Select Undo Hyperlink. The text remains, but the hyperlink is removed.

    Now let's add a hyperlink with a Screen Tip.

18. Click cell A7. Click the Ribbon's Insert tab. In the Links group, you will see the Links button. The Links button has an upper and lower half. Clicking the lower half of the Links button will show the links that you have recently used. Clicking the upper half of the links button brings up the Insert Hyperlink dialog box.

Note   The Insert Hyperlink window can also be brought up by right-clicking an empty cell and selecting Link from the context menu.

---

19. In the **Link to** pane, select Existing File or Web Page.

20. Type **Professor Office web site** in the Text to display text box.

21. Type www.professoroffice.com in the Address: box. Excel automatically adds the prefix http:// to the address.

22. Click the ScreenTip … button. Enter **Microsoft Office and other training** for the ScreenTip text. See Figure 10-5. Press OK.

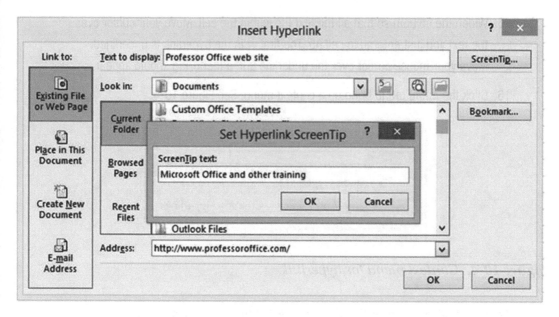

Figure 10-5. *Create a ScreenTip for the hyperlink*

Click the OK button for the Insert Hyperlink dialog box.

23. You should now have the hyperlink Professor Office web site in cell A7. Move your cursor over the hyperlink. The ScreenTip **Microsoft Office and other training** displays. Click the hyperlink. Close the web page.

24. Right-click cell A7. Select Remove Hyperlink.

EXERCISE 10-2: HYPERLINKING TO A FILE

There may be a file that contains more information that you don't want stored in Excel, but you want to have quick access to it. This exercise shows you how to hyperlink to a file. When you click that hyperlink, the file will open.

1. Create a new worksheet in your Hyperlinks workbook.

2. You can use the linkedtext.txt file provided with this book, or you can create your own linkedtext.txt file using the following steps:

   a. Start the Notepad program.

b. Type the text **This is text that I am linking to**. Save the file with a name of linkedtext. See Figure 10-6.

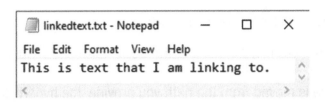

Figure 10-6. *Create text in Notepad*

c. Save the file to your My Documents folder or a location you can easily find the path to.

d. Close the Notepad program.

3. Right-click inside cell A5 on your spreadsheet and select Link.

4. Enter **Notepad file** in the Text to display text box. See Figure 10-7.

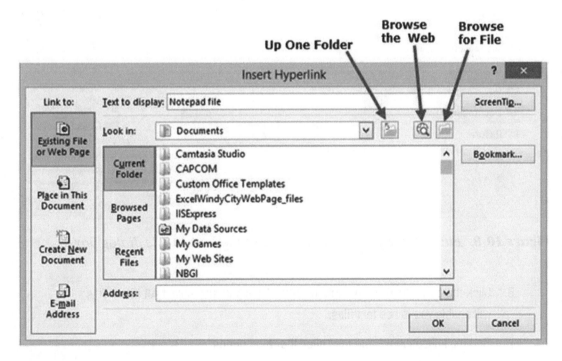

Figure 10-7. *Adding a link to a file in the Insert Hyperlink dialog box*

- You can type the path directly.

- You can click the down arrow to the right of the Look in drop-down box to get to the correct folder and then select the linkedtext file.

- You can click the **Browse for File** button.

There are several ways to get the address (path) and file name in the Address box:

Note   If your file is moved from the path you provide, the hyperlink will not work.

---

5. Click the **Browse for File** button. This brings up the Link to File dialog box in Figure 10-8. Excel thinks you are searching for an Office file, but you are searching for a text file.

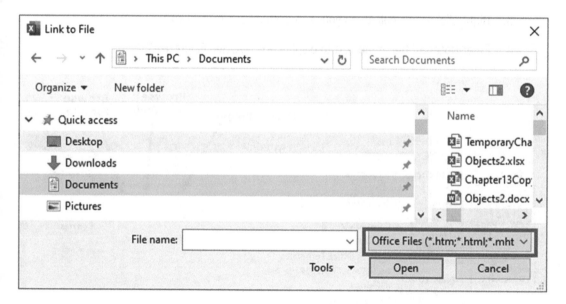

Figure 10-8. *Find the file you want to link to in the Link to File dialog box*

6. Click the down arrow to the right of Office Files and then select **All Files**. This will enable you to see text files.

7. Find the linkedtext file and then click the Open button.

8. The address with the linkedtext file should now appear in the Insert Hyperlink dialog box. See Figure 10-9. Click the OK button.

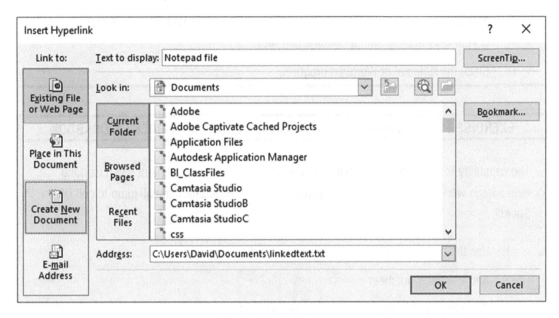

Figure 10-9. *The file name with its path is displayed in the Address field*

9. Click the hyperlink in cell A5. If you get the message in Figure 10-10, click the Yes button.

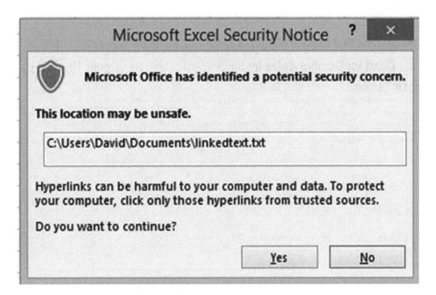

Figure 10-10. *Ignore the message and click the Yes button*

The `linkedtext.txt` file will open either in Notepad or WordPad, whichever is your default program for opening text files.

10. Close the Notepad or WordPad program.

EXERCISE 10-3: HYPERLINKING TO A LOCATION IN THE CURRENT WORKBOOK

The capability to hyperlink to any cell in your workbook is really convenient when working with a large workbook. In this exercise, you will create a hyperlink that will jump to cell DA on Sheet2.

1. Use the same workbook as the previous practice.

2. Add another worksheet.

3. On Sheet3, right-click inside cell D7 and select Link.

4. Select **Place in This Document** in the "Link to" area.

5. Enter **Oct Sales** in the Text to display text box.

6. Enter **DA35** for the cell address in the Type the cell reference text box.

7. Select **Sheet2** for the Cell Reference.

8. Let's assume that there is a chart in cell DA35. Click the ScreenTip… button. Enter **Chart for October Sales** for the ScreenTip text. See Figure 10-11. Click the OK button.

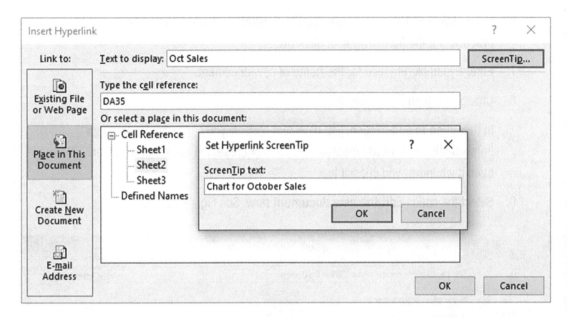

Figure 10-11. *Enter a ScreenTip for the chart*

9. Click the OK button in the "Insert Hyperlink" dialog box.

10. Move your cursor over the hyperlink. The ScreenTip that you created displays.
    Click the hyperlink. The active cell is now DA35 on Sheet2.

EXERCISE 10-4: CREATING A HYPERLINK TO A NEWLY CREATED EXCEL WORKBOOK

You can create a hyperlink that, when clicked for the first time, creates a new Excel workbook
and then opens that workbook. Every time you click it thereafter, it will open the workbook.
The Excel workbook will be given a name that you have assigned within the Hyperlink options.

1. Use the same workbook as the previous practice.

2. On Sheet2, enter the text "Hyperlink to a New Document" in cell A1.
   Press Ctrl + Enter.

3. Right-click cell A1 and select Link.

4. The Insert Hyperlink window appears. In the **Link to**: area, click Create New
   Document.

5. Click the ScreenTip… button.

6. Enter "Takes you to a newly created Excel workbook named Hyperlink_ Practice" for the ScreenTip text. Click OK.

7. Enter Hyperlink_Practice for the Name of new document.

8. Change the path if you need to.

    In the **When to edit** section, you have two options: you can either select to go immediately to the newly created workbook or create the workbook and then open it whenever you choose to.

9. Select the option **Edit the new document now**. See Figure 10-12.

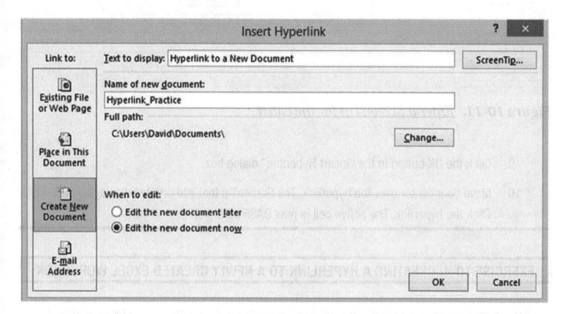

Figure 10-12. *Select Edit the new document now on the Insert Hyperlink window*

10. Click the OK button.

11. Your cursor should now be in cell A1 of the newly created workbook named Hyperlink_Practice. Go back to your workbook named Hyperlinks. Click Sheet2 tab cell A1. Since Hyperlink_Practice has already been created, Excel will not try to recreate it, but it will take you to the Hyperlink_Practice workbook. Before opening the Hyperlink_Practice workbook, Excel may display a Security Notice window. Since you created the hyperlink yourself, you know it is safe, so click the Yes button.

You have learned how to create and remove hyperlinks. You can create hyperlinks to open documents and take you to various websites on the Internet, different locations in Excel workbooks, and so on. If you want a user to get information from another location, then use hyperlinks. It saves the user a lot of time in locating the information. Next, you will see how to combine data from different cells (e.g., combining the first, middle, and last names into one column).

Concatenation and Flash Fill

Joining the data from multiple cells or joining cell data with additional text into a single text item is called *concatenation*. The symbol used to perform concatenation is the & (ampersand). An ampersand must be placed between each cell address and each additional string of text that you're adding.

For example, if you wanted to combine the contents of cells A1 and B1 and place the results in cell C1, then you would need to place the following formula in cell C1: =A1&B1. If you wanted to place a space between the data in cell A1 and cell B2, you would need to change the formula to =A1&" "&B1. A space between two quotation marks creates a blank space. If you wanted two spaces between the data, you would place two spaces between the quotation marks.

Figure 10-13 displays a concatenation starting in cell A8 that includes the contents from columns A, B, and C along with additional text. The formula entered in cell A8 is

```
= A2 & " " & B2 & " used " & "$" & C2 & " of fuel this week"
```

◢	A	B	C	D	E
1	**First Name**	**Last Name**	**Fuel**		
2	Mike	Romanchok	28.95		
3	Donnie	Welch	63.85		
4	Isaiah	Parmley	45.75		
5	Jake	Halsey	19.2		
6	Marsha	Landau	25.25		
7					
8	Mike Romanchok used $28.95 of fuel this week				
9	Donnie Welch used $63.85 of fuel this week				
10	Isaiah Parmley used $45.75 of fuel this week				
11	Jake Halsey used $19.2 of fuel this week				
12	Marsha Landau used $25.25 of fuel this week				

Figure 10-13. *Combining data from columns A, B, and C*

The formula was copied down through cell A12 using the AutoFill Handle.

EXERCISE 10-5: USING CONCATENATION

In this exercise, you will use the & character to combine column data along with other characters.

1. Open the workbook Chapter 10.

2. Click the worksheet Concatenation.

3. Figure 10-14 shows the data in the worksheet.

⏴	A	B	C	D
1	City	State	Zip	Combined
2	Fort Wayne	IN	46804	
3	Hobart	IN	46404	
4	Chicago	IL	60601	
5	Springfield	IL	60205	

Figure 10-14. *Data from Concatenation worksheet*

4. In this exercise, you will combine the City, State, and Zip into a single cell. In cell D2, enter the following formula and then press Ctrl + Enter:

    =A2 & ", " & B2 & " " & C2

    The formula displays a comma after the city followed by a space and provides two spaces between the state and zip.

5. Widen column D so that the city, state, and zip all fit inside the column.

6. Double-click the AutoFill Handle to copy the formula down through cell D5. The result should be as shown in Figure 10-15.

⏴	A	B	C	D
1	City	State	Zip	Combined
2	Fort Wayne	IN	46804	Fort Wayne, IN 46804
3	Hobart	IN	46404	Hobart, IN 46404
4	Chicago	IL	60601	Chicago, IL 60601
5	Springfield	IL	60205	Springfield, IL 60205

Figure 10-15. *Result of using AutoFill Handle*

Now that you have practiced using concatenation, let's look at a feature called **Flash Fill**, which can do concatenation without having to use the & sign. Flash Fill can speed up concatenation when you want to concatenate values that are in a row. You enter an example of how you want the data to appear. Excel checks for that pattern in the other rows and using Flash Fill copies that pattern down through the column.

It is still necessary to know how to concatenate using the & sign when

1. You are concatenating to a single cell

2. You are using it in a formula

3. What you are concatenating isn't in the same row

4. You are using it in programming code

Let's look at the same sample that was used for concatenation. The Flash Fill does a lot more than just concatenation. It can be used to find patterns in your columns and then repeat those patterns in another column. For example, you could create a column that uses the first letter from a column containing middle names or create a column that uses the day portion from a date column.

For this example, we will put the concatenation on the same line as the data it is assembling.

EXERCISE 10-6: USING FLASH FILL

In this example, you will use Flash Fill to combine multiple columns of data along with additional text.

1. Open workbook Chapter 10 if it isn't already opened.

2. Click worksheet FlashFill A.

3. Figure 10-16 shows the data in the worksheet.

	A	B	C
1	**First Name**	**Last Name**	**Fuel**
2	Mike	Romanchok	28.95
3	Donnie	Welch	63.85
4	Isaiah	Parmley	45.75
5	Jake	Halsey	19.2
6	Marsha	Landau	25.25

Figure 10-16. *Data in worksheet FlashFill A*

4. In cell D2, type **Mike Romanchok used $28.95 of fuel this week**. Press Enter. Your cursor should be in cell D3.

5. Click the Ribbon's Home tab. In the Editing group, click the Fill button and then select Flash Fill. See Figure 10-17.

Figure 10-17. *Select Flash Fill*

Flash Fill used the three columns of data along with the additional text you typed. It even put the dollar sign in front of each of the amounts. See Figure 10-18.

	A	B	C	D	E	F	G	H
1	First Name	Last Name	Fuel					
2	Mike	Romanchok	28.95	Mike Romanchok used $28.95 of fuel this week				
3	Donnie	Welch	63.85	Donnie Welch used $63.85 of fuel this week				
4	Isaiah	Parmley	45.75	Isaiah Parmley used $45.75 of fuel this week				
5	Jake	Halsey	19.2	Jake Halsey used $19.2 of fuel this week				
6	Marsha	Landau	25.25	Marsha Landau used $25.25 of fuel this week				

Figure 10-18. *Result of Flash Fill*

EXERCISE 10-7: USING MORE FLASH FILL FEATURES

This exercise will show different features as well as showing a potential problem that can arise when using Flash Fill.

1. Open workbook Chapter 10 if it isn't already opened.

2. Click worksheet 3 Names.

3. Figure 10-19 shows the data in the worksheet.

	A	B	C
1	**First Name**	**Middle Name**	**Last Name**
2	John	James	Studebaker
3	Carrie	Mae	Romanchok
4	Russell	Johnson	Carson
5	Annette	Florence	Carson
6	Jake	Wolf	Shaw

Figure 10-19. *Data on the worksheet named 3 Names*

We want to combine the individual's first and last name.

4. In cell D2, type **John Studebaker** and then press Enter. Widen the column to fit the name.

5. Click inside cell D3.

   Instead of going through the process of clicking the Home tab, then clicking the Fill button in the Editing group, and then selecting Flash Fill, we will use the shortcut key for Flash Fill: Ctrl + e.

◢	A	B	C	D
1	**First Name**	**Middle Name**	**Last Name**	
2	John	James	Studebaker	John Studebaker
3	Carrie	Mae	Romanchok	Carrie Romanchok
4	Russell	Johnson	Carson	Russell Carson
5	Annette	Florence	Carson	Annette Carson
6	Jake	Wolf	Shaw	Jake Shaw

Figure 10-20. *Result of using Flash Fill*

6. Press Ctrl + e. Figure 10-20 shows the result.

    You need to be careful about your spelling which will be illustrated in the following steps.

7. Select the data in column D. Right-click the selected data and select Clear Contents.

8. In cell D2, type **John Sludebaker** and then press Enter.

9. Press Ctrl + e.

    Since Excel couldn't find a match name in the row, it created the new name of Sludebaker for all the copied cells. See Figure 10-21.

◢	A	B	C	D
1	**First Name**	**Middle Name**	**Last Name**	
2	John	James	Studebaker	John Sludebaker
3	Carrie	Mae	Romanchok	Carrie Sludebaker
4	Russell	Johnson	Carson	Russell Sludebaker
5	Annette	Florence	Carson	Annette Sludebaker
6	Jake	Wolf	Shaw	Jake Sludebaker

Figure 10-21. *All of the last names in column D were changed to Sludebaker*

    If you make a change to the end of the text, that change is reflected in the rows below.

10. Clear the contents in column D.

11. In cell D2, type **John Studebaker's** and then press Enter.

12. Press Ctrl + e. Figure 10-22 shows the results.

	A	B	C	D
1	**First Name**	**Middle Name**	**Last Name**	
2	John	James	Studebaker	John Studebaker's
3	Carrie	Mae	Romanchok	Carrie Romanchok's
4	Russell	Johnson	Carson	Russell Carson's
5	Annette	Florence	Carson	Annette Carson's
6	Jake	Wolf	Shaw	Jake Shaw's

Figure 10-22. *Result of changing cell D2 to John Studebaker's*

Note   If we hadn't bolded or centered the column headers in columns A, B, and C, Excel wouldn't have known they were column headers. This would have created a problem when we applied Flash Fill. Excel would have combined the text in cell A1 and cell C1 and placed it into cell D1. This problem could also be avoided by putting a column header in cell D1.

You don't need to enter the data from the columns in the same order. We can easily use the last name, followed by a column, then a space, and then the first name.

13. Clear the contents in column D.

14. In cell D2, type **Studebaker, John** and then press Enter.

15. Press Ctrl + e. Figure 10-23 shows the results.

	A	B	C	D
1	**First Name**	**Middle Name**	**Last Name**	
2	John	James	Studebaker	Studebaker, John
3	Carrie	Mae	Romanchok	Romanchok, Carrie
4	Russell	Johnson	Carson	Carson, Russell
5	Annette	Florence	Carson	Carson, Annette
6	Jake	Wolf	Shaw	Shaw, Jake

Figure 10-23. *Result of changing cell D2 to Studebaker, John*

The number of characters you type will be the number of characters that will be displayed from that column. If you type the first two letters in the name Studebaker, then use Flash Fill, Excel will display the first two letters of the other last names. See Figure 10-24.

	A	B	C	D
1	First Name	Middle Name	Last Name	
2	John	James	Studebaker	St
3	Carrie	Mae	Romanchok	Ro
4	Russell	Johnson	Carson	Ca
5	Annette	Florence	Carson	Ca
6	Jake	Wolf	Shaw	Sh

Figure 10-24. *Typed two letters in cell D2 so all the other cells in column 2 display with two letters*

The last name Shaw only contains four characters, so if you enter more than four characters of the last name, the name Shaw will not appear at all. See Figure 10-25.

	A	B	C	D
1	First Name	Middle Name	Last Name	
2	John	James	Studebaker	Stude
3	Carrie	Mae	Romanchok	Roman
4	Russell	Johnson	Carson	Carso
5	Annette	Florence	Carson	Carso
6	Jake	Wolf	Shaw	

Figure 10-25. *Shaw doesn't appear in column D because it is only four characters*

Let's try an example where we combine the individual's first initial and last name.

16. Clear the contents in column D.

17. In cell D2, type **J. Studebaker** and then press Enter.

18. Press Ctrl + e. The results are shown in Figure 10-26.

	A	B	C	D
1	**First Name**	**Middle Name**	**Last Name**	
2	John	James	Studebaker	J. Studebaker
3	Carrie	Mae	Romanchok	C. Romanchok
4	Russell	Johnson	Carson	R. Carson
5	Annette	Florence	Carson	A. Carson
6	Jake	Wolf	Shaw	J. Shaw

Figure 10-26. *Result of entering J. Studebaker in cell D2*

Potential Problem

In the second row, John and James both start with a "J," which can cause some confusion. When there is a duplicate of text, Excel uses the left-most column. If we type John J. Studebaker and then use Flash Fill, Excel will use the initial from column A rather than column B. See Figure 10-27.

	A	B	C	D
1	**First Name**	**Middle Name**	**Last Name**	
2	John	James	Studebaker	John J. Studebaker
3	Carrie	Mae	Romanchok	Carrie C. Romanchok
4	Russell	Johnson	Carson	Russell R. Carson
5	Annette	Florence	Carson	Annette A. Carson
6	Jake	Wolf	Shaw	Jake J. Shaw

Figure 10-27. *Result of entering John J. Studebaker in cell D2*

If we want to use the middle initials, we have to teach it the correct pattern. We can do this by typing the data from the next row(s) until the pattern meets our requirement. Since the middle name in row 3 doesn't have the same initial as the first name, it will meet our needs.

1. Clear the contents in column D.

2. In cell D3, type **Carrie M. Romanchok** and then press Enter.

3. Press Ctrl + e. Figure 10-28 shows the results.

◢	A	B	C	D
1	**First Name**	**Middle Name**	**Last Name**	
2	John	James	Studebaker	John J. Studebaker
3	Carrie	Mae	Romanchok	Carrie M. Romanchok
4	Russell	Johnson	Carson	Russell J. Carson
5	Annette	Florence	Carson	Annette F. Carson
6	Jake	Wolf	Shaw	Jake W. Shaw

Figure 10-28. *Result of entering Carrie M. Romanchok in cell D3*

Using the Status Bar

The status bar at the bottom of your Excel window can be used to display lots of useful information. You can view computational results on different ranges without ever having to enter any formulas. What is displayed on the status bar is up to you. Right-clicking the status bar brings up all the available options, as shown in Figure 10-29. You merely select the options you want to see and deselect those you don't want to see. The selected items appear with check marks to the left of them. Clicking anywhere on a line that has a check mark on it will remove the check mark. Clicking anywhere on a line that doesn't have a check mark will place a check mark on it.

In this section, we'll look at most of the options. We'll omit Signatures, Information Management Policy, and Permissions because these options are used only when dealing with SharePoint.

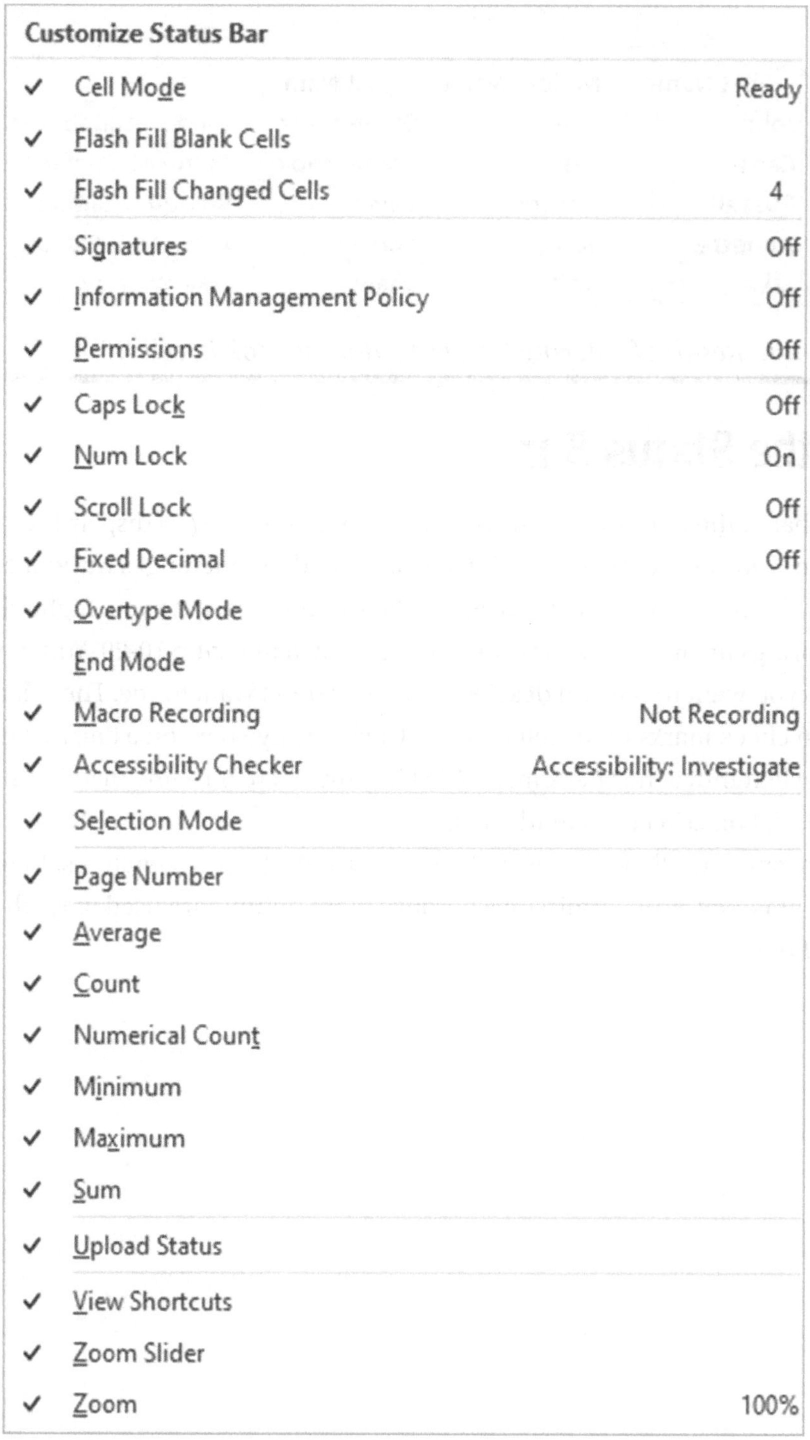

Customize Status Bar

✓	Cell Mode	Ready
✓	Flash Fill Blank Cells	
✓	Flash Fill Changed Cells	4
✓	Signatures	Off
✓	Information Management Policy	Off
✓	Permissions	Off
✓	Caps Lock	Off
✓	Num Lock	On
✓	Scroll Lock	Off
✓	Fixed Decimal	Off
✓	Overtype Mode	
	End Mode	
✓	Macro Recording	Not Recording
✓	Accessibility Checker	Accessibility: Investigate
✓	Selection Mode	
✓	Page Number	
✓	Average	
✓	Count	
✓	Numerical Count	
✓	Minimum	
✓	Maximum	
✓	Sum	
✓	Upload Status	
✓	View Shortcuts	
✓	Zoom Slider	
✓	Zoom	100%

Figure 10-29. *Available options on the Status Bar*

Cell Mode

The Cell Mode is displayed in the left-most position of the status bar. It displays one of three words: **Ready**, **Enter**, or **Edit**. **Ready** indicates that the cell is ready for you to enter data into. If you are entering data into a cell, the word **Enter** appears. If you double-click a cell to put it in Edit mode, then the word **Edit** will appear.

Flash Fill Blank Cells and Flash Fill Changed Cells

If these options are turned on whenever you do a Flash Fill, the status bar will display the number of Blank cells and the number of cells Flash Fill changed.

Using the data in Figure 10-30, if you typed **Rhianna Slager** in cell D1 to show Excel the pattern, selected the cell range D1:D4, and then pressed Ctrl + e to start Flash Fill, there would be one blank cell and two cells would have changed which would show on the status bar as   Flash Fill Blank Cells: 1   Flash Fill Changed Cells: 2

	A	B	C	D
1	Rhianna	Cailin	Slager	
2	Marten	David	Slager	
3				
4	Rebecca	Denise	Parmley	

Figure 10-30. *Original data entered*

The resulting spreadsheet would look like Figure 10-31.

	A	B	C	D
1	Rhianna	Cailin	Slager	Rhianna Slager
2	Marten	David	Slager	Marten Slager
3				
4	Rebecca	Denise	Parmley	Rebecca Parmley

Figure 10-31. *Result of using Flash Fill*

Caps Lock, Num Lock

The words **Caps Lock** or **Num Lock** will display if they are turned on.

Scroll Lock

The words **Scroll Lock** will display if Scroll Lock is turned on. Scroll Lock is helpful when you want to scroll through your worksheet, but you want to keep track of your current location. When Scroll Lock is on, using your arrow keys moves your worksheet in the direction of the arrow, but the active cell doesn't change.

Fixed Decimal

This indicates when Fixed Decimal mode is on. Fixed Decimal mode can be turned on by going to File ➤ Options ➤ Advanced and then placing a check mark in front of **Automatically insert a decimal point**. You can set the number of decimal places. Fixed Decimal mode enters decimals automatically for you. If you have set the decimal places at 2 and you type 2375, Excel will change it to 23.75. If you type a 3, it will be changed to 0.03.

Overtype Mode

Pressing the Insert key turns on Overtype Mode. If Overtype Mode is turned on and you are editing data in a cell, characters that you type will replace existing characters; otherwise, existing characters will move over as you insert new characters. Press the Insert key again to turn off Overtype Mode.

End Mode

This displays when the End key has been pressed. The End key is used as part of a two-key combination with one of the arrow keys.

Macro Recording

The Macro Recording displays the button ☐ on the status bar when a macro is being recorded. Clicking the button stops the recording. Macros are explained in Chapter 19.

Accessibility Checker

The accessibility checker checks your document while you are creating it and provides information on how to make your document more accessible for those with visual impairments.

Selection Mode

The Selection Mode is used for selecting cells with your arrow keys rather than dragging across them with your mouse.

Page Number

The Page Number displays the current page number and total number of pages. The Page Number only displays in Page Layout view.

Average, Count, Numerical Count, Minimum, Maximum, Sum

These options instantly display the result of calculations performed on the cells you select. The selected cells do not have to be adjacent. These math functions are called Aggregate functions. Aggregate functions perform a calculation on a set of values and return a single value.

View Shortcuts

These three buttons appear on the right side of the task bar. They control how the worksheet is viewed. In order, they are the Normal view, Page Layout view, and Page Break Preview.

Zoom and Zoom Slider

The **Zoom** option shows the current zoom percentage. The **Zoom Slider** controls the amount of Zoom.

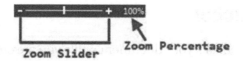

Zoom Slider   Zoom Percentage

The Zoom tool is located on the status bar at the bottom right corner of your Excel window. You can drag the Zoom tool's slider to the right to zoom in on objects (make them larger), or you can drag the slider to the left to zoom out on objects (make them smaller). The zoom range goes from 10% to 400%. Not only is this a great tool when you are developing your worksheets, but it also makes a great presentation tool.

In addition to dragging the slider to change the zoom ratio, you can also do the following:

- Click anywhere in the slide range and the slider will move to that position.

- Each time you click the Zoom out button ▬, objects will decrease by 10% in size.

- Each time you click the Zoom in button ➕, objects will increase by 10% in size.

Note   Changing the zoom ratio only affects your worksheet. It doesn't change the size of your Ribbon.

EXERCISE 10-8: USING THE STATUS BAR

This exercise covers the status bar options.

1. Open the workbook Chapter 10 if it isn't already open.

2. Click the worksheet Status.

3. Figure 10-32 shows the data in the worksheet.

◢	A	B
1		
2	300	500
3	400	600
4	180	900
5	700	400
6	200	500
7	abc	400
8	150	200

Figure 10-32. *Data in the worksheet named Status*

Computations

1. Right-click the status bar. Select Average, Count, Numerical Count, Minimum, Maximum, and Sum.

2. Click an empty cell to close the status bar menu.

3. Drag across cells A2 through A8. Look at the result on your status bar. Your results should look like those in Figure 10-33.

Average: 321.6666667   Count: 7   Numerical Count: 6   Min: 150   Max: 700   Sum: 1930

Figure 10-33. *Computations displayed on status bar*

Note   Empty fields are excluded from computations and cell counts. The difference between Count and Numerical Count is that Numerical Count only counts cells that contain numeric data; therefore, it excluded counting the cell that contained the text **abc**.

---

4. Right-click the status bar. Notice that the results are also displayed on the menu. See Figure 10-34.

✓	A̲verage	321.6666667
✓	C̲ount	7
✓	Numerical Count̲	6
✓	Mi̲nimum	150
✓	Ma̲ximum	700
✓	S̲um	1930

Figure 10-34. *Results of selection are displayed on the Status Bar menu*

Note   The cells you select do not need to be vertical; in fact, they don't even need to be adjacent to each other.

---

5. Select cells A2:B8. Look at the results.

6. Click a blank cell to remove the selection.

7. Hold down the Ctrl key and select cells A2, A4, and B6. The status bar now shows the results for only these three cells.

Selection Mode

1. Click cell A2. Press the F8 key. This makes cell A2 the starting or anchor cell. The status bar displays **Extend Selection**.

2. Click cell B9. This highlights all cells between the anchor cell and the selected cell.

3. Press the right arrow twice. This extends the selection two columns.

4. Press the down arrow key twice. This extends the selection two rows.

5. Click cell F15. This extends the selection from the anchor point to cell F15.

6. Use your arrow keys to move the selection back so that the range A2:B9 is selected.

7. The Shift + F8 key can be used to add additional ranges or individual cells to your current selection. Press Shift + F8. The text on the status bar changes from **Extend Selection** to **Add or Remove Selection.**

8. Drag across cells E2:G9.

9. Drag across cells B13:B16.

10. Randomly click individual cells. They will all become part of the selection.

11. Press Shift + F8 to return to normal mode.

Cell Mode

1. Click cell A17. If Cell Mode is selected, you should see the word **Ready** on the far left side of the status bar. If you don't, right-click the status bar and select Cell Mode.

2. Start typing in the cell. The text should change from **Ready** to **Enter**. After you finished entering something in the cell, press the Enter key.

3. Double-click cell A17 to go into Edit mode. The text on the status bar should now change to **Edit**. You can now make any changes you want to the text in cell A17.

Caps Lock, Num Lock, Scroll Lock

1. Right-click the status bar and select Caps Lock, Num Lock, and Scroll Lock if they are not already selected. Click a cell to close the Status Bar menu.

    The status bar will show when these keys are turned on. When they aren't turned on, nothing will appear. These keys usually have lights associated with them on the keyboard to give you a visual clue as to whether they are on or not. Selecting these gives you one more visual clue.

2. Click each of these keys to see how they appear on the status bar and then click the keys again to turn them off.

Overtype Mode

1. Right-click the status bar and then select Overtype Mode if it isn't already selected.

2. Misspell the word **Excel** in both cell A10 and cell A11 by leaving out the letter c (Exel). If you see the word **Overtype** on the status bar, click the Insert key to turn it off.

3. Double-click inside cell A10 to go into Edit mode. Place your cursor between letters x and e. Now type the letter c. The letter c should insert and the letters e and I are moved to the right. Press Enter.

4. Double-click inside cell A11 to go into Edit mode. Place your cursor between the letters x and e. Press the Insert key. You should now see **Overtype** on the status bar. Now type the letter c. It overtypes the letter e. The letter I doesn't move over. Overtyping prevents the insertion of characters. Press the Insert key. You should no longer see **Overtype** on the status bar.

5. Type the letter e and press Enter.

Zoom, Zoom Slider

1. Drag the Zoom Slider bar to the right to increase the size of items on your worksheet.

2. Drag the Zoom Slider bar to the left to decrease the size of items on your worksheet.

3. Click the plus sign several times. Notice it increases the size by 10% every time you click it.

4. Click the minus sign several times. It decreases the size by 10% every time you click it.

5. Click the Ribbon's View tab. In the Zoom group, click the Zoom button. Select Custom. Enter 147 for the percent. Click the OK button.

6. In the Zoom group, click the 100% button.

7. Select cells A2:B8. Click the Ribbon's View.

8. In the Zoom group, click the **Zoom to Selection** button. Using the Zoom to Selection makes whatever is selected take up the size of the window either vertically or horizontally whichever meets its limit first.

9. Click the center of the Zoom Slider to return to a 100% view.

Summary

Hyperlinks can be used to create quick links to files, web pages, a specific cell, a new document, or an email address.

You can combine data by using either Concatenation or Flash Fill. The & (ampersand) sign is used to tie different strings of data together when using concatenation. Flash Fill can connect data automatically without having to manually type the formula.

Users often forget about the status bar. This is probably because the options aren't visible unless you right-click it. The status bar is the fastest way to view the results of multiple functions applied to multiple ranges at the same time. The results displayed on the status bar are temporary; they are not stored in cells.

In the next chapter, you will learn the different methods of copying and pasting data. You will also see how to switch row and column headings. You'll learn how to use the Microsoft Office Clipboard and how to key data into multiple worksheets at the same time.

CHAPTER 11

Transferring and Duplicating Data to Other Locations

There are times when you place data into cells and then later decide that you would rather display that data in a different location, or maybe you want that data duplicated elsewhere. Excel provides a variety of methods for doing this as well as allowing you to select the format of the newly transferred or duplicated data. This transferred or duplicated data can be copied to locations within other data, in other worksheets, and in other workbooks.

After reading and working through this chapter, you should be able to

- Move and copy cells using the drag-and-drop method

- Copy and paste manually

- Copy and paste using the keyboard

- Copy data to other worksheets

- Copy data from one workbook to another

- Use Paste Special to transpose rows and columns

- Use the Paste button gallery

- Use the Add and Multiply option of Paste Special

- Use the Skip Blanks option of Paste Special

- Use the Microsoft Office Clipboard

- Insert copied or moved cells

© David Slager and Annette Slager 2020
D. Slager and A. Slager, *Essential Excel 2019*, https://doi.org/10.1007/978-1-4842-6209-2_11

Moving and Copying Data

You can move data using Excel's **move** feature also known as the **cut** feature. **Moving** a cell moves the value in one cell to another cell. After the move, the original cell location will be empty. Any existing values in the receiving location will be overwritten by the moved data.

You may want to have the same data in multiple locations. Rather than typing that same data repeatedly, you can use Excel's **copy** feature. **Copying** a cell involves copying the value in a cell to another cell. After the copy, the value will exist in both the original location and the location to where the value was copied. Whenever you cut or copy data, a duplicate of that data is placed in the Microsoft Office Clipboard. The clipboard is shared with other Microsoft Office products. If you copy cells and then start Word and open its clipboard, you will see the cells that you copied in Excel. You can then paste that cell data into your Word document.

Moving a cell can be accomplished by using drag and drop or using the **cut-and-paste** method. Copying can be accomplished by using drag and drop while holding down the Ctrl key or using the copy-and-paste method.

Moving and Copying Cells Using the Drag-and-Drop Method

When moving data using the drag-and-drop method, follow these steps:

1. Select an area to be moved.

2. Move the mouse cursor on one of the selected areas sides. The cursor should turn into a move arrow, as shown in Figure 11-1.

◢	A	B
1	123	235
2	475	195
3	190	280

Figure 11-1. *Select area to be moved*

3. Drag the data to its new location. As you drag the data, a box the size of the selected area moves with the cursor so that you always know where the data will be moved. See Figure 11-2.

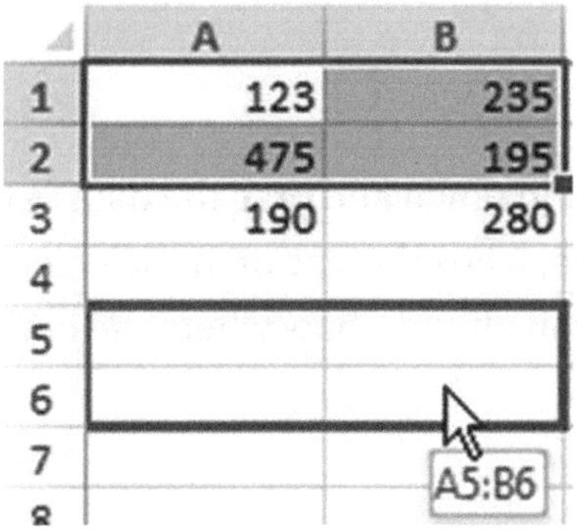

	A	B
1	123	235
2	475	195
3	190	280
4		
5		
6		
7		A5:B6
8		

Figure 11-2. *As you drag the data, a box the size of selected area moves with the cursor*

4. Release the mouse button. The data is moved to its new location. Any formatting that was applied to a cell is moved with the value. See Figure 11-3.

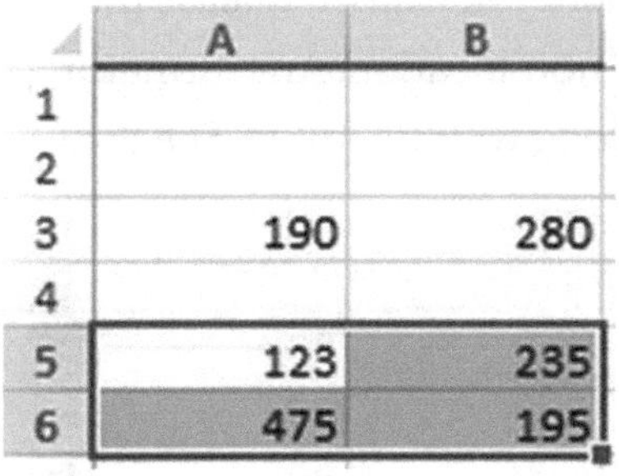

	A	B
1		
2		
3	190	280
4		
5	123	235
6	475	195

Figure 11-3. *Result of moving data*

When copying data using the drag-and-drop method, follow these steps:

1. Select an area to be copied.

2. Hold down the Ctrl key and move the mouse cursor on one of the selected areas sides. The cursor should turn into an arrow with a plus sign. The plus sign means you are copying the selected data. See Figure 11-4.

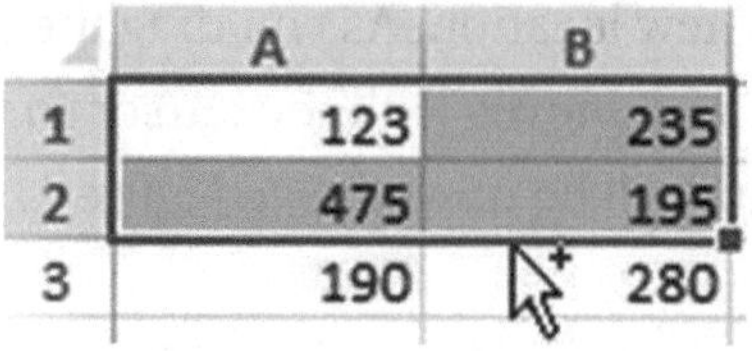

Figure 11-4. *Select area to be moved*

3. Keep the Ctrl key held down and drag the data to its new location. As you drag the data, a box the size of the selected area moves with the cursor so that you always know where the data will be copied. See Figure 11-5.

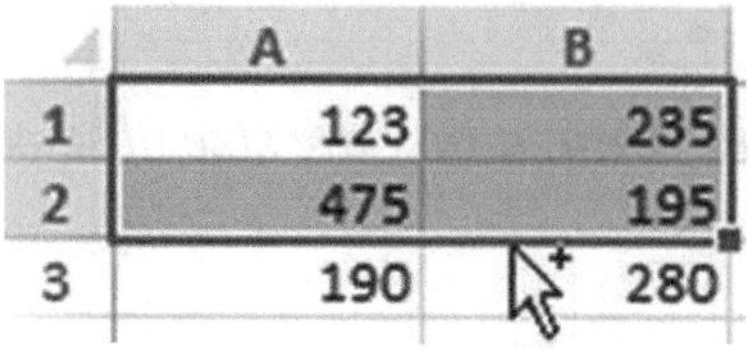

Figure 11-5. *As you drag the data, a box the size of selected area moves with the cursor*

4. When the box is where you want it, release the mouse button. The data is copied to its new location. Any formatting that was applied to a cell is copied along with the value. See Figure 11-6.

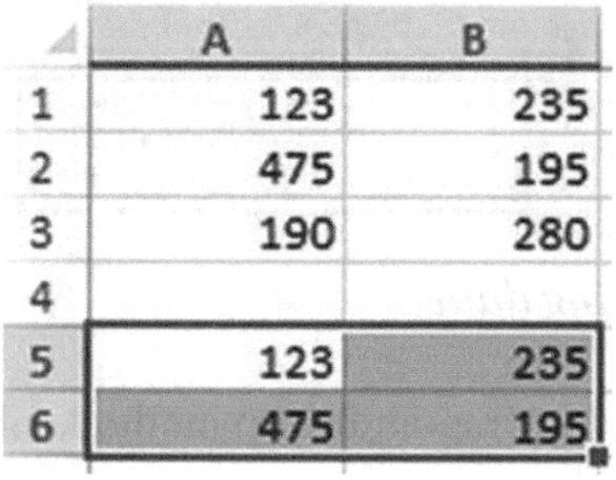

Figure 11-6. *Result of copying data*

Note   Cell references are automatically adjusted when copying cells but not when moving them. You will need to manually adjust cell references for moved cells and any cells that pointed to them.

Moving and Copying Cells Using the Cut and Copy Buttons

The buttons for copying and pasting are located on the Home tab in the Clipboard group. See Figure 11-7.

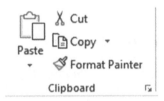

Figure 11-7. *Ribbon's Clipboard group*

To move cell values

1.   Select the cells you want to move.

2.   Click the Cut button ✂.

3.   Select the upper left cell of where you want the contents moved. Click the Paste button.

To copy cell values

1.   Select the cells you want to move.

2.   Click the Copy button ▣.

3.   Select the upper left cell of where you want the contents copied. Click the Paste button.

Rather than using Cut, Copy, and Paste buttons on the Ribbon, you can select them from the context menu by right-clicking the selected cells.

Note   Excel displays an animated marquee around cells that have been cut or copied. After you have finished pasting the contents to their new location, cancel the animated marquee by pressing the Esc key.

Moving and Copying Cells Using the Keyboard

It's impossible to remember all the keyboard shortcuts, but you should remember the ones for copying, cutting, and pasting since you will be using these all the time. See Table 11-1.

Table 11-1. *Keyboard Shortcuts*

Function	Shortcut Keys
Cut	Ctrl + X
Copy	Ctrl + C
Paste	Ctrl + V

These same shortcuts are used in most other programs as well.

EXERCISE 11-1: COPYING, PASTING, AND MOVING CONTENT

In this exercise, you duplicate cell content by using copy and paste and then move content using cut and paste.

1. Start a new workbook named Chapter 11. Name the worksheet "CopyPaste."

2. Enter the data in Figure 11-8 into a worksheet.

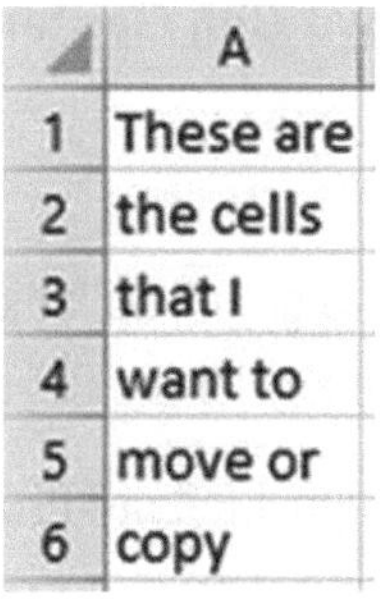

◢	A
1	These are
2	the cells
3	that I
4	want to
5	move or
6	copy

Figure 11-8. *Enter this data*

Duplicate Cell Contents

1. Select cells A1:A6.

2. On the Home tab in the Clipboard group, click the Copy button ⧉ to copy the contents of cells A1:A6. Pressing Ctrl + C is the shortcut key for the Copy command.

3. An animated marquee appears around cells A1:A6 to indicate that these cells are being copied.

4. Click inside cell C3. This is the upper left cell of where the content will be copied to.

5. On the Home tab in the Clipboard group, click the top half of the Paste button ⧉. A duplicate copy of the contents in cells A1:A6 now exists in cells C3:C8.

6. As long as the animated marquee appears around the cells being copied, you can continue to paste them in other locations. Click inside cell C5. This time use the paste shortcut Ctrl + V instead of using the Paste button. When you paste content into cells that already have content, the original contents are copied over.

7. Press the Esc key. The animated marquee is removed.

Move Cell Contents

1. Select cells A1:A6.

2. On the Home tab in the Clipboard group, click the Cut button ✂ to cut the cells. Pressing Ctrl + X is the shortcut key for the Cut command.

3. Click inside cell E1. This is the upper left cell of where the content will be moved to.

4. Press Ctrl + V to paste the contents to their new location. The cell contents have been moved to cells E1:E6. The animated marquee has been removed.

You have learned how to move and copy cell data by using the drag-and-drop method, using keyboard shortcuts, and clicking the Cut and Copy command buttons. These methods didn't provide any options for specifying what to paste and how to format it. If you want a choice as to how the pasted data is to appear, you will need to make your selection from the Paste button gallery. The Paste button gallery is accessed by clicking the bottom half of the Paste button.

Paste Button Gallery

The Paste button is actually two separate buttons. If you let your cursor set on the top half of the Paste button, you will see the tooltip shown on the left side of Figure 11-9. If you let your cursor set on the bottom half of the Paste button (the half with the down arrow), you will see the tooltip on the right side of Figure 11-9. Clicking the bottom portion of the Paste button highlights the entire button and presents a gallery of paste options.

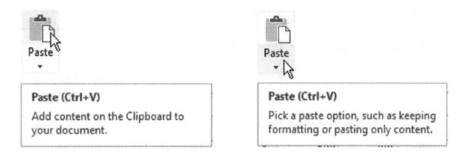

Figure 11-9. *Paste by clicking the top of the button, or open the Paste gallery by clicking the lower half of the button*

The paste options in the gallery can perform such functions as pasting only the formatting rather than the data or pasting the results of formulas and not the formulas themselves. Figure 11-10 shows the Paste gallery options. Not all the options appear every time; it depends upon what you copied. Table 11-2 displays each paste option and a description for it.

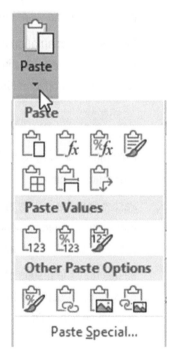

Figure 11-10. *Paste options*

Table 11-2. *The Paste Gallery Options with Their Descriptions*

Button	Name with Shortcut Key	Description
	Paste (P)	Pastes everything exactly as is from the original copy. Keeps all the formatting, formulas, etc. Doesn't adjust column widths.
	Formulas (F)	Pastes everything except the formatting.
	Formula & Number Formatting (O)	Pastes formulas & number formatting such as dollar signs, commas, negatives in parentheses, etc., but does not copy text formatting.
	Keep Source Formatting (K)	Pastes all the data along with *all* the formatting.
	No Borders (B)	Pastes everything except any borders that were applied to the copied cells.
	Keep Source Column Widths (W)	Pastes everything and adjusts the column widths to match those that were copied.
	Transpose	Changes the orientation of the copied cells. For example, if the copied cells were in a vertical format, they would be changed to a horizontal format.

Paste Values

	Values (V)	Only the values are pasted not the formulas. No formatting is pasted.
	Values & Number Formatting (A)	Pastes only the results of formulas, not the formulas themselves. Pastes any numeric formatting such as $ and commas, etc., but not any text formatting.
	Values & Source Formatting (E)	Pastes only the results of formulas, not the formulas themselves. Pastes all the formatting applied to the original copied cells.

(continued)

Table 11-2. (*continued*)

Button	Name with Shortcut Key	Description
Other Paste Options		
	Formatting (R)	Pastes only the formatting applied to the cells, not the data itself.
	Paste Link (N)	Creates a link to the copied item. Pasting a link means the pasted cell will always be updated when you change the original cell.
	Picture (U)	Pastes your data as a picture rather than individual cells of data. The pasted data can't be edited.
	Linked Picture (I)	Pastes everything. Any change you make to the original data or its formatting is automatically updated in the pasted data.

EXERCISE 11-2: USING THE PASTE GALLERY

In this exercise, you will apply the various paste options to see how they affect your data.

1. Use the Chapter 11 workbook you created. Create a new worksheet named "Pasting."

2. Type Jan in cell A1. Press Ctrl + Enter.

3. Drag the AutoFill Handle to the right until you have the months Jan through May.

4. Select Cells A1:E1. On the Home tab in the Font group, click the **Increase Font Size** button three times. Change the font color to Red. Change the fill color to (Green, Accent 6, Lighter 80%). Bold and center the headings.

5. Enter the rest of the data in Figure 11-11.

▲	A	B	C	D	E
1	**Jan**	**Feb**	**Mar**	**Apr**	**May**
2	1234875	85729.29	18495.73	501.27	389
3	3800.25	2500.5	1850.01	201.23	275.25
4	-1400.75	100.3	1450.2	130.49	129.14

Figure 11-11. *Enter the rest of this data*

6. Apply AutoSum to cells A5:E5.

7. Select cells A2:E4. On the Home tab, in the Number group, click the Comma Style button.

8. Select cells A5:E5. On the Home tab, in the Number group, click the **$** button.

9. On the Home tab, in the Font group, click the down arrow for borders. Select **Top and Double Bottom** border.

10. Select column heads A through E. Double-click between column heads E and F to AutoFit the column width as shown in Figure 11-12.

▲	A	B	C	D	E
1	**Jan**	**Feb**	**Mar**	**Apr**	**May**
2	1,234,875.00	85,729.29	18,495.73	501.27	389.00
3	3,800.25	2,500.50	1,850.01	201.23	275.25
4	(1,400.75)	100.30	1,450.20	130.49	129.14
5	$ 1,237,274.50	$ 88,330.09	$ 21,795.94	$ 832.99	$ 793.39

Figure 11-12. *AutoFit the column width*

11. Select cells K1 and K2. Apply bold, italic, and underline to the cells.

View How Paste Options Affect the Copied Data

1. Select cells A1:E5. Press Ctrl + C.

2. Click inside cell G1. Click the bottom half of the Paste button so that you can see the gallery of paste options.

3. Move your cursor over the paste options. The results should appear as those in Table 11-3.

Table 11-3. *The Paste Gallery Options with Examples*

Paste Option	Example

📋 Paste

Pastes all the data and formatting but doesn't adjust the columns widths to match those being copied. This is the same as pressing Ctrl + V.

Jan	Feb	Mar	Apr	May
#######	#######	#######	501.27	389.00
#######	#######	#######	201.23	275.25
#######	100.30	#######	130.49	129.14
#######	#######	#######	$ 832.99	$ 793.39

📋ƒₓ Formulas

Pastes the data and the formulas. Doesn't paste any formatting. The formatting that exists in the pasted area will affect the copied cells (notice the formatting in cells K1 and K2).

Jan	Feb	Mar	Apr	*May*
1234875	✪729.29	18495.73	501.27	***389***
3800.25	2500.5	1850.01	201.23	275.25
-1400.75	100.3	1450.2	130.49	129.14
1237275	88330.09	21795.94	832.99	793.39

📋%ƒₓ Formula & Number Formatting

Pastes formulas and only number formatting, no other formatting. The formatting that exists in the pasted area will affect the copied cells.

Jan	Feb	Mar	Apr	*May*
#######	#######	#######	501.27	***389.00***
#######	#######	#######	201.23	275.25
#######	100.30	#######	130.49	129.14
#######	#######	#######	$ 832.99	$ 793.39

📋 Keep Source Formatting

Preserves the look of the original data except it doesn't adjust column widths.

Jan	Feb	Mar	Apr	May
#######	#######	#######	501.27	389.00
#######	#######	#######	201.23	275.25
#######	100.30	#######	130.49	129.14
#######	#######	#######	$ 832.99	$ 793.39

📋 No Borders

Notice the top and double bottom borders weren't pasted.

Jan	Feb	Mar	Apr	May
#######	#######	#######	501.27	389.00
#######	#######	#######	201.23	275.25
#######	100.30	#######	130.49	129.14
#######	#######	#######	$ 832.99	$ 793.39

(continued)

Table 11-3. (*continued*)

Paste Option	Example
⬛ Keep Source Column Widths Looks like the original including maintaining the column widths.	Jan Feb Mar Apr May 1,234,875.25 85,729.29 18,495.73 501.27 389.00 3,800.25 2,500.50 1,850.01 201.23 275.25 (1,400.75) 100.30 1,450.20 130.49 129.14 $1,237,274.75 $88,330.09 $21,795.94 $ 832.99 $ 793.39
⬛ Transpose Changes the orientation of the rows and columns.	Jan ###### ###### ###### ###### Feb ###### ###### 100.30 ###### Mar ###### ###### ###### ###### Apr 501.27 201.23 130.49 $ 832.99 May 389.00 275.25 129.14 $ 793.39
⬛ Values	None of the formatting is copied over. It displays the results of the formulas but doesn't keep the formulas.
⬛ Values & Number Formatting	Looks the same as Formula & Number Formatting. It doesn't preserve formulas; rather, it displays the results of the formulas.
⬛ Values & Source Formatting	Looks the same as Keep Source Formatting. It doesn't preserve formulas; rather, it displays the results of the formulas.

4. With the copy area A1:E5 still selected, click inside cell G1. Click the down arrow portion of the Paste button. In the **Other Paste Options** area, click the Formatting button ⬛. All the formatting has been pasted but none of the cell contents. See Figure 11-13.

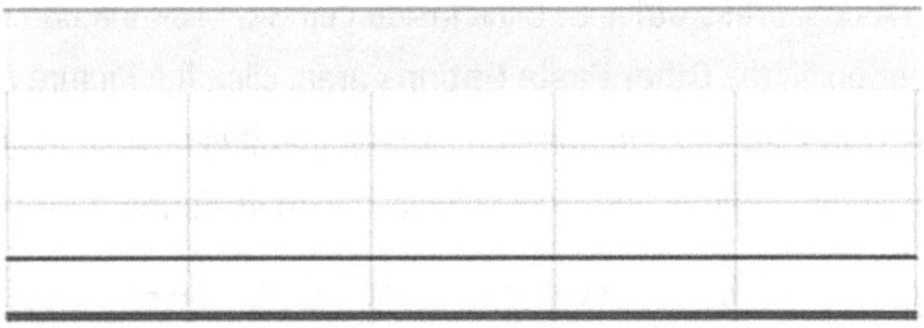

Figure 11-13. *Formatting pasted without cell contents*

5. If you type data in one of the cells in the pasted area, the data will take on the formatting of the original copied cell:

   a. Type June in cell G1.

   b. Type 395 in cell G5.

6. Select cells G1:K5. On the Home tab, in the Editing group, click the Clear button and select Clear All. Selecting Clear All clears the data and all formatting.

7. Select cells A1:E5. Press Ctrl + C. Click inside cell G1. Click the bottom half of the Paste button. In the **Paste Values** area, click the Values button 🗋. Click inside cell G5. Notice in the formula bar that there is no formula but only the result of the formula used in the copied cell.

8. Select cells G1:K5. On the Home tab, in the Editing group, click the Clear button and select Clear All.

9. Select cells A1:E5. Press Ctrl + C. Click inside cell G1. Click the bottom half of the Paste button. In the **Other Paste Options** area, click the Paste Link button. The pasted area is linked to the copied area. Any changes you make in the copied area will be reflected in the pasted area.

10. Click inside cell G2. Cell G2 contains the formula =A2. Cell G3 contains the formula =A3.

11. Change the value in cell A3 from 3,800.25 to 6,725.50. Press Enter. The change is reflected in cell G3. The total in cell G5 was changed so that it has the same value as cell A5.

12. Select cells G1:K5. On the Home tab, in the Editing group, click the Clear button and select Clear All.

13. Select cells A1:E5. Press Ctrl + C. Click inside cell G1. Click the bottom half of the paste button. In the **Other Paste Options** area, click the Picture button. The pasted area is treated as an image. You can tell that it is now an image because the contextual tab group Picture Tools displays with its Format tab.

14. Click the Format tab. You can move your cursor across the Picture Styles to see how they would affect the image. Click the down arrow for the Picture Styles to see more choices.

15. Press the Delete key to remove the image in cell G1.

16. With cells A1:E5 still selected, click inside cell G1. Click the bottom half of the paste button. In the **Other Paste Options** area, click the Linked Picture button. Notice the contextual tab group Picture Tools is visible again. Since the image is linked to the copied cells, any changes you make to the original cells will be reflected in the pasted image.

17. Change the value in cell D3 to 87.75. Press Enter. The change is reflected in the pasted image.

18. Change the font color in cell A1 to blue. Italicize cell A1. Notice the change in the pasted area.

19. Close the Chapter 11 workbook.

You have used Paste options to copy cell values, cell formats, formulas only, and other formatting. Next, you will learn how to copy data from one worksheet to another and then later how to copy data from one workbook to another.

Copy Data to Other Worksheets Using Fill Across Worksheets

You can copy all or a portion of the data on one worksheet to other worksheets by using the Fill Across Worksheets copy command. The following steps are used to perform the Fill Across Worksheets copy command:

1. Select the cells that you want copied to the other worksheets. You can only select the entire worksheet or a single range. If you want to copy multiple ranges, then you will need to repeat the steps.

2. Select the worksheet tabs of the worksheets where you want the data copied.

3. On the Ribbon, on the Home tab, in the Editing group, click the Fill button. Choose Across Worksheets. The Fill Across Worksheets dialog box opens. Here you can select if you want to copy everything, only the data, or only the formatting.

EXERCISE 11-3: USING THE FILL ACROSS WORKSHEETS COPY COMMAND

In this exercise, you will copy data from one worksheet to other worksheets.

1. Create a new workbook named Chapter 11B.

2. Create two new worksheets so that you have tabs for Sheet1, Sheet2, and Sheet3.

3. Enter the Data in Figure 11-14 in Sheet1 and format as follows:

   a. Widen column B so that all the text fits in the column.

   b. Bold and italicize the title **Chart of Accounts**. Change the size of the title font to 16. Select cells A1 and B1. Merge and center the title.

   c. Apply a fill color of **Green, Accent 6, Lighter 40%** for the title.

   d. Click cell B3. Hold down the Ctrl key while you click cells B12 and B17. Click Bold and then Center for the Alignment.

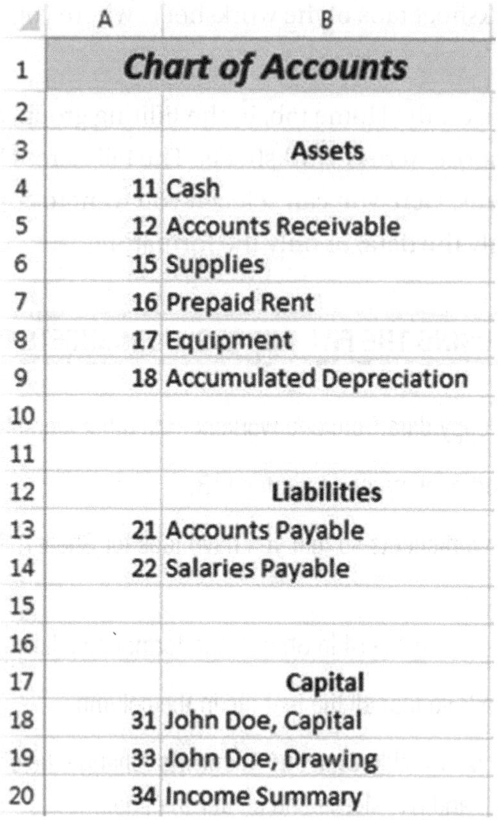

	A	B
1		*Chart of Accounts*
2		
3		**Assets**
4	11	Cash
5	12	Accounts Receivable
6	15	Supplies
7	16	Prepaid Rent
8	17	Equipment
9	18	Accumulated Depreciation
10		
11		
12		**Liabilities**
13	21	Accounts Payable
14	22	Salaries Payable
15		
16		
17		**Capital**
18	31	John Doe, Capital
19	33	John Doe, Drawing
20	34	Income Summary

Figure 11-14. *Enter this data*

4. Select cells A3:B9.

5. Hold down the Ctrl key and click the Sheet2 tab and Sheet3 tab at the bottom of the worksheets.

6. On the Ribbon, on the Home tab, in the Editing group, click the Fill button. Choose Across Worksheets. See Figure 11-15.

Figure 11-15. *Select Across Worksheets*

The Fill Across Worksheets dialog box opens. See Figure 11-16.

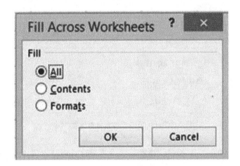

Figure 11-16. *Fill Across Worksheets dialog box*

You can select if you want to copy everything, only the data, or only the formatting.

7. Select All and click the OK button.

8. Right-click one of the worksheet tabs. Select **Ungroup Sheets**.

9. Click the Sheet2 tab. You will notice that the column widths didn't automatically adjust to the copied data.

10. Adjust the column widths on Sheet2 and Sheet3.

Copy Data from One Workbook to Another

You can copy data from one workbook to another by copying the data from one workbook and then pasting it into another workbook using one of the paste methods.

EXERCISE 11-4: COPYING DATA FROM ONE WORKBOOK TO ANOTHER

This exercise covers using the various paste options when copying data.

1. Use the workbook Chapter 11B that you created in the previous exercise.

2. Press Ctrl + N. This is the shortcut key to create a new workbook.

3. Click the Ribbon's View tab. In the Window group, click the **Arrange All** button. The Arrange Windows dialog box displays. See Figure 11-17.

Figure 11-17. *Arrange Windows dialog box*

4. Select Vertical. Click the OK button.

5. Select all of the data on Sheet1 of the Chapter 11B workbook. Press Ctrl + C to copy the data.

6. Click inside cell A1 of the new workbook. Press Ctrl + V to paste the data into the new workbook.

   The column widths of the pasted data do not match the column widths of the copied data.

7. Click the Paste Options button ![Paste Options icon] (Ctrl) ▾ that appears at the bottom right of the pasted data, or click the Ctrl key to bring up the Paste Options. See Figure 11-18.

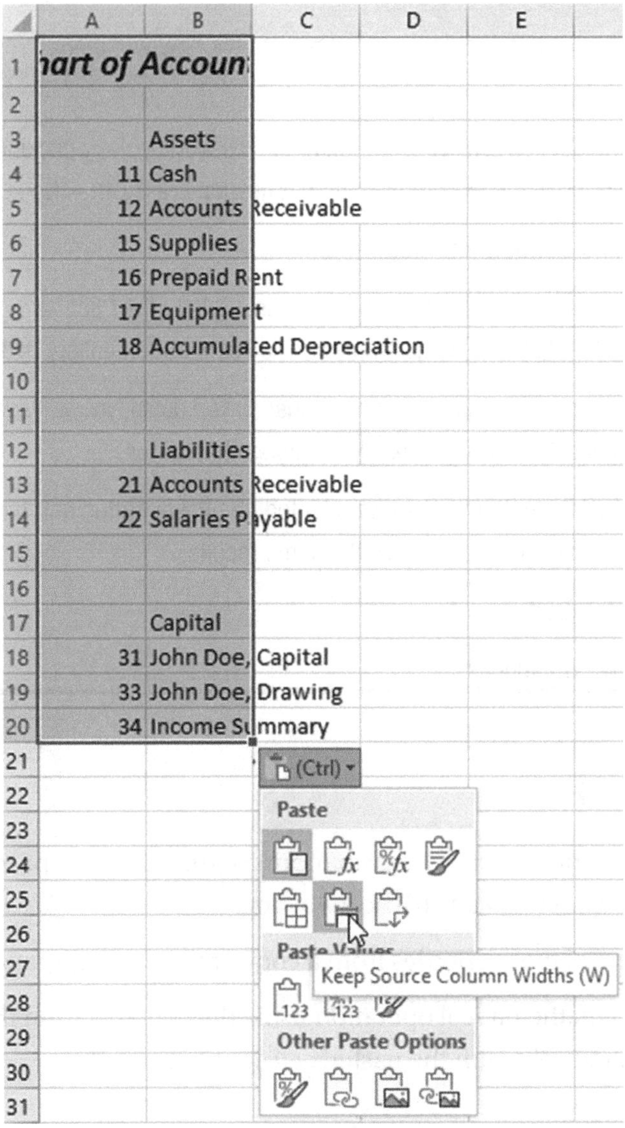

Figure 11-18. *Paste options*

8. Click the Keep Source Column Widths button. The columns adjust to their original width.

   Let's repeat what we have just done, but this time, we will copy the entire worksheet.

9. Let's start by clearing the data we pasted into the new workbook. Press Ctrl + Z.

10. Select all the data on Sheet1 of the Chapter 11B workbook by clicking the Select All button to the left of the column A header. Press Ctrl + C.

11. Click cell A1 of the new workbook.

12. On the new workbook, click the Ribbon's Home tab. In the Clipboard group, click the bottom half of the Paste button. The Paste button gallery is displayed.

13. Move your cursor over the different buttons; as you do so, you will see exactly how the spreadsheet would look if you clicked that button.

    Since you had the entire worksheet selected, the Paste button and the Keep Source Column Widths button give the same results.

14. Click the Paste button.

15. Close the two workbooks.

Paste Special

The Paste Special command provides even more control over your pasting. It should be used whenever you want to do any of the following:

- Paste some of the properties from a copy or a cut cell operation.

- Perform a mathematical operation using the contents from the copied or cut cells with the pasted cell contents.

- Ignore cell blanks.

- Transpose row data into column data or column data into row data.

- Link the pasted cells to the copied cells.

The Paste Special command is located at the bottom of the Paste gallery. See Figure 11-19.

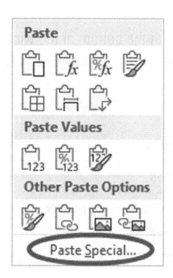

Figure 11-19. *Select Paste Special*

Using Paste Special

Selecting Paste Special brings up the Paste Special dialog box. See Figure 11-20.

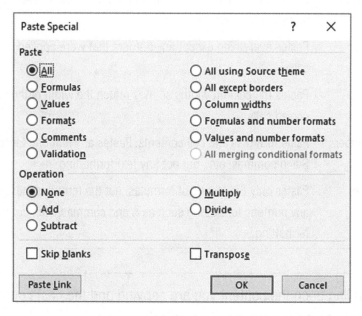

Figure 11-20. *Paste Special dialog box*

Note   The Paste Link button at the bottom of the window is only available when All or All Except Borders is selected. It is grayed out when anything else is selected.

Table 11-4 describes the Paste options. Table 11-5 describes the Operation options.

Table 11-4. *Paste Special Options*

Paste Special Option	Description
All	This performs the same operation as pressing Ctrl + V. It pastes all cell contents and formatting. It does not adjust the column widths.
Formulas	Pastes the formulas used in the copied cells.
Values	Pastes only the results of formulas, not the formulas themselves.
Formats	Pastes only the formatting, not the data.
Comments	Pastes only comments attached to the cell.
Validation	Pastes data validation rules for the copied cells.
All Using Source Theme	Pastes all cell contents and formatting using the theme that was applied to the copied cells.
All Except Borders	Pastes everything except any borders that were applied to the copied cells.
Column Widths	Pastes the column widths so they match the width of the columns of the copied cells.
Formulas and Number Formats	Paste formulas and cell contents. Pastes any numeric formatting such as $ and commas, etc., but not any text formatting.
Values and Number Formats	Pastes only the results of formulas, not the formulas themselves. Pastes any numeric formatting such as $ and commas, etc., but not any text formatting.

Note   Depending upon the content you are copying and the paste options that you have selected, some options may be unavailable.

Table 11-5. *Operation Options*

Operation Option	Description
None	Do not perform any mathematical operation to the pasted cells.
Add	Add the values in the copied cells to those in the pasted cells.
Subtract	Subtract the values in the copied cells from those in the pasted cells.
Multiply	Multiply the values in the copied cells to those in the pasted cells.
Divide	Divide the values in the copied cells into those in the pasted cells.

Note   Mathematical operations can only be applied to values. If you are performing one of the mathematical operations (Add, Subtract, Multiply, or Divide), you must select either All, All except borders, or Values and number formats from the Paste options.

To use Paste Special

1. Select the cells that contain the data or attributes that you want to copy.

2. Press Ctrl + C.

3. Select the upper left cell of the paste area. If you are performing a mathematical operation, you will need to select all the cells in the paste area.

4. On the Home tab, in the Clipboard, click the bottom half of the Paste button. Select Paste Special.

5. Select the Paste option you wish to perform.

6. Select the Operation option you want to perform. If you aren't performing a mathematical operation, select **None**.

7. If you want to paste static data (not linked), click the Paste option you want to use and then click the OK button.

Or

- If you want to paste linked data, you must select either **All** or **All except borders** from the Paste options and then click the Paste Link button.

- Avoid copying blank cells to your paste area by selecting **Skip blanks**.

- Change row data to columns or vice versa by selecting **Transpose**.

Helpful Hint   If you find yourself using Paste Special frequently you can add it to your Quick Access Toolbar.

Using Paste Special to Transpose Rows and Columns

The Paste Special command has a Transpose option that is used for converting row data to column data and vice versa. A likely scenario in which you would use the Transpose function is that you have set up your row and column headings and then decide that what you really wanted was to have the row headings as the column headings and the column headings as the row headings. Rather than manually retyping the headings, you can flip-flop them by using Paste Special's Transpose option.

EXERCISE 11-5: USING THE PASTE SPECIAL TRANSPOSE OPTION

In this exercise, you will transpose the rows and columns of the original copied data.

1. Create a new workbook named Chapter 11C. Name the worksheet "Transpose."

2. Enter the data in Figure 11-21. Bold the column headings.

⁄	A	B	C	D	E
1		**Product 1**	**Product 2**	**Product 3**	**Product 4**
2	January	80	75	68	78
3	February	41	45	67	60
4	March	61	22	56	29
5	April	74	19	89	28
6	May	90	26	13	41
7	June	32	80	69	73

Figure 11-21. *Enter this data*

3. Select cells A1:E7.

4. Right-click the selected data and select Copy.

5. Right-click inside cell A10. Select Paste Special from the context menu.

6. In the Paste area, select **Values** (Figure 11-22). Selecting **All** would leave what will now become are row headings bolded.

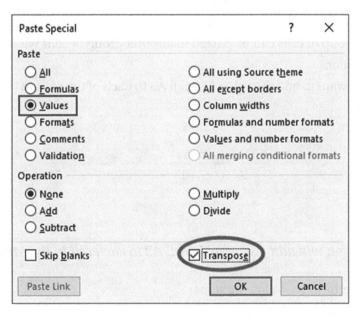

Figure 11-22. *Select Values and Transpose from the Paste Special dialog box*

7. Select Transpose.

8. Click the OK button.

9. Bold and center the column headings.

The result should be as shown in Figure 11-23.

10		January	February	March	April	May	June
11	Product 1	80	41	61	74	90	32
12	Product 2	75	45	22	19	26	80
13	Product 3	68	67	56	89	13	69
14	Product 4	78	60	29	28	41	73

Figure 11-23. *Result of transposing row and column heads*

Using Paste Special to Perform Calculations

Paste Special can perform calculations while pasting. Paste Special has operation options for Add, Subtract, Multiply, and Divide.

A single cell value can be pasted to a group of cells while performing an arithmetic operation, or a group of cells can be pasted to another group of cells while performing an arithmetic operation.

Let's say you want to add the value 5 in cell A3 to each of the cells in the range A1:D1. See Figure 11-24.

◢	A	B	C	D
1	10	15	20	25
2				
3	5			

Figure 11-24. *You will add the value in cell A3 to each cell in the range A1:D1*

You would perform the following steps:

1. Select cell A3 and then do a copy.

2. Select the cells where you want to paste the operation. In this case, that would be range A1:D1.

3. Right-click one of the selected cells and then select Paste Special.

4. Select the Add operation.

The result would be as shown in Figure 11-25 where 5 has been added to each value in the pasted range.

◢	A	B	C	D
1	15	20	25	30
2				
3	5			

Figure 11-25. *5 was added to each cell in row 1*

You can combine a group or multiple groups while performing an arithmetic operation.

If you wanted to add the individual values in row 3 to the individual values in row 1, as shown in Figure 11-26, you would perform the following steps:

1. Select the cell range A3:D3 and then do a copy.

2. Select cells A1:D1.

3. Right-click one of the selected cells and then select Paste Special.

4. Select the Add operation.

◢	A	B	C	D
1	10	15	20	25
2				
3	5	8	10	5

Figure 11-26. *Values in row 3 will be added to those in row 1*

Figure 11-27 shows the result.

◢	A	B	C	D
1	15	23	30	30

Figure 11-27. *Result of adding the values in row 3 to row 1*

If you were copying fewer values than you were pasting, then the copied values would repeat their pattern.

In Figure 11-28, the value in cell A3 would be added to the value in cell A1. The value in cell B3 would be added to the value in B1.

Then the value in cell A3 would be added to the value in C1, and the value in cell B3 would be added to the value in D1.

◢	A	B	C	D
1	10	15	20	25
2				
3	5	10		

Figure 11-28. *Value in cell A3 will be added to cells A1 and C1. Value in cell B3 will be added to cells B1 and D1*

Figure 11-29 shows the result.

◢	A	B	C	D
1	15	25	25	35

Figure 11-29. *Value in A3 was added to A1 and C1. Value in B3 was added to B1 and D1*

EXERCISE 11-6: USING THE PASTE SPECIAL ADD OPTION

The Paste Special option **Add** can be beneficial when you want to combine groups of values. The **Add** option adds the values you copied in the *source* range with those in the *destination* range and then replaces the destination range with the new values.

In this example, we will add the values in Groups B and C to the values in Group A. We will then delete the rows that contain the Group B and Group C values.

1. Open workbook Chapter 11C. Create a new worksheet named "PasteAdd."

2. Create the spreadsheet as shown in Figure 11-30.

	A	B	C	D	E	F	G
1	Group A	25	50	60	80	90	120
2	Group B	47	58	54	39	40	71
3	Group C	64	32	27	18	49	51
4	Group D	14	69	27	85	43	29

Figure 11-30. *Enter this data*

3. Select the cell range B2:G2 and do the following:

   a. Press Ctrl + C.

   b. Click inside cell B1. Click the lower half of the Paste button.

   c. Select Paste Special.

   d. Under Operation, select Add. Click OK.

4. Select the cell range B3:G3. Press Ctrl + C. Click inside cell B1. Click the lower half of the Paste button. Select Paste Special. Under Operation, select Add. Click OK.

5. Select row heads 2 and 3 and then right-click and select Delete from the context menu. See Figure 11-31.

	A	B	C	D	E	F	G
1	Group A	136	140	141	137	179	242
2	Group D	14	69	27	85	43	29

Figure 11-31. *Result of deleting rows 2 and 3*

The values in Group A are now the sum of the values that were previously in Group A plus Group B plus Group C.

EXERCISE 11-7: USE THE MULTIPLY OPTION OF PASTE SPECIAL

We have a list of prices for the items we sell. We want to increase the price for each product by 10%.

1. Open the workbook Chapter 11C. Create a new worksheet named "PasteMultiply."

2. Enter the data as shown in Figure 11-32.

	A	B
1	**Price List**	
2	Product 1	5.75
3	Product 2	8.95
4	Product 3	12.75
5	Product 4	14.85
6	Product 5	1.92
7	Product 6	4.75
8	Product 7	13.92
9	Product 8	0.99
10	Product 9	1.75
11	Product 10	10.5
12		
13	Increase	1.1

Figure 11-32. Enter this data in worksheet PasteMultiply

3. Click cell B13. Press Ctrl + C.

4. Select cell range B2:B11. Right-click the selected cells and then select **Paste Special** from the context menu.

5. Select Multiply under Operation. Click the OK button.

   Excel multiplies each value in the pasted area by 1.1. Figure 11-33 shows the result.

◢	A	B
1	**Price List**	
2	Product 1	6.325
3	Product 2	9.845
4	Product 3	14.025
5	Product 4	16.335
6	Product 5	2.112
7	Product 6	5.225
8	Product 7	15.312
9	Product 8	1.089
10	Product 9	1.925
11	Product 10	11.55
12		
13	Increase	1.1

Figure 11-33. *Result of multiplying each value in the pasted area by 1.1*

Experiment with the other Operation options.

EXERCISE 11-8: USING THE PASTE SPECIAL SKIP BLANKS OPTION

When you copy a range of cells, any cells that are blank in the copied range will overwrite any cells in the pasted range, thus making those cells also blank.

However, if you use the Skip Blanks option for Paste Special, the blank cells in the copied range will not overwrite the data in the cells in the pasted range.

1. Use the workbook Chapter 11C. Create a new worksheet named "Skip."

2. Type the data as shown in Figure 11-34.

◢	A	B	C	D	E
1	20		30		50
2					
3	90	75	85	140	200

Figure 11-34. *Enter this data*

3. Drag across cells A1:E1. Press Ctrl + C to copy them.

4. Click cell A3. Press Ctrl + V. See Figure 11-35.

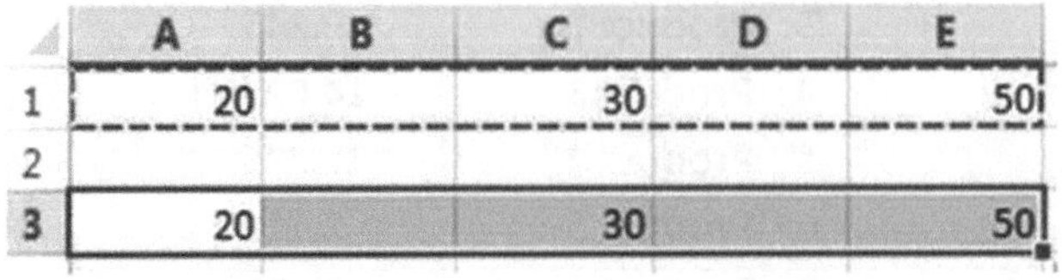

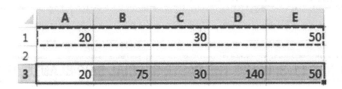

Figure 11-35. *Cell range A1:E1 pasted to range A3:E3*

5. Notice that the copied blank cells replaced those in cells B3 and D3.

6. Press the Undo button on the Quick Access Toolbar or press Ctrl + Z to undo the paste.

7. Click the lower half of the Paste button. Select Paste Special.

8. Select **Skip Blanks**. Click OK. See Figure 11-36.

Figure 11-36. *Used Paste Special to prevent blank cells from copying over nonblank cells*

The values in cells B3 and D3 have not been changed.

Inserting Copied or Moved Cells

Rather than having a move or copy override the values in the receiving location, you can insert the copied or moved cells so that the values in the receiving location are moved to accommodate them.

Insert Copied Cells

The three steps to copying the cells in the range A1:B2 of Figure 11-37 to the cell range A6:B7 are as follows.

	A	B	C
1	123	235	
2	475	195	
3	190	280	
4			
5	48	85	90
6	75	95	60
7	20	80	43

Figure 11-37. *Copy cells in range A1:B2 to the cells in range A6:B7*

Step 1

First, select the cells that you want to copy. Right-click the selection and then select Copy from the context menu. See Figure 11-38.

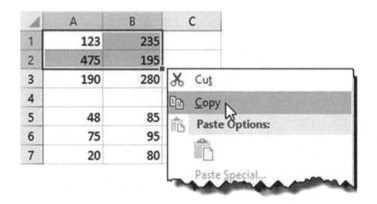

Figure 11-38. *Right-click the selected cells and then select Copy*

Step 2

Right-click cell B6 and then select Insert Copied Cells from the context menu. See Figure 11-39.

Figure 11-39. *Select Insert Copied Cells*

Step 3

Select the direction in which you want the current cell values moved.
See Figure 11-40.

Figure 11-40. *Select the direction in which you want the current cell values moved*

Figure 11-41 displays the result from Step 3 if you selected Shift cells down.

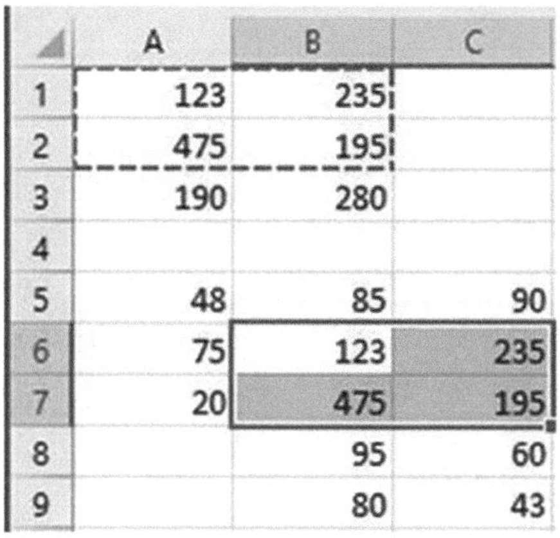

Figure 11-41. *Results if you selected Shift cells down*

Figure 11-42 displays the result from Step 3 if you selected Shift cells right.

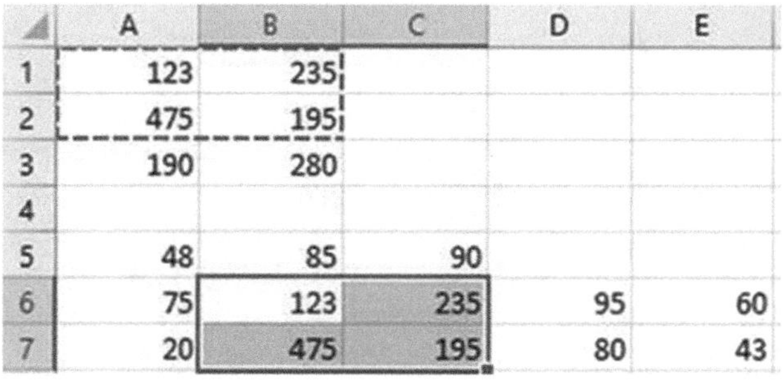

Figure 11-42. *Results if you selected Shift cells right*

Insert Cut Cells

Inserting cut cells works similar to inserting copied cells.

EXERCISE 11-9: INSERTING CUT CELLS

In this exercise, you will move data to another location forcing the data that is currently in the receiving location to move down into the cells below.

1. Use the Chapter 11C workbook you created. Create a new worksheet named "CutCells."

2. Enter the data in Figure 11-43 into a worksheet.

◢	A	B	C
1	123	235	
2	475	195	
3	190	280	
4			
5	48	85	90
6	75	95	60
7	20	80	43

Figure 11-43. *Enter this data*

3. Select the cell range A1:B2. Press Ctrl + X to cut the cells for a move.

4. Right-click inside cell B5. Select **Insert Cut Cells**. Select **Shift Cells Down**. Figure 11-44 shows the result.

◢	A	B	C
1			
2			
3	190	280	
4			
5	48	123	235
6	75	475	195
7	20	85	90
8		95	60
9		80	43

Figure 11-44. *Copied data over existing data*

5. Close the Chapter 11C workbook.

Using the Microsoft Office Clipboard

Whenever you do a Copy, the item you copied is placed into the Office Clipboard. The Office Clipboard is a temporary storage area to which you can copy as many as 24 items. The items can be text or graphical such as charts, images, clip art, shapes, and so on. The clipboard is shared with the other Microsoft Office products. This means that you can copy and paste items between Excel, Access, PowerPoint, Word, and so on. The clipboard makes it quick and easy to rearrange data in your documents or to merge elements from other office products into your documents. Items that are on the clipboard can be pasted individually or all at once.

The items stay on the clipboard until you delete them from the clipboard or until you have closed all the Microsoft Office applications. The last item you added to the clipboard remains even after you have closed all the Microsoft Office applications. The last item copied is added to the top of the clipboard. If you try to add more than 24 items to the clipboard, the first item you added will be removed. All items are cleared when you restart your computer.

The clipboard can be opened by clicking the Dialog Box Launcher for the clipboard group. This opens the Clipboard Task Pane.

EXERCISE 11-10: USING THE CLIPBOARD

This exercise covers copying text, shapes, images, and charts using the clipboard.

1. Open workbook Clipboard.xlsx.

2. Click the Clipboard worksheet tab.

3. Click the Dialog Box Launcher ⌐ in the lower right corner of the Clipboard group to open the clipboard pane.

4. Click the Clear All button.

5. Make the following selections:

   a. Select cells B3:C3. Press Ctrl + C.

   b. Select cells B4:C4. Press Ctrl + C.

   c. Select cells B5:C5. Press Ctrl + C.

   d. Select cells B6:C6. Press Ctrl + C.

6. Click the Blank worksheet tab and complete the following steps:

   a. Click cell A2. Click Wolves 7 in the Clipboard pane.

   b. Click cell A3. Click Elephants 5.

    c. Click cell A7. Click the Paste All button.

    d. Click the Clear All button.

7. Select the Clipboard worksheet tab and complete the following steps:

    a. Click the edge of the Shape. Press Ctrl + C.

    b. Click the frame of the Chart. Press Ctrl + C.

    Notice that the Paste All button is grayed out. The Paste All button doesn't work with Charts and Images.

8. Click the Image worksheet tab. Click the Automobile picture. Press Ctrl + C.

    The image, the shape, and the chart should appear as three separate items in your clipboard.

9. Click the Blank tab. Copy the shape, chart, and image to different areas on the Blank worksheet.

10. Click the Clear All button.

11. Close the Clipboard pane.

12. Close the workbook.

You have learned a wide variety of methods for copying and pasting data. You have seen how to insert copied data between other data. Next, you will learn to save data entry time by entering the same data in multiple worksheets at the same time.

Entering Data into Multiple Worksheets at the Same Time

When you have multiple worksheets selected, whatever you enter in one worksheet will be entered into the other worksheets at the same time. Any formatting you apply will be applied to all the worksheets.

EXERCISE 11-11: KEYING THROUGH MULTIPLE WORKSHEETS

In this exercise, you will enter text and formulas into one worksheet which will be duplicated through three other worksheets. The formatting you apply to one worksheet will also be duplicated in the other worksheets.

1. Create a new workbook with four worksheets. Name the worksheets "Qtr1," "Qtr2," "Qtr3," and "Qtr4."

2. Hold down your Ctrl key and click the Qtr1, Qtr2, Qtr3, and Qtr4 worksheet tabs.

3. Enter Windy City Comics in cell A1 and format as follows:

    a. Merge and center the cell range A1:E1.

    b. Change the font to Harlow Solid Italic and change the font size to 14.

    c. Use Fill Color to apply a light gold background color.

    d. Apply a thick box border.

4. Enter Quarter 1 Sales in cell A2. Merge and center the cell range A2:E2. Change the font size to 14.

5. Enter Store 1 in cell A4. Use the Autofill Handle on cell A4 and drag through cell A9.

6. Enter the rest of the worksheet so that it looks like Figure 11-45.

◢	A	B	C	D	E	F	G	H
1			*Windy City Comics*					
2			Quarter 1 Sales					
3					Totals		Average Store Sales	
4	Store 1						Highest Store Sale	
5	Store 2						Lowest Store Sale	
6	Store 3							
7	Store 4							
8	Store 5							
9	Store 6							
10	Totals							
11								
12	Avg							
13	High							
14	Low							

Figure 11-45. *Make your worksheet look like this worksheet*

7. Select the cell range B3:E3. Bold the text. Select the Center button in the Alignment group.

8. Select the cell range B10:D10. Hold down your Ctrl key while you drag across cells E4 through E9. Apply the lightest green background color.

9. Apply a thick outside border to cell E10. Apply the second lightest green background color.

10. Select the cell range B3:D3. Apply a thick bottom border. Your worksheet should now look like Figure 11-46.

◢	A	B	C	D	E	F	G	H
1		*Windy City Comics*						
2		Quarter 1 Sales						
3					Totals		Average Store Sales	
4	Store 1						Highest Store Sale	
5	Store 2						Lowest Store Sale	
6	Store 3							
7	Store 4							
8	Store 5							
9	Store 6							
10	Totals							
11								
12	Avg							
13	High							
14	Low							

Figure 11-46. *Your worksheet should now look like this*

11. Enter a 1 in cell B4. Use the Autofill Handle to copy the 1 down through cell B9. Drag the Autofill Handle in cell B9 to cell D9 to copy the 1s to columns C and D.

12. Select the cell range B4:E10. Click the AutoSum button.

13. Enter the formula =Average(B4:B9) in cell B12.

14. Enter the formula =Max(B4:B9) in cell B13.

15. Enter the formula =MIN(B4:B9) in cell B14.

16. Select the cell range B12:B14. Use the Autofill Handle to copy the formulas through cell D14.

17. Enter the formula =Average(E4:E9) in cell I3.

18. Enter the formula =Max(E4:E9) in cell I4.

19. Enter the formula =Min(E4:E9) in cell I5.

20. Select cell range B12:D12. Hold down the Ctrl key and select cell I3. On the Ribbon's Home tab in the Number group, click the down arrow of the Number Format and select Number. We are selecting the Number format so that the averages have two decimal positions.

▲	A	B	C	D	E	F	G	H	I
1		*Windy City Comics*							
2		Quarter 1 Sales							
3					Totals		Average Store Sales		3.00
4	Store 1	1	1	1	3		Highest Store Sale		3
5	Store 2	1	1	1	3		Lowest Store Sale		3
6	Store 3	1	1	1	3				
7	Store 4	1	1	1	3				
8	Store 5	1	1	1	3				
9	Store 6	1	1	1	3				
10	Totals	6	6	6	18				
11									
12	Avg	1.00	1.00	1.00					
13	High	1	1	1					
14	Low	1	1	1					

Figure 11-47. *Your worksheet should now look like this worksheet*

Your spreadsheet should now look like Figure 11-47.

21. Select the cell range B4:D9. Right-click the selection and select Clear Contents. You will get divide by zero errors for the Averages because there is no data.

    Our template is completed. Now, you need to ungroup the spreadsheets so that you can enter data into each individual worksheet.

22. Right-click a selected worksheet and select Ungroup Sheets.

23. Click each of the worksheets. They should be identical.

24. Click the Qtr1 worksheet. Enter Jan in cell B3, Feb in cell C3, and Mar in cell D3.

25. Select the cell range B4:B9. Right-click one of the selected cells and select Clear Contents. Enter any numbers you want to in the cell range B4:D9.

Your worksheet should look similar to the one in Figure 11-48.

◢	A	B	C	D	E	F	G	H	I
1		*Windy City Comics*							
2		Quarter 1 Sales							
3		Jan	Feb	Mar	Totals		Average Store Sales		4259.33
4	Store 1	1127	1305	1409	3841		Highest Store Sale		4885
5	Store 2	1285	1502	1604	4391		Lowest Store Sale		3841
6	Store 3	1303	1208	1407	3918				
7	Store 4	1205	1309	1706	4220				
8	Store 5	1401	1695	1205	4301				
9	Store 6	1807	1606	1472	4885				
10	Totals	8128	8625	8803	25556				
11									
12	Avg	1354.67	1437.50	1467.17					
13	High	1807	1695	1706					
14	Low	1127	1208	1205					

Figure 11-48. Your worksheet should now look like this worksheet

26. Click the Qtr2 worksheet. Change the heading from Quarter 1 Sales to Quarter 2 Sales. Enter Apr in cell B3, May in cell C3, and Jun in cell D3.

27. Enter any numbers you want to in the cell range B4:D9.

Summary

The difference between a move and a copy is that after you have performed a copy, the data will exist in both its original location and in the pasted location, while a move clears the data from its original location and stores it in the pasted location. The Paste Special command has much more capability than just pasting data into a cell. While it is pasting, it can perform calculations at the same time. It has options for copying only the cell values without copying the formatting or copying the formatting without the cell values.

The Transpose option for Paste Special converts row data into column data or column data into row data. The Fill Across Worksheets option copies data from one worksheet into multiple worksheets.

Whenever you do a copy, the copy is temporarily stored in the Office Clipboard. You can store 24 items on the Office Clipboard at a time. This clipboard is shared with other office products so you can easily copy and paste between them.

In the next chapter, you will see how to convert your data into a table. You can apply multiple sorts and complex filtering to tables so you can view only the data you want to see and in the order you want to see it.

CHAPTER 12

Working with Tables

A table is a structured collection of data consisting of a header row and multiple data rows. A table is used to contain data of a single type such as orders, sales, inventory, and so on. A table makes it easier to filter, sort, analyze, summarize, and format your data.

After reading and working through this chapter, you should be able to

- Create and format a table

- Sort and filter a table

- Add to a table

- Resize a table

- Add calculated fields to a table

- Create slicers

- Apply themes

- Apply styles

- Create conditional formatting

You can pull data from multiple tables to create another table. Data in a table can be kept separate from other data in your worksheet. You can take an existing range of cells and format them into a table. If you no longer want the data to be in a table format, you can convert it back to a range.

© David Slager and Annette Slager 2020
D. Slager and A. Slager, *Essential Excel 2019*, https://doi.org/10.1007/978-1-4842-6209-2_12

The last part of this chapter covers conditional formatting. Conditional formatting improves the readability of your data by allowing you to apply formats or display icons based on cell values. For example, you can have cell data in a column appear in bold red if the value is less than 0, blue if the value falls between 0 and 1000, and green if the value is greater than 1000. You can decide how the data will be formatted and create the rules for those formats.

Creating and Formatting Tables

The process of creating and formatting a table is very simple. Select the data you want to be in a table and pick a format you would like it to be in. That's all there is to it.

You format the data you have selected on a worksheet as a table by clicking the Format as Table button. The Format as Table button is located on the Home tab in the Styles group. Clicking the Format as Table button brings up a gallery of styles for you to select from. See Figure 12-1. The gallery is divided into the categories Light, Medium, and Dark. Some of the styles have colored column headings; some have cell borders and others do not; some have a different color for every other row and others do not. If you don't like any of the styles Excel has provided for you, then you can create your own style.

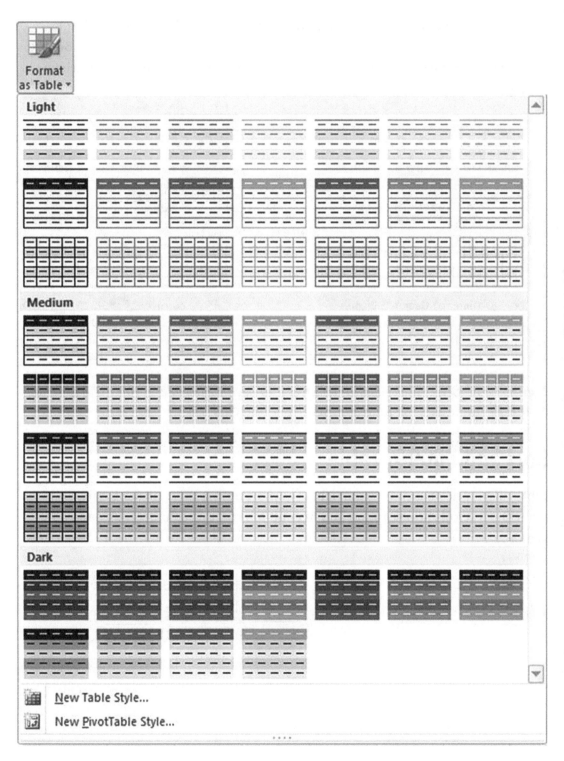

Figure 12-1. *Gallery of styles. Select the one you want to format your table*

When you select a style, Excel brings up the **Format As Table** dialog box. If you haven't already selected the cells to become a table, you can do so at this point. Be sure the **My table has headers** check box is selected if your selected data contains headers. See Figure 12-2.

	A	B	C	D	E	F	G	H	I	J
1										
2	Student ID	Last Name	First Name	Major	Degree	Sex	City	State	School Fees	Dorm Fees
3	10	Carson	Annette	English					3952.25	
4	14	Sterk	Carrie	Math					4327.73	1495.5
5	39	Bundy	Peggy	History					4832.25	1275.85
6	49	Smith	Margaret	Physics					5927.72	1495.5
7	55	Hough	Marten	Science					3178.7	
8	56	Kamp	Brianna	Biology					4835.7	1495.5
9	63	Thomas	George	English					4075.85	1500
10	66	Carlson	Rhiana	Biology					5050.29	1500
11	69	Ulrich	Robert	Biology	BS	M	Chicago	IL	6175.8	1475
12	70	Lupner	Lisa	English	MA	F	Toledo	OH	5439.5	1495.5

Dialog box overlay:
Format As Table   ?   ×
Where is the data for your table?
=$A$2:$J$12
☑ My table has headers
OK   Cancel

Figure 12-2. *Select the cells that you want to become a table. If you are including headings, make sure My table has headers is selected*

Sort and Filter a Table

Tables display down arrows to the right of each column header. Clicking one of these down arrows brings up sort and filter options for that column as illustrated in Figure 12-3.

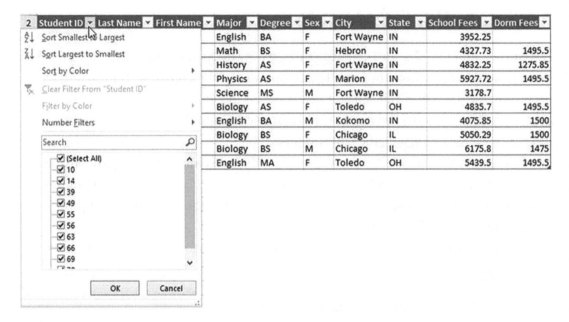

Figure 12-3. *Select down arrow to the right of the header to select filters*

Every unique value that appears in the column is displayed. If there are duplicate values in a column, they are not repeated in the list.

Removing a check mark in front of a value will prevent the row that contains that value from being displayed. If you have removed multiple check marks, you can quickly reselect all of them by selecting (Select All). If you only want to select a couple of items for filtering, you can deselect (Select All) and then just click the items that you want included in your table.

Filtering does not rearrange or alter your data; it temporarily hides rows that don't match the criteria you specify. Removing filters restores your table to the way it was before the filters were applied.

Columns that have been filtered can easily be identified because the down arrow button changes to a filtered button ⏷. Excel displays a tool tip that shows what will be filtered when you hover your cursor over the filtered button.

Adding to the Excel Table

You can add additional records to the table by typing them in the row after the last record in the table. After you press the Enter key at the end of the record, it will automatically become part of the table. You can add new columns in a similar manner by keying a new column to the right of the last column. You can also expand the table to include more rows or columns by dragging the table expansion handle located at the bottom right-hand corner of the table 🔲. You can then type additional data into the expanded area. You can also insert a row or column anywhere in the table. If you insert a new column, the column heading will be named Column. Double-click the column heading and change the name to what you want.

Automatic expansion of the table can be stopped by performing the following steps:

1. Click File ➤ Options.

2. Click Proofing in the left pane.

3. Click the AutoCorrect Options button in the right pane.

4. Click the **AutoFormat As You Type** tab.

5. Deselect **Include new rows and columns in table Automatically as you work**. See Figure 12-4.

6. Click OK.

7. Click OK.

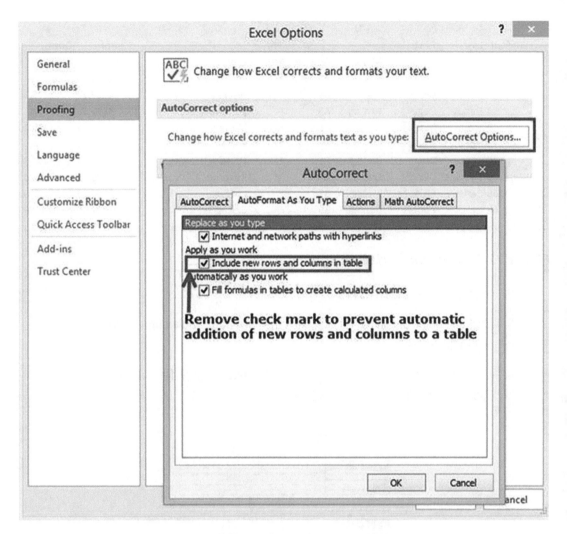

Figure 12-4. *Remove check mark from Include new rows and columns in table*

EXERCISE 12-1: CREATING AND USING TABLES

In this exercise, you will create a table and then apply filters to different columns so that you can view only the data that you want to see.

1. Open the workbook named **Chapter12.** Click the worksheet named "Students." See Figure 12-5.

	A	B	C	D	E	F	G	H	I	J
1	Student ID	Last Name	First Name	Major	Degree	Sex	City	State	School Fees	Dorm Fees
2	10	Carson	Annette	English	BA	F	Fort Wayne	IN	3952.25	
3	14	Sterk	Carrie	Math	BS	F	Hebron	IN	4327.73	1495.5
4	39	Bundy	Peggy	History	AS	F	Fort Wayne	IN	4832.25	1275.85
5	49	Smith	Margaret	Physics	AS	F	Marion	IN	5927.72	1495.5
6	55	Hough	Marten	Science	MS	M	Fort Wayne	IN	3178.7	
7	56	Kamp	Brianna	Biology	AS	F	Toledo	OH	4835.7	1495.5
8	63	Thomas	George	English	BA	M	Kokomo	IN	4075.85	1500
9	66	Carlson	Rhianna	Biology	BS	F	Chicago	IL	5050.29	1500
10	69	Ulrich	Robert	Biology	BS	M	Chicago	IL	6175.8	1475
11	70	Lupner	Lisa	English	MA	F	Toledo	OH	5439.5	1495.5

Figure 12-5. *Data on worksheet Students*

2. Select the cell range A1:J11.

3. Click the Home tab on the Ribbon. Click the **Format as Table** button in the Styles group.

4. Click the first style in the third row of the **Medium** category. This will be White, Table Style Medium 15. See Figure 12-6.

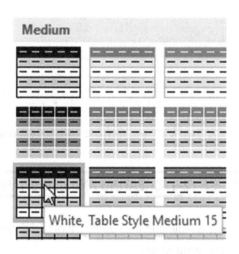

Figure 12-6. *Select Table Style Medium 15*

5. Since you have already selected the range, Excel just wants you to verify it. Since you included your headers in the table, you need to select **My table has headers**. See Figure 12-7. Click the OK button.

Figure 12-7. *Select My table has headers*

6. Click the down arrow to the right of Student ID. Click **Sort Largest to Smallest**. The table is rearranged so that the record with the highest Student ID appears first. See Figure 12-8.

Student ID ↓	Last Name	First Name	Major	Degree	Sex	City	State	School Fees	Dorm Fees
70	Lupner	Lisa	English	MA	F	Toledo	OH	5439.5	1495.5
69	Ulrich	Robert	Biology	BS	M	Chicago	IL	6175.8	1475
66	Carlson	Rhianna	Biology	BS	F	Chicago	IL	5050.29	1500
63	Thomas	George	English	BA	M	Kokomo	IN	4075.85	1500
56	Kamp	Brianna	Biology	AS	F	Toledo	OH	4835.7	1495.5
55	Hough	Marten	Science	MS	M	Fort Wayne	IN	3178.7	
49	Smith	Margaret	Physics	AS	F	Marion	IN	5927.72	1495.5
39	Bundy	Peggy	History	AS	F	Fort Wayne	IN	4832.25	1275.85
14	Sterk	Carrie	Math	BS	F	Hebron	IN	4327.73	1495.5
10	Carson	Annette	English	BA	F	Fort Wayne	IN	3952.25	

Figure 12-8. *Table has been sorted so that the highest Student ID appears first*

A down arrow icon has been added to the button to the right of Student ID. This is to inform you that the column is sorted from largest to smallest.

7. Click the down arrow to the right of City. The list displays all the values in the City column. We have a very small table, but if the column consisted of thousands of different cities, we could use the Search box to find the city we wanted to use.

8. Click the check box for (Select All) to deselect all items in the list.

9. Select the Fort Wayne check box.

10. Click the OK button. Only the records that have Fort Wayne for a city are displayed. See Figure 12-9.

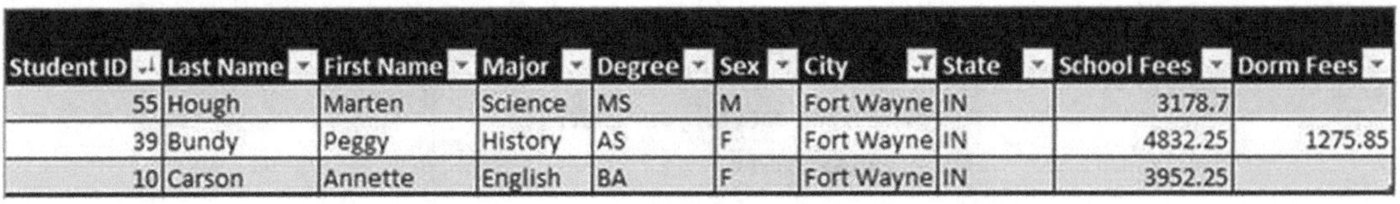

Student ID	Last Name	First Name	Major	Degree	Sex	City	State	School Fees	Dorm Fees
55	Hough	Marten	Science	MS	M	Fort Wayne	IN	3178.7	
39	Bundy	Peggy	History	AS	F	Fort Wayne	IN	4832.25	1275.85
10	Carson	Annette	English	BA	F	Fort Wayne	IN	3952.25	

Figure 12-9. *The City column has been filtered to only show records that have Fort Wayne for a city*

11.   The button to the right of City now displays a filter icon so that we can easily see that this column is being filtered. Move your cursor over the filter icon. The tool tip displays the filter that is being applied to the column. See Figure 12-10.

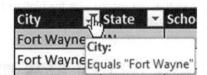

Figure 12-10. *Move cursor over filter icon to view tool tip*

12.   Click the Filter button to the right of City. A highlighted check mark appears to the left of the column data. See Figure 12-11. This is to inform you that the column is being filtered by one or more of its values.

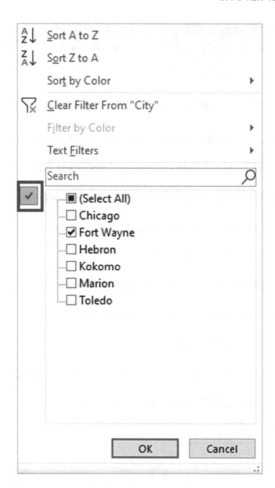

Figure 12-11. *Highlighted check mark means that the field is currently being filtered*

13. Click **Clear Filter from "City."** All the data is now displayed.

14. Let's view only the students who are getting a BS degree in Biology:

    a. Click the down arrow to the right of Major.

    b. Deselect (Select All).

    c. Select only **Biology**.

    d. Click the OK button.

15. Click the down arrow to the right of Degree. Select only **BS**. Click the OK button. Both buttons should have a filter icon on them. Only those records that meet both conditions are displayed. See Figure 12-12.

Student ID	Last Name	First Name	Major	Degree	Sex	City	State	School Fees	Dorm Fees
69	Ulrich	Robert	Biology	BS	M	Chicago	IL	6175.8	1475
66	Carlson	Rhianna	Biology	BS	F	Chicago	IL	5050.29	1500

Figure 12-12. *Result of filtering Biology for Major and BS for Degree*

16. Click the Filter button to the right of Major. Click **Clear Filter from "Major."**

17. Click the Filter button to the right of Degree. Click **Clear Filter from "Degree."**

18. Let's find students who have **ann** in their first name but not the letters tt:

    a. Click the down arrow to the right of First Name.

    b. Select Text Filters and then click Custom Filter.

    c. Click the down arrow for the first drop-down box and select **contains**.

    d. Press the Tab key and enter **ann** in the next box.

    e. Press the Tab key. Select **And**.

    f. Click the down arrow in the next box and select **does not contain**.

    g. Press the Tab key. Enter **tt** in the box. See Figure 12-13.

    h. Click the OK button.

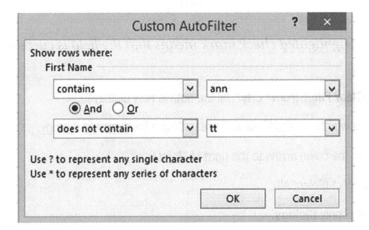

Figure 12-13. *Filter for First Name that contains ann but doesn't contain the letters tt*

Records for Rhianna and Brianna are displayed because they both contain ann. The record for Annette doesn't display because even though the name contains Ann it also contains tt. See Figure 12-14.

Student ID	Last Name	First Name	Major	Degree	Sex	City	State	School Fees	Dorm Fees
66	Carlson	Rhianna	Biology	BS	F	Chicago	IL	5050.29	1500
56	Kamp	Brianna	Biology	AS	F	Toledo	OH	4835.7	1495.5

Figure 12-14. *Result of filtering First Name*

19. Move your cursor over the Filter button to the right of First Name. The tool tip displays the filter that is being applied to the column. See Figure 12-15.

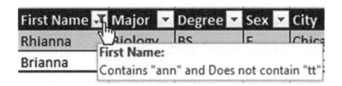

Figure 12-15. *Tool tip for filter*

20. Click the Filter button to the right of First Name. Click **Clear Filter from "First Name"**.

21. Let's find the three students who are paying the most for their school fees:

   a. Click the down arrow to the right of School Fees.

   b. Select **Number Filters**.

   c. Click **Top 10**. You are not limited to the top 10. You can enter however many items you want displayed. You can select items from the bottom by selecting Bottom from the first drop-down box.

   d. Enter the data shown in Figure 12-16.

Figure 12-16. *Filter for top 3 items*

e. Click OK. The result is the top three highest school fees as shown in Figure 12-17.

Student ID	Last Name	First Name	Major	Degree	Sex	City	State	School Fees	Dorm Fees
70	Lupner	Lisa	English	MA	F	Toledo	OH	5439.5	1495.5
69	Ulrich	Robert	Biology	BS	M	Chicago	IL	6175.8	1475
49	Smith	Margaret	Physics	AS	F	Marion	IN	5927.72	1495.5

Figure 12-17. *Filtered for top three highest school fees*

22. Click the Filter button to the right of School Fees. Click **Clear Filter from "School Fees."**

23. Let's find the 30% of the students who are paying the least for their school fees:

    a. Click the down arrow to the right of School Fees. Select Number Filters.

    b. Click Top 10.

    c. Select **Bottom** in the first Box. Enter 30 in the middle box. Select Percent in the third box. See Figure 12-18.

Figure 12-18. *Filter for bottom 30%*

d. Click OK. The result is the bottom 30% of the school fees as shown in Figure 12-19.

Student ID	Last Name	First Name	Major	Degree	Sex	City	State	School Fees	Dorm Fees
63	Thomas	George	English	BA	M	Kokomo	IN	4075.85	1500
55	Hough	Marten	Science	MS	M	Fort Wayne	IN	3178.7	
10	Carson	Annette	English	BA	F	Fort Wayne	IN	3952.25	

Figure 12-19. *Result of filtering School Fees for bottom 30%*

24. Click the Filter button to the right of School Fees. Click **Clear Filter from "School Fees."**

25. Let's find the students who are paying above average for their school fees compared to the other students in the table. Click the down arrow to the right of School Fees. Select Number Filters. Click **Above Average**. Figure 12-20 shows the results.

Student ID	Last Name	First Name	Major	Degree	Sex	City	State	School Fees	Dorm Fees
70	Lupner	Lisa	English	MA	F	Toledo	OH	5439.5	1495.5
69	Ulrich	Robert	Biology	BS	M	Chicago	IL	6175.8	1475
66	Carlson	Rhianna	Biology	BS	F	Chicago	IL	5050.29	1500
56	Kamp	Brianna	Biology	AS	F	Toledo	OH	4835.7	1495.5
49	Smith	Margaret	Physics	AS	F	Marion	IN	5927.72	1495.5
39	Bundy	Peggy	History	AS	F	Fort Wayne	IN	4832.25	1275.85

Figure 12-20. *Result of selecting Above Average for School Fees*

26. Click the Filter button to the right of School Fees. Click **Clear Filter from "School Fees."**

27. Right-click any cell in the table. Select **Table** from the context menu and then click **Totals Row**.

28. Click the total in cell J12. Click the down arrow to the right of the cell. Select **Average**.

29. Click inside cell I12. Click the down arrow to the right of the cell. Select **Average**.

30. Change the text in cell A12 from **Total** to **Averages**. See Figure 12-21.

Student ID	Last Name	First Name	Major	Degree	Sex	City	State	School Fees	Dorm Fees
70	Lupner	Lisa	English	MA	F	Toledo	OH	5439.5	1495.5
69	Ulrich	Robert	Biology	BS	M	Chicago	IL	6175.8	1475
66	Carlson	Rhianna	Biology	BS	F	Chicago	IL	5050.29	1500
63	Thomas	George	English	BA	M	Kokomo	IN	4075.85	1500
56	Kamp	Brianna	Biology	AS	F	Toledo	OH	4835.7	1495.5
55	Hough	Marten	Science	MS	M	Fort Wayne	IN	3178.7	
49	Smith	Margaret	Physics	AS	F	Marion	IN	5927.72	1495.5
39	Bundy	Peggy	History	AS	F	Fort Wayne	IN	4832.25	1275.85
14	Sterk	Carrie	Math	BS	F	Hebron	IN	4327.73	1495.5
10	Carson	Annette	English	BA	F	Fort Wayne	IN	3952.25	
Averages								4779.579	1466.60625

Figure 12-21. Result of change to averaging

You have learned how to filter data from a table by clicking the down arrow to the right of the column head you want to filter and then selecting only the values you want to view. Next, you will learn an alternative method of filtering by using a slicer.

Filtering Data with a Slicer

Rather than filtering a table by using the filter buttons next to the column heads, you can create slicers. Slicers don't provide all of the options that table filter buttons provide, but they do make it visually easier to see what has been selected.

EXERCISE 12-2: FILTERING DATA WITH A SLICER

In this exercise, you will create two slicers to filter a table.

1. Continue working with workbook Chapter12 that you used in Exercise 12-1. Click anywhere within the table.

2. Click the Design tab. In the Tools group, click the **Insert Slicer** button. This brings up the Insert Slicers dialog box. See Figure 12-22.

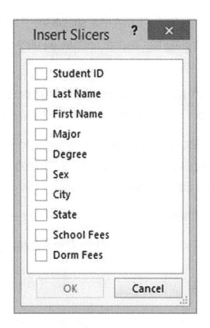

Figure 12-22. *Insert Slicers dialog box*

3.   Select Major and City. Click the OK button. See Figure 12-23.

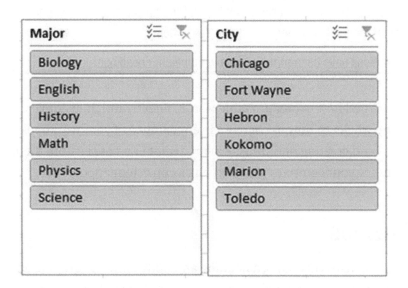

Figure 12-23. *Slicers for Major and City*

4. Click Biology in the Major slicer. Excel only shows the students with a Biology major. The City slicer highlights Chicago and Toledo because the Biology majors are from these two cities.

5. In the Major slicer, click the Multi-Select button ⊞ in the upper right corner; this allows you to select more than one item from the slicer.

6. Select English. This adds English to the already selected Biology. The City slicer now highlights Chicago, Fort Wayne, Kokomo, and Toledo because these are where the Biology and English majors are from.

7. Click the Clear Filter button in the upper right-hand corner of the Major slicer.

8. In the City slicer, click Hebron. The record for Carrie Sterk appears.

Convert Table Back to a Range

1. Convert the Table back to a range by right-clicking any cell in the table and then selecting **Table**.

2. Click **Covert to Range**.

3. A message box displays asking you to confirm that you really do want to convert the table to a range. Click the **Yes** button.

All the filter buttons have been removed from your column headings.

You have learned how to create a table, sort it, filter it, and then convert it back to a range. Next, you will learn how to change the appearance of your workbooks by using existing themes and creating new ones. You may want to create custom themes to give your documents or your company's documents a consistent appearance.

Using Themes

A theme in Excel consists of colors, a heading and body font, and graphic effects applied to a worksheet. Themes are located on the Page Layout tab in the Themes group. See Figure 12-24. You can also create your own themes. You can use themes to enhance the look of your spreadsheets as well as to give them a consistent look. You could, for instance, apply different themes to different projects, or maybe your company elects to always use the same theme. Microsoft Word and PowerPoint have the same themes available as those found in Excel.

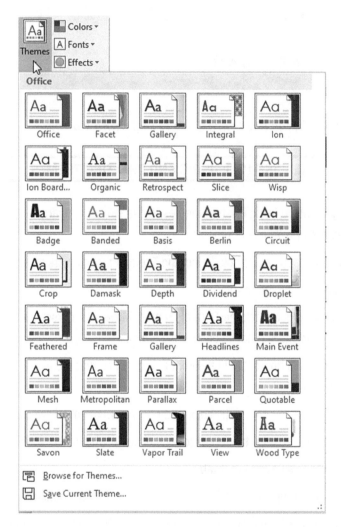

Figure 12-24. *Themes gallery*

Office is the Default theme.

If you click the down arrow for Font on the Home tab's Font group, you will see the two theme fonts above all of the other fonts. See Figure 12-25.

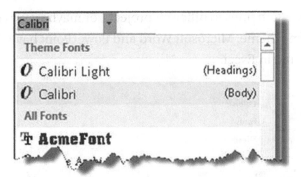

Figure 12-25. *The Theme Fonts are Calibri Light and Calibri*

EXERCISE 12-3: USING AND CREATING THEMES

In this exercise, you will be using existing themes to change the appearance of a workbook. You will also create your own custom theme.

1. Open the Chapter12 workbook. Select the Theme Practice worksheet.

2. Click the Ribbon's Page Layout tab.

3. In the Themes group, click the Theme's button. Move your cursor across the different themes as you notice each one's effect.

4. Click the Ribbon's Home tab. In the Font group, notice that the font is Calibri with a size of 11.

5. Click the down arrow for the fill color and observe what colors are available.

6. Click the down arrow for the font color and observe what colors are available.

7. Click the Ribbon's Page Layout tab.

8. In the Themes group, click the Theme button. Click the Savon theme.

9. Click the Ribbon's Home tab. Click one of the cells. Notice that the default font is now Century Goth size 11. Click the down arrow for fill color. Notice that the theme colors have changed. Click the down arrow for the font color. The theme colors have changed here as well.

10. Click the Ribbon's Page Layout tab. Click the Theme button. Select Office.

11. A theme consists of colors, fonts, and effects. If you didn't want to change an entire theme, you can change part of it, such as just changing the colors. In the Themes group, click the Colors button. Select Grayscale.

12. Click the Ribbon's Home tab. You still have the Office Theme's default font of Calibri, but if you look at the colors available from the color fill or the font color, they will be from the grayscale palette.

13. Click the Ribbon's Page Layout tab. In the Themes group, click the Colors button.

14. Click Customize Colors. The Create New Theme Colors window displays where you can select the colors you want to use for your customized colors. After you have selected some colors, enter a name. Use your name followed by CustomColors. See Figure 12-26. Click the Save button.

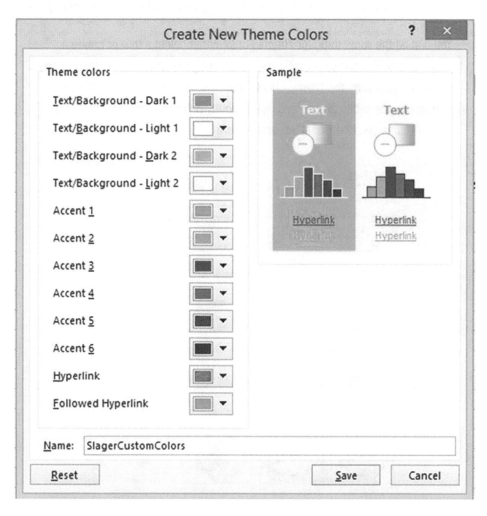

Figure 12-26. *Theme Color options*

15. Click the Theme Colors button. Your custom colors should appear at the top. See Figure 12-27.

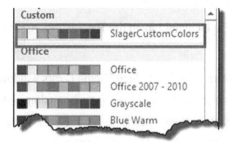

Figure 12-27. *The theme colors you selected appear at the top under Custom*

You can do the same thing for Theme Fonts; either select a font or create a Custom Font. You can select a Theme effect, but you can't create a custom one. Figure 12-28 shows the Theme Effects gallery for the default Office Theme. Effects provide subtle differences like a slightly different edge to your Shapes, SmartArt, Images, Charts, Tables, and WordArt.

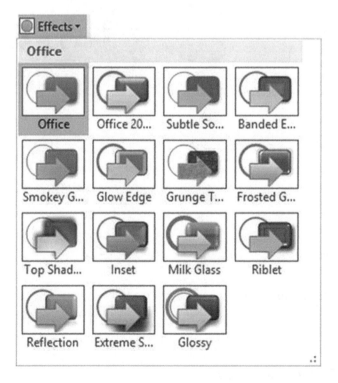

Figure 12-28. *Theme Effects for the Office Theme*

16. Click the Themes button. Click Save Current Theme. Enter a name for your theme and click the Save button. The theme you saved contains whatever is currently selected in Theme Colors, Theme Fonts, and Theme Effects.

17. Click the Theme button. Your theme should appear in the Custom group at the top as shown in Figure 12-29.

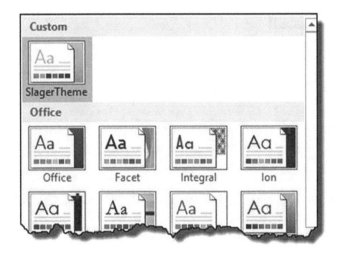

Figure 12-29. *The theme you created appears above the other themes in the Theme Gallery*

18. Click the Office theme.

You have seen how to apply formatting through a theme. Next, you will learn how apply styles to cells.

Applying and Defining Cell Styles

Cell styles provide a way of improving the look and readability of your spreadsheet. Figure 12-30 shows the cell styles. They can be accessed from the Home tab's Styles group by clicking the Cell Styles button. The Cell Styles are divided into five categories:

- Good, Bad and Neutral

- Date and Model

- Titles and Headings

- Themed Cell Styles

- Number Format

Figure 12-30. Cell styles

Styles provide you with an easy way to give a consistent format to certain cells, such as those that contain what the company would consider good news and those that contain what the company would consider bad news, cells that are used for input and those that are used for output, those cells that contain links, and so on. You can also use Themed Cell Styles, which provide different amounts of transparency using the same color. These predefined styles are located at Home ➤ Styles ➤ Cell Styles. Themed Cell Styles are derived from the theme you are currently using. If you change your document theme, your themed cell styles will be different. You apply styles to your worksheet by first selecting those cells you want to apply the style to and then clicking the appropriate style from the Cell Styles gallery. A nice feature of the styles is that you can see exactly how a style will affect your spreadsheet before you ever apply it. Move your mouse over any style to get a visual preview of how it will affect your spreadsheet.

What if there is no style in the Cell Styles gallery that meets your needs? No problem; just create your own style. To do this, click **New Cell Style...** at the bottom of the Cell Styles gallery. This brings up the Style dialog box where you can assign a name to your new style.

Styles can be deleted by displaying the Cell Styles gallery, right-clicking the style, and then pressing the Delete key.

EXERCISE 12-3: USING AND CREATING CELL STYLES

In this exercise, you will apply existing cell styles, and then you will create a new cell style and apply it.

Note   If your Excel window is maximized, you will see a portion of the Cell Styles gallery. If not, you will see a Cell Styles button. Clicking the Cell Styles button will show the gallery. If you only see a portion of the gallery, then click the More button .

---

1. Open Workbook Chapter12. Click the Cell Styles worksheet tab.

2. Click cell A1. Open the Cell Styles gallery (see the preceding Note). Maximize your Excel window.

3. Under the Titles and Headings group, click Heading 1.

4. Select cells A3:J3. Click the More button for Cell Styles. Select the cell style 20% - Accent5.

5. Select cells I14:J14. Click the More button for Cell Styles. Select the cell style Total.

6. Select cell J8. Click the More button for Cell Styles. Select the Bad cell style.

<u>Create a New Style</u>

1. Click the More button for Cell Styles. Click New Cell Style.

2. Click the Format button.

3. Select the Alignment tab. Make the text 45 degrees.

4. Select the Border tab. Select a light blue color. Select Outline in the Presets.

5. Select the Fill tab. Select a light gold color.

6. Select the Font tab. Select the color Black. Click the OK button.

7. Change the style name on the Style window to your last name followed by Style, for example, "SlagerStyle."

8. Click the OK button.

9. Select cells A3:J3. Click the More button for Cell Styles. The style you created should be in the Custom group. See Figure 12-31.

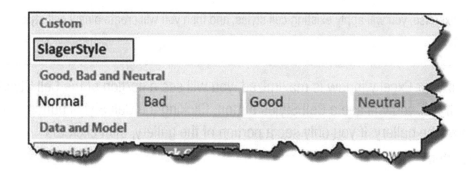

Figure 12-31. Style you created should be in the Custom group

10. Click the style you created. See Figure 12-32.

2016 Student List for Westwood Dorms									
Student ID	Last Name	First Name	Major	Degree	Sex	City	State	School Fees	Dorm Fees
10	Carson	Annette	English	BA	F	Fort Wayne	IN	3952.25	
14	Sterk	Carrie	Math	BS	F	Hebron	IN	4327.73	1495.5
39	Bundy	Peggy	History	AS	F	Fort Wayne	IN	4832.25	1275.85
49	Smith	Margaret	Physics	AS	F	Marion	IN	5927.72	1495.5
55	Hough	Marten	Science	MS	M	Fort Wayne	IN	3178.7	1850.5
56	Kamp	Brianna	Biology	AS	F	Toledo	OH	4835.7	1495.5
63	Thomas	George	English	BA	M	Kokomo	IN	4075.85	1500
66	Carlson	Rhianna	Biology	BS	F	Chicago	IL	5050.29	1500
69	Ulrich	Robert	Biology	BS	M	Chicago	IL	6175.8	1475
70	Lupner	Lisa	English	MA	F	Toledo	OH	5439.5	1495.5
								47795.79	13583.35

Figure 12-32. Formatted with the style you created

You have learned how to apply consistent formatting by using themes and cell styles. Next, you will learn how to format based on a condition (e.g., applying a format to all cells in a column that have a value over 500).

Conditional Formatting

Conditional formatting formats cells based on the values they contain. The formatting is conditional based on criteria that you determine. Cells that are of special interest to you can be highlighted. You can apply conditional formatting to tables or any range of data.

If you have a large amount of data in your spreadsheet, it can make those cells of importance stand out, or it can give you a quick visual overview about the range of the values in a column.

EXERCISE 12-4: CREATING CONDITIONAL FORMATTING

In this example, you will create formatting based on whether certain conditions are met. You will learn how to edit, delete, and create new formatting rules.

1. Open workbook Chapter12.

2. Click the Orders tab. Figure 12-33 shows the data.

	A	B	C	D	E
1	Order Amt	Order Placed	Due to complete	Date Completed	Took Order
2	5185.25	1/3/2016	1/10/2016	1/12/2016	Gina
3	7212.38	1/5/2016	1/17/2016	1/15/2016	Larry
4	2919.95	1/8/2016	1/19/2016	1/20/2016	Larry
5	8753.27	1/8/2016	1/21/2016	1/17/2016	Larry
6	1275.75	1/9/2016	1/23/2016	1/20/2016	Gina
7	4875.5	1/10/2016	1/24/2016	1/22/2016	Larry
8	10172.22	1/10/2016	1/25/2016	1/26/2016	Gina
9	882.32	1/11/2016	1/14/2016	1/13/2016	Gina
10	3975.52	1/12/2016	1/17/2016	1/21/2016	Larry
11	6205.28	1/14/2016	1/19/2016	1/18/2016	Gina
12	935.75	1/15/2016	1/20/2016	1/23/2016	Larry
13	1495.35	1/16/2016	1/22/2016	1/19/2016	Gina

Figure 12-33. *Data on the Orders tab*

Top/Bottom Rulesv

1. Select the cell range A2:A13.

2. On the Home tab in the Styles group, click Conditional Formatting.

3. Select Top/Bottom Rules.

4. Select Top 10%. You are not limited to the Top 10%. You can enter any value. Change 10 to 25 to get the top 25%.

    The top 25% will be highlighted with whatever formatting you select. It is currently showing with **Light Red Fill with Dark Red Text**.

    Click the down arrow and then click each option as you observe how it affects the selected range. Keep repeating this until you have seen the effect of each option. Finally, click Custom Format.

    This brings up the Format Cells window. You can make selections from any of the tabs.

5. Click the Font tab. Select Bold. Click the down arrow for color. Select Purple. Click the Border tab. Select Outline.

6. Click the OK button. Figure 12-34 shows the results. Click the OK button.

	A	B	C	D	E
1	Order Amt	Order Placed	Due to complete	Date Completed	Took Order
2	5185.25	1/3/2016	1/10/2016	1/12/2016	Gina
3	7212.38	1/5/2016	1/17/2016	1/15/2016	Larry
4	2919.95	1/8/2016	1/19/2016	1/20/2016	Larry
5	8753.27	1/8/2016	1/21/2016	1/17/2016	Larry
6	1275.75	1/9/2016	1/23/2016	1/20/2016	Gina
7	4875.5	1/10/2016	1/24/2016	1/22/2016	Larry
8	10172.22	1/10/2016	1/25/2016	1/26/2016	Gina
9	882.32	1/11/2016	1/14/2016	1/13/2016	Gina
10	3975.52	1/12/2016	1/17/2016	1/21/2016	Larry
11	6205.28	1/14/2016	1/19/2016	1/18/2016	Gina
12	935.75	1/15/2016	1/20/2016	1/23/2016	Larry
13	1495.35	1/16/2016	1/22/2016	1/19/2016	Gina

Figure 12-34. *Formatted with top/bottom rules*

Clear Rules

Next, you will clear the rules you set for cell range A2:A13.

1. With the cell range still selected, click Conditional Formatting, select Clear Rules, and then select Clear Rules from Selected Cells. See Figure 12-35.

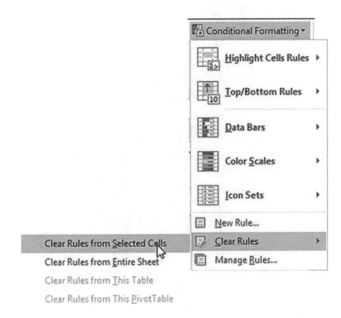

Figure 12-35. *Select Clear Rules from Selected Cells*

2. With cell range A2:A13 still selected, select Conditional Formatting, then select Top/Bottom Rules, and then select Below Average. Leave the format as **Light Red Fill with Dark Red Text**.

3. Click the OK button. The smaller values should be highlighted. Click a blank cell so that you can see the results.

4. Let's clear the rules again. The only rules we currently have for this worksheet are those for the cell range A2:A13. Click Conditional Formatting, then select Clear Rules, and then select **Clear Rules from Entire Sheet**.

Highlight with Data Bars

1. Select the cell range A2:A13, select Conditional Formatting, and then select Data Bars.

2. Move your cursor over each of the Gradient Fill and Solid Fill Colors. See Figure 12-36.

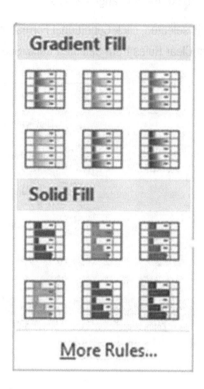

Figure 12-36. *Conditional Formatting Data Bars*

The larger the value in the cell, the longer the data bar will be. See Figure 12-37. Don't select any.

Note   Some of the conditional formats have more of an impact if the column widths are wider.

	A	B	C	D	E
1	**Order Amt**	**Order Placed**	**Due to complete**	**Date Completed**	**Took Order**
2	5185.25	1/3/2016	1/10/2016	1/12/2016	Gina
3	7212.38	1/5/2016	1/17/2016	1/15/2016	Larry
4	2919.95	1/8/2016	1/19/2016	1/20/2016	Larry
5	8753.27	1/8/2016	1/21/2016	1/17/2016	Larry
6	1275.75	1/9/2016	1/23/2016	1/20/2016	Gina
7	4875.5	1/10/2016	1/24/2016	1/22/2016	Larry
8	10172.22	1/10/2016	1/25/2016	1/26/2016	Gina
9	882.32	1/11/2016	1/14/2016	1/13/2016	Gina
10	3975.52	1/12/2016	1/17/2016	1/21/2016	Larry
11	6205.28	1/14/2016	1/19/2016	1/18/2016	Gina
12	935.75	1/15/2016	1/20/2016	1/23/2016	Larry
13	1495.35	1/16/2016	1/22/2016	1/19/2016	Gina

Figure 12-37. *The larger the value in the cell, the longer the data bar*

3. With cell range A2:A13 still selected, select Conditional Formatting ➤ **Highlight Cells Rules** ➤ **Greater than**. Enter 5000 in the text box and select Yellow Fill with Dark Yellow Text. See Figure 12-38.

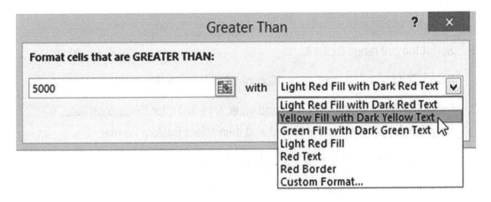

Figure 12-38. *Conditional formatting*

4. All values in the range that are greater than 5000 should be highlighted. Click the OK button.

Highlight Duplicate or Unique Values

1. Select the cell range B2:B13.

2. Select Conditional Formatting ➤ Highlight Cell Rules ➤ Duplicate Values.

3. You can highlight duplicate cells, or you can highlight those that don't have a duplicate by selecting Unique. See Figure 12-39. Select Duplicate.

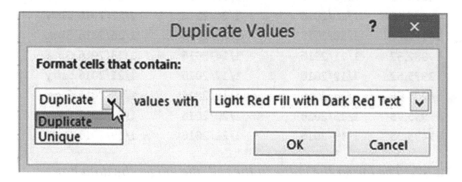

Figure 12-39.   *Formatting for Duplicate Values*

4. Click the OK button.

Highlight Values Between Other Values

1. Select the cell range C2:C13.

2. Select Conditional Formatting ➤ Highlight Cell Rules ➤ Between.

3. Enter 1/15/2016 for the first date and enter 1/19/2016 for the second date. Click the down arrow for the format and then select Custom Format.

4. Click the Font tab and then select Bold for the Font style.

5. Click the Fill tab and then select a light green color.

6. Click the OK button for the Format Cells dialog box.

7. Click the OK button for the Between dialog box.

Highlight Values Greater Than Another Value in Another Column

We want to highlight the dates where we didn't complete an order by its due date. Therefore, we want to highlight those cells in column D whose date is greater than the same row in column C.

1. Select cell range D2.

2. Select Conditional Formatting ➤ Highlight Cell Rules ➤ **Greater Than**.

3. With the date from D2 currently highlighted in the Greater Than window, click Cell C2. The text box displays the cell address in absolute form. We will need to copy the formatting, so select the text and change it to **=C2.** See Figure 12-40.

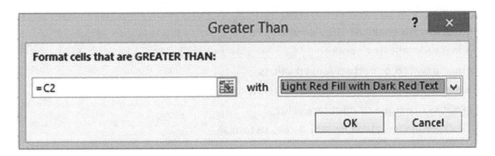

Figure 12-40. *Highlight values that are greater than those in cell C2*

4. Click the OK button.

5. Drag the Autofill Handle from D2 down through D13. Click the Autofill options button below the selected data. Select **Fill Formatting Only**.

Highlight Specified Text

1. Select cell range E2:E13.

2. Select Conditional Formatting ➤ Highlight Cell Rules ➤ Text that Contains. Leave Gina in the first text box. Click the down arrow for the format and then select Red Text.

3. Click the OK button.

Edit, Delete, and Create New Rules

1. Click Conditional Formatting and then select Manage Rules.

2. Click the down arrow for **Show formatting rules for** and then select **This Worksheet**.

3. Select the Rule for Cell Value contains "Gina" and then click the Edit Rule button. You have more options here to work with. Select **beginning with**. See Figure 12-41.

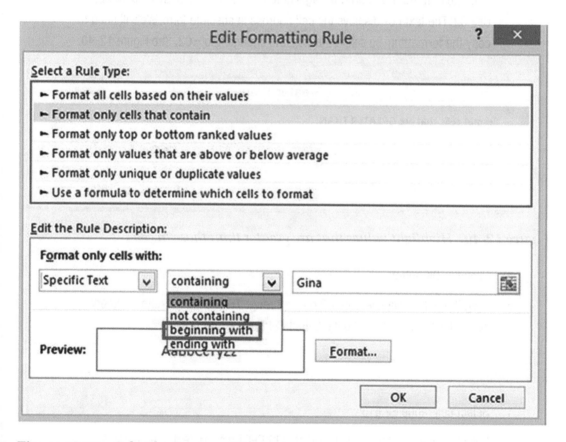

Figure 12-41. *Select beginning with*

4. Change Gina to just G. (Now anybody whose name starts with G will be highlighted.)

5. Click the Format button.

6. Click the Font tab. Select Bold Italic from the Font style.

7.   Click the OK button for Format Cells.

8.   Click the OK button for the Editing Formatting Rule dialog box.

9.   Click the OK button for the Conditional Formatting Rules Manger dialog box.

10.   Change the name Larry in cell E3 to Gary. It should now be highlighted because it starts with a G.

Applying Conditional Formatting to a Table

You applied conditional formatting, but what if you added additional rows of data? The conditional formatting will apply to those cells immediately below the selected rule range except for dates. For date columns, use the Autofill handle to copy and then click the Autofill options button and select Fill Formatting Only.

1.   Select cell range A2:E14. Drag the Autofill handle down to E20. Click the Autofill options button and select **Fill Formatting Only**. Add an additional row of data that meets the conditional rules for each column. See row 14 in Figure 12-42.

	A	B	C	D	E
1	**Order Amt**	**Order Placed**	**Due to complete**	**Date Completed**	**Took Order**
2	5185.25	1/3/2016	1/10/2016	1/12/2016	*Gina*
3	7212.38	1/5/2016	**1/17/2016**	1/15/2016	*Gary*
4	2919.95	1/8/2016	**1/19/2016**	1/20/2016	Larry
5	8753.27	1/8/2016	1/21/2016	1/17/2016	Larry
6	1275.75	1/9/2016	1/23/2016	1/20/2016	*Gina*
7	4875.5	1/10/2016	1/24/2016	1/22/2016	Larry
8	10172.22	1/10/2016	1/25/2016	1/26/2016	*Gina*
9	882.32	1/11/2016	1/14/2016	1/13/2016	*Gina*
10	3975.52	1/12/2016	**1/17/2016**	1/21/2016	Larry
11	6205.28	1/14/2016	**1/19/2016**	1/18/2016	*Gina*
12	935.75	1/15/2016	1/20/2016	1/23/2016	Larry
13	1495.35	1/16/2016	1/22/2016	1/19/2016	*Gina*
14	5285	1/10/2016	**1/18/2016**	1/19/2016	*George*

Figure 12-42. *Result of adding an additional row*

2.   Click the Lab Scores worksheet tab.

3.   Select cell range B2:B8.

4. Select Conditional Formatting ➤ Color Scales. Move your cursor over each option to see how it affects the cell background colors.

5. Select one.

   We want to highlight the student averages based on their scores. The grade scale is as follows:

   90–100 A

   80–89 B

   70–79 C

   60–69 D

   0–59 F

6. Select the cell range F2:F8.

7. Click Conditional Formatting and then select Icon Sets. Select the first option in the Indicators group. See Figure 12-43.

Figure 12-43. *Indicators group in the Icon Sets*

We want to use an icon with a green circle and a check mark in it for those who have an A or a B, a yellow circle icon with an exclamation mark in it for students who have a C, and a red circle with an X through it for those students who have a D or an F.

8. Click Conditional Formatting and then select Manage Rules.

9. Click the Icon Set and then click the Edit Rule button.

   You don't have to use all the icons from one set; you can pick among the different icons. See Figure 12-44.

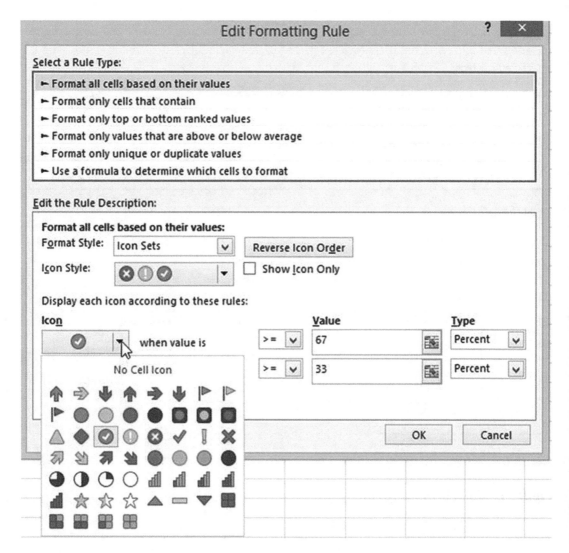

Figure 12-44. *Select an icon*

10. Change the Type for the green and yellow icon to Number. The greater than or equal to sign is already selected. Enter .8 for the green icon Value.

11. Enter .7 for the value for the yellow icon. The greater than or equal to sign should already be selected. The type should be Number. Press the Tab key.

    The red icon should show that it will display for any value less than .7. See Figure 12-45.

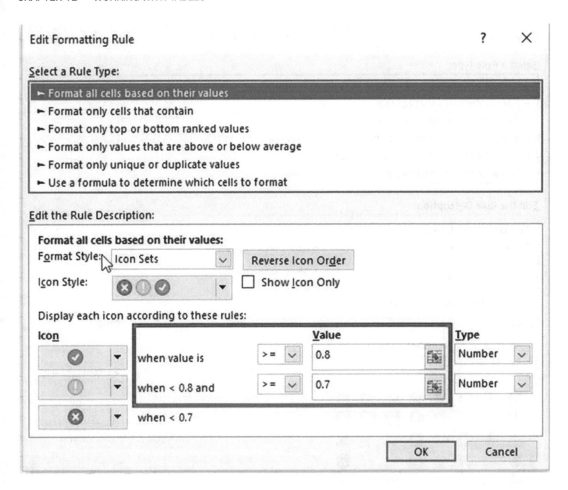

Figure 12-45. *Formatting rule*

    12.   Click OK. Figure 12-46 shows the results. Click OK for the Conditional
                Formatting Rules Manager dialog box.

⎘	A	B	C	D	E	F
1	Student	Lab 1	Lab 2	Lab 3	Totals	Average
2	Abe, Honest	60	45	52	157 ⓘ	79%
3	Apple, Barney	50	38	49	137 ⊗	69%
4	Parmley, Rebecca	70	50	63	183 ✓	92%
5	Parmley, Isaiah	68	49	59	176 ✓	88%
6	Howard, Curly	70	50	55	175 ✓	88%
7	Fine, Larry	66	48	53	167 ✓	84%
8	Sweeney, Mike	48	35	35	118 ⊗	59%
9	Possible	75	55	70	200	

Figure 12-46. *Result of applying formatting rule*

13.   Change Cell D7 from 53 to 44. Your icon should change to the yellow icon.

Summary

A theme in Excel consists of colors, a heading and body font, and graphic effects applied to a worksheet. You can also create your own themes. You can use themes to enhance the look of your spreadsheets as well as to give them a consistent look. You could, for instance, apply different themes to different projects, or maybe your company elects to always use the same theme. Microsoft Word and PowerPoint have the same themes available as those found in Excel.

Microsoft Office Themes consist of galleries of colors, heading and body fonts, and graphic effects. You can click the various themes and at the same time view how they affect your worksheet. You can select the one that is most pleasing to you, or perhaps your company takes it a step further and creates their own custom company or corporate theme; this is done both to create a consistent sense of professionalism within the organization and to give external communications a sense of branding. If you can't find a theme that matches your needs, you can create your own by making selections from the color, heading and body font, and graphic effect galleries.

Rather than manually formatting your cells, you select from groups of predesigned cell styles. The groups are Good, Bad and Neutral; Data and Model; Titles and Headings; and Number Format. Like themes, cell styles can give your worksheets a consistent look, such as always giving your headings, titles, and totals the same appearance.

You can convert the data in your worksheet to a table and then convert it back again to its original format. You create a table by selecting the data you want to convert into a table and then clicking the Format as Table button located in the Ribbon's Styles group. Converting data to a table allows you to filter data by clicking the drop-down arrow to the right of a column head. You can then either select the values that you want to use for your filter or create a custom filter; if the data is text, you can select a Text Filter, and if the data is Numeric, you can select a Numeric Filter. Another way of filtering table data is to create slicers. Slicers let you see exactly what is being filtered without having to use the table filters.

You can format cells with a variety of colors or icons to highlight cells that are of particular interest. For example, you could highlight cells whose value is in the top 20% with one icon, a value within the middle 60% with another icon, and a value in the bottom 20% with yet another icon.

The next chapter covers charts. Excel makes it very simple to take your spreadsheet data and turn it into a chart. Charts make it easy to find trends in your data that you might not otherwise detect.

CHAPTER 13

Working with Charts

Where would we be without charts? Not only are charts more appealing to our eye than raw data, but they also make it much simpler for us to spot trends. The old saying "a picture is worth a thousand words" also applies to charts. Managers don't always want to look at every record in a spreadsheet; they would rather make their decisions based on trends, whether they be good or bad. By looking at trends, management can decide what actions the company should take in the future. Charts can summarize vast amounts of data and show relationships between data. Charts make it easier to see such things as what lines of product are more profitable, what the most profitable time of year is, whether the profitability of the company is on the rise or going down, what departments are doing well and which are not, and on and on.

After reading and working through this chapter, you should be able to

- Use shortcut keys to create charts

- Change chart types

- Add a data point

- Change the data source

- Create a pie chart

- Create a pie of a pie chart

- Create a pie of a bar chart

- Create a chart with multiple data chart types (Combination Chart)

- Create a Treemap chart

- Create a Sunburst chart

- Create a Funnel chart

- Create a Map Chart

© David Slager and Annette Slager 2020

D. Slager and A. Slager, *Essential Excel 2019*, https://doi.org/10.1007/978-1-4842-6209-2_13

- Create a Bing Map

- Create Sparklines

Chart Types

Excel provides a wide variety of chart types to choose from to best display your data. What type of chart you should use depends upon the data you are charting. Before creating a chart, you need to decide what it is you want the chart to say. Once you have a chart selected, you can change the format of the entire chart by changing its theme, or you can change the individual components of the chart. Excel provides a wide variety of different chart options. As you will see later, Excel looks at the data you have selected, makes a **quick analysis** of it, and then provides **recommended** chart types for your data.

The Ribbon's Insert tab in the Charts group displays the available chart types. Clicking a down arrow next to one of the chart types will display available variations of that chart type. See Figure 13-1.

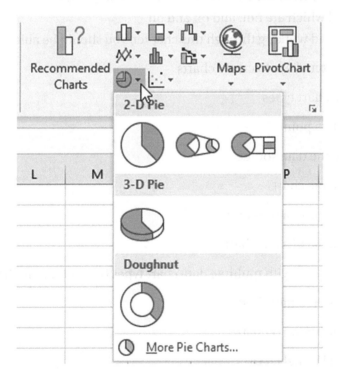

Figure 13-1. *Variations of the pie chart*

Clicking the Charts group dialog box launcher as shown in Figure 13-2 brings up the Insert Chart dialog box (Figure 13-3).

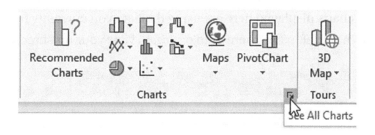

Figure 13-2. *Charts group*

The Insert Chart dialog box has two tabs. The Recommended Charts tab displays the Recommended Charts for the data you currently have selected in your worksheet. See Figure 13-3. The charts displayed here are using the data that you have selected. Clicking a chart type gives a description of what that chart type can be used for.

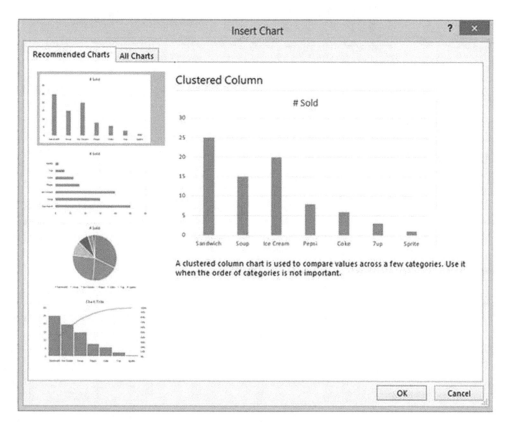

Figure 13-3. *Insert Chart dialog box with Recommended Charts tab selected*

The Recommended Charts are just a suggestion. You don't have to use Excel's suggestion. The All Charts tab shows all the chart types in the left pane. Clicking a chart type in the left pane will display all the variations of that type in the right pane. See Figure 13-4. The charts displayed here are using the data you currently have selected on your worksheet. In the left pane, there is an option to view your most recent chart.

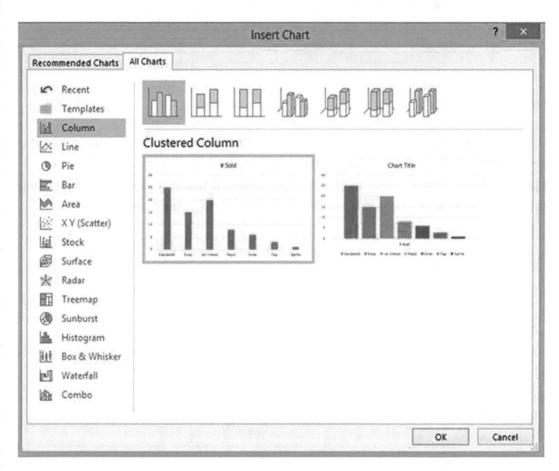

Figure 13-4. *Variations for the Column chart*

Creating and Modifying Charts

There are several ways to create a chart, including using the options on the Insert tab shown in the preceding section. The easiest way to create a default 2D Clustered bar chart is by selecting the data you want to display in your chart and then

- Pressing the Alt + F1 key if you want to display the chart on the same worksheet as the data, or

- Pressing F11 if you want the chart to appear on a separate worksheet

Once you have created your chart, you can change it in so many ways. Some of the changes you can make are the following:

- The type of the chart

- Its location

- Its style

- What is displayed in the chart

- Reversing the columns and rows

- Adding or removing items from the chart

- Changing the color and appearance of individual bars

- Adding or removing legends, titles, and data tables

- The incremented value on the plot area

Selecting the chart adds two contextual tabs to the Ribbon. The contextual tabs Design and Format appear under the heading Chart Tools. The Design tab is mostly used for making changes that affect the entire chart, while the Layout tab is used for changing the look of individual elements in the chart.

Note   Don't get carried away with trying to make your charts too fancy and overdoing colors and so on. The most important thing is how well your chart conveys the information that you want it to and how easy it is for the intended audience to read and understand. The audience should be able to quickly grasp what the chart is conveying.

EXERCISE 13-1: CREATING AND MODIFYING EXCEL CHARTS

In this exercise, you will create and modify column and bar charts from existing Game Sales data.

1. Open workbook Chapter 13. Click the Game Sales tab. Figure 13-5 shows the data.

◢	A	B	C	D	E	F	G
1				Game Sales			
2		January	February	March	April	May	June
3	Risk	43	51	40	34	33	36
4	Stratego	29	29	30	23	22	28
5	Clue	40	50	35	33	30	35
6	Sorry	25	24	23	19	20	23
7	Monopoly	48	53	46	39	38	43

Figure 13-5. *Data on Game Sales tab*

2. Select the cell range A2:G7. Press Alt + F1.

    Your chart is in a moveable frame. Move your cursor on the chart frame. Your cursor should display as four arrows ✥.

3. Drag and drop the chart to whatever location you want.

4. Increase the size of the chart by placing your cursor on the bottom right of the chart frame and then hold down the Shift key while you drag the cursor toward the bottom right. Holding down the Shift key makes the chart retain the same width-to-length ratio.

    When a chart is selected, Excel displays two new tabs on the Ribbon: Design and Format. The two tabs are under the Chart Tools grouping.

Note   A chart consists of many different individual elements. What the elements are depends upon the type of chart and the data being charted.

On the Ribbon, click the Format tab. In the Current Selection group, the Chart Elements drop-down box displays the current item selected in the chart.

5.   Click the down arrow of the Chart Elements drop-down box. See Figure 13-6.

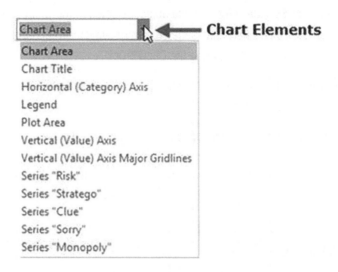

Figure 13-6.   *Chart Elements drop-down box*

Selecting an item from the Chart Elements list will select that item on the chart.

6.   Select Plot Area from the Chart Elements list. The Plot Area is now selected on the chart.

7.   Select Horizontal (Category) Axis from the Chart Elements list. The Months on the Chart are now selected.

Selecting an item from the Chart Elements list will select that item on the chart and vice versa.

8.   Click the Chart Title on the Chart. Look at the Chart Elements drop-down box. It shows that Chart Title is selected.

9.   Drag across the Chart Title text on the chart to select it. Type **Game Sales**.

10. Drag across Game Sales to select it. Right-click Game Sales and then select Font. Change the font size to 18. Click the OK button.

11. On the Ribbon's Format tab, in the Current Selection group, click **Format Selection**.

Clicking **Format Selection** formats the object selected in the Chart Elements drop-down Box. Since Chart Title is the current object, clicking Format Selection brings up the options for formatting the Chart Title:

- The **Title Options** menu has options that affect the area inside the Chart Title area but outside the text itself. See Figure 13-7.

- The **Text Options** menu has options that affect only the text.

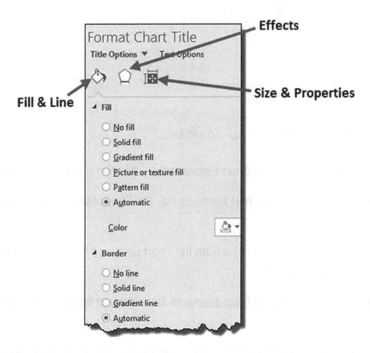

Figure 13-7. *Chart options*

12. Click the Fill & Line paint bucket.

13. Expand the Fill if it isn't already. Click Solid fill. Clicking Solid fill brings up options for Color and Transparency.

14. Pick the color Orange, Accent 2, Lighter 60% from the Color option. See Figure 13-8.

15. Change the Transparency to 50%.

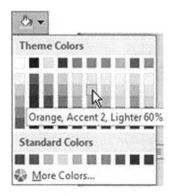

Figure 13-8. *Select Orange, Accent 2, Lighter 60% from the Theme Colors*

16.   Collapse the **Fill**. Expand the **Border** if it isn't already. Selecting an option other than **No line** brings up numerous options.

17.   Select Solid line. Change the color of the border to red.

18.   Increase the size of the border around the Chart Title by changing the width to 2.5.

19.   Change the border to a double line by changing the Compound Type to Double.

20.   Make the border line corners rounded by changing the Join type to Round.

21.   Click the Effects button ⬠. There are four basic effects: Shadow, Glow, Soft Edges, and 3-D format.

22.   Expand the 3-D Format options.

23.   Click the down arrow for Top bevel. Select the Slant bevel. See Figure 13-9.

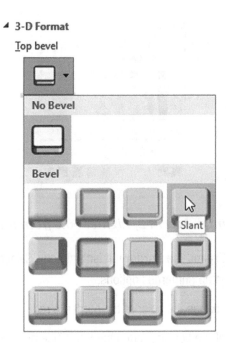

Figure 13-9. *Select Slant*

The chart title should now display with a 3-D appearance.

Figure 13-10. *The chart title now has a 3-D appearance*

24.   Collapse the 3-D Format options.

25.   Expand the Shadow options. Creating a Shadow can make an object appear to come off the page:

a. Click the down arrow for Presets.

b. There are three groups: Outer, Inner, and Perspective. Click the first option in the first row (Offset Bottom Right).

c. Drag the Distance slider to 21 pt.

d. Drag the blur slider to 16 pt.

26.   Click the Size & Properties button ▥.

27.   Click the down arrow for Custom angle until the angle is –3.

28.   Drag the Game Sales chart title to the left side of the chart.

29.   Click Text Options on the Format Chart Title pane. See Figure 13-11.

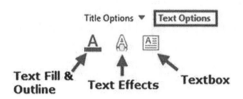

Figure 13-11.   *Text options*

30.   Click Text Fill & Outline and complete the following steps:

a. Expand the Text Fill options.

b. Select Pattern Fill.

c. Change the Foreground color to Dark Red.

d. Change the Background color to Black.

e. Select the last pattern in the last row (Solid Diamond). See Figure 13-12.

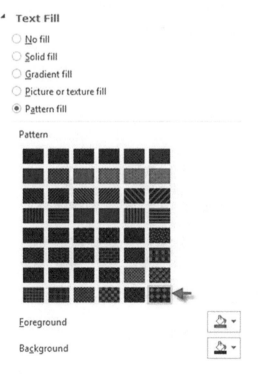

Figure 13-12.   *Pattern Fill options*

31. On the Ribbon's Format tab, in the Current Selection group, click the down arrow for Chart Elements and then select Chart Area.

32. In the Current Selection group, click the Format Selection button.

33. In the Format Chart Area pane, select Gradient fill. See Figure 13-13.

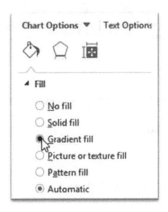

Figure 13-13. *Select Gradient fill*

34. Click the down arrow for the Preset gradients.

35. Select Light Gradient – Accent 2. See Figure 13-14.

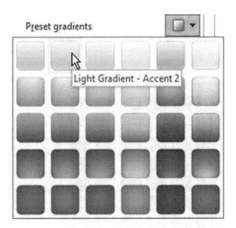

Figure 13-14. *Select Light Gradient – Accent 2*

Figure 13-15 shows the result.

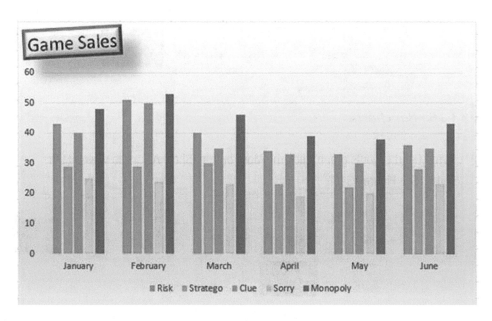

Figure 13-15. *Your chart should now look like this*

36. Close the Format Selection pane by clicking the X in the upper right corner of the pane.

37. Select Legend from the Chart Elements drop-down box. A border appears around the legend at the bottom of the chart. You can move the legend by dragging its border, or you can resize it by using its handles.

Applying Chart Styles

1. On the Ribbon under the Chart Tools grouping, there is a Design tab and a Format tab. Click the Design tab.

   Excel has a series of predefined styles that you can select from. The style buttons reflect the type of the current chart. Since we are using a bar chart, the style buttons display styles for bar charts. See Figure 13-16. While working on a pie chart, the style buttons would display pie charts.

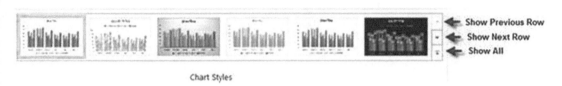

Chart Styles

Figure 13-16. *Bar charts*

2. In the Chart Layouts group, click the More button (Show All) ⩝ to view all the styles. See Figure 13-17.

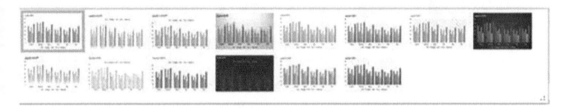

Figure 13-17. *Charts available after clicking the More button*

The number of available predefined layouts varies for the different types of charts.

3. Move your mouse over the style buttons while noticing how it changes your chart. A tool tip displays the name assigned to each button such as Style 1, Style 2, and so on.

Click the **Style 6** button. When the chart is selected, you will see three buttons to the right of it. See Figure 13-18.

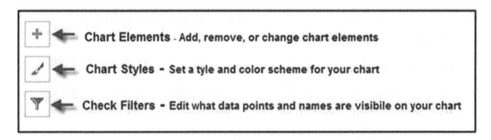

Figure 13-18. *These three buttons display to the right of the chart when it is selected*

If you click anywhere off the chart, these three buttons will not display.

4. Click the Chart Elements button. See Figure 13-19.

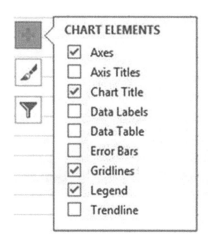

Figure 13-19. *Options for Chart Elements button*

5. Select Data Labels. The Data Labels values appear above each column.

6. Select Data Table. The Data Table displays the individual values in a spreadsheet form. These are the same values that are being displayed by the Data Labels above the column. Eliminate the Data Labels by removing the check mark in front of it.

Changing the Chart Style

1. Click the Chart Styles button. What appears here depends upon what you have in your current chart. Since we currently have a data table in our chart, all of the charts being presented here have a Data Table.

2. Move your cursor over each of the charts as you scroll through them to see how each option will affect your chart. These are the same styles that appear on the Ribbon's Data Tab in the Chart Styles group.

3. Click Style 14. Figure 13-20 shows the results.

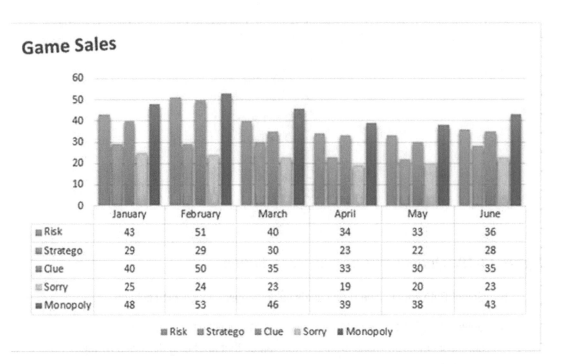

Figure 13-20. *Results after selecting Style 14*

4. Click the Chart Elements button. Move your cursor over the options. Notice that there is an arrow to the right of each of the Chart Elements. Clicking the arrow allows you to choose various options for that Element. The Axis Titles and the Data Labels are not currently selected. Moving your cursor over any unselected item will show what that item will look like on your chart. Move your cursor over the Axis Title. There is a horizontal Axis Title and a vertical Axis Title. See Figure 13-21.

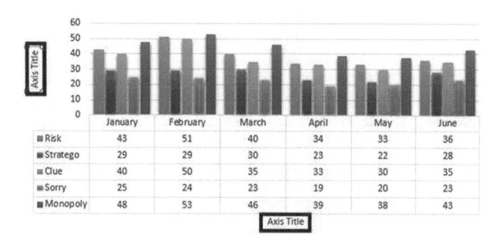

Figure 13-21. *Horizontal and vertical axes*

5. Click the arrow to the right of Axis Title.

6. We only want to use the horizontal Axis Title so select **Primary Horizontal**.

7. Drag across the text **Axis Title** on the chart and type **Games Sold per Month**.

Quick Layouts

1. On the Ribbon under the Chart Tools grouping, click the Design tab. In the Chart Layouts group, click the Quick Layout button. See Figure 13-22.

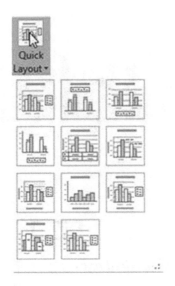

Figure 13-22. *Options for Quick Layout*

Excel has many predefined layouts that contain different arrangements of your chart's elements.

2. Move your cursor over each of the layouts to see how they alter the look of your chart.

3. Select Layout 3.

Changing Chart Types

1. On the Ribbon's Design tab, in the Type group, click the Change Chart Type button. A list of chart categories appears in the left pane of the Change Chart Type window. The upper right pane contains variations of the chart type selected in the left pane.

    Clicking one of the variations displays the results in the bottom right pane. See Figure 13-23. There are two available charts in the lower right pane. The difference between these is that one uses the column headings for the horizontal axis and the other chart uses the row headings for the horizontal axis. Move your cursor over the two charts; they will enlarge so you can get a better view of them.

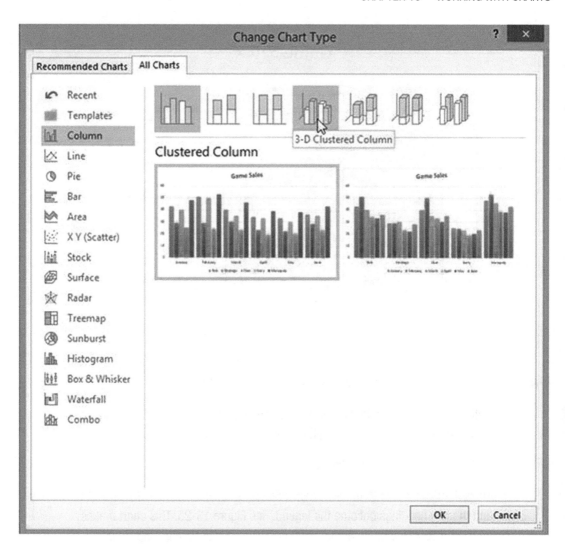

Figure 13-23. *Charts available for the selected chart appear in the bottom right pane*

2. Click Bar in the left pane. Click the first bar chart in the upper right pane which is a clustered bar chart.

3. Two clustered bar charts appear in the pane below. Click the chart on the right side.

4. Click the OK button. See Figure 13-24.

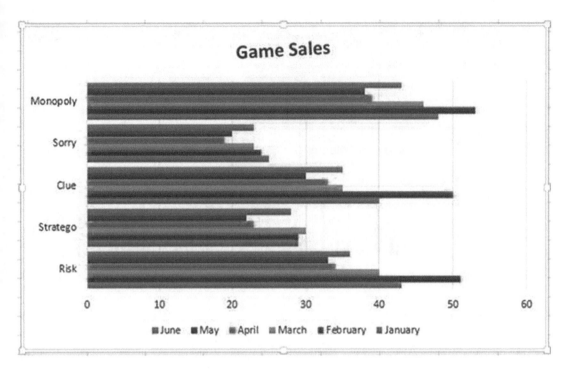

Figure 13-24. *Result of selecting the Clustered Bar Chart*

Notice that the buttons in the Chart Styles group and those in the Quick Layout have changed to reflect that the chart is now a bar chart.

5.   In the Data group, click the Switch Row/Column button.

The column headings from the spreadsheet are now used for the vertical axis and the row headings become the legend. See Figure 13-25. This chart makes it easier to see how much difference there was in sales for each individual game during the six-month period.

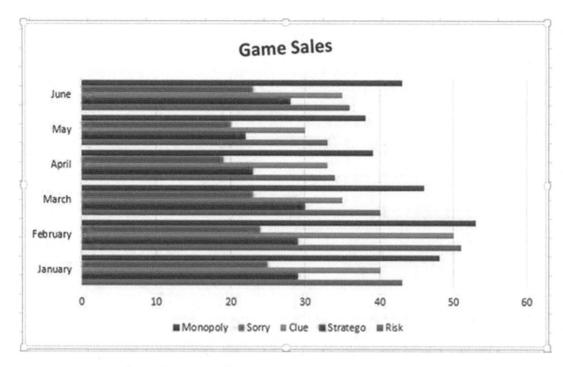

Figure 13-25. *The column headings from the spreadsheet are the vertical axis and the row headings are the legend*

6. Right-click the legend at the bottom of the chart and then select Format Legend.

7. Select Right from the Format Legend pane.

8. On the Ribbon's Design tab, in the Data group, click the Select Data button. This brings up the Select Data Source dialog box shown in Figure 13-26.

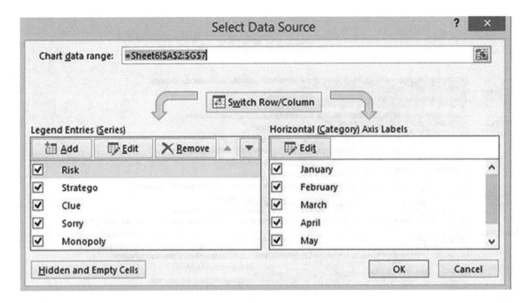

Figure 13-26. *Select Data Source dialog box*

The Select Data Source dialog box provides you with an option for changing the range of cells that the chart is based on. There is also a button here for switching the rows and columns just as you did in the previous step.

There is an up and down arrow to the right of the Remove button. These buttons, as shown in Figure 13-27, are for moving the current legend up or down along with its series.

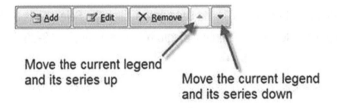

Figure 13-27. *Use up and down buttons to move the current legend*

Originally, we had a column chart, and the Risk game was represented by the first column in each month's group. When we changed to a bar chart, the series was reversed, and the Risk game became the last bar in each month's group. It seems strange, but in the Select Data Source dialog box, Risk is at the top even though it appears at the bottom of a bar chart.

9. With Risk selected, keep clicking the down arrow shown in Figure 13-27 until Risk becomes the last item in the series.

    Risk now appears at the top of each month's group in the bar chart.

10. Click the check mark in front of Stratego to remove it.

11. Click the check mark in front of June to remove it.

12. Click the OK button.

    Notice that no changes have been made to the spreadsheet data, only to the chart. The legend and chart series for Stratego have been removed, and all the June data has been removed from the chart.

13. Let's add another game to the spreadsheet. Add the line for the game Battleship as shown in Figure 13-28.

◢	A	B	C	D	E	F	G
1				Game Sales			
2		January	February	March	April	May	June
3	Risk	43	51	40	34	33	36
4	Stratego	29	29	30	23	22	28
5	Clue	40	50	35	33	30	35
6	Sorry	25	24	23	19	20	23
7	Monopoly	48	53	46	39	38	43
8	Battleship	40	48	48	36	35	38

Figure 13-28. *Row 8 data added to worksheet*

14. Click the chart to select it. Click the Design tab on the Ribbon. In the Data group, click the Select Data button.

15. Click the Add button. The Edit Series dialog box appears.

16. Click the Series name text box.

17. Click the word **Battleship** in cell A8.

18. Delete the text that is currently in the Series values: text box.

19. Drag across the cell range B8:G8 to get the sales data for the Battleship game. See Figure 13-29.

Figure 13-29. *Data entered for the Edit Series window*

20. Click the OK button.

    The chart should now show the monthly sales for Battleship.

21. Click the OK button for the Select Data Source dialog box.

Changing a Series Color

The series color for Risk and Monopoly are very close, and it is hard to distinguish one from another. We will change the color of the Risk series to red.

1. On the Ribbon's Format tab, in the Current Selection group, click the down arrow for Chart Elements and then select Series "Risk."

2. In the Current Selection group, click the Format Selection button.

3. In the Format Data Series pane, click the Fill & Line button ◇. You can make a bar color a solid color, a gradient, or fill it with a pattern or a picture.

4. Select Solid.

5. Select the red color from the Color button. See Figure 13-30.

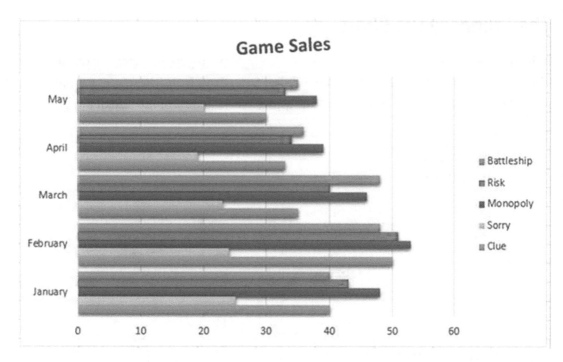

Figure 13-30. *The bars representing Risk have been changed to the color red*

You have learned how to create column and bar charts. Column and bar charts work well for multiple columns of numeric data. Next, you will learn how to create pie charts. Pie charts should only use a single column of numeric data.

Pie Charts

Pie charts are used to show percentages. The entire circle represents 100%. Each item in a data series is represented by a slice of the pie. The size of each slice shows what part of the 100% it represents. Sometimes you may find that the information in one slice needs to be broken out into percentages itself. You can use the pie of pie subtype in such circumstances. In this section, we'll first look at the standard pie chart and then cover the pie of pie subtype.

The Standard Pie Chart

In Figure 13-31, we have ten pieces of fruit. The entire pie represents 100% of our fruit. There are five bananas so they represent 5/10 or 50% of our fruit. Therefore, the bananas represent one-half of our pie. The slice for Apples takes up 30% of the size of our pie.

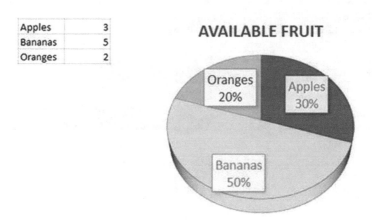

Apples	3
Bananas	5
Oranges	2

Figure 13-31. *Pie chart with labels*

EXERCISE 13-2: CREATING A PIE CHART

In this exercise, you will create a pie chart from Dave's Lunch Wagon data. A pie chart works well with this data because there is only one column of label data and only one column of numeric data.

1. Open workbook Chapter 13. Select the worksheet named Pie.

2. Select cells A2:B6. Click the **Quick Analysis** button at the bottom right of the data. See Figure 13-32.

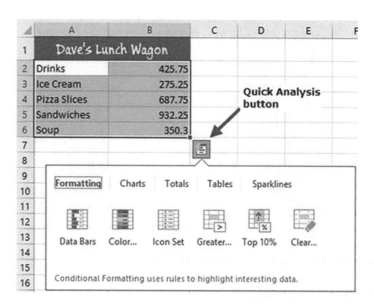

Figure 13-32. *Options for Quick Analysis button*

3. Click Charts.

   Excel recommends that you use a clustered column chart, a pie chart, or a clustered bar chart for the data you have selected. If you were to click the Clustered Column button, Excel would create a clustered column chart using the selected data on the current worksheet.

4. Move your cursor over the chart buttons to see how the chart would look with your selected data.

5. Click the pie chart. Click the chart.

6. Click the Chart Styles button ⌀ to the right of the chart.

   The Chart Styles box shows various ways of displaying the current chart type. Let's select a chart that displays the label and percent on each slice of the pie. Having a label on each slice of the pie makes having a legend unnecessary.

7. Click Pie Chart Style 10, which you can see in Figure 13-33.

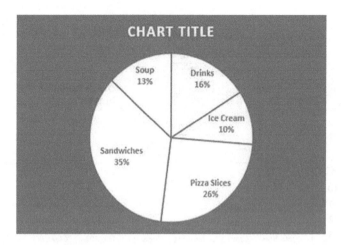

Figure 13-33. *Pie chart with Pie Chart Style 10 applied*

8. Click CHART TITLE to select it.

9. On the Ribbon's Home tab, in the Font group, change the font size to 28.

10. Drag across CHART TITLE text to select it. Type Dave's Chuck Wagon.

    If the title appears on two lines, then drag the bottom right sizing handle of the chart frame (not the chart title) to the bottom right while you hold down the Shift key. Do this until Dave's Chuck Wagon displays on only one line. See Figure 13-34.

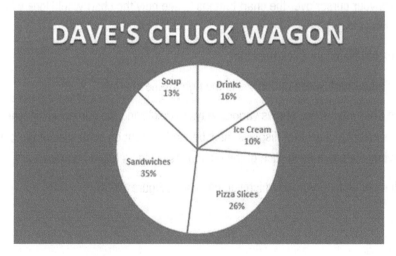

Figure 13-34. *Title should appear on one line*

11. Click the new chart title text.

12. Click the Chart Elements button ⊞ to the right of the chart.

13. Move your cursor on Chart Title. Click the arrow to the right of it. Select **More Options**. See Figure 13-35.

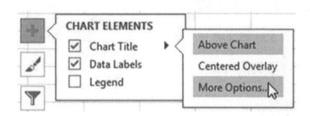

Figure 13-35. *Select More Options from the Chart Title options*

The Format Chart Title pane opens. There are two options to choose from. The **Title Options** affects the area inside the Chart Title area but outside the text itself. The **Text Options** affects only the text.

14. Click Text Options and complete the following:

a. Click the Text Effects button 🄰.

b. Expand the Shadow option.

c. Click the down arrow for Presets.

d. Click the second shadow in the Outer group (Offset Bottom).

e. Drag the Distance slider to 12 pt.

15. Click the pie portion of the pie chart. The right pane should change to Format Data Series.

16. Click the Series Options button (has a bar graph on it). Drag **Angle of first slice** slider to 26. This will rotate the pie 26 degrees.

17. To create some separation between the pie slices, change Pie Explosion to 2%. See Figure 13-36.

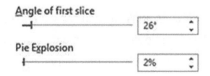

Figure 13-36. *Change Angle of first slice to 26 and the Pie Explosion to 2%*

The result should look like Figure 13-37.

Figure 13-37. *Your pie chart should now look like this*

Pie of Pie Subtype

A pie of pie chart takes a slice of a pie chart and then breaks that slice into another pie chart. One reason for doing this is that your pie chart contains too many items and some of the pie slices are so small that you can't tell what they are. Another reason might be that one of the pie slices is a category that you want further broken down.

Annette owns a soup and sandwich shop. She wants to see the breakdown of her drink category. She wants to see what types of drinks customers purchased this month. This can be done by pulling the drink category piece of the pie away from the others. This is called exploding the pie. You can explode the entire pie or individual slices.

EXERCISE 13-3: CREATING PIE SUBTYPES

In this exercise, you will be creating a pie chart for a sandwich shop to show what percent of sales are represented by sandwiches, soup, ice cream, and drinks. The shop decided that it would also like to see a breakout of its drink sales to see what percentage of its drink sales is represented by Pepsi, Coke, 7Up, and Sprite. You will create a pie of pie chart to display each of the different types of drinks, and then you will represent the same drink data with a bar of pie chart.

1. Open workbook Chapter 13. Click the worksheet tab Pie of Pie.

   Figure 13-38 shows the worksheet data.

◢	A	B
1	**Product**	**# Sold**
2	Sandwich	35
3	Soup	17
4	Ice Cream	20
5	Pepsi	13
6	Coke	10
7	7Up	3
8	Sprite	2

Figure 13-38. *Data on the worksheet Pie of Pie*

When creating a pie of pie chart, Excel takes the items at the end of your selection for the second pie chart so your data needs to be entered accordingly.

2. Select cell range A1:B8.

3. On the Ribbon's Insert tab, in the Charts group, click the Pie Chart button.

4. Select the second option (Pie of Pie) in the 2-D Pie area. See Figure 13-39.

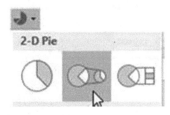

Figure 13-39. *Select the Pie of Pie option*

Excel took the last three items by default and made them the second chart. See Figure 13-40. We are going to need to change that to the last four items.

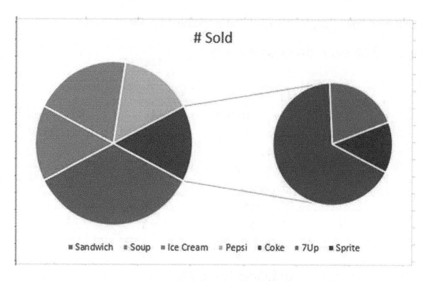

Figure 13-40. *Your pie charts should now look like this*

5.   Click the Chart. Click the Chart Elements button ⊞ .

6.   Move over Data Labels and then click the arrow to the right of it.

7.   Select Data Callout. See Figure 13-41.

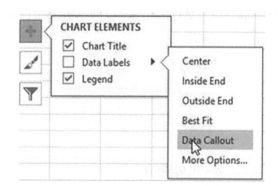

Figure 13-41. *Select Data Callout from the Data Labels options*

Figure 13-42 shows the result.

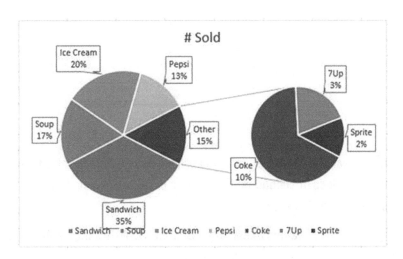

Figure 13-42. *Your charts should now look like this*

8. Double-click one of the pie slices from either chart. This displays the Format Data Series pane on the right side of the spreadsheet.

9. Click the Series Options button as shown in Figure 13-43.

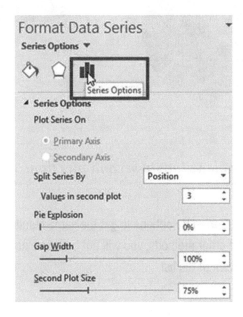

Figure 13-43. *Select Series Options*

The Split Series By combo box has four options. See Figure 13-44.

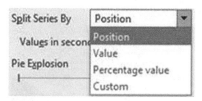

Figure 13-44. *Options for Split Series By*

The default value for **Split Series By** is Position, and the default Value for the combo box **Values in second plot** is 3. These defaults place the last three items in the list on the second chart.

10. Change the value in **Values in second plot** from 3 to 4. This moves the Pepsi slice to the second chart. See Figure 13-45.

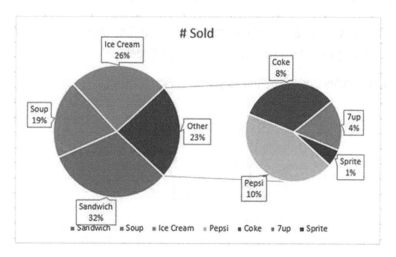

Figure 13-45. *Changing the **Values in second plot** from 3 to 4 moved Pepsi to the pie of pie chart*

You can use the other **Split Series By** options to get the same results as displayed in Figure 13-45. If you use the other methods, you will not have to place the items you want in the second chart at the bottom of the list.

Changing the Value of Split Series By to Value

1. Click the Undo button or press Ctrl + Z to undo moving Pepsi to the second chart.

For this example, you can get the same result by changing the **Split Series By** to Value. When you select Value, Excel displays a combo box asking for **Values less than**. Since Pepsi has a value of 13, you will need to change the value to 13.1 to move the Pepsi slice to the second chart.

2.   Select Value for the **Split Series By**.

3.   Enter 13.1 for the **Values less than**. Press Enter. See Figure 13-46.

Figure 13-46. *Select Value for Split Series By and then enter 14 for Values less than*

Changing the Value of Split Series By to Percentage Value

1.   Click the Undo button or press Ctrl + Z to remove the drinks from the second chart.

Changing the option for the **Split Series By** combo box to Percentage value brings up the **Values less than** combo box just as it does when selecting Value. Pepsi on the chart displays a percentage value of 13%. For this example, the values and the percentages are the same because percentage values were used for the spreadsheet data. To move the Pepsi slice to the second chart, we need to change the value in the **Values less than**.

2.   Select Percentage value for the **Split Series By**.

3.   Click the up arrow for **Values less than** until it is at 14. See Figure 13-47.

Figure 13-47. *Select Percentage value for Split Series By and then enter 14 for Values less than*

Changing the Value of Split Series By to Custom

1. Click the Undo button to undo moving Pepsi to the second chart.

   You can move slices from one chart to the other by selecting Custom from **Split Series By** and then clicking the pie slice you want moved to the second chart and then selecting which chart you want to move it to from the **Point Belongs to** combo box.

2. Select Custom for the **Split Series By**.

3. Double-click the Pepsi pie slice on the first chart. In the Point Belongs to drop-down box, select Second Plot. See Figure 13-48.

Figure 13-48. *Select Custom for the Split Series By*

618

4.   The chart shows that the four drinks are part of the **Other** slice. The word **Other** doesn't tell us anything. Select the word **Other** and change it to **Drinks**. See Figure 13-49. Click inside a blank area.

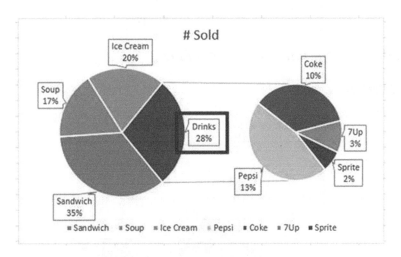

Figure 13-49. *Changed the word* ***Other*** *to* ***Drinks***

5.   Click one of the charts. Then in the Format Data Series pane, try the following options:

- The **Pie Explosion** is used for putting spacing between the pie slices. The higher the number, the more the charts explode out.

- The **Gap Width** is used for changing the amount of space between the two charts.

- The **Second Plot Size** is used for changing the size of the second chart.

Changing Pie of Pie to Bar of Pie

The bar of pie chart works the same way as the pie of pie chart.

1.   Select the Chart frame.

2.   On the Ribbon's Insert tab, in the Charts group, click the Pie Chart button.

3.   Select the third option (Bar of Pie) in the 2-D Pie group. See Figure 13-50.

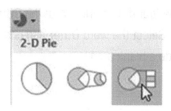

Figure 13-50. *Select the third option Bar of Pie*

Figure 13-51 shows the result.

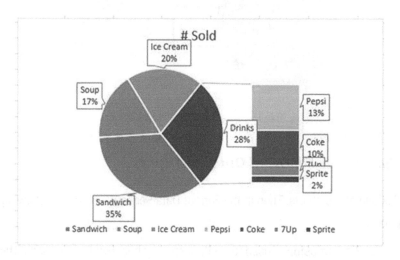

Figure 13-51. *Your charts should look like this*

You have learned how to create pie charts and even learned how to break down a slice of the pie chart into another pie chart or a bar chart. Next, you will learn how to combine two chart types into a single chart called a Combination Chart.

Combination Chart

A combination chart combines more than one chart type in a single chart. The most common combination chart consists of a bar chart for one series and a line chart for another series. Figure 13-52 shows a combination where the sales data and predicted sales data are shown on the same chart. The sales data is displayed in a column format and the predicted sales data in a line format. This makes it easy to see which predicted sales were higher and which predicted sales were lower than the actual sales. Another reason for creating a combination occurs when there is a large disparity between the values in the different series.

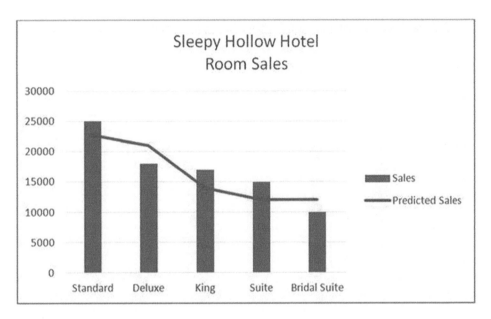

Figure 13-52. *Combination chart consisting of a column and a line chart*

EXERCISE 13-4: CREATING A COMBINATION CHART

In this example, you will create a combination chart consisting of a column and a line chart. Looking at Figure 13-53, you can see that there is a large disparity in the values between the two columns. As you will see this, disparity prevents us from creating an accurate column chart.

1. Open workbook Chapter 13. Select the worksheet named Combination.
   Figure 13-53 shows the data.

	A	B	C
1		**Homes Sold**	**Average Price**
2	Jan	6500	175,235
3	Feb	7000	176,015
4	Mar	7200	177,500
5	Apr	7400	174,118
6	May	8100	170,025
7	June	8400	175,920

Figure 13-53. *Data in the worksheet named Combination*

2. Select the cell range A1:C7 and complete the following:

   a. Click the Quick Analysis button at the bottom right of the selected data.

   b. Click the Charts tab.

   c. Click Clustered Column. Figure 13-54 shows the result.

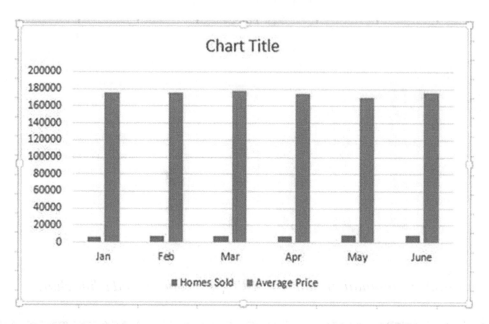

Figure 13-54. *Clustered Column chart*

You can see that this chart isn't very useful. There is far too much disparity between the values for Homes Sold and the Average Price.

3. Click one of the Average Price columns in the chart to select the Average Price series.

4. On the Ribbon under Chart Tools, click the Design tab. In the Type group, click Change Chart Type.

5. In the Change Chart Type window, select Combo in the left pane. Both series are displayed. The Homes Sold and the Average Price are both clustered columns.

6. Click the down arrow for Clustered Column to the right of Average Price. Under the Line group, select Line with Markers. See Figure 13-55.

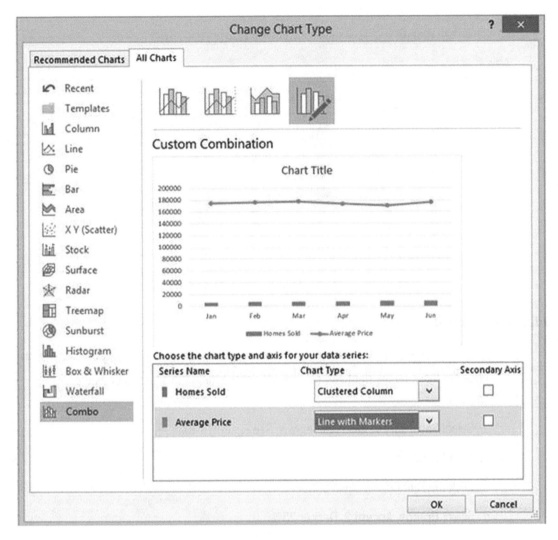

Figure 13-55. *Select **Line with Markers***

7. We need to create a second vertical axis that will consist of values related to the Average Price. The left vertical axis will then show values related to Homes Sold:

    a. Check the Secondary Axis box to the right of Average Price.

    b. Click the OK button.

8. Click Chart Title. Change the text to **Number of Homes Sold and their Average Price**. See Figure 13-56.

Figure 13-56. *Chart with Title change*

9.   So that the user can easily see that the values associated with the Average Price are those on the right vertical axis, let's make them the same color as the line chart:

   a.   Select Secondary Vertical (Value) Axis. See Figure 13-57.

   b.   Click the Home tab.

   c.   Click the Font Color's down button.

   d.   Select **Orange, Accent 2, Darker 25%**.

   e.   Click the Bold button. See Figure 13-57.

Figure 13-57. *The vertical axis has* **Orange, Accent 2, Darker 25%** *applied*

You have learned how to combine two chart types in a single combination chart. Next, you will learn how to create the two new hierarchical charts, the Treemap chart and the Sunburst chart.

Hierarchical Charts

Sometimes you will want to show the hierarchical relationships among data. In this section, we'll look at two common types of hierarchical charts: the Treemap and Sunburst chart types.

Treemap Chart

A Treemap is used for representing hierarchical data. The chart consists of rectangles whose size represents the value associated with a label. Figure 13-59 shows a Treemap chart.

EXERCISE 13-5: CREATING A TREEMAP CHART

In this exercise, you will create a Treemap chart using the data in Figure 13-58. Since the largest value in the spreadsheet is 1000, this item will be represented by the largest rectangle. The rectangles representing the products will be organized within the states.

1. Open workbook Chapter 13. Select the worksheet named Treemap.

   Our company sells snack products to gas stations in five states. Figure 13-58 shows our profit for the past week.

◢	A	B	C
1	State	Product	Profit
2	Indiana	cookies	$ 500.00
3	Indiana	muffins	$ 650.00
4	Indiana	donuts	$ 800.00
5	Indiana	candy	$ 550.00
6	Illinois	muffins	$ 1,000.00
7	Illinois	donuts	$ 300.00
8	Michigan	cookies	$ 900.00
9	Michigan	donuts	$ 300.00
10	Michigan	muffins	$ 100.00
11	Georgia	cookies	$ 800.00
12	Florida	donuts	$ 275.00

Figure 13-58. Data in the worksheet named Treemap

   A Treemap chart will give us a quick view of the distribution of our profit for the various products across the five states.

Note   The headings don't need to be selected when creating the chart.

---

2. Select the range A2:C12. On the Ribbon's Insert tab, click the **See all Charts** button ◪ in the bottom right corner of the Charts group.

3. Select the All Charts tab and then select Treemap in the left pane. Click the OK button. See Figure 13-59. The largest rectangle represents the largest value.

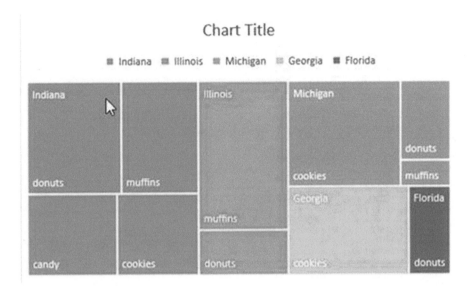

Figure 13-59. *Treemap chart*

It's usually easier to read a Treemap chart if you turn the headings (the state names in this case) to banners.

4.   Right-click any one of the rectangles on the chart and then select **Format Data Series**. Select **Banner** from the Label Options. See Figure 13-60.

Figure 13-60. *Select Banner from the Label Options*

Figure 13-61 shows the result.

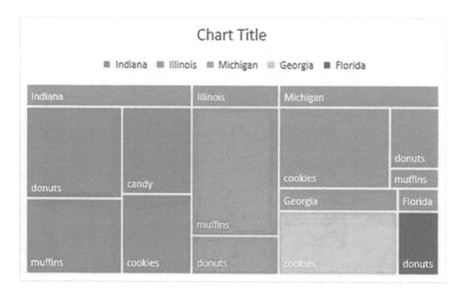

Figure 13-61. *Result of changing the headings to banners*

It's easier to see the state headings now. Since the Michigan Items are in gray, they blend in with the state headings and are hard to read.

5.   Change colors of the rectangles:

a. Click the Ribbon's Design tab. In the Chart Styles group, click the Change Colors button.

b. Move your cursor over the different color palettes to see how they affect your chart. We want one that doesn't have a gray color. Select the third row in the Colorful group. See Figure 13-62.

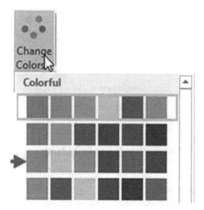

Figure 13-62. *Select the third row in the Colorful group*

You might want to see the values along with the labels.

6. Click the Chart frame. Click on the Ribbon's Format tab. In the Current Selection group click on Format Selection. Click the down arrow for Chart Options in the Format Chart Area pane and then select Data Labels. See Figure 13-63.

Figure 13-63. *Select Data Labels from the Chart Options*

This adds another option button, Label Options. See Figure 13-64.

Figure 13-64. *Label Options*

7. Click the Label Options button. Expand Label Options so that you can see the options as shown in Figure 13-65.

Figure 13-65. *Label Options*

8. Select **Value**. Select **New Line** from the Separator drop-down box. Selecting New Line places the label and the value on two separate lines as shown in Figure 13-66.

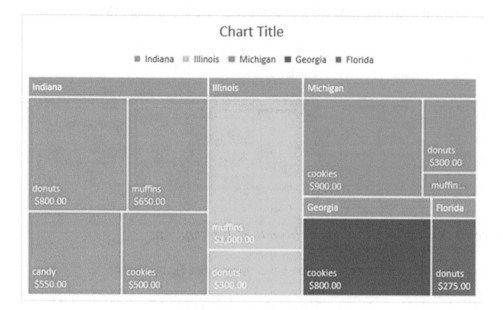

Figure 13-66. *The labels and values should be on two separate lines*

9. The white text can be hard to read:

    a. Click the Ribbon's Home tab.

    b. Click the down arrow for Font Color and then select the black color.

10. With the labels still selected, press Ctrl + B to bold the text.

11. Click Chart Title. Drag across it to select it. Type **Sales by State**.

    If you click on one of the state's rectangles and then click on the same rectangle then all the rectangles not related to that state will be dimmed. In Figure 13-67, the legend item Michigan was clicked twice. Once a group is selected, you can single-click an individual rectangle to select it or you can select another group by clicking its header. If you click one of the labels in a rectangle, everything becomes selected again.

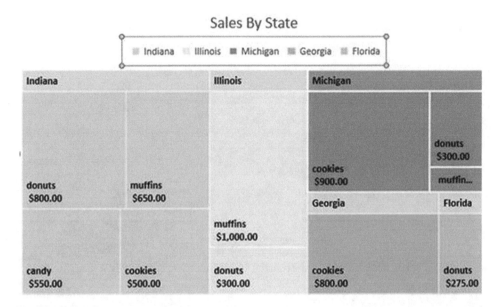

Figure 13-67. *Michigan is highlighted because the Michigan legend item was clicked once and then clicked again*

12.   Click on one of the Michigan rectangles and then click on it again. Click on the donuts rectangle in the Michigan group. All rectangles should be dimmed except the Michigan donuts. Click an empty cell to undim the rectangles.

When a group of rectangles or an individual rectangle is selected, you can change its background to a different color, gradient, picture, or texture. Figure 13-68 shows that a texture was selected for the Michigan group of rectangles. You can also apply various effects and alter its borders by clicking the Ribbon's Format tab and then selecting one of the options from the Shape Styles group.

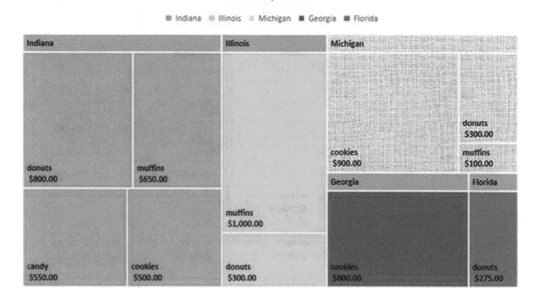

Figure 13-68. *Texture background was selected for Michigan*

13. Click twice on one of the Michigan rectangles to reselect all of the Michigan rectangles. In the Format Data Point pane, click the Fill & Line button (the paint bucket).

14. Expand the Fill group.

15. Select **Picture or texture fill**.

16. Click the drop-down box for Texture shown in Figure 13-69. Select whatever texture you want to use.

Figure 13-69. *Select a texture*

17.   Try changing the color or texture of individual rectangles.

If not all your data values are showing, you may need to expand the size of the chart.

Sunburst Chart

Like the Treemap chart, the Sunburst chart is also a hierarchical chart. Instead of using the rectangle sizes to represent a value, it uses the size of pie pieces.

An advantage of the Sunburst chart over the Treemap chart is that it can represent more columns of labels. In the example for the Treemap chart, the first column (State) was used for the headings and the next column (Product) was used to represent the individual rectangles. If we wanted to break things down farther by going State, then Counties, and then Products within those counties, the Counties wouldn't display in the Treemap chart, but they would in a Sunburst chart.

Using the same data we used for the Treemap chart and selecting the same options, let's look at the result for the Sunburst chart in Figure 13-70.

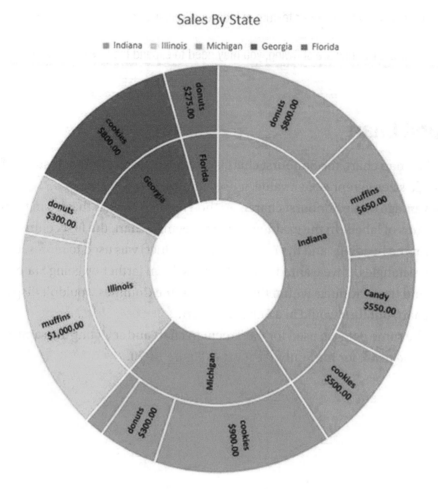

Figure 13-70. *Sunburst chart*

The Sunburst chart clearly shows that Indiana is our most profitable state. It's easy to see from the chart that donuts are the most profitable item we sell in Indiana. The chart bursts from the inside out. The largest grouping is in the center of the chart. The details are farthest from the center.

The pie slice for muffins in Michigan isn't large enough to display the label and the value. If we remove the values, the label for the muffins becomes visible. See Figure 13-71.

Sales By State

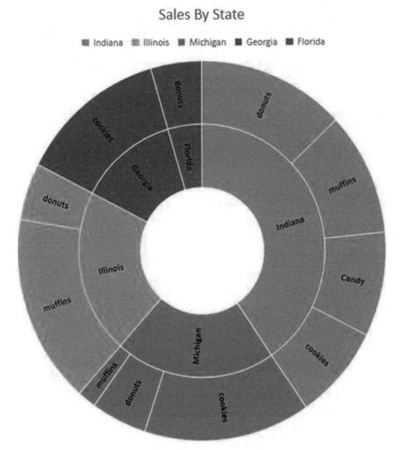

Figure 13-71. *Removing the values gives more room for the labels to display*

The label for the muffins profit in Michigan is now visible. If you want to know the value associated with each slice, just move your cursor over one of the items. See Figure 13-72.

Figure 13-72. *Move your cursor over a slice to see the associated value*

EXERCISE 13-6: CREATING A SUNBURST CHART

This exercise covers selecting what data you want represented in the Sunburst chart as well as selecting portions of the chart and applying styles.

1. Open workbook Chapter 13. Select the worksheet named Sunburst.

   Figure 13-73 shows the data.

	A	B	C	D	E	F
1	Region	State	Product	Dozens	Price Per Dozen	Gross
2	North	Illinois	cookies	2000	2.95	$ 5,900
3	North	ILlinois	donuts	1800	3.5	$ 6,300
4	North	Indiana	cookies	2200	2.88	$ 6,336
5	North	Indiana	donuts	2275	3.4	$ 7,735
6	North	Michigan	cookies	2850	3.1	$ 8,835
7	North	Michigan	donuts	2717	3.45	$ 9,374
8	South	Georgia	cookies	1500	2.9	$ 4,350
9	South	Georgia	donuts	1325	3.4	$ 4,505
10	South	Florida	cookies	1210	3.25	$ 3,933
11	South	Florida	donuts	1085	3.5	$ 3,798
12	South	Alabama	cookies	1105	2.85	$ 3,149
13	East	New York	cookies	1200	9.5	$ 11,400
14	East	Maryland	donuts	1250	4.3	$ 5,375
15	East	Vermont	cookies	1000	4.25	$ 4,250

Figure 13-73. *Data in the worksheet named Sunburst*

2. Select the cell range A2:D15 and complete the following:

   a. On the Ribbon's Insert tab, click the dialog box launcher (See All Charts) for the Charts group.

   b. Click the All Charts tab.

   c. Select Sunburst in the left pane. Click the OK button.

   d. Enlarge the chart. See Figure 13-74.

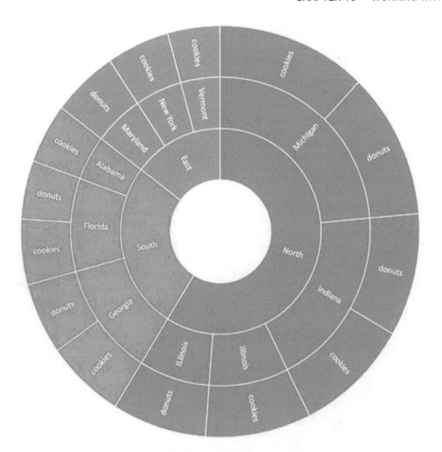

Figure 13-74. *Sunburst chart using the range A2:D15*

The Sunburst chart makes it easy to see that we sell the most items in the North region and that Indiana and Michigan are the biggest selling states.

What if we wanted the chart to display the pie slices based on the Gross column?

3. Delete the Sunburst chart.

4. Select the cell range A2:C15, hold down the Ctrl key, and select the cell range F2:F15. Let's apply a cell style to the range so that we can easily identify which columns are being used for the chart.

5. On the Ribbon's Home tab in the Styles group, click the Cell Styles button. Select Accent5. See Figure 13-75.

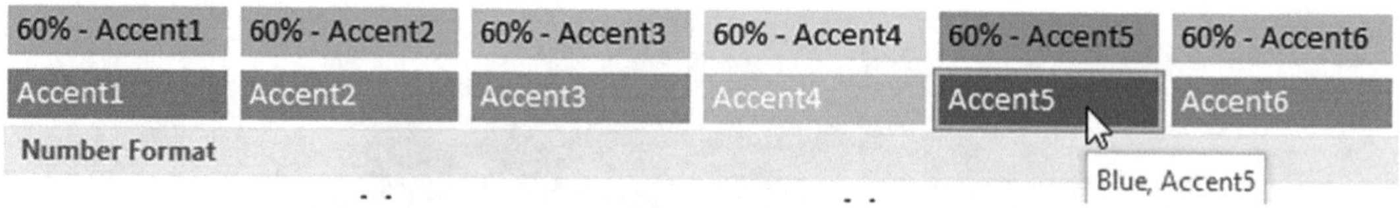

60% - Accent1	60% - Accent2	60% - Accent3	60% - Accent4	60% - Accent5	60% - Accent6
Accent1	Accent2	Accent3	Accent4	Accent5	Accent6
Number Format					

Blue, Accent5

Figure 13-75. *Select Accent5 from the Styles group*

Your spreadsheet should look like Figure 13-76.

	A	B	C	D	E	F
1	**Region**	**State**	**Product**	**Dozens**	**Price Per Dozen**	**Gross**
2	North	Illinois	cookies	2000	2.95	$ 5,900
3	North	ILlinois	donuts	1800	3.5	$ 6,300
4	North	Indiana	cookies	2200	2.88	$ 6,336
5	North	Indiana	donuts	2275	3.4	$ 7,735
6	North	Michigan	cookies	2850	3.1	$ 8,835
7	North	Michigan	donuts	2717	3.45	$ 9,374
8	South	Georgia	cookies	1500	2.9	$ 4,350
9	South	Georgia	donuts	1325	3.4	$ 4,505
10	South	Florida	cookies	1210	3.25	$ 3,933
11	South	Florida	donuts	1085	3.5	$ 3,798
12	South	Alabama	cookies	1105	2.85	$ 3,149
13	East	New York	cookies	1200	9.5	$ 11,400
14	East	Maryland	donuts	1250	4.3	$ 5,375
15	East	Vermont	cookies	1000	4.25	$ 4,250

Figure 13-76. *Your spreadsheet should now look like this*

6. On the Ribbon's Insert tab, click the dialog box launcher (See All Charts) for the Charts group.

7. Click the All Charts tab. Select Sunburst in the left pane. Click the OK button.

8. Enlarge the chart. See Figure 13-77.

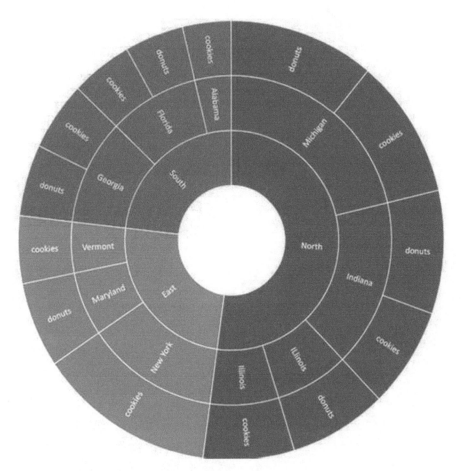

Figure 13-77. *Sunburst chart created from range A2:C15 and F2:F15 selected*

We sell our product in New York City in a very upscale area at a higher price, which makes our New York pie slice much larger than it was for the number of dozens sold.

9. Click the Ribbon's Design tab. In the Chart Layouts group, click the Quick Layout button. Select Layout 5. This adds the Gross values to the Chart.

10. In the Chart Styles group, click the Change Colors button. Select Colorful Palette 3.

11. In Excel 2016, selecting the last chart in the Chart Styles group gave the chart a 3D appearance. This no longer works, but we will give it kind of a 3D appearance by formatting it. Select the chart and then, from the Ribbon's Design tab, select the last chart choice from the Chart Styles group. Right-click the chart. Select Format Data Series. Click on the Effects button. Expand Shadow. Enter 132 for the angle. See Figure 13-78.

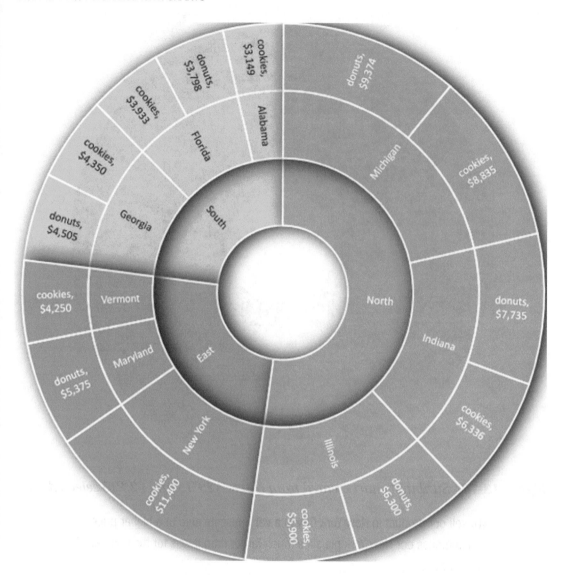

Figure 13-78. *Sunburst chart with 3D effect*

Just like with the Treemap chart, you can select a group by clicking two
separate times on one of the pieces of that group. You can then select individual
pieces or other groups by using a single click. If you purposely or accidently
click a label, everything gets reselected. Figure 13-79 shows the North group
selected.

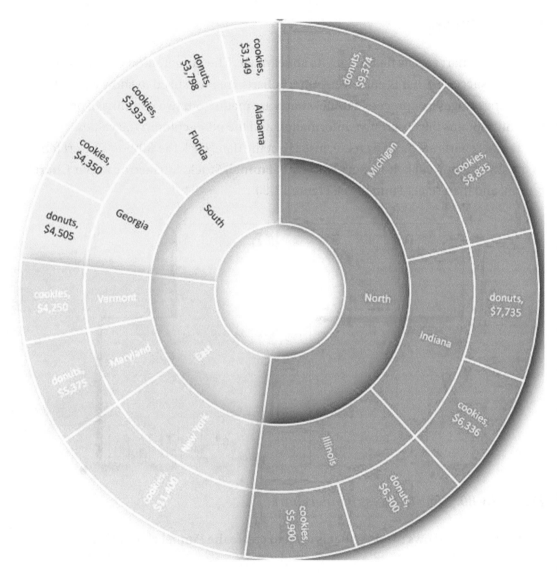

Figure 13-79. *Sunburst chart with North selected*

You have learned how to create the Treemap and Sunburst hierarchical charts. Hierarchical charts work well for data where natural groupings exist and the data is organized in some level of order.

Funnel Chart

The Funnel chart gets its name from its appearance. The funnel starts with a wide bar, and each successive bar is narrower. Funnel charts are often used to show values for the various stages in a process. Funnel charts are made from two columns of data. One column contains the descriptive information and the other column the values.

Funnel charts are easy to create; just select the data, click the Ribbon's Insert tab, in the Chart group click the **Insert Waterfall, Funnel, Stock, Surface or Radar Chart** button, and then select Funnel. See Figure 13-80.

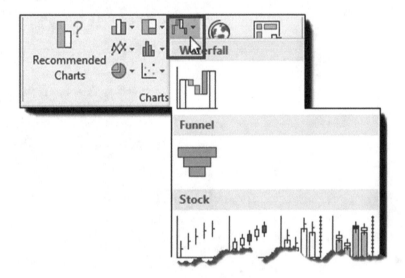

Figure 13-80. *How to access Funnel chart*

Figure 13-81 shows the data that is used to create the Funnel chart in Figure 13-82.

▲	A	B
1		
2	Prospective students in mass mailing	80000
3	Students attending on-campus sessions	55000
4	Students applying to enroll	40000
5	Students completing enrollment	30000
6	Students Accepted	27000
7	Students attending freshman year	25000
8	Students attending Sophmore year	22000
9	Students attending Junior Year	20000
10	Students attending Senior Year	19000
11	# of graduates in 4 years	18000

Figure 13-81. *Funnel chart data*

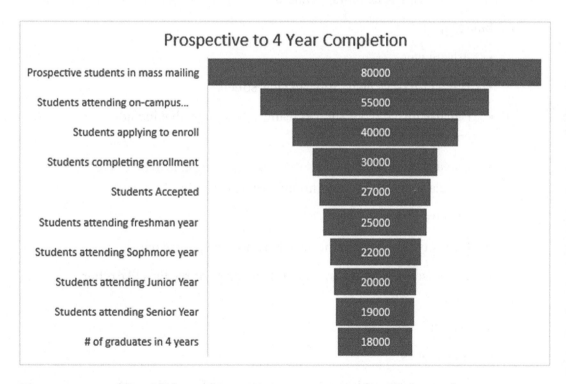

Figure 13-82. *Funnel chart shows stages to get to # of students graduating*

You have learned how to create a Funnel chart to show values for the various stages in a process. Next, you will look at Excel's two mapping charts: the Map Chart and the Bing Chart. The Bing Chart is not part of the Chart group; rather, it is an Add in. They are both discussed here so that you can compare the two to see when you need to use one or the other.

Map Chart

Excel's new Map Chart is used for showing geographical data by countries/regions, states, counties, or postal codes. You can use the Map Chart to display Numerical or Categorical data. It is used for showing a location and one variable such as a location's population, sales for that area, hospitals in that area, and so on. The Bing Map is discussed after this cart. The advantages and disadvantages of each are listed as follows:

- Bing Map

  - Advantages

  - Has a bird's-eye view. You can view streets.

  - Can use mixed types such as using a data range that includes zip codes, states, and countries.

  - Can use more than one variable such as being able to chart the population, revenue, and number of hospitals for the states.

  - Disadvantages

  - Can only represent Numerical data, not Categorical data.

  - Can't use commands from the Design and Format tabs like other charts.

- Map Chart

  - Advantages

  - Big advantage is that you can use all the formatting and design tools like other charts you have work with use.

  - Can use Categorical data as well as Numerical data.

- Disadvantages

- Doesn't have a bird's-eye view and you can't view any streets.

- Can't use mixed types.

- Can only use one variable.

Let's look at a map that shows a problem that has occurred while writing this book. That is the coronavirus. Figure 13-83 shows some states and their current number of coronavirus cases.

	A	B
1	**State**	**Cases**
2	Alabama	6
3	Arizona	13
4	Indiana	15
5	Illinois	93
6	Kentucky	21
7	Ohio	37

Figure 13-83. *Coronavirus data*

We can chart this data by selecting the cell range A1:B7 and then clicking the Ribbon's Insert tab. In the Charts group, there is a Maps button. See Figure 13-84. Currently, there is only one type of Map Chart and that is the Filled Map Chart. Clicking More Map Charts just lets you select from one of the available chart types.

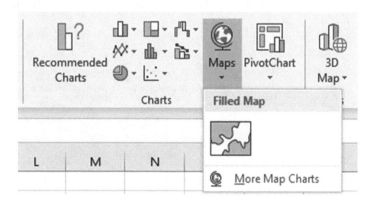

Figure 13-84. *Select Filled Map*

After selecting the Map Data and clicking the Filled Map button, Excel creates a chart of the United States with the states from the chart filled with a gradient color. See Figure 13-85. The darker the color, the more cases of coronavirus. The Legend shows the number of cases. It shows from the minimum (6) to the maximum (93).

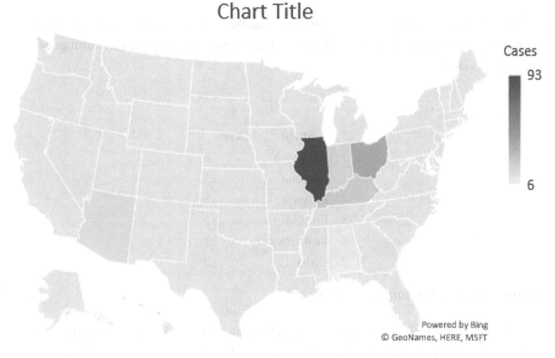

Figure 13-85. *Map of coronavirus cases*

You can use the Map Chart to display categories rather than a value. Each category will be assigned its own color. Figure 13-86 shows a portion of a table used to create the map in Figure 13-87.

Country	Primary Religion
United States	Protestant
Canada	Catholic
Mexico	Catholic
Venezula	Catholic
Uganda	Cattolic
Singapore	Buddhist
Burma	Buddhist
India	Hindu
Turkey	Muslim

Figure 13-86. *Category data for map*

Figure 13-87. *Map Chart using Categories*

You can change the color of all the areas represented by a single category by doing the following:

1.   Click the Legend to select it.

2.   Right-click the category whose color you want to change.

3.   Click the Fill Color button.

4.   Select a Color.

See Figure 13-88.

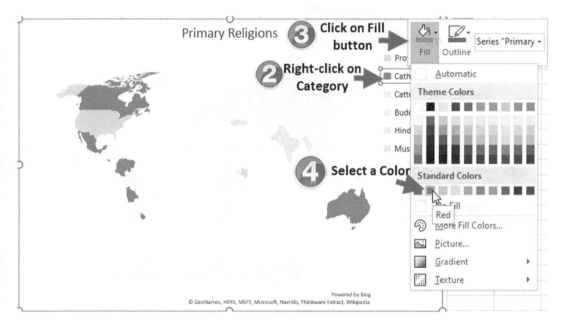

Figure 13-88.   *Change Category Color*

Using Bing Maps

Using Excel's Bing Maps, you can create a visual representation of your data in any location in the world. The first column of the data must be an address, city, state, country, zip, latitude/longitude, or a combination. Any columns following that must be numeric data. The problem with using a city is that there may be many cities with the same name. You can concatenate a city with a state or country name.

EXERCISE 13-7: USING BING MAPS

In this exercise, you will create maps that reflect the revenue created in different cities.

1. Open the Chapter 13 workbook. Select the worksheet named Maps.

   The examples in this worksheet use Revenue for the numeric amount. The values could be anything (e.g., city or state populations, growth rates, death rates, number of employees in a company working in that location, etc.).

2. Click the Ribbon's Insert tab. In the Add-ins group, click the Bing Maps button. The Bing Map displays. See Figure 13-89.

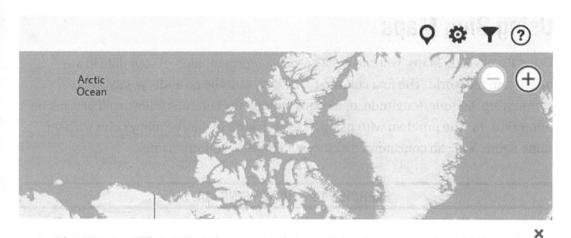

Figure 13-89. *Bing Maps*

3. Click the X to close the opening message.

4. Place your cursor on the border of the map above the icons, and it will turn into four arrows. You can now drag the map to wherever you want.

5. If you click while on the border, the sizing handles display. When you move your cursor over a sizing handle, it changes to a double arrow. Hold down the Shift key while you drag one of the corner sizing handles so that the map keeps it proportions.

6. Select the cell range A2:B4. Click the Show Location button 📍 in the upper right corner.

Bing displays a map with a circle in each of the three states. See Figure 13-90. The circle represents the amount of the numeric data. Illinois had the largest amount, so its circle is the largest.

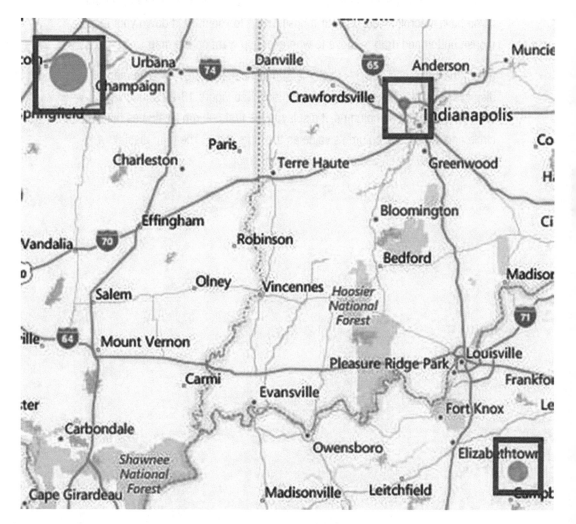

Figure 13-90. *Circles represent the amount of revenue for that location*

7. Click one of the circles. The data associated with that circle displays. See Figure 13-91. Click the close button.

Indiana

Revenue: 500

close

Figure 13-91. *Data for Indiana*

8. Click the plus and minus buttons at the top right of the map to zoom in and out. If you have a scroll wheel on your mouse, use it to zoom. Hold down your left mouse button and drag to move to wherever you want on the map.

9. Select the cell range A6:C11. Click the Show Location button ♀. The map displays with a circle at each of our locations. See Figure 13-92. Since we provided two numeric columns, it displayed the first column's value as the outer circle and the second column's value as the inner circle. The map displays a legend to show which value goes with which color.

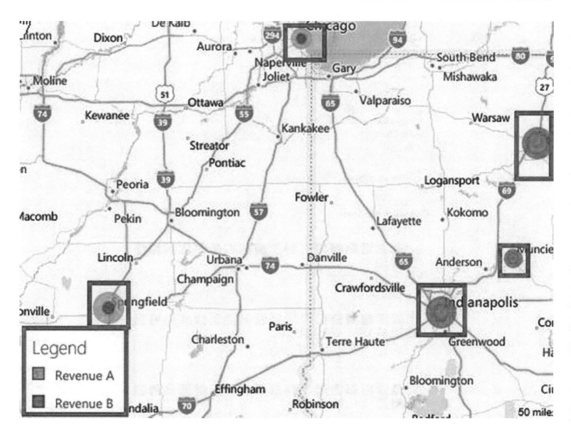

Figure 13-92. *Legend identifies the colors for the different revenue columns*

The circles are okay when viewing two numeric values, but when using more than two columns, it is really difficult to identify the colors. Instead of using circles, you can use a pie chart.

10.   Select the cell range A13:D17. Click the Show Location button 📍.

11.   Click the Settings button ⚙. Select Pie Chart under **Multiple Data Point Display**. See Figure 13-93. Excel lets you select a different color for each of the three pie slices.

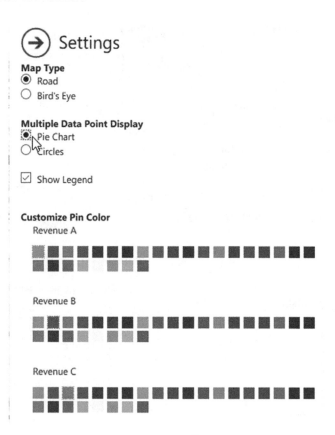

Figure 13-93. *Select Pie Chart*

12. Pick whatever colors you want. When you have finished picking your colors, click the arrow within the circle to return to the map.

The pie charts should display using the colors you selected. See Figure 13-94.

Figure 13-94. *Values are represented by pie charts*

13.   Click one of the pie charts. The data associated with that pie chart shows. See
Figure 13-95. Click the close button.

Leerdam, Netherlands
Revenue A: 5000
Revenue B: 4000
Revenue C: 5000

close

Figure 13-95. *Data for Pie Chart*

14.   Select the cell range A19:B22. Click the Show Location button ⛳. Because all
the locations are in the same city, the map is zoomed in.

15.   Click the Settings button ⚙. Under Map Type, select Bird's Eye. Click the arrow
within the circle to return to the map. Zoom in. See Figure 13-96.

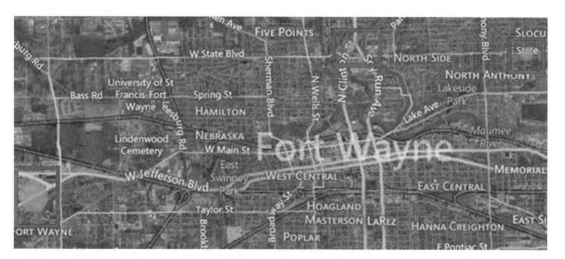

Figure 13-96. *Zoomed in Bird's Eye view*

The locations don't have to be of the same type. You can have a mix of cities, states, countries,
zip codes, addresses, and so on, as shown in Table 13-1.

Table 13-1. *Locations of Mixed Types*

Location	Revenue
Kentucky	7500
Chicago	4500
France	5000
Germany	8000
46341	7000
550 N. Clinton St., Fort Wayne	6000

16.   Select the cell Range A24:B30. Click the Show Location button ⬤.

17.   Delete the map by clicking the map border and then press the Delete key.

Next, you will learn how to create charts that represent a row of data and are stored in a single cell.

Sparklines

Sparklines are small charts displayed in individual cells. They are used for highlighting trends within a single row or single column. There are three types of sparklines: Line, Column, and Win/Loss. Sparklines are created by first selecting the cells where you want the sparklines to be located and then selecting the type of sparkline you want to create. The Sparklines group is located on the Insert tab. There are three buttons in the group: Line, Column, and Win/Loss. See Figure 13-97.

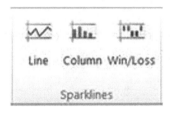

Figure 13-97. *Ribbon's Sparklines group*

Because the Sparklines are so small, if there isn't a wide variance in your values, the column charts may appear disproportionate in size compared to the actual values. It is much easier to see the trends if you enlarge the cell that contains the Sparklines.

Column sparklines that are negative are displayed in the bottom half of the cell. Zero values are not given a sparkline; instead, a space is placed in that position. See Figure 13-98.

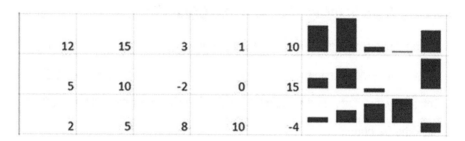

Figure 13-98. *Sparklines with negative values are displayed at the bottom of the cell*

Win/Loss sparklines show whether your data is positive (win) or negative (loss). See Figure 13-99.

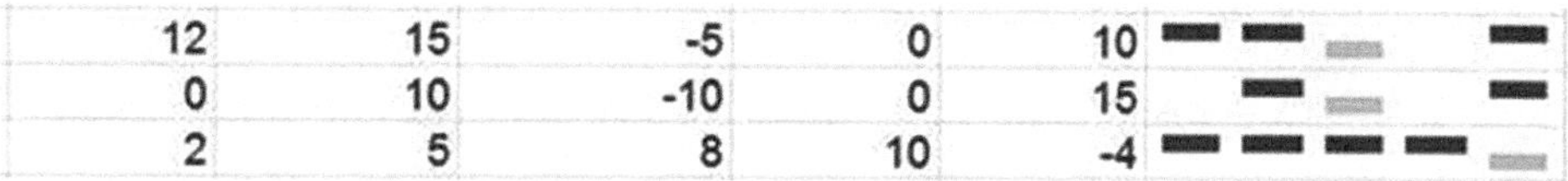

12	15	-5	0	10	
0	10	-10	0	15	
2	5	8	10	-4	

Figure 13-99. *Sparkline that identifies wins and losses*

EXERCISE 13-8: CREATING SPARKLINES

In this exercise, you will create sparklines for the row data in Figure 13-100. You will group and ungroup the sparklines so that you can format some of your sparklines one way and some of them another way.

1. Open the Chapter 13 workbook. Select the worksheet named Sparklines. Figure 13-100 shows the data.

⊿	A	B	C	D	E
1		**Qtr 1** **% Change**	**Qtr 2** **% Change**	**Qtr 3** **% Change**	**Qtr 4** **% Change**
2	Stock A	4	8	10	13
3	Stock B	-3	0	-2	2
4	Stock C	4	3	1	3
5	Stock D	0	0	-2	-4
6	Stock E	3	-2	5	4

Figure 13-100. *Data in the worksheet named Sparklines*

2. On the Ribbon, on the Insert tab, in the Sparklines group, click the Line button.

3. Delete any text that is in the Data Range text box. Select the cell range B2:E6.

4. Delete any text that is in the Location Range text box. Select the cell range F2:F6.

5. Click the OK button.

6. Under the Sparkline Tools contextual tab, click the Design tab. In the Style group, click the down arrow for more line color options. Select the Sparkline Style Colorful #4 in the last row. See Figure 13-101.

Figure 13-101. *Select Sparkline Style Colorful #4*

7. Drag across row heads 2, 3, 4, 5, and 6. On the Ribbon, on the Home tab, in the Cells group, click the Format button.

8. Click Row Height. Change the row height to 22. Click the OK button.

9. Change the column width for column F to 17.43.

10. Select cells F2:F6.

11. Click the Ribbon's Design tab. In the Show group, click the High Point and Low Point check boxes. The Sparkline lines should now have dots representing the highest and lowest points on the line as shown in Figure 13-102.

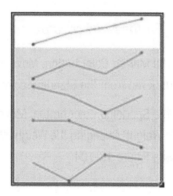

Figure 13-102. *Dots represent highest and lowest points*

12. Remove the check marks from the High Point and Low Point. Check the Negative Points. Now the points on the line represent those cells that have negative values.

13. Check the Markers. Now all the cells are represented by points on the line.

14. In the Style group, click the Marker Color button. Select purple from the Standard Colors. See Figure 13-103.

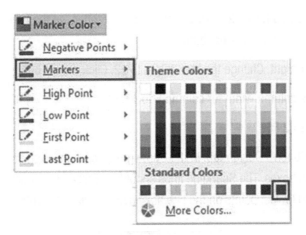

Figure 13-103. *Select purple for the Marker color*

15. In the Style group, click the Marker Color button. Move your cursor over Negative Points. Select dark red from the Standard colors.

16. In the Style group, click the Sparkline Color button. Move your cursor over Weight. Select the 4½-pt Weight. Changing the Weight changes both the size of the line and the points. See Figure 13-104.

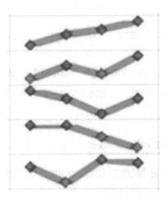

Figure 13-104. *Changing the weight changes the line and font size*

17. Click the Sparkline Color button again. In the Theme colors, select Blue, Accent 5, Lighter 40%.

18. In the Type group, click Column. Let's change the style. Click the More button for Style. Click Sparkline Style Accent 5, Darker 50%. See Figure 13-105.

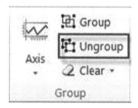

Green, Sparkline Style Accent 5, Darker 50%

Choose a visual style for the selected sparkline group.

Figure 13-105. *Select Sparkline Style Accent 5, Darker 50%*

<u>Ungrouping Sparklines</u>

1. You may want some of your sparkline cells formatted one way and some of the others formatted another way. You can do this by grouping and ungrouping the sparkline cells.

2. Select cells F2 and F3.

3. Click Ungroup. See Figure 13-106.

Figure 13-106. *Click Ungroup*

4. In the Type group, click Line.

5. In the Show group, check off Markers and check on Last Point. See Figure 13-107.

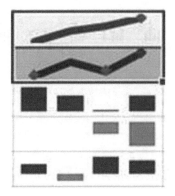

Figure 13-107. *Last Point selected*

Grouping Sparklines

1. Select cells F2:F6. Click the Group button.

2. Remove the check from Last Point.

3. In the Type group, click Win/Loss.

The points representing positive values (wins) are displayed in the top half of the cell. The points representing negative values (losses) are displayed in the bottom half of the cell.

Clearing Sparklines

1. Click cell F4.

2. Click the down arrow to the right of the Clear button and then click Clear Selected Sparklines. See Figure 13-108.

Figure 13-108. *Select Clear Selected Sparklines*

Changing the Data Location and Location of the Sparklines

1. Right-click column head B. Select Insert.

2. Select cells G2:G6.

3. Under Sparkline Tools, click the Design tab.

4. In the Sparkline group, click the top half of the Edit Data button.

5. Click the Collapse Dialog button to the right of the Data Range text box. See Figure 13-109.

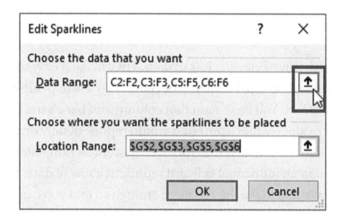

Figure 13-109.   *Click the Collapse Dialog button*

6. Select cells C2:F6. Click the Expand Dialog button (the down arrow).

7. Click the Collapse Dialog button to the right of the Location Range text box.

8. Select cells B2:B6. Click the Expand Dialog button.

9. Click the OK button. See Figure 13-110.

	A	B	C	D	E	F
1			Qtr 1 % Change	Qtr 2 % Change	Qtr 3 % change	Qtr 4 % Change
2	Stock A		4	8	10	13
3	Stock B		-3	0	-2	2
4	Stock C		4	3	1	3
5	Stock D		0	0	-2	-4
6	Stock E		3	-2	5	4

Figure 13-110. *Your spreadsheet should now look like this*

Summary

Charts provide a quick way of viewing lots of data. They make it easy to see if you are meeting your goals and to spot problems. What chart you use depends upon the data that you want to represent. You have seen that column and bar charts can be used for many rows of numeric data, while a pie chart would represent only one row of numeric data. The Treemap and Sunburst charts work well for representing hierarchal data. Sparklines are stored in an individual cell and represent a row of data. If you are using Sparklines, you will want to make the cell large enough so that you can easily see what the chart is displaying.

Pie charts represent data in pie slices. The pie slices represent a percent of the entire pie. The Treemap chart uses rectangles to represent values, while a Sunburst chart represents values with pie pieces. Sparklines can use one of three chart types: Line, Column, or Win/Loss. A pie chart slice can be broken down into another pie or bar chart so that you can see all of the items that were used to create it.

A Combination chart combines two charts into one. It usually consists of a bar chart for one series and a line chart for another.

The Bing Map and Map Chart allow you to chart data related to a geographical area.

The fastest way to create a 2D clustered bar chart is to use the following shortcut keys:

- Press the Alt + F1 key if you want to display the chart on the same worksheet as the data.

- Press F11 if you want the chart to appear on a separate worksheet.

The next chapter details how to import data from various sources.

CHAPTER 14

Importing Data

Excel is the premier analysis tool. This is the reason for bringing data from other sources into Excel. Microsoft has added even more sources and ways of importing data for Excel 2019.

After reading and working through this chapter, you should be able to

- Import delimited text and fixed-width text files

- Import data from an Access database

- Import data as is from a web page and modify and transform it before importing it

- Create connections between a workbook and a data source

Importing Data into Excel

The buttons for importing and transforming data are located on the Ribbon's Data tab. Excel 2016 had a group for getting external data and another for transforming the data. Excel 2019 has combined these into one group named Get & Transform Data. See Figure 14-1. Excel has added a lot more types of databases that you can access data from. They have also extended the number of data sources and file types that you can download.

Figure 14-1. *Get & Transform Data group*

© David Slager and Annette Slager 2020
D. Slager and A. Slager, *Essential Excel 2019*, https://doi.org/10.1007/978-1-4842-6209-2_14

Excel can import data from a wide variety of sources. The buttons in the Get & Transform Data group lets you import data from text files, an Access database, the Web, XML files, SQL server, and so on. It can even connect to other programs using an ODBC (Open Database Connectivity) driver or an OLEDB (Object Link Embedded) driver. The different file types do not all have the same capabilities in Excel. Some options work with some data types and not with others.

Importing Text Files

Text files do not contain any formatting. This makes it easy for different programs to exchange text file data. There are two types of text files: delimited and fixed-width. The data in a file is usually divided into fields. Delimited files have fields that have variable lengths. All values in a fixed-width field are the same length. Spaces are used to make the length of text fields the same size; numeric fields can be padded with zeroes to make them the same length.

Delimited Text Files

Fields in text files are usually separated by commas, but they can also be separated by tabs, spaces, or any other character. The character used to separate the fields is called the delimiter. A text file that has its fields separated by commas is called a comma-delimited text file. Most text files have an extension of .txt. Excel spreadsheets can be saved as text files with an extension of .csv. The extension stands for comma-separated values.

Following is the content from a comma-delimited file. The file has headings in the first row. Not all text files contain a header row.

```
EmployeeID,DeptID,LastName,FirstName,MidInitial
A3251,05,Carson,Annette,F
A3291,04,Sterk,Carrie,M
A1450,04,Uhrhammer,Dorothy,L
A1425,05,Carson,Dick,J
```

EXERCISE 14-1: IMPORTING A COMMA-DELIMITED TEXT FILE

In this exercise, you will import data from a delimited text file into a worksheet. You will create two connections to the text file. One connection will automatically refresh the worksheet with any updated data from the text file when the file is open, and the other connection will not. You will also learn to import just the data that you want to be imported.

1. Create a new blank workbook. Click the Ribbon's File tab. Click Save As. Name the workbook "ImportingData." Return to the worksheet.

2. Click the Ribbon's Data tab. In the Get & Transform Data group, click **From Text/CSV**.

3. This brings up the Import Data window where you browse until you find the file you want. Find and then select the `Employee.txt` file. Click the Import button.

4. This brings up the dialog box shown in Figure 14-2. You can pick the delimiter by clicking the Delimiter Text box and selecting one of the separators or you could select Custom or Fixed Width. Select Comma for the Delimiter.

Note   There is a delimiter option for Fixed-Width. This currently doesn't work for Fixed-Width text files. No delimiter text box will display if you try to load a Fixed-Width text file. You will learn how to import Fixed-Width text files in the next exercise.

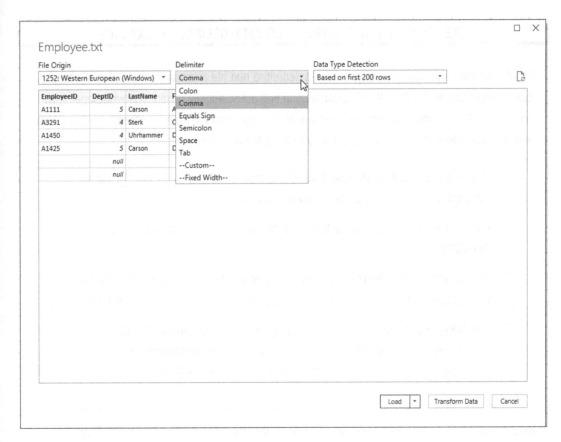

Figure 14-2. *Import Data dialog box*

5. You can now load the data directly into a new worksheet by clicking the Load
    button, or you can click the down arrow of the Load button and then select
    Load To which brings up the Import Data dialog box shown in Figure 14-3.
    Click the down arrow for the Load button and select **Load To**.

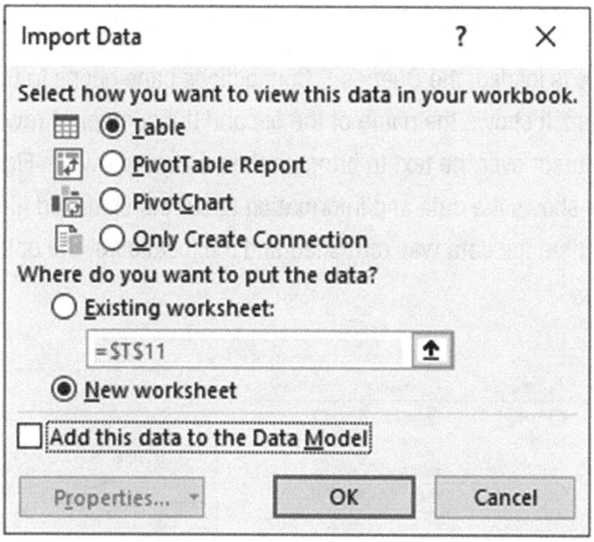

Figure 14-3. *Import Data dialog box*

The Import Data dialog box has options for letting you load the text into a worksheet that you select or a new worksheet. Unlike Excel 2016, you can also load the text file into a PivotTable or PivotChart. We will be looking at PivotTables and PivotCharts in the next chapter. You can add the file to a data model. Data models are used for integrating data from multiple tables.

Click the Option button for **Existing worksheet**. Click cell A1. Click the OK button. The data from the text file is loaded into a table in the current worksheet. See Figure 14-4.

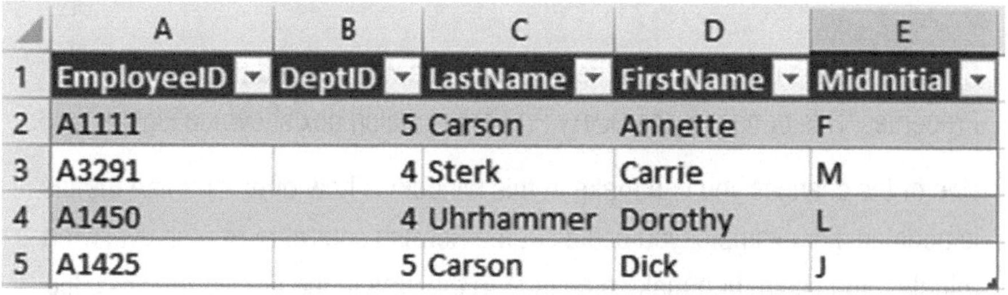

▲	A	B	C	D	E
1	EmployeeID ▼	DeptID ▼	LastName ▼	FirstName ▼	MidInitial ▼
2	A1111	5	Carson	Annette	F
3	A3291	4	Sterk	Carrie	M
4	A1450	4	Uhrhammer	Dorothy	L
5	A1425	5	Carson	Dick	J

Figure 14-4. *Imported Data loaded into worksheet table*

Creating a Connection to the External Data

1. When the file is loaded, the Queries & Connections pane opens to the right of the worksheet. It shows the name of the file and the number of rows loaded. Move your cursor over the text to bring up the window shown in Figure 14-5. The window shows the data and information about the uploaded file. You can see the last time the data was refreshed and the location of the uploaded file.

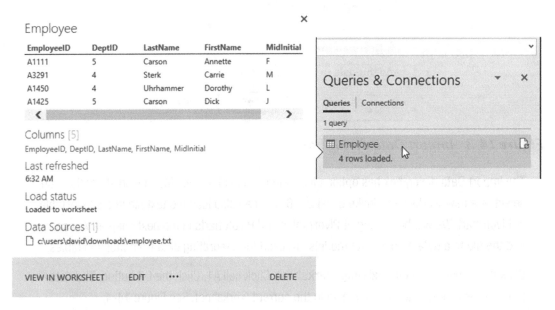

Figure 14-5. *Uploaded File Information*

2. Click the ellipsis button (…) at the bottom of the window and then select Properties. This brings up the Query Properties dialog box shown in Figure 14-6.

   One of the changes you can make in this window is how often or when the file is updated. For example, you could set the Refresh control to refresh every ten minutes. You could then make a change to the data in the Employee.txt file, wait ten minutes, and then see the changes displayed in the worksheet, or you could set it to refresh when you reopen the workbook.

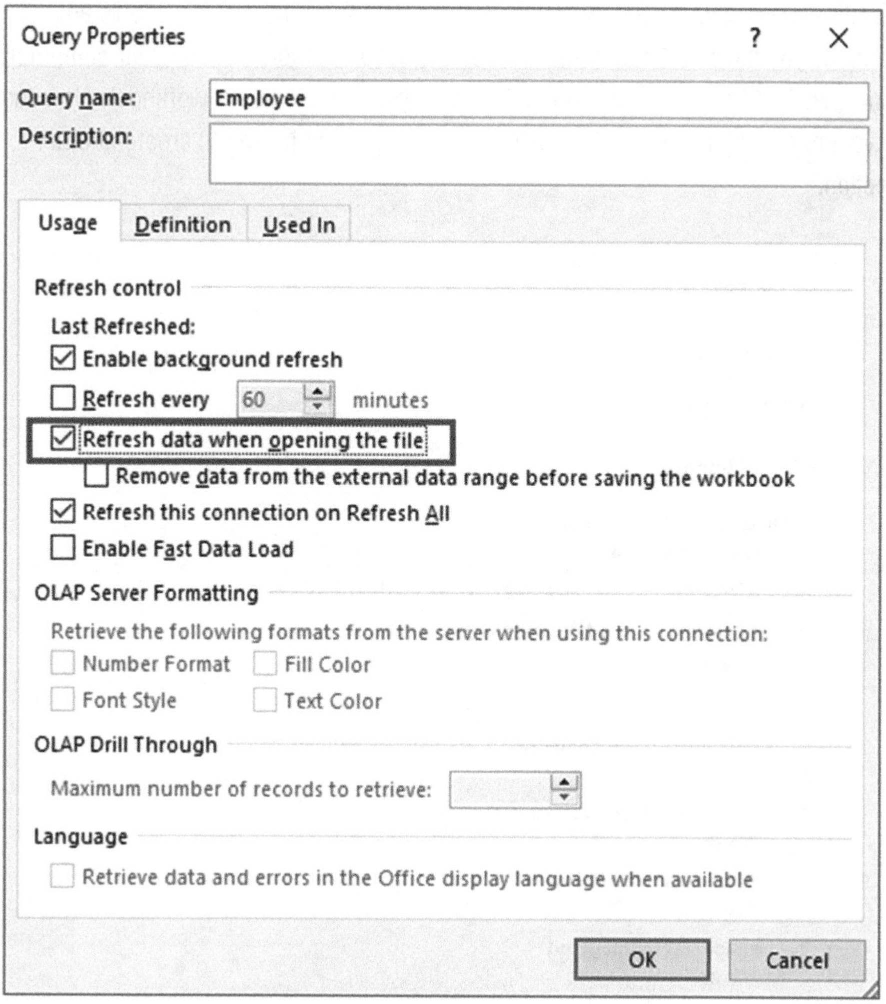

Figure 14-6. *Query Properties dialog box*

3. Select **Refresh data when opening the file**. Click the OK button.

4. Click cell F7.

5. Click the Ribbon's Data tab; in the Get & Transform Data group, click Existing Connections.

The Existing Connections window opens. See Figure 14-7. This window shows any connections you have in the current workbook, on your network, and on your computer. Currently, we just have a connection to the Employee text file in our workbook. Now that we have a connection, we can keep using this data without having to create another connection.

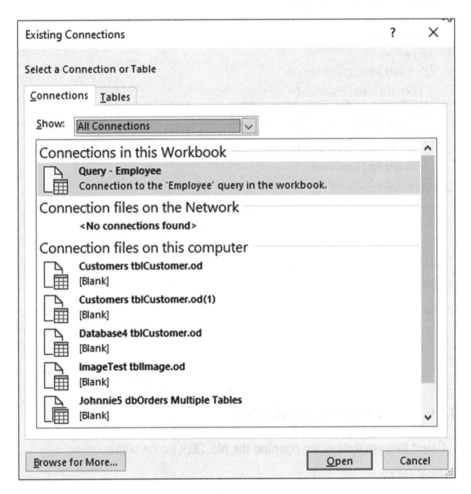

Figure 14-7. *Existing Connections dialog box*

Creating a Second Connection to the Employee.txt File

1. Double-click the Query – Employee connection or click the Open button.

2. The Import Data window opens. See Figure 14-8. You can place the data in another location or another worksheet.

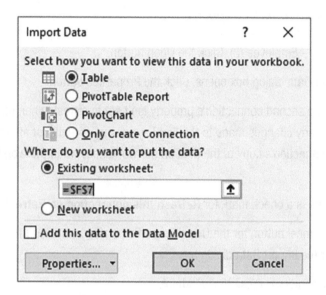

Figure 14-8. *Specify where you want the imported data placed*

3. Leave the Existing worksheet cell at F7.

    Click the OK button. Another copy of the file from the first connection displays starting in cell F7.

4. Click the Ribbon's Data tab; in the Get & Transform Data group, click Existing Connections.

    The Existing Connections dialog box opens. Excel named the second connection Query – Employee(2). See Figure 14-9.

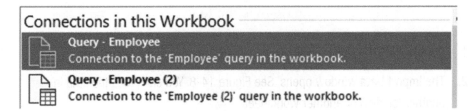

Figure 14-9. *Connections to imported files*

5. Click Query – Employee (2). Click the Open button.

6. The Import Data dialog box opens. Click the Properties button.

    Because the second connection's property isn't set to refresh when the file is opened, any changes made to the Employee.txt file will not affect the second connection's copy of the data on the worksheet when the workbook is opened.

    Notice there is a check mark for **Refresh this connection on Refresh All**.

7. Click the Cancel button for the Query Properties dialog box. Click the Cancel button for the Import Data dialog box.

8. Save this workbook. Close the workbook.

9. Use Notepad or WordPad to change the Employee.txt file so that the first EmployeeID is A9999 rather than A1111. Save the Employee.txt file.

10. Reopen the ImportingData workbook.

    You may get a security warning above your worksheet. If you do, click the Enable Content button.

    Figure 14-10 shows that the EmployeeID in cell A2 has been changed to the same value you updated it to in the Employee.txt file. The value in cell F8 kept the original value. The reason this occurred was because in the first connection we selected the property for **Refresh data when opening the file**.

◢	A	B	C	D	E	F	G	H
1	EmployeeID ▾	DeptID ▾	LastName ▾	FirstName ▾	MidInitial ▾			
2	A9999	5	Carson	Annette	F			
3	A3291	4	Sterk	Carrie	M			
4	A1450	4	Uhrhammer	Dorothy	L			
5	A1425	5	Carson	Dick	J			
6								
7						EmployeeID ▾	DeptID ▾	LastName
8						A1111	5	Carson
9						A3291	4	Sterk
10						A1450	4	Uhrhammer
11						A1425	5	Carson

Figure 14-10. *EmployeeID in cell A2 was changed to the value you updated it to in the Employee.txt file*

11. If you want to refresh the data for all the connections immediately, click the Refresh All button. See Figure 14-11.

Figure 14-11. *Refresh All button*

Click the Ribbon's Data tab; in the Connections group, click the Refresh All button. The data from both connections should now match.

Breaking the Connection Between the Imported File and Your Workbook

What if you didn't want Excel to update the data anymore when a change was made to the file?

1. Click the table starting in cell A1. The Ribbon has a context group for Table Tools and Query Tools. Click Query Tools. Click the Design tab. On the Ribbon's Design tab, in the External Table Data group, click the Unlink button. The dialog box shown in Figure 14-12 displays.

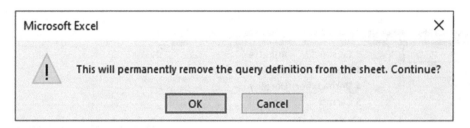

Figure 14-12. Permanently remove query definition

2. Click the OK button.

3. Change the Employee.txt file so that the first EmployeeID is A0000.
    Save the file.

4. Click the Ribbon's Data tab. In the Queries & Connections group, click the
    Refresh All button.

Because the link to the first table was removed, there was no change to the EmployeeID, but
the EmployeeID was updated in the other table. See Figure 14-13.

	A	B	C	D	E	F	G	H
1	EmployeeID ▾	DeptID ▾	LastName ▾	FirstName ▾	MidInitial ▾			
2	A9999	5	Carson	Annette	F			
3	A3291	4	Sterk	Carrie	M			
4	A1450	4	Uhrhammer	Dorothy	L			
5	A1425	5	Carson	Dick	J			
6								
7						EmployeeID ▾	DeptID ▾	LastName
8						A0000	5	Carson
9						A3291	4	Sterk
10						A1450	4	Uhrhamm
11						A1425	5	Carson

Figure 14-13. Data no longer updates for first table because it was unlinked

Changing What Data Is Being Imported

1. Click the table starting in cell F7. Click the Query menu. In the Edit group, click
    the Properties button.

2. The Connection Properties window displays. See Figure 14-14.

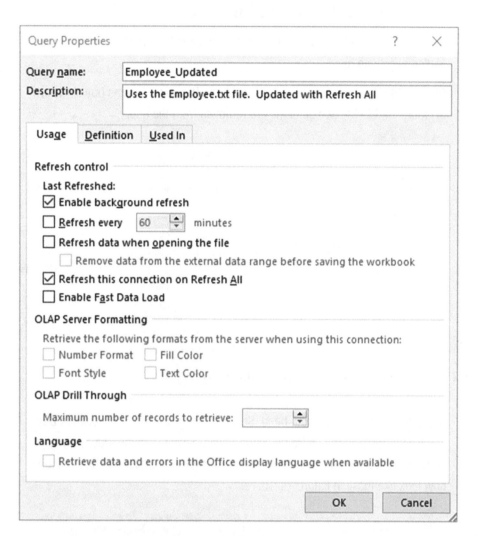

Figure 14-14. *Connection Properties dialog box*

3. Employee(2) isn't a very good name; change it to Employee_Updated.

4. For the Description, enter **Uses the Employee.txt file. Updated with Refresh All**.

Notice that Refresh this connection on Refresh All is selected.

5. Click the Definition tab. Click the Edit Query button.

6. Click the EmployeeID column head, then hold down the Ctrl Key, and click the MidInitial column head.

7. Click on the Home tab. Click the top half of the Remove Columns button shown in Figure 14-15. The two columns are removed from the query.

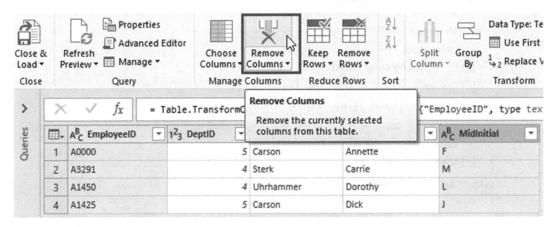

Figure 14-15. *Remove two columns from the query*

8. Right-click the DeptID column head. Select Rename. Put a space between Dept and ID. Right-click the LastName column head. Select Rename. Put a space between Last and Name. Do the same for FirstName. Your column heads should now look like those in Figure 14-16.

1²₃ Dept ID	A⁸꜀ Last Name	A⁸꜀ First Name
1	5 Carson	Annette
2	4 Sterk	Carrie
3	4 Uhrhammer	Dorothy
4	5 Carson	Dick

Figure 14-16. *Remove two columns from the query and change column names*

9. Click the Close & Load button. The table in cell F7 changes to reflect the result of the edited query. See Figure 14-17.

Dept ID ▼	Last Name ▼	First Name ▼
5	Carson	Annette
4	Sterk	Carrie
4	Uhrhammer	Dorothy
5	Carson	Dick

Figure 14-17. *Table changed according to changes in query*

You have learned how to import delimited files. The fields can be delimited by commas, semicolons, and so on. There doesn't need to be spaces between fields in a delimited file. Next, you will learn how to import fixed-width text files. In a fixed-width text file, there is no delimiter. The fields are separated by spaces.

Fixed-Width Text Files

In a fixed-width text file, all fields take up the same amount of space. Headings aren't usually used because they would rarely line up with the data.

```
A3251 05 Carson       Annette F
A3291 04 Sterk        Carrie  M
A1450 04 Uhrhammer    Dorothy L
A1425 05 Carson       Dick    J
```

EXERCISE 14-2: IMPORTING A FIXED-WIDTH TEXT FILE

In this exercise, you will import data from a text file stored in fixed-width format into a location that you specify in your worksheet.

Before we import a fixed-width text file, we will need to bring back the legacy (previous) method for doing this. When you try to import a fixed-width text file using **From Text/CSV**, as we did in the previous exercise, it will not give you an option for a delimiter.

1. Click the Ribbon's File tab. Click Options in the left pane.

2. Click Data in the left pane. Select From Text (Legacy). Click OK.

3. Create a new worksheet named "Fixed-Width."

4. Click the Ribbon's Data tab; in the Get & Transform Data group, click the **Get Data** button. There is an option for Legacy Wizards which you just created. Click Legacy Wizards and then click From Text (Legacy).

5. This brings up the Import Text File dialog box where you browse until you find the file. Select the file Employee_Fixed.txt and then select the Import button.

6. This brings up the Text Import Wizard. See Figure 14-18. Excel recognized this as a fixed-width file and automatically selected that option.

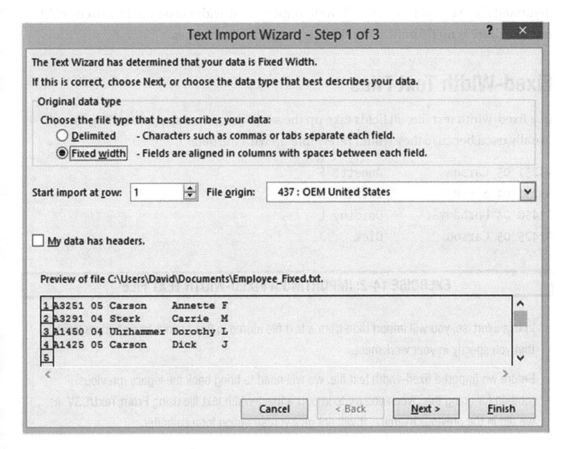

Figure 14-18. Text Import Wizard – Step 1 of 3

7.   Click the Next button. This brings up the second step. See Figure 14-19.

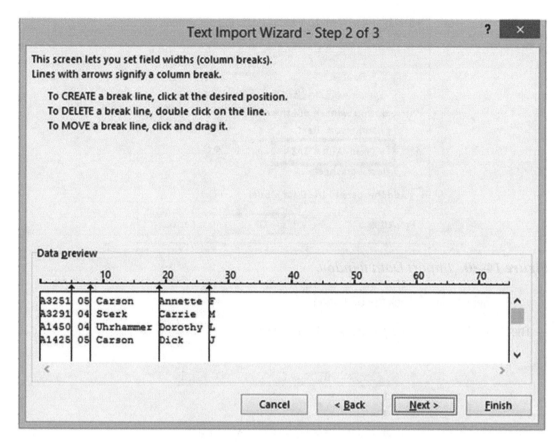

Figure 14-19. *Text Import Wizard – Step 2 of 3*

The fields in a fixed-width file can be separated wherever you want them to be, by dragging the lines. You could combine fields by double-clicking the line between them to remove it. You could create another column between columns by clicking where you want to create the extra lines.

8.   Click the Next button for Step 2.

9.   Click the Finish button for Step 3.

The Import Data dialog box displays. See Figure 14-20. Here you can specify the location of where you want the data loaded in the current workbook or if you want it placed in a new worksheet.

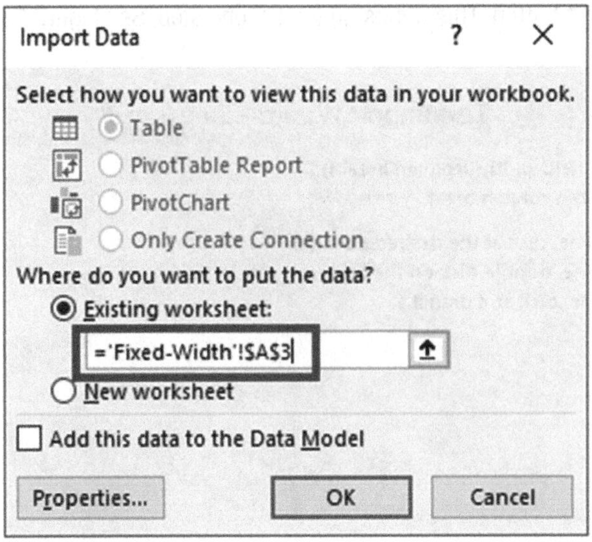

Figure 14-20. *Import Data window*

10. Select cell A3. Click the OK button.

The data is imported starting in cell A3. See Figure 14-21.

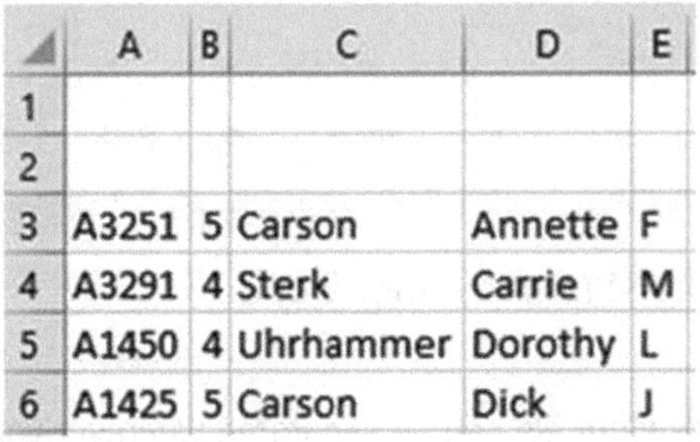

	A	B	C	D	E
1					
2					
3	A3251	5	Carson	Annette	F
4	A3291	4	Sterk	Carrie	M
5	A1450	4	Uhrhammer	Dorothy	L
6	A1425	5	Carson	Dick	J

Figure 14-21. *Imported data*

You have learned how to import text files. Next, you will learn how to import data from an Access database.

Importing Data from an Access Database

Microsoft Access is a part of the Microsoft Office suite. An Access database can contain multiple linked tables and queries. A query contains only the fields from a table that you requested. The records it contains are limited to any filters you placed on it. You can import data from tables and queries that you have created in Access.

EXERCISE 14-3: IMPORTING DATA FROM AN ACCESS DATABASE

In this exercise, you will import the data from a table named tblCustomer, which is in the Access database named Customers.

1. Create a new worksheet named "Access."

2. Click the Ribbon's Data tab. In the Get & Transform Data group, click the Get Data button. Click **From Database**. Click From Microsoft Access Database. The Import Data dialog box opens.

3. Find and Select the Access database named Customers and then click the Import button. See Figure 14-22.

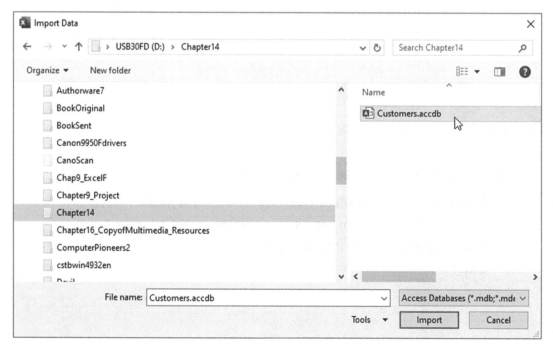

Figure 14-22. *Select the Access database to be opened*

4.   The Customers Access Database contains two tables: tblCustomer and
     tblOrderinfo. It also contains a query named qryCustomerbyState. See
     Figure 14-23. Click the table named tblCustomer from the Navigator dialog
     box. The table displays in the right pane. If you wanted to import more than one
     table or query, you would need to check the **Select multiple items**.

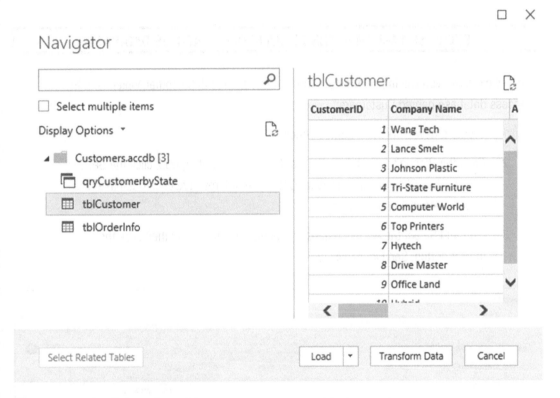

Figure 14-23. *Select the table or query you want to import*

5.   Click the down arrow for the Load button and select Load To.

6.   In the Import Data dialog box, you need to specify how you want to view the
     data and where you want to locate it in your workbook. Click the option button
     for Existing worksheet and then click cell A2. See Figure 14-24. Click the OK
     button.

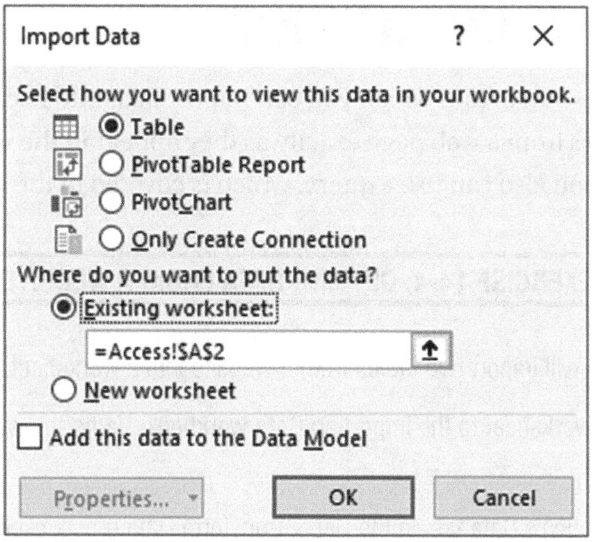

Figure 14-24. *Select how you want to view the data and where to put it*

The data you selected is placed in a table in your workbook. See Figure 14-25. If you want to update the imported data in your Excel workbook after you have made a change to the data in Access, then right-click anywhere in the table and select Refresh.

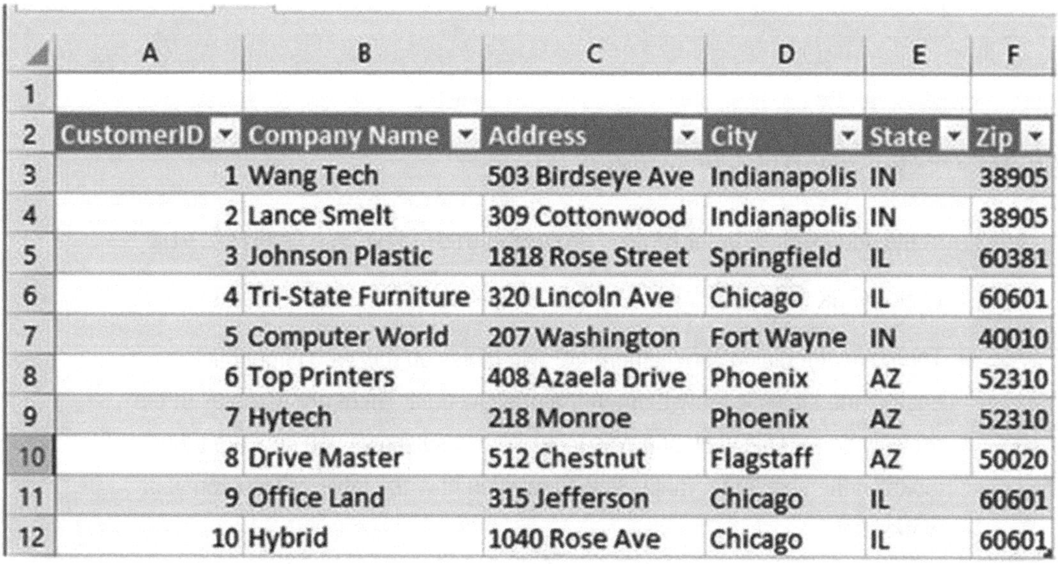

⊿	A	B	C	D	E	F
1						
2	CustomerID ▼	Company Name ▼	Address ▼	City ▼	State ▼	Zip ▼
3	1	Wang Tech	503 Birdseye Ave	Indianapolis	IN	38905
4	2	Lance Smelt	309 Cottonwood	Indianapolis	IN	38905
5	3	Johnson Plastic	1818 Rose Street	Springfield	IL	60381
6	4	Tri-State Furniture	320 Lincoln Ave	Chicago	IL	60601
7	5	Computer World	207 Washington	Fort Wayne	IN	40010
8	6	Top Printers	408 Azaela Drive	Phoenix	AZ	52310
9	7	Hytech	218 Monroe	Phoenix	AZ	52310
10	8	Drive Master	512 Chestnut	Flagstaff	AZ	50020
11	9	Office Land	315 Jefferson	Chicago	IL	60601
12	10	Hybrid	1040 Rose Ave	Chicago	IL	60601

Figure 14-25. *Data has been imported into a table*

Importing Data from a Website

A website is another source from which you can import data into Excel. In this section, you will import tables from a web page exactly as they appear in the web page. As with other data sources, you also can use a query, which is covered in the next section.

EXERCISE 14-4: GETTING DATA FROM A WEBSITE

In this exercise, you will import two tables from a website into a worksheet.

1. Add a new worksheet to the Importing Data workbook. Name the worksheet ImportWeb.

2. Click the Ribbon's Data tab. In the Get & Transform Data group, click the **From Web** button. The From Web dialog box opens. See Figure 14-26.

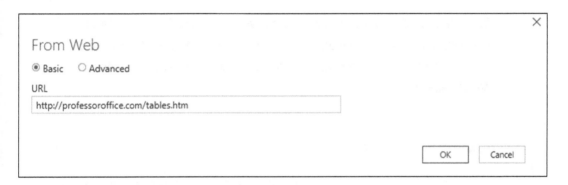

Figure 14-26. *New Web Query dialog box*

3. In the Address box, enter `http://professoroffice.com/tables.htm`.

4. Click the OK button.

   The Navigator dialog shows there are two tables on this web page. Click each of the tables to view them in the preview area. There are two tabs in the preview area: Table View and Web View. Web View shows you how the table looks on the web page. Table View shows you how the table will look on your worksheet.

5. Check the Select multiple items check box and then check each of the tables. See Figure 14-27.

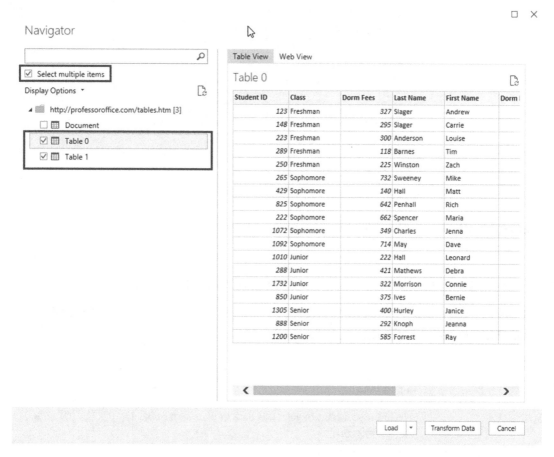

Figure 14-27. Both tables are selected

6. Click the Load button. The Queries & Connections pane displays to the right of the worksheet. Excel named the first table Table 0 and the second one Table 1. See Figure 14-28.

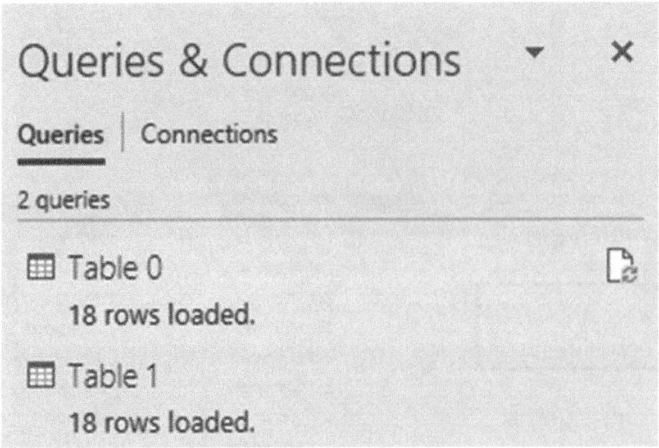

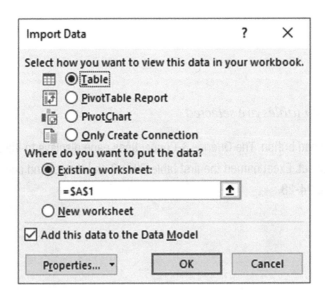

Figure 14-28. *The two tables have been loaded into queries*

7. Right-click Table 0 in the Queries & Connections pane. Select Load To.

Excel asks you where you want to import the data. See Figure 14-29. Notice that you could load the data into a PivotTable or PivotChart.

Figure 14-29. *Select where you want to store the imported data*

8. Click the Table option and Existing worksheet option. Click cell A1. Click the OK button.

The data has now been imported into a table, and you can manipulate it any way you wish.

9. Right-click Table 1 in the Queries & Connections pane. Select Load To.

10. In the Import Data dialog box, click the Table option and Existing worksheet option. Click cell A23. Click the OK button. Both tables have been loaded into the worksheet. See Figure 14-30.

	A	B	C	D	E	F	G	H	I
1	Student ID	Class	Dorm Roo	Last N	First Name	Dorm Fees	Parking Fees	School Fees	Book Fees
2	123	Freshman	327	Slager	Andrew	787.25	75.25	1785.5	585
3	148	Freshman	295	Slager	Carrie	695.5	75.25	1785.5	585
4	223	Freshman	300	Anderso	Louise	750.3	25	1885.3	670.75
5	289	Freshman	118	Barnes	Tim	700.4	50	1650	499.75
6	250	Freshman	225	Winston	Zach	730.35	68	1735.4	600.1
7	265	Sophomore	732	Sweene	Mike	828.32	75.25	1909.25	550.35
8	429	Sophomore	140	Hall	Matt	850.25	68	1825.5	499.75
9	825	Sophomore	642	Penhall	Rich	750.3	50	1885.3	670.75
10	222	Sophomore	662	Spencer	Maria	730.35	75.25	1885.3	670.75
11	1072	Sophomore	349	Charles	Jenna	730.35	50	1883.3	670.75
12	1092	Sophomore	714	May	Dave	695.5	75.25	1785.5	585
13	1010	Junior	222	Hall	Leonard	700.4	68	1705	575
14	288	Junior	421	Mathew	Debra	850.25	25	1600	600
15	1732	Junior	322	Morrisor	Connie	901.1	68	1650	499.75
16	850	Junior	375	Ives	Bernie	705.25	25	1785.5	585
17	1305	Senior	400	Hurley	Janice	730.35	68	1825.5	499.75
18	888	Senior	292	Knoph	Jeanna	705.25	70	1903.3	585
19	1200	Senior	585	Forrest	Ray	530.25	75.25	1705	575
20									
21									
22	Student ID	Address	City	State					
23	123	211 Belmont	Fort Wayne	IN					
24	148	408 Jackson	Muncie	IN					
25	223	319 Washington	Redford	IN					

***Figure 14-30.** Imported data*

You have learned how you can import data directly into Excel from tables you find on the Internet. Next, you will learn how to use Excel's Power **Query** feature. This feature lets you select only the fields and data that you want to import. You can even add your own calculated fields.

Importing Data Using Transform

Excel's Power Query feature can bring in data from many sources and has the ability to do a great deal of modification to your data. Among the sources from which you can import data are web pages and databases. Excel enables you to import data from a variety of database types by default. You can use any data source that is available from the Get & Transform Data group.

Note   You also can import data from other sources, such as various database management systems (DBMSs) via the ODBC standard. You can get ODBC and other data source drivers from third parties.

In this section, you will import data from a web page using a query from the Get & Transform Data group and then transform that data to the way you want.

EXERCISE 14-5: TRANSFORM DATA USING POWER QUERY

In this exercise, we will use the same data as in the previous example. This time, rather than loading the data directly into the worksheet, we will use the Power Query to choose which fields you want to import. You will filter the data, create a calculated column, merge the queried data, and change the order of the columns.

1. Start a new blank workbook.

2. Click the Ribbon's Data tab. In the Get & Transform group, click the Get Data button.

3. Click **From Other Sources** and then select **From Web**.

4. In the URL (uniform resource locator) text box, enter http:// professoroffice.com/tables.htm. See Figure 14-31.

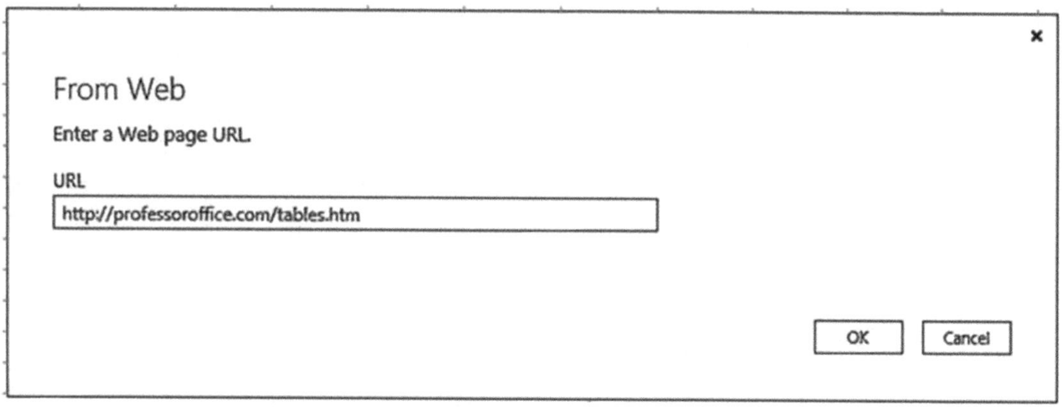

Figure 14-31. *Enter the web address in the URL text box*

5.   Click the OK button.

The Navigator window displays. See Figure 14-32. The two tables from the web page are now available. Clicking a table name in the left pane displays that table in the right pane. If you want to select more than one table, you need to check **Select multiple items**.

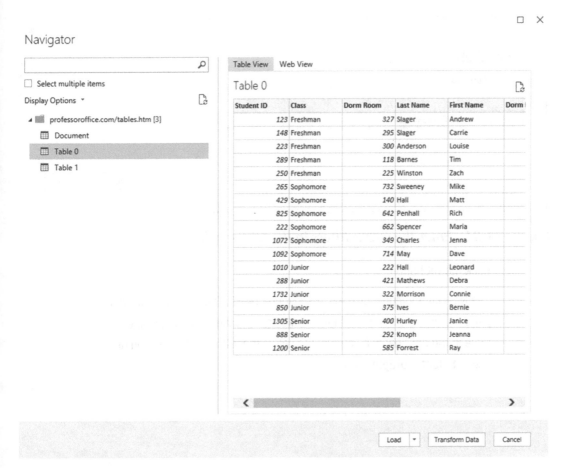

Figure 14-32. *Selecting a table from the left pane displays it in the right pane*

At the bottom of the window, there is a Load and Transform Data button.
Clicking the Transform Data button gives you the opportunity to modify the data
and limit what data is brought in before it is imported into your worksheet.

6. Check **Select multiple items**. Check Table 0 and Table 1. See Figure 14-33.

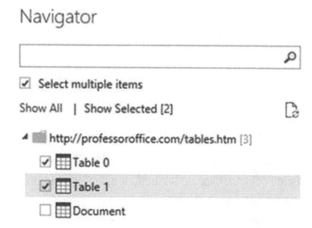

Figure 14-33. *Select multiple items and Table 0 and Table 1*

7. Click the Transform Data button.

Naming the Tables

1. The Power Query Editor displays. The Query Settings pane is on the right side of the screen. Change the table name from Table 0 to tblStudentFees. Press Enter. See Figure 14-34.

Figure 14-34. *Change the name to tblStudentFees*

2. Click Table 1 on the left side of the screen. Change its name in the Query Settings pane to tblStudentAddress. Press Enter.

Choose Fields to Import

The icons in the column heads identify the type of data that is in the columns. ABC is text data and 123 is Numeric data. We don't need the Address column in the tblStudentAddress table.

1. Click the Address column head to select the entire column. Click the Ribbon's Home tab.

   In the Power Query Editor's Ribbon in the Manage Columns group, there is a Remove Columns button. Clicking the top half of the button removes the currently selected column. Clicking the bottom half of the button provides two options. One option is Remove Columns, which does the same thing as clicking the top half of the button. Clicking the second option, Remove Other Columns, removes all the columns except the currently selected ones.

2. Click the top half of the Remove Columns button.

3. On the left side of the window, click table tblStudentFees. We don't need the column Dorm Room from this table.

4. Click the Dorm Room column head.

5. In the Ribbon's Manage Columns group, click the top half of the Remove Columns button.

Applying a Filter

We only want to import records of students whose parking fees are greater than $25.

1. Click the down arrow to the right of the Parking Fees column heading.

2. Click Number Filters. Select Greater Than. The Filter Rows window appears.

3. Click the down arrow for the second drop-down box. This drop-down box displays all the values in the column. See Figure 14-35.

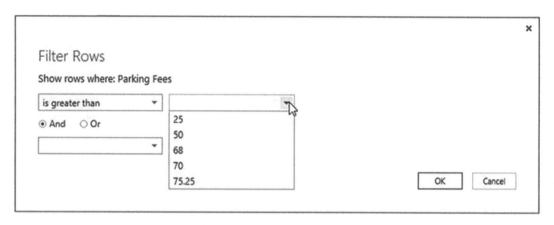

Figure 14-35. *Select 25 after selecting is greater than*

4. Select 25. Click the OK button. All records with 25 or less for the parking fees are removed. Three records were removed.

5. On the left side of the window, click the tblStudentAddress table.

6. Click the down arrow to the right of the State column heading.

7. Uncheck IL. Click the OK button. The four records that contained the state of IL are removed.

Undoing and Redoing Query Steps

The Applied Steps area in the Query Settings pane keeps track of all the changes you make to the query. If you want to see what your query looked like before your last step, you can click the step above it. As you move up in the Applied Steps, you are basically performing an Undo. Click the line below the current one if you want to Redo the step.

1. In the APPLIED STEPS area, click Removed Columns. See Figure 14-36.

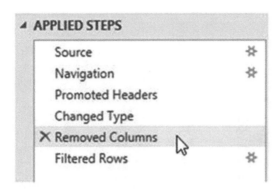

Figure 14-36. *Select Removed Columns*

The records that contained the state of IL are displayed again.

2. In the left pane, click the tblStudentFees table. In the APPLIED STEPS area, click Removed Columns.

The records that contained parking fees of 25 or less are displayed again.

3. Click Filtered Rows. The filter is reapplied, and those records that contained parking fees of 25 or less are no longer displayed.

    - Right-clicking an applied step brings up the menu options shown in Figure 14-37. From here, you can elect to do the following:

        i.   Edit the settings you created. For example, this step could be changed from only showing records where the parking fee is greater than 25 to showing those greater than 50 or whatever you want.

        ii. Rename a step.

        iii. Delete a step. If you delete it, you can't get the step back.

        iv. Delete all the steps from the current step to the last step.

        v.   Move steps up and down to change their order.

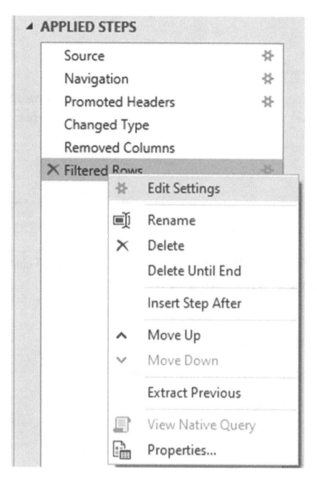

Figure 14-37. *Right-click a step to view its options*

Creating a Calculated Column

1. Click the Power Query Editor Ribbon's Add Column tab. In the General group, click the **Custom Column** button.

2. You may get the message in Figure 14-38. If you do, click the Insert button.

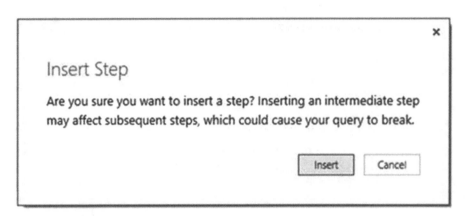

Figure 14-38. *Insert Step message*

3. Enter Total Fees for the New Column name in the Add Custom Column dialog box and complete the following steps. See Figure 14-39.

    a. In the Available columns list box, double-click Dorm Fees.

    b. Enter a plus sign (+).

    c. Click Parking Fees. Click the Insert button.

    d. Enter a plus sign (+).

    e. Click School Fees. Click the Insert button.

    f. Enter a plus sign (+).

    g. Click Book Fees. Click the Insert button.

    h. Click the OK button.

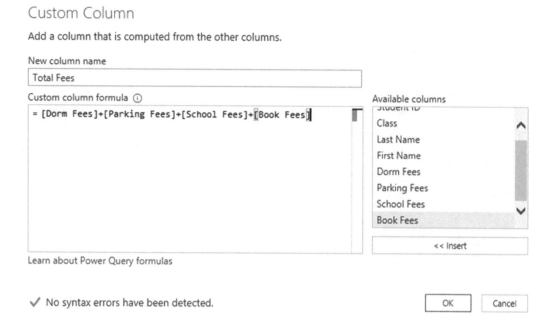

Figure 14-39. *Add a custom column*

4. Click the down arrow to the right of the Total Fees field. Select Sort Descending.

The reason Excel creates the applied steps is so that the next time you want to do an import from the web page, you will not have to go through this editing process again. The steps will be saved and automatically run for you.

Merging Queries

1. Click the Ribbon's Home tab. In the Combine group, click the Merge Queries button.

    The Merge window displays with the data from tblStudentFees displaying at the top.

2. Select tblStudentAddress from the drop-down box as shown in Figure 14-40. We will link the tables together, and then we will add the fields we want from the tblStudentAddress field.

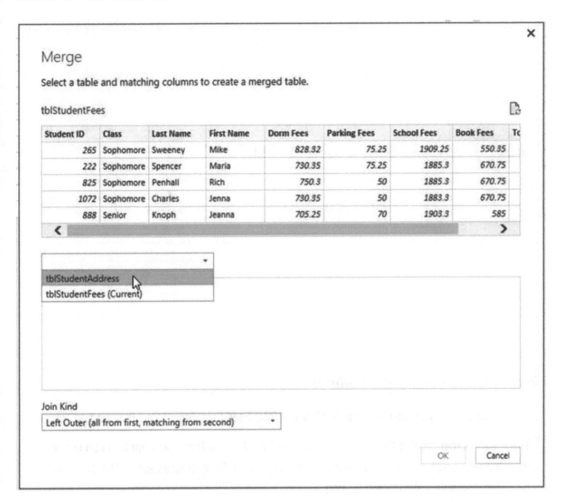

Figure 14-40. *Select tblStudentAddress*

3. Click the Student ID column head in both tables. This links the two tables. Both tables have the same Student ID numbers. Linking tables together requires that your tables have a common field.

4. Click on the down arrow for **Join Kind** and then select **Inner (only matching rows)**. See Figure 14-41.

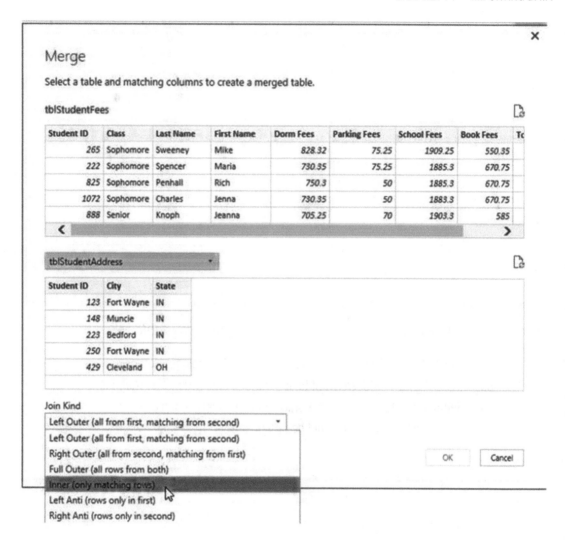

Figure 14-41. *Click the Student ID column head in both tables to link the two tables*

Twelve of the records in tblStudentFees have a matching record in tblStudentAddress. Because you selected Inner (only matching rows), only the records with matching Student IDs will become part of the query.

5. Click the OK button.

A new column appears at the end of the tblStudentFees table. Each row in the column contains the word **Table**. See Figure 14-42.

2 School Fees	▼	1.2 Book Fees	▼	ABC 123 Total Fees	▼	⊞ tblStudentAddress	⬌
1785.5		585		3233		Table	
1885.3		670.75		3361.65		Table	
1785.5		585		3141.25		Table	
1885.3		670.75		3356.35		Table	
1883.3		670.75		3334.4		Table	
1735.4		600.1		3133.85		Table	
1825.5		499.75		3243.5		Table	
1785.5		585		3141.25		Table	
1705		575		3048.4		Table	
1825.5		499.75		3123.6		Table	
1650		499.75		3118.85		Table	
1705		575		2885.5		Table	

Figure 14-42. *A new column is displayed to the right of the table*

To the right of the tblStudentAddress header, there is a column expansion button. See Figure 14-43. We want to add the City and State fields from the tblStudentAddress table but not the Student ID. The merged table already contains a Student ID so that would be redundant.

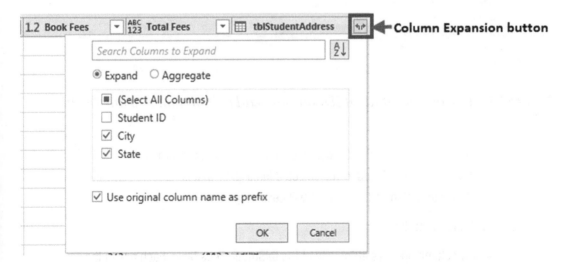

Figure 14-43. *Click the Column Expansion button and uncheck Student ID*

6.   Click the Column Expansion button. Uncheck Student ID.

7.   Click the OK button.

   The City and State columns have been added to the table. We would like to
   remove the table name from the City and State columns. See Figure 14-44.

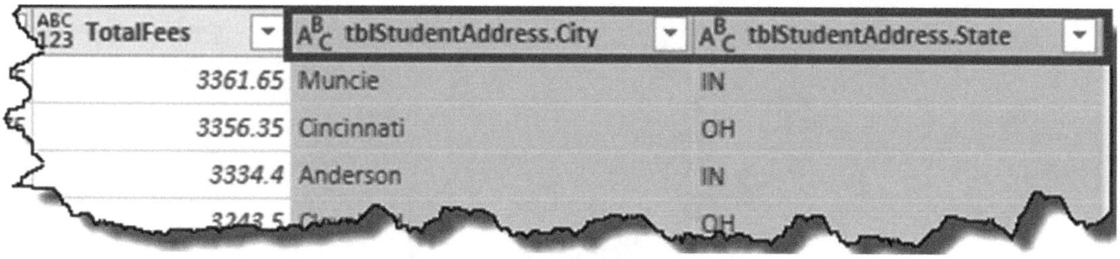

Figure 14-44. *Remove the table names from the City and State column
headings*

8.   In the formula bar, remove tblStudentAddress. in front of City.

9.   In the formula bar, remove tblStudentAddress. in front of State.

Moving Column Heads

1.   Click the City column head and then hold down the Ctrl key and click the State
   column head.

2.   Drag the column heads so that they are just to the right of the Class column.
   See Figure 14-45.

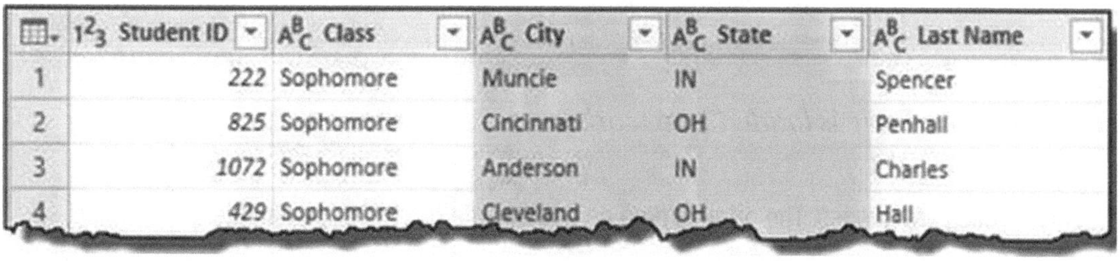

Figure 14-45. *Drag the City and State columns to the right of Class*

Closing Query and Loading Results to Excel

1. Click the Power Query Editor's Ribbon. In the Close group, click the bottom half of the Close & Load button.

2. Click Close & Load. See Figure 14-46.

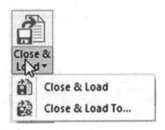

Figure 14-46. *Select Close & Load*

The Query Editor window closes, and the tblStudentFees and the tblStudentAddress tables are loaded into separate worksheets. The tblStudentAddress table is shown in Figure 14-47.

Student ID	Class	City	State	Last Name	First Name	Dorm Fees	Parking Fees	School Fees	Book Fees	TotalFees
222	Sophomore	Muncie	IN	Spencer	Maria	730.35	75.25	1885.3	670.75	3361.65
825	Sophomore	Cincinnati	OH	Penhall	Rich	750.3	50	1885.3	670.75	3356.35
1072	Sophomore	Anderson	IN	Charles	Jenna	730.35	50	1883.3	670.75	3334.4
429	Sophomore	Cleveland	OH	Hall	Matt	850.25	68	1825.5	499.75	3243.5
123	Freshman	Fort Wayne	IN	Slager	Andrew	787.25	75.25	1785.5	585	3233
148	Freshman	Muncie	IN	Slager	Carrie	695.5	75.25	1785.5	585	3141.25
1092	Sophomore	Fort Wayne	IN	May	Dave	695.5	75.25	1785.5	585	3141.25
250	Freshman	Fort Wayne	IN	Winston	Zach	730.35	68	1735.4	600.1	3133.85
1305	Senior	Gas City	IN	Hurley	Janice	730.35	68	1825.5	499.75	3123.6
1732	Junior	Muncie	IN	Morrison	Connie	901.1	68	1650	499.75	3118.85
1010	Junior	Muncie	IN	Hall	Leonard	700.4	68	1705	575	3048.4
1200	Senior	Fort Wayne	IN	Forrest	Ray	530.25	75.25	1705	575	2885.5

Figure 14-47. *Table is loaded into worksheet*

These tables are queries. The original data on the website has not been changed.

3. Move your cursor over tblStudentFees in the Queries & Connections pane. See Figure 14-48.

From this window, you can

    a. See the last time the query was refreshed

    b. Load the query into another worksheet (from the ellipsis (…) menu)

    c. Click EDIT which will take you back into the Query Editor

    d. Click DELETE to delete the query

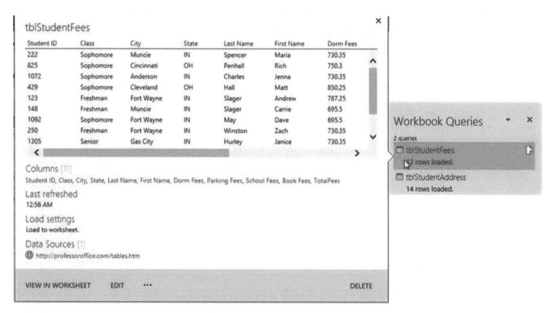

Figure 14-48. *tblStudentFees table*

Summary

Data can be imported into Excel from many different data sources. You can use a single data source or multiple data sources. Once the data is imported, it can take advantage of Excel's wonderful analysis features.

Power Query let's you extract, transfer, manipulate, link and load data into Excel from many different sources.

There are many different file types such as text files, Access files, Microsoft query files, and so on. Different file types can use different Excel options.

Text files are files that contain no formatting, which makes them easy to import. They can be delimited or fixed-width. The fields in a text-delimited file are separated by some character such as a comma or semicolon. All fields have the same length in a fixed-width text file. Spaces are added to fields to make them the same length.

The next chapter covers Pivot Tables and Pivot Charts. A pivot table has much more capability than a regular Excel table. Data can be dragged into an area if you want it to be rows, another area if you want it to be columns, and another area if you want it to be summarized. Pivot Charts have the advantage over normal Excel charts because they can be filtered. A user can select which items he or she wants to appear in the Pivot Chart.

CHAPTER 15

Using PivotTables and PivotCharts

Few people use PivotTables, but they are missing out on probably the strongest component in Excel. It takes a little time to learn, but once you have conquered it, you will become an Excel power user capable of grouping and summarizing data in almost any form you can think of. Your boss will be amazed at what you can produce. If you are the boss, you will be even more amazed at yourself. PivotTables can create the types of reports managers want to see—those that don't contain every detail but rather contain only the data they want to see when they want to see it.

After reading and working through this chapter, you should be able to

- Create a PivotTable

- Create a PivotChart from a data source

- Create a PivotChart from a PivotTable

- Rearrange and filter PivotTables

- Create multiple PivotTables from the data source

- Clear, move, and delete PivotTables

- Use the slicer tool

- Use the time interval tool

- Filter the data

- Create a relational database

© David Slager and Annette Slager 2020
D. Slager and A. Slager, *Essential Excel 2019*, https://doi.org/10.1007/978-1-4842-6209-2_15

Working with PivotTables

It's called a PivotTable because you can pivot your data, switching between rows and columns. You can dynamically rearrange your existing data into an almost endless variety of different views. You can rapidly create these views by dragging and dropping fields to different locations. Excel then automatically recalculates the results to match the view you created.

You can create a PivotTable from a named table or from a range of data that has not been declared as a table. The benefit of creating a PivotTable from a named table is that you can take advantage of table features such as automatically extending the table when adding rows of data.

Excel analyzes your data and then displays "Recommended" PivotTables from which you can select the PivotTable that matches your needs. We will look at this feature later, but in our first practice, we will start with a simple example in which we will create a PivotTable manually.

The Work & Leisure Company is a chain of four stores located in different areas of the country. It sells sports products, outdoor furniture, and tools. The spreadsheet in Figure 15-1 shows the products the companies sell under those categories and the number of units sold of that product during the previous month.

◢	A	B	C	D
1		**Work & Leisure World**		
2	**Store**	**Category**	**Product**	**Unit Sales**
3	Store 1	Sports	Fishing Poles	125
4	Store 2	Sports	Fishing Poles	227
5	Store 3	Sports	Fishing Poles	320
6	Store 1	Outdoor Furn	Lounge Chairs	118
7	Store 2	Outdoor Furn	Lounge Chairs	220
8	Store 3	Outdoor Furn	Lounge Chairs	150
9	Store 4	Outdoor Furn	Lounge Chairs	139
10	Store 1	Sports	Bats	300
11	Store 2	Sports	Bikes	250
12	Store 2	Outdoor Furn	Umbrellas	190
13	Store 4	Tools	Saws	510
14	Store 3	Tools	Hammers	95
15	Store 2	Outdoor Furn	Tables	73
16	Store 3	Sports	Roller Skates	45
17	Store 4	Tools	Drills	135
18	Store 4	Sports	Fishing Pole	199

Figure 15-1. *Data to be used in PivotTable*

EXERCISE 15-1: MANUALLY CREATE A PIVOTTABLE

In this exercise, you will create a PivotTable. You will practice rearranging the column and row data, adding and removing fields, changing the order of the data, looking at different ways of summarizing the data, filtering the data, and changing the data source location.

1. Open the workbook Chapter15_PivotTables.

2. Select the worksheet named PivotTable1.

3. Click anywhere within the data. Click the Ribbon's Insert tab. In the Tables group, click the PivotTable button.

4. The Create PivotTable dialog box displays with a range already selected. The range should consist of cells $A$2:$D$18. See Figure 15-2. If this isn't the range showing in the text box, then change it.

5. Change the location of the PivotTable report to **Existing Worksheet**.

6. Enter F2 in the Location text box.

7. Click the OK button.

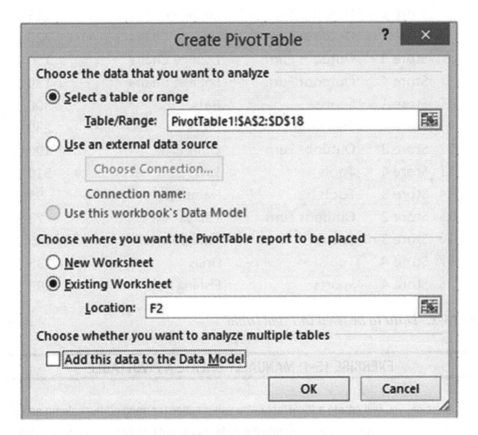

Figure 15-2. *Select Existing Worksheet*

8. The PivotTable Fields list consists of five separate areas.

    • A Fields area which lists all the column names in the table along with a check box next to each column name

    • Filters area

    • Columns area

- Rows area

- Values area

9. Click the check box for the Store field. See Figure 15-3. The Store field contains text data. Columns that contain text are placed in the Rows area by default.

10. Click the check box for Unit Sales. The Unit Sales field contains numeric data. Columns that contain numeric data are placed in the Values area by default.

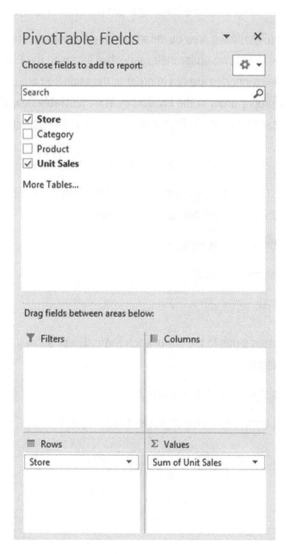

Figure 15-3. *Select fields to be used in PivotTable*

The PivotTable now shows each store in its own row, and it has summed all the Sales for each one of those stores. It also displays a grand total that consists of all Sales from all the stores.

Note   You may need to use the Zoom tool on the status bar to view all of the spreadsheet and PivotTable at the same time while doing this practice.

---

11.   In the PivotTable Fields list, click the check box next to Category.

Category goes to the Rows area by default because it contains text. The categories appear as rows under their respective stores in the PivotTable. Even though the categories are not in order in the table you are using, they are placed in their proper order in the PivotTable. The PivotTable now shows the total sales for each store, the sales for each category, and a grand total.

Hiding and Unhiding the PivotTable Fields List

The PivotTable Fields list only displays when you select a cell in the PivotTable. Select a cell anywhere outside the PivotTable and the pane disappears.

1.   Click cell A2. The pane disappears.

2.   Click cell F2. The pane reappears.

Collapsing and Expanding Groupings

Notice that there are minus signs to the left of each Store. These allow you to collapse what appears underneath a store.

1.   Click the minus sign to the left of Store 2.

The minus sign turns into a plus sign. See Figure 15-4.

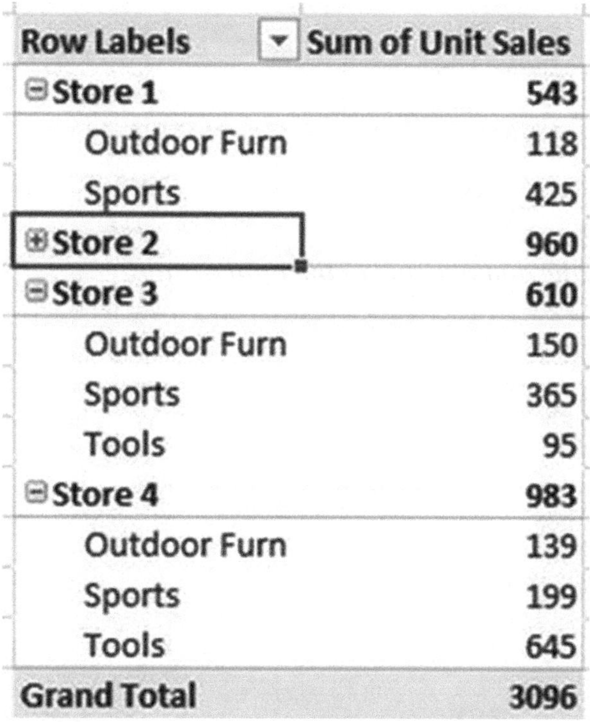

Row Labels ▼	Sum of Unit Sales
⊖ Store 1	543
Outdoor Furn	118
Sports	425
⊞ Store 2	960
⊖ Store 3	610
Outdoor Furn	150
Sports	365
Tools	95
⊖ Store 4	983
Outdoor Furn	139
Sports	199
Tools	645
Grand Total	**3096**

Figure 15-4. *The plus and minus signs can be used to collapse and expand the categories*

2. Click the plus sign to expand the details for Store 2.

Viewing Field Values

You can view all of the different values in a field (column) by moving your mouse over a field in the PivotTable Fields list and then clicking the down arrow.

1. Move your mouse over the Store field in the PivotTable Fields list. Click the down arrow to the right of Store. See Figure 15-5. This displays all of the different values in the Store field. See Figure 15-6.

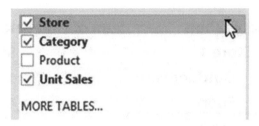

Figure 15-5. *Click the down arrow to the right of Store*

Figure 15-6. *All of the different values in the Store field are displayed*

2.   Click the Cancel button.

Filtering the Data

1. Move your mouse cursor over the word **Product** in the Fields List. Click the down arrow to the right of Product. A list of all the products in the table is displayed.

2. Click the check boxes next to bats and bikes to remove their check marks.

3. Click the OK button.

   A filter icon appears on the line (Figure 15-7) to inform you that if you use this field, not all the items will be displayed.

Figure 15-7. *The filter icon to the right of a field informs you that the field has been filtered so not all the items for that field will be displayed*

   Because the Product field hasn't been placed in the PivotTable, performing a filter has no effect on what is currently displayed in the PivotTable.

4. Click the check box next to Product in the PivotTable Fields list. The Products appear under their prospective categories in the PivotTable. Widen the columns if you need to. Notice that there are no bats or bikes.

   A Filter button appears next to the Row Labels heading in the PivotTable. See Figure 15-8.

5. Move your mouse over the Filter button. A message appears informing you that you have manually applied a filter to the PivotTable based on the Product field.

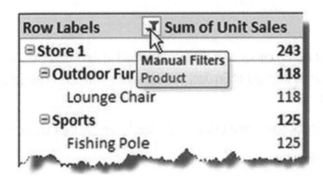

Figure 15-8. *A filter button appears next to the Row Labels heading*

6. Click the Filter button.

7. Click the down arrow for the Select field and then select Product. See Figure 15-9.

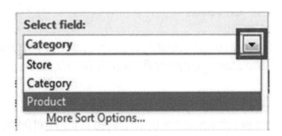

Figure 15-9. *Select Product*

8. Click **Clear Filter from "Product."** This adds the bats and bikes values to the PivotTable.

9. Click Lounge Chairs under Outdoor Furn in the PivotTable.

10. Click the down arrow to the right of the Row Labels column heading. See Figure 15-10.

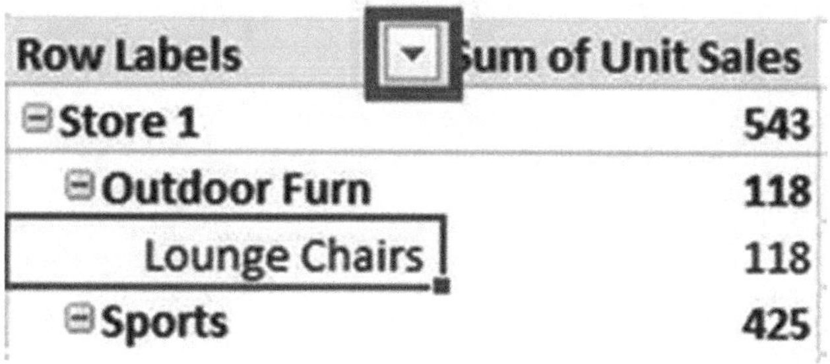

Row Labels	▼	um of Unit Sales
⊟ Store 1		543
⊟ Outdoor Furn		118
Lounge Chairs		118
⊟ Sports		425

Figure 15-10. *Click the down arrow to the right of the Row Labels column heading*

11. Since you have a product currently selected, all of the available products are listed. Click the Cancel button.

12. Click Outdoor Furn in the PivotTable under Store 1.

13. Click the down arrow to the right of the column heading **Row Labels**. Since you have a category selected, all of the available categories are listed. Click the Cancel button.

14. Click Store 1. Click the down arrow to the right of the column heading **Row Labels**. Since you have a store selected, all the available stores are listed. We only want to display the sales for Store 2. Click the check marks next to all of the stores except for Store 2 to remove them. Click the OK button. The down arrow to the right of Row Labels changes to a filter icon.

15. Bring back all the stores by clicking the filter icon next to Row Labels and then click the check box for Select All. (You could have accomplished the same thing by clicking **Clear Filter from "Store"**.) Click the OK button.

16. There may be a large blank area in the PivotTable Fields list between the fields and the drop areas. This area could prevent you from seeing all the fields in the drop areas. Move your cursor on the dotted line just above the text "Drag fields between areas below:". When the cursor changes to a double-headed arrow, move the mouse cursor up to remove some of the space. See Figure 15-11.

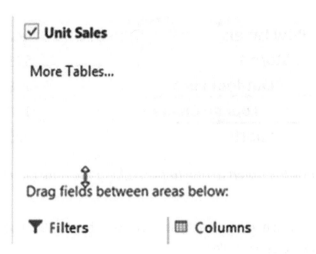

Figure 15-11. *Remove space between PivotTable Fields list and the drop areas*

<u>Changing the Field Order</u>

The order in which fields are placed in the Rows drop area in the PivotTable Fields list affects the order in which they are displayed in the PivotTable. Therefore, you need to be careful about the order in which you drop items into the drop area.

1.   In the Rows area, drag Product above Category. See Figure 15-12.

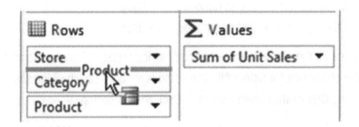

Figure 15-12. *Drag Product above Category*

2.   Look at the PivotTable. Sports now falls under bats, and Outdoor Furniture falls under Lounge Chairs. This is incorrect.

3.   Drag Category from the Rows area into the Columns area. The categories are now displayed in the PivotTable as columns rather than rows. Notice how the PivotTable automatically adjusted the totals.

4. Click the down arrow to the right of Column Labels in the PivotTable. See Figure 15-13.

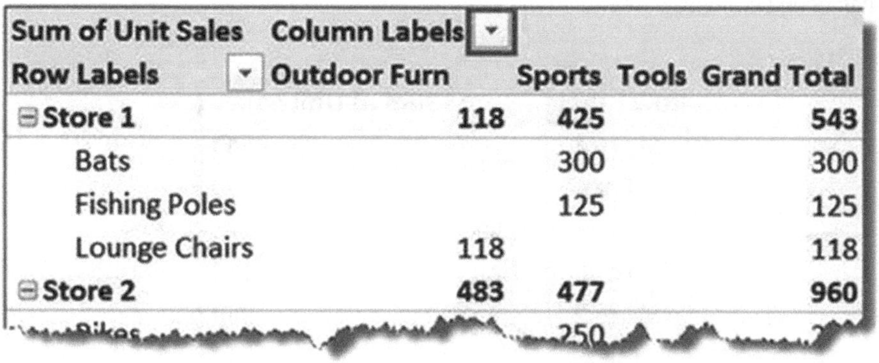

Sum of Unit Sales	Column Labels ▾			
Row Labels ▾	Outdoor Furn	Sports	Tools	Grand Total
⊟ Store 1	118	425		543
Bats		300		300
Fishing Poles		125		125
Lounge Chairs	118			118
⊟ Store 2	483	477		960
Bikes			250	

Figure 15-13. *Click the down arrow to the right of Column Labels*

5. Here, you can filter which columns you want to display. Click the check box next to Tools to remove it. Click the OK button. The down arrow to the right of Column Labels changes to a filter icon.

Removing a Field from the PivotTable

Removing a field from one of the drop areas of the PivotTable Fields list removes that field from the PivotTable.

1. Drag the Category field from the Columns drop area to the spreadsheet.

2. As you are dragging the field over the spreadsheet, an X displays under the field name. See Figure 15-14. Stop dragging at this point.

Figure 15-14. *As you are dragging the field over the spreadsheet, an X displays under the field name*

The Category field information is removed from the PivotTable.

Summarizing Values

Fields in the Values area are totaled using the Sum function. Figure 15-15 shows that Unit Sales has been summed.

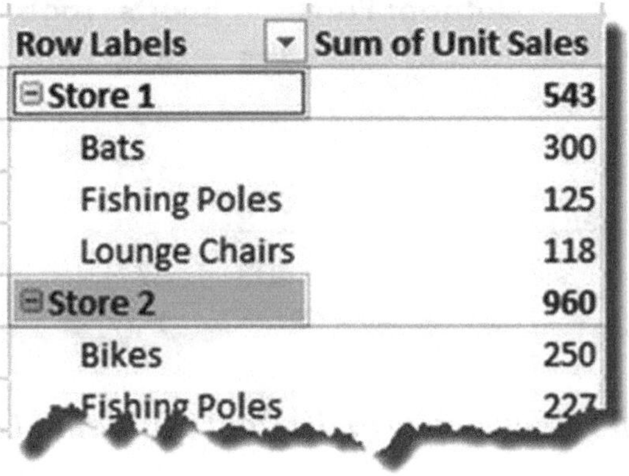

Figure 15-15. *The values in Store 1 are summed. The total is 543*

The type of calculation can be changed by right-clicking **Sum of Unit Sales** on the Pivot Table and then selecting **Summarize Values By**. See Figure 15-16.

Here, in addition to the Sum function, you can elect to use the Count, Average, Max, Min, or Product functions.

It would be rare that you would use the Product function. It multiplies all the values under a grouping. For example, in Figure 15-15, it would determine the product of Store 1 by multiplying 300 * 125 * 118. You can view other available functions such as variance and standard deviation by clicking More Options....

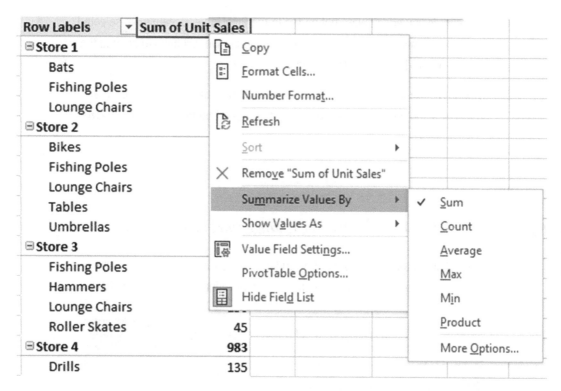

Figure 15-16. *The type of calculation can be changed by right-clicking **Sum of** **Unit Sales** and then selecting **Summarize Values By***

1. Right-click **Sum of Unit Sales** in the PivotTable and then select Summarize Values By.

2. Click Average. Each of the Unit Sales is now an average and the column heading changed to **Average of Unit Sales**.

3. Try the other options.

<u>Changing the Data Source Location</u>

1. Add the records in Figure 15-17 to the spreadsheet.

| 19 | Store 2 | Tools | Hammers | 120 |
| 20 | Store 2 | Tools | Saws | 450 |

Figure 15-17. *Add these two records*

The records you added were not added to the PivotTable. Because we selected a data range instead of a table for our data source, we will have to change the location of our data source to include the new records.

2. Click anywhere in the PivotTable. Click the Ribbon's Analyze tab, in the Data group, click the top half of the Change Data Source button. See Figure 15-18.

Figure 15-18. *Select the top half of the Change Data Source button*

3. Reselect your data range so that it includes the two records you added (don't forget to include your column headings in the selection). See Figure 15-19.

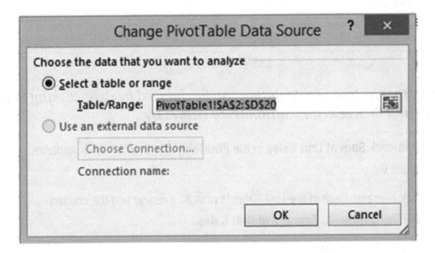

Figure 15-19. *Reselect your data range so that it includes the two records you added*

4. Click the OK button. Your PivotTable should now include the records you added.

EXERCISE 15-2: USING A RECOMMENDED PIVOTTABLE

In this exercise, you will look at Excel's recommended PivotTable formats based on the table data you have selected.

1. Open workbook **Chapter15_PivotTables** if it isn't opened already.

    Select the worksheet named Recommended.

2. Select the range A2:D20. Click the Quick Analysis button ⊞ at the bottom right corner of the selected range.

3. Select Tables from the tabs at the top. Move your cursor over the four recommended PivotTables. If one of them matched your needs, you could click it and Excel would create that PivotTable on a separate worksheet.

4. Click the More button.

    The left pane contains six recommended PivotTables. Clicking a PivotTable view in the left pane displays an enlarged view of that PivotTable in the right pane. See Figure 15-20.

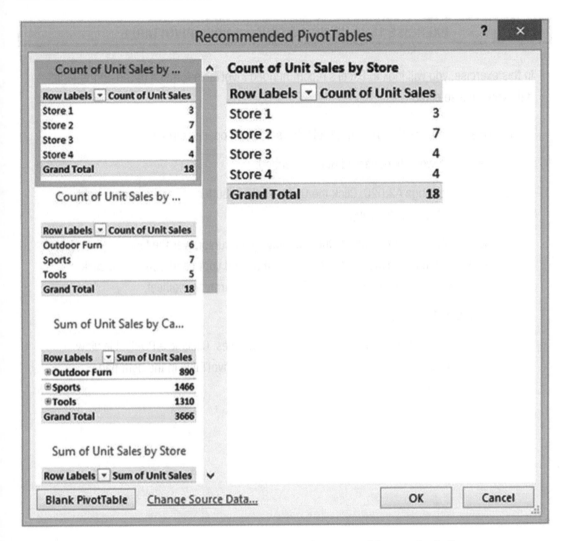

Figure 15-20. *Click one of the Recommended PivotTables in the left pane to get an enlarged view of it in the right pane*

5. Click the Cancel button.

6. Click the Insert tab on the Ribbon. In the Tables group, click the Recommended PivotTables button. This brings up the same Recommended PivotTables dialog box that you saw when you clicked the More option in the Quick Analysis.

7. Select the last PivotTable in the left pane. Click the OK button.

8. The new PivotTable is created on a new worksheet.

EXERCISE 15-3: HANDLING DATES AND TIMES IN PIVOTTABLES

In this exercise, you will convert the data from a spreadsheet into a table and then convert the table to a PivotTable. You will practice grouping and ungrouping the fields. You will also learn how to analyze the data by using the Slicer and TimeLine.

Our production line starts manufacturing radios throughout the year whenever inventory runs low. A gauge on the line displays the number of units produced. An employee takes a reading from the gauge every 20 minutes and then enters the results into a spreadsheet. The spreadsheet covers the years 2012 and 2013. See Figure 15-21.

	A	B	C
1	**Date**	**Time**	**Units**
2	6/4/2012	8:20 AM	640
3	6/4/2012	8:40 AM	530
4	6/4/2012	9:00 AM	500
5	6/4/2012	9:20 AM	480
6	6/4/2012	9:40 AM	500
7	6/4/2012	10:00 AM	520
8	6/4/2012	10:20 AM	550

Figure 15-21. *Inventory data*

1. Open the workbook **Chapter15_PivotTables**.

2. Select the DateTime worksheet.

3. Select all the data in the Spreadsheet by doing the following:

   Click inside cell A1.

   Hold down the Ctrl + Shift keys while you press the right arrow key and then the down arrow key.

Converting the Data into a Table

1. Click the Ribbon's Home tab, and then in the Styles group, click the Format As Table button.

2. Select the second table in the second row of the Light category (Table Style Light 9).

    The Format As Table dialog box displays with the selected data for the location. See Figure 15-22.

Figure 15-22. *Format As Table dialog box*

3. Be sure that you have checked **My table has headers**. Click the OK button.

Changing the Name of the Table

1. Click the Ribbon's Design tab.

2. In the Properties group, change the name from Table1 to DateTimeTable.

Creating the PivotTable

1. Click anywhere within the data.

2. Click the Ribbon's Insert tab; in the Tables group, click the PivotTable button.

3. We will place the PivotTable on the same sheet as our data:

    a. Click the Existing Worksheet radio button.

    b. Place your cursor in the Location field. Click cell E3. See Figure 15-23.

    c. Click the OK button.

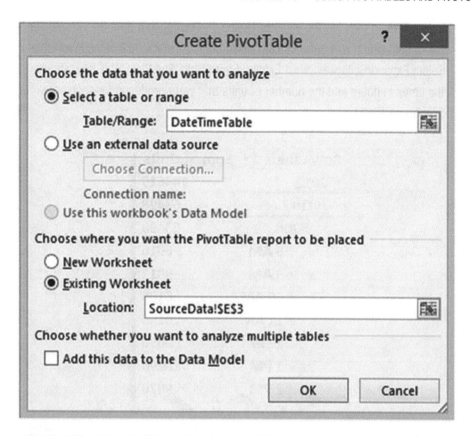

***Figure 15-23.** The PivotTable will start in cell E3 of the current worksheet*

4. Select Date in the PivotTable Fields list. Notice that there are two new fields
    in the PivotTable Fields list that weren't columns in the table. Selecting a Date
    field adds the Quarters and Years fields and makes them selected as well.

5. Right-click 2012 in the PivotTable. Select Expand/Collapse. Select Expand
    Entire Field.

6. Right-click Qtr2 under 2012 in the PivotTable. Select Expand/Collapse. Select
    Expand Entire Field. You can see that Excel breaks down (ungroups) date fields
    into years and quarters. You can see that Excel has the intelligence to know
    what months fall within what quarters.

7. Select Units in the PivotTable Fields list. Because Units is a numeric field, Units
    is placed in the Values pane. Since summing is done on numeric fields, it shows
    as **Sum of Units**.

8.   Select Time in the PivotTable Fields. Notice that selecting a time field creates an Hours field which was automatically selected. Right-click Qtr2 in the PivotTable. Select Expand/Collapse. Select Expand Entire Field. The PivotTable now shows the times in hours and the number of units that were produced for each hour. See Figure 15-24.

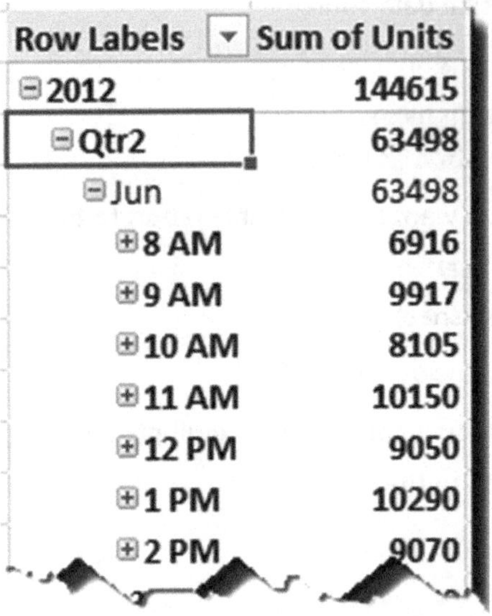

Row Labels ▼	Sum of Units
⊟2012	144615
⊟Qtr2	63498
⊟Jun	63498
⊞8 AM	6916
⊞9 AM	9917
⊞10 AM	8105
⊞11 AM	10150
⊞12 PM	9050
⊞1 PM	10290
⊞2 PM	9070

Figure 15-24. *PivotTable shows the times in hours and number of units that were produced for each hour*

9.   Click the plus sign to the left of 8 AM in the Pivot Table to expand it. The time 8 AM under each month expands. The 20-minute intervals display with the number of units produced during that time. See Figure 15-25.

Row Labels ▼	Sum of Units
⊟ 2012	144615
⊟ Qtr2	63498
⊟ Jun	63498
⊟ 8 AM	6916
:20	3563
:40	3353
⊞ 9 AM	9917
⊞ 10 AM	8105
⊞ 11 AM	10150
⊞ 12 PM	9050
⊞ 1 PM	10290

Figure 15-25. *Number of units produced within 20-minute intervals*

10.   Click the minus sign to the left of 8 AM to collapse it. The time under 8 AM collapses for each month.

11.   Right-click any Qtr under 2012. Select Expand/Collapse and then select Collapse Entire Field. The result is shown in Figure 15-26.

Row Labels ▼	Sum of Units
⊟ 2012	144615
⊞ Qtr2	63498
⊞ Qtr3	62950
⊞ Qtr4	18167
⊟ 2013	80549
⊞ Qtr1	16249
⊞ Qtr2	17547
⊞ Qtr3	28216
⊞ Qtr4	18537
Grand Total	225164

Figure 15-26. *Collapsed Quarters*

12. Click anywhere in the PivotTable. Drag Years from the Rows area to the Columns area in the PivotTable Fields list.

You can now easily compare the number of units produced in each quarter of 2012 to those produced in 2013. See Figure 15-27.

Sum of Units	Column Labels ▼		
Row Labels ▼	2012	2013	Grand Total
⊞ Qtr1		16249	16249
⊞ Qtr2	63498	17547	81045
⊞ Qtr3	62950	28216	91166
⊞ Qtr4	18167	18537	36704
Grand Total	144615	80549	225164

Figure 15-27. *Compare 2012 values to those from 2013*

13. Click Year 2012 in the PivotTable.

14. Click the Ribbon's Analyze tab. In the Active Field group, click the Expand Field button. See Figure 15-28.

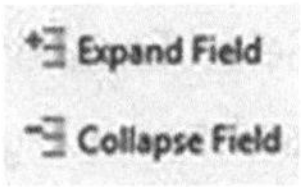

Figure 15-28. *Select Expand Field*

15. Click Date in the Show Detail dialog box. See Figure 15-29.

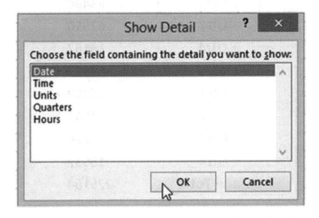

Figure 15-29. *Select Date*

16. Click the OK button. Figure 15-30 shows the result.

Sum of Units	Column Labels															
	2012						2012 Total	2013							2013 Total	Grand Total
Row Labels	Jun	Jul	Aug	Sep	Oct	Dec		Jan	Mar	Apr	Jun	Jul	Sep	Oct		
Qtr1								8240	8009						16249	16249
Qtr2	63498						63498			8496	9051				17547	81045
Qtr3		18734	18390	25826			62950					8300	19916		28216	91166
Qtr4					10047	8120	18167							18537	18537	36704
Grand Total	63498	18734	18390	25826	10047	8120	144615	8240	8009	8496	9051	8300	19916	18537	80549	225164

Figure 15-30. *Your PivotTable should now look like this*

17. Click Qtr1 in the PivotTable. On the Ribbons' Analyze tab, in the Active Field group, click the Expand Field button. See Figure 15-31.

Sum of Units	Column Labels															
	2012						2012 Total	2013							2013 Total	Grand Total
Row Labels	Jun	Jul	Aug	Sep	Oct	Dec		Jan	Mar	Apr	Jun	Jul	Sep	Oct		
Qtr1								8240	8009						16249	16249
8 AM								1076	771						1847	1847
9 AM								1372	722						2094	2094
10 AM								1003	1378						2381	2381
11 AM								1614	1538						3152	3152
12 PM								1594	1598						3192	3192
1 PM								705	990						1695	1695
2 PM								876	1012						1888	1888
Qtr2	63498						63498			8496	9051				17547	81045
8 AM	6916						6916			977	966				1943	8859
9 AM	9917						9917			1455	1154				2609	12526
10 AM	8105						8105			1454	1439				2893	10998
11 AM	10150						10150			1095	1142				2237	12387
12 PM	9050						9050			1233	1218				2451	11501
1 PM	10290						10290			1376	1611				2987	13277
2 PM	9070						9070			906	1521				2427	11497
Qtr3		18734	18390	25826			62950					8300	19916		28216	91166
8 AM		1477	1728	3648			6853					1214	1965		3179	10032
9 AM		2440	2593	3640			8673					1147	2994		4141	12814
10 AM		2730	2378	3243			8351					1245	2439		3684	12035
11 AM		3002	3039	3074			9115					1038	3325		4363	13478
12 PM		2855	3218	3696			9769					1177	2919		4096	13865
1 PM		3002	2457	4736			10195					1374	3208		4582	14777
2 PM		3228	2977	3789			9994					1105	3066		4171	14165
Qtr4					10047	8120	18167							18537	18537	36704
8 AM					1085	685	1770							1382	1382	3152
9 AM					1561	972	2533							2729	2729	5262
10 AM					1624	1275	2899							2703	2703	5602
11 AM					1236	1548	2784							2911	2911	5695
12 PM					1619	1129	2748							3147	3147	5895
1 PM					1480	1137	2617							2985	2985	5602
2 PM					1442	1374	2816							2680	2680	5496
Grand Total	63498	18734	18390	25826	10047	8120	144615	8240	8009	8496	9051	8300	19916	18537	80549	225164

Figure 15-31. *Expanded Quarter view*

18. We want to collapse the times so that we don't see them. Click any 8 AM in the PivotTable. You can click any of the times. On the Ribbons' Analyze tab, in the Active Field group, click the Collapse Field button.

19. Drag the Years and Date fields from the Columns area back to the Rows in the PivotTable Fields pane in the order shown in Figure 15-32.

Your PivotTable should now look like Figure 15-33.

Figure 15-32. *Rows pane*

Row Labels ▼	Sum of Units
⊟ **2012**	
⊞ **Qtr2**	63498
⊞ **Qtr3**	62950
⊞ **Qtr4**	18167
⊟ **2013**	
⊞ **Qtr1**	16249
⊞ **Qtr2**	17547
⊞ **Qtr3**	28216
⊞ **Qtr4**	18537
Grand Total	225164

Figure 15-33. *Your PivotTable should now look like this*

732

Grouping and Ungrouping Fields

1. Expand Qtr2 under 2012 by clicking the plus sign.

2. Excel added the Years and Quarters fields when you selected the Date field in the PivotTable Fields list. They can be removed by using Ungroup:

    a. Click the year 2012 in the PivotTable.

    b. On the Ribbon's Analyze tab, in group, click the Ungroup button.

    The Year and Quarter have been removed from the PivotTable. The dates now show as they do in the source data. Notice that Years and Quarters are no longer in the Rows area.

3. Click one of the times in the PivotTable.

4. Click the Ribbon's Analyze tab. In group, click the Ungroup button.

    The Hours data (separated minutes) no longer appears in the PivotTable. The Time now shows as it does in the source data. Notice that Hours is no longer in the Rows area or in the PivotTable Fields list.

5. Click the Ribbon Design tab. In the Layout group, click the **Report Layout** button. See Figure 15-34.

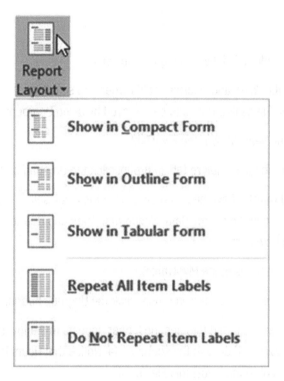

Figure 15-34. *Options under Report Layout button*

The PivotTable is currently displayed in the Compact Form.

6. Click **Show in Outline Form**. The Outline Form separates the Date and Time field into separate columns. The Sum of Units displays at the top of the data being summed.

7. Click the **Report Layout** button again. Click **Show in Tabular Form**.

    The first Time and Units appear on the same row as the Date. The total of the Units appears at the end of each Date rather than at the top. Figure 15-35 shows how each of the Report Layout types treats the same data.

Compact Form

Row Labels ▾	Sum of Units
⊟6/4/2012	10176
8:20 AM	640
8:40 AM	530
9:00 AM	500
9:20 AM	480
9:40 AM	500
10:00 AM	520
10:20 AM	550
10:40 AM	460
11:00 AM	490
11:20 AM	500
11:40 AM	499
12:00 PM	507
12:20 PM	490
12:40 PM	500
1:00 PM	660
1:20 PM	650
1:40 PM	590
2:00 PM	490
2:20 PM	400
2:40 PM	220
⊟6/5/2012	10185

Outline Form

Date ▾	Time ▾	Sum of Units
⊟6/4/2012		10176
	8:20 AM	640
	8:40 AM	530
	9:00 AM	500
	9:20 AM	480
	9:40 AM	500
	10:00 AM	520
	10:20 AM	550
	10:40 AM	460
	11:00 AM	490
	11:20 AM	500
	11:40 AM	499
	12:00 PM	507
	12:20 PM	490
	12:40 PM	500
	1:00 PM	660
	1:20 PM	650
	1:40 PM	590
	2:00 PM	490
	2:20 PM	400
	2:40 PM	220
⊟6/5/2012		10185

Tabular Form

Date ▾	Time ▾	Sum of Units
⊟6/4/2012	8:20 AM	640
	8:40 AM	530
	9:00 AM	500
	9:20 AM	480
	9:40 AM	500
	10:00 AM	520
	10:20 AM	550
	10:40 AM	460
	11:00 AM	490
	11:20 AM	500
	11:40 AM	499
	12:00 PM	507
	12:20 PM	490
	12:40 PM	500
	1:00 PM	660
	1:20 PM	650
	1:40 PM	590
	2:00 PM	490
	2:20 PM	400
	2:40 PM	220
6/4/2012 Total		10176

Figure 15-35. *Three types of forms: Compact, Outline, and Tabular*

Notice the Compact Form doesn't display a separate Date column as do the Outline and Tabular forms.

8. Click the **Report Layout** button and then click **Repeat All Item Labels**. See Figure 15-36.

Date ▾	Time ▾	Sum of Units
⊟6/4/2012	8:20 AM	640
6/4/2012	8:40 AM	530
6/4/2012	9:00 AM	500
6/4/2012	9:20 AM	480
6/4/2012	9:40 AM	500
6/4/2012	10:00 AM	530

Figure 15-36. *Dates are repeated down the column*

9. Remove the repeated dates by clicking the Report Layout button and then clicking **Do Not Repeat Item Labels**.

10. Click the Ribbon's Design tab, and in the Layout group, click the Subtotals button. Select **Do Not Show Subtotals**. The Subtotals are removed from the PivotTable.

11. Click the Subtotals button again. Select **Show all Subtotals at Bottom of Group**. The subtotals appear at the bottom of each date.

12. Click the Ribbon's Design tab, and in the Layout group, click the Blank Rows button. Select **Insert Blank Line after Each Item**. A blank line appears between each date.

13. Click the date 6/4/2012 in the PivotTable.

14. Click the Ribbon's Analyze tab. In the Active Field group, click the Collapse Field button.

15. On the Ribbon's Design tab, in the Layout group, click the Blank Rows button. Select **Remove Blank Line after Each Item**.

16. On the Ribbon's Design tab, in the PivotTable Styles group, click one of the styles you like.

Adding Records to the Source Data and Refreshing the PivotTable

When you add new records, the PivotTable doesn't automatically update to reflect the new records. The data source doesn't need to change since we are using a table; however, the PivotTable still needs to be refreshed.

1. Add the three records in Figure 15-37 to the bottom of the source data.

502	10/7/2013	8:20 AM	327
503	10/7/2013	8:40 AM	432
504	10/7/2013	9:00 AM	500

Figure 15-37. *Add this data*

2. Click any cell in the PivotTable. Click the Ribbon's Analyze tab. In the Data group, click the top half of the Refresh button.

    The records you added should now appear at the bottom of the PivotTable. See Figure 15-38.

Date ▼	Time ▼	Sum of Units
⊕ 6/4/2012		10176
⊕ 6/5/2012		10185
⊕ 6/6/2012		9522
⊕ 6/7/2012		7566
⊕ 6/8/2012		8642
⊕ 6/18/2012		9427
⊕ 6/22/2012		7980
⊕ 7/5/2012		9606
⊕ 7/25/2012		9128
⊕ 8/3/2012		9147
⊕ 8/27/2012		9243
⊕ 9/5/2012		9195
⊕ 9/11/2012		8845
⊕ 9/27/2012		7786
⊕ 10/8/2012		10047
⊕ 12/14/2012		8120
⊕ 1/5/2013		8240
⊕ 3/2/2013		8009
⊕ 4/5/2013		8496
⊕ 6/13/2013		9051
⊕ 7/8/2013		8300
⊕ 9/1/2013		9860
⊕ 9/2/2013		10056
⊕ 10/5/2013		9278
⊕ 10/6/2013		9259
⊖ 10/7/2013	8:20 AM	327
	8:40 AM	432
	9:00 AM	500
10/7/2013 Total		1259
Grand Total		226423

Figure 15-38. *Records you added appear at the bottom of the PivotTable*

3. Click the minus sign next to 10/7/2013. Click anywhere in the pivot table.

4. Click the Ribbon's Analyze tab. In the Filter group, click Insert Slicer.

5. Check Date and Time. Click the OK button.

6. Click 6/6/2012 in the Date slicer; only the data for 6/6/2012 displays. All of the other dates in the slicer become unselected.

7. Click 6/4/2012 in the Date slicer. Only the data for 6/4/2012 is displayed, and only this date appears highlighted in the slicer.

8. If you want to select more than one date at a time, you will need to click the Multi-Select button on the top right of the slicer. Click the Multi-Select button.

9. Select the other four dates shown in Figure 15-39. Click on the Multi-Select button to turn it off.

Figure 15-39. *Date slicer*

Note   If you click an item in the slicer and then click the Multi-Select button, all other items become unselected and then you click the rest of the items you need. If you haven't clicked an item before you click the Multi-Select button, then items you click actually become unselected.

10.   Click the Filter Icon 𝖸̲𝗑. Click the Multi-Select button. Now click 6/4/2012. Notice that because you clicked the Multi-Select button while no dates were selected, 6/4/2012 is in white which means it is unselected. The date 6/4/2012 no longer appears in your Pivot Table. Click the other dates in white in Figure 15-40 to remove their data from the PivotTable.

Figure 15-40. *Date slicer*

11.   Right-click one of the dates in the PivotTable. Select Expand/Collapse and then select Expand Entire Field.

12.   Click 10:00 AM in the Time slicer. Click the Multi-Select button. Select times 10:20 AM, 10:40 AM, and 11:00 AM. See Figure 15-41.

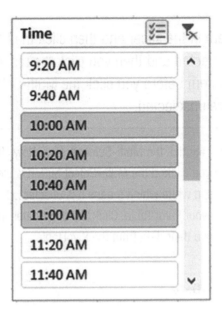

Figure 15-41. *Time slicer*

13. Click anywhere on the PivotTable. Click the Ribbon's Analyze tab. In the Filter group, Click Insert Timeline.

14. Check Date. Click the OK button.

15. The Timeline has a horizontal scroll bar. Scroll all the way to the left. Under 2012, click June.

16. All the June 2012 dates display in the PivotTable.

Drag across June, July, and August. See Figure 15-42. All the June, July, and August 2012 dates display in the PivotTable.

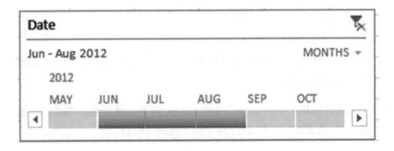

Figure 15-42. *Timeline*

All the June, July, and August dates appear in the PivotTable.

You have learned how to create PivotTables. You have seen that you can move fields to different areas in the PivotTable Fields pane to control rows, columns, and summarizing. You have also seen how easy it is to view only data from a particular time or date using a slicer or a timeline. Next, you will learn how to create a chart that can be filtered.

Creating a PivotChart

PivotCharts can be created directly from a data source or from a PivotTable. The difference between a PivotChart and other charts is that a PivotChart has filter buttons associated with it that the user can use to view just what he or she wants on the chart.

EXERCISE 15-4: CREATING A PIVOTCHART

In this exercise, we first will create a PivotChart from a data source. Then we will create one from a PivotTable.

1. Open the Chapter15_PivotTables workbook.

2. Select the worksheet named PivotChart.

Creating a PivotChart Based on Source Data

1. Click anywhere within the data. Click the Ribbon's Home tab. In the Styles group, click Format As Table.

2. Select the second style in the second row (Table Style Light 9).

3. Click the Ok button for Format As Table. See Figure 15-43.

Figure 15-43. *Format As Table dialog box*

4. Click the Ribbon's Insert tab. In the Charts group, click the top half of the PivotChart button. The Create PivotChart dialog box displays.

5. Select Existing Worksheet. Click inside the Location text box. Click inside cell F2. See Figure 15-44.

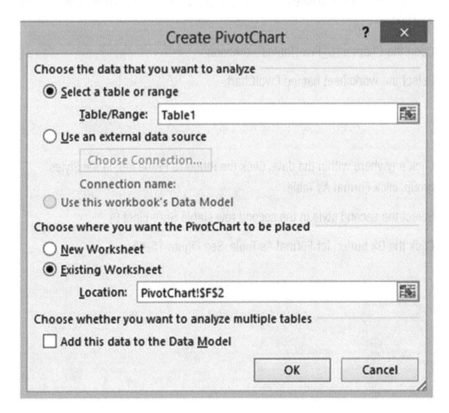

Figure 15-44. *Create PivotChart dialog box*

6. Click the OK button.

7. You create a PivotChart the same way that you create a PivotTable. Select State, Category, Product, and Sales in the PivotChart Fields list. State, Category, and Product appear in the Axis (Categories) area, and because Sales is a numeric field, it appears in the Values area as **Sum of Sales**.

8. Expand the size of the chart so that you can see everything. There are three filter buttons below the chart. See Figure 15-45.

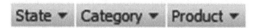

Figure 15-45. *Filter buttons*

9. Click the State filter button. Remove the check in front of Illinois. Click the OK button. Now, just the data for Indiana and Kentucky shows.

10. Click the Category filter button. Remove the check in front of Books_Mag. Click the OK button.

11. Click the Product filter button. Remove the check in front of Beverages. Click the OK button. Your PivotChart should look like Figure 15-46.

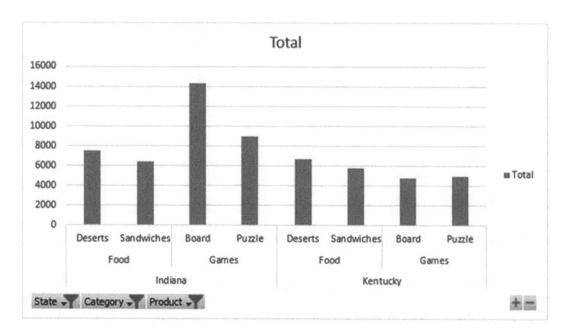

Figure 15-46. *Your PivotChart should now look like this*

Note   When you created a PivotChart, a PivotTable was created at the same time. When you make changes to the PivotChart, the same changes are made to the PivotTable.

12.   Click the State filter button. Select **Clear Filter from "State."**

13.   Click the Category filter button. Select **Clear Filter from "Category."**

14.   Click the Product filter button. Select **Clear Filter from "Product."**

Below the chart on the right, there is a plus and minus button. The plus button is the Expand Entire Field button. The minus button is the Collapse Entire Field button.

15.   Click the minus button. This collapses all of the Product data. See Figure 15-47.

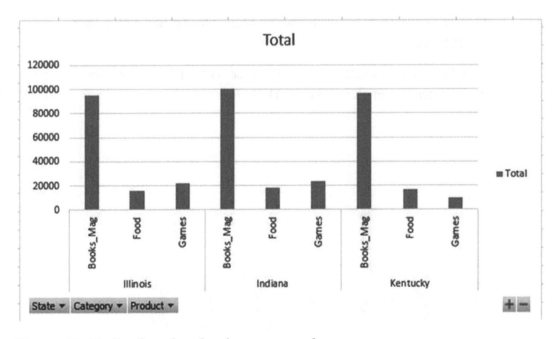

Figure 15-47.   *Product data has been removed*

16.   Click the plus button. This brings back all the Products.

17.   Press the Delete key to remove the PivotChart.

18. Click anywhere in the PivotTable. Click the Ribbon's Analyze tab. In the Actions group, click the Select button.

19. Select **Entire PivotTable**. Press the Delete key.

Creating a PivotChart Based on a PivotTable

1. Click anywhere in the table data. Click the Ribbon's Insert tab.

2. In the Tables group, click PivotTable. The Create PivotTable dialog box displays.

3. Select Existing Worksheet. Click inside the Location text box. Click inside cell F2.

4. Click the OK button.

5. Select the State, Category, Product, and Sales fields in the PivotChart Fields list.

6. Move the Category and Product fields from the Rows area to the Columns area. See Figure 15-48.

Figure 15-48. *Moved Category and Product to the Columns area*

The PivotTable should look like Figure 15-49.

Sum of Sales	Column Labels											
	⊟Books_Mag		Books_Mag Total	⊟Food			Food Total	⊟Games		Games Total	Grand Total	
Row Labels ▼	Books	Magazines		Beverages	Deserts	Sandwiches		Board	Puzzle			
Illinois	65800	29000	94800	3900	4250	6000	14150	14000	7800	21800	130750	
Indiana	72500	28300	100800	4200	7500	6400	18100	14350	9000	23350	142250	
Kentucky	68000	29000	97000	4000	6700	5800	16500	4800	5000	9800	123300	
Grand Total	206300	86300	292600	12100	18450	18200	48750	33150	21800	54950	396300	

Figure 15-49.   Your PivotTable should now look like this

7.   Click anywhere on the PivotTable. Click the Ribbon's Analyze button. In the Tools group, click the PivotChart button.

The Insert Chart dialog box displays. See Figure 15-50.

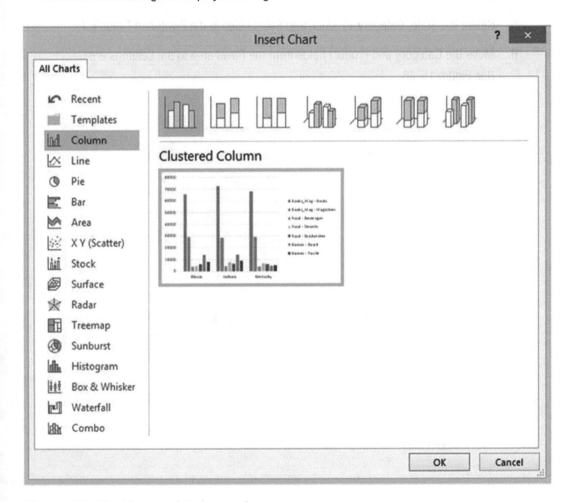

Figure 15-50.   Clustered Column chart

8. Leave the Clustered Column chart selected. Click the OK button.

9. The PivotChart displays. See Figure 15-51. It is based on the PivotTable. It takes on the same format. The results here are much different than the first PivotChart you created from the source data.

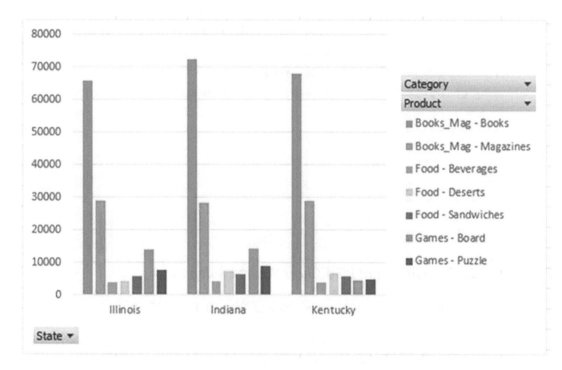

Figure 15-51. *PivotChart*

Creating Multiple PivotTables and PivotCharts from the Same Data Source

It is possible to create multiple PivotTables with different structures from the same data source. You may want to create multiple PivotTables so you can look at various summarized views of your data at the same time.

1. Click anywhere in the source data.

2. Bring up the PivotTable and PivotChart Wizard by pressing the Alt key, then the D key, and then the P key. (Do not press any of them at the same time.) If you are not at Step 1 of 3 of the Wizard, press the Back button until you are.

3. Leave **Microsoft Excel list or database** selected. Select **PivotChart report (with PivotTable report)**. See Figure 15-52. Click the Next button.

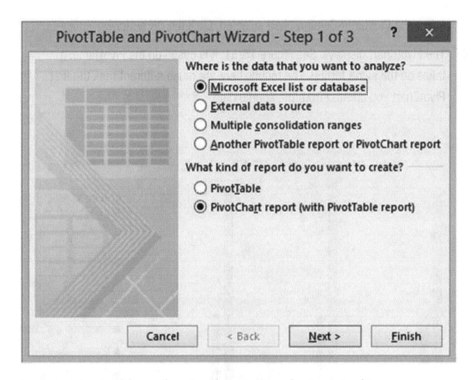

Figure 15-52. *PivotTable and PivotChart Wizard – Step 1 of 3*

4. Step 2 of the Wizard should have the range as shown in Figure 15-53.

Figure 15-53. *PivotTable and PivotChart Wizard – Step 2 of 3*

5. Click the Next button.

6. Select Existing worksheet. Click cell F35. See Figure 15-54.

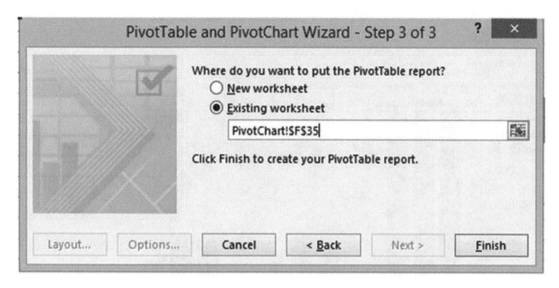

Figure 15-54. *PivotTable and PivotChart Wizard – Step 3 of 3*

7. Click the Finish button.

8. In the PivotTable Fields list, select the State, Category, and Sales fields.

9. Drag the State field from the Axis (Categories) area to the Legend (Series) area.

10. Close the PivotChart Fields list.

The PivotTable should now look like Figure 15-55.

Sum of Sales	Column Labels			
Row Labels	Illinois	Indiana	Kentucky	Grand Total
Books_Mag	94800	100800	97000	292600
Food	14150	18100	16500	48750
Games	21800	23350	9800	54950
Grand Total	**130750**	**142250**	**123300**	**396300**

Figure 15-55. *Your PivotTable should now look like this*

The PivotChart should now look like Figure 15-56.

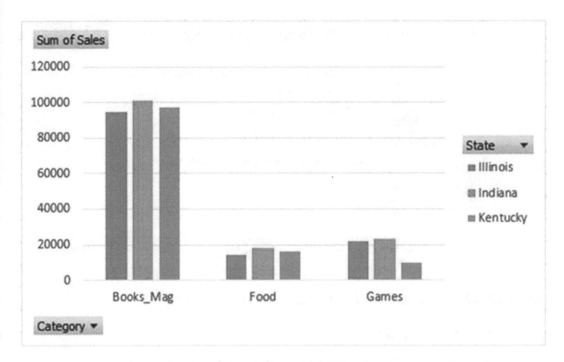

Figure 15-56. *Your PivotChart should now look like this*

Moving the PivotChart and PivotTable

You can move a PivotChart by moving the cursor to its border at which time the cursor changes to a four-headed arrow. You can then drag the PivotChart to wherever you want. There are two ways to move a PivotTable. You can use drag and drop if you are moving the PivotTable to another location on the same worksheet, or you can use the Move PivotTable button on the Ribbon.

1.   Click anywhere on the PivotTable you just created.

2.   On the Ribbon, click the Analyze tab. In the Actions group, click the Select button and then select Entire PivotTable. See Figure 15-57.

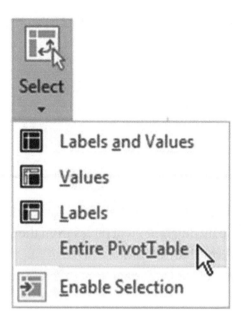

Figure 15-57. *Select Entire PivotTable from the Select button*

3. Move the cursor to any border of the PivotTable. The cursor changes to a four-headed arrow to signify that the PivotTable is now able to be dragged.

4. Drag the PivotTable so that it starts in cell G33. You may need to widen the columns.

5. Method 2 for moving a PivotTable. Click anywhere on the PivotTable. On the Analyze tab in the Actions group, click the Move PivotTable button. The Move Table dialog box displays. With the cursor in the Location text box, click cell D34. Click the OK button.

Clearing a PivotTable

1. In the Ribbon's Action group, click the Clear button.

2. Select Clear All. This clears all the data and labels, but the PivotTable still exists.

Deleting a PivotTable

1. Press Ctrl + Z to undo the Clear.

2. Click inside any cell in the PivotTable.

3. In the Actions group, click the Select button. Select Entire PivotTable.

4. Press the Delete key.

Deleting a PivotChart

1. Click the PivotChart.

2. Press the Delete key.

You have learned how to create a PivotChart from source data and from a PivotTable. You have learned how to create different PivotTables from the same data source and how to use the PivotTable and PivotChart Wizard.

Next, you will learn how you can take data from different sources and link them together by a key field.

Creating PivotTable on a Relational Database

PivotTables become even more powerful when you can apply them to multiple files that are linked together. You can do this through a relational database.

When we think of databases, we usually think of programs such as Access, SQL Server, Oracle, and so on. We can import data from these programs into Excel. We can, however, create small relational databases directly in Excel. A relational database is created by linking multiple tables through a common field. The tables can be linked together to create a PivotTable.

Relational databases consist of Master and Detail tables. If you have previously used Access, you are probably familiar with the concept. There must be a common field in the master and detail files on which the tables are linked together. There is one Master record for one or more detail records.

Figure 15-58 shows a Customer table, and Figure 15-59 shows an Orders table. The two tables can be linked by the common field CustomerID. Each customer in the Customer table is assigned a unique CustomerID. Each customer can make many orders; therefore, the CustomerID field in the Orders table is not unique. We call this type of relationship between the two tables a one-to-many relationship. The fields used to link tables are called **keys**. The common field that is unique is called the **Primary Key**.

The common field on the many side is called the **Foreign Key**. Looking at the two tables, you can see that a CustomerID must already exist in the Customer table before it can be used in the Orders table.

CustomerID	Compay Name	Address	City	State	Zip
1	Wang Tech	503 Birdseye Ave	Indianapolis	IN	38905
2	Lance Smelt	309 Cottonwood	Indianapolis	IN	38905
3	Johnson Plastic	1818 Rose Street	Springfield	IL	60381
4	Tri-State Furniture	320 Lincoln Ave	Chicago	IL	60601
5	Computer World	207 Washington	Fort Wayne	IN	40010
6	Top Printers	408 Azaela Drive	Phoenix	AZ	52310
7	Hytech	218 Monroe	Phoenix	AZ	52310
8	Drive Master	512 Chestnut	Flagstaff	AZ	50020
9	Office Land	315 Jefferson	Chicago	IL	60601
10	Hybrid	1040 Rose Ave	Chicago	IL	60601

Figure 15-58. *Customer table*

◢	A	B	C	D	E
1	OrderID	CustomerID	OrderDate	RequiredDate	Order Amount
2	1231	5	3/14/2015	5/15/2015	3812.64
3	1232	8	3/15/2015	5/18/2015	4875.85
4	1233	4	3/17/2015	4/18/2015	2750.32
5	1234	5	3/22/2015	6/1/2015	7395.5
6	1235	2	3/23/2015	5/17/2015	3817.5
7	1236	4	3/26/2015	6/1/2015	14975.32
8	1237	3	3/30/2015	5/21/2015	8327.42
9	1238	5	4/1/2015	5/23/2015	4832.17
10	1239	8	4/4/2015	5/23/2015	10731.15
11	1240	3	4/8/2015	5/25/2015	6753.25
12	1241	2	4/9/2015	5/26/2015	3975.75

Figure 15-59. *Orders table*

EXERCISE 15-5: CREATING A PIVOTTABLE ON A RELATIONAL DATABASE

In this exercise, you will first create the Master table and then the Detail table. You will then link the two tables together. You will then create a PivotTable that uses the two linked tables.

Creating the Master Table

1. Open the Chapter15_PivotTables workbook. Select the worksheet named Master. Click anywhere in the data.

2. Click the Ribbon's Home tab. In the Styles group, click the Format As Table button. Select any style you wish.

   The Format As Table dialog box displays. See Figure 15-60.

Figure 15-60. *Format As Table dialog box*

3. Be sure that you select **My table has headers** and then click the OK button.

4. The Design tab is active when you select the table. In the Properties group, change the Table Name to **Customer** and then press Enter.

Creating the Detail Table

1. Click the worksheet tab named Detail. Click anywhere in the data.

2. Click the Ribbon's Home tab. In the Styles group, click the Format As Table button. Select any style you wish.

3. Be sure that you select **My table has headers** in the Format As Table dialog box and then click the OK button.

4. The Design tab is active when you select the table. In the Properties group, change the Table Name to **Orders** and then press Enter.

Linking the Tables Together

1. Click the Ribbon's Data tab. In the Data Tools group, click the Relationships button.

2. Click the New button.

3. The Customer table and the Orders table both contain a CustomerID field by which they can be linked together. The CustomerID field in the customer table is the Primary key, and the CustomerID in the Orders table is the Foreign key. Make the entries shown in Figure 15-61 and then click the OK button.

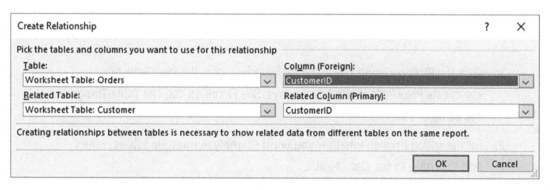

Figure 15-61. *Create Relationship dialog box*

4. The Manage Relationships dialog box shows the relationship you just created. Click the Manage Relationships Close button. See Figure 15-62.

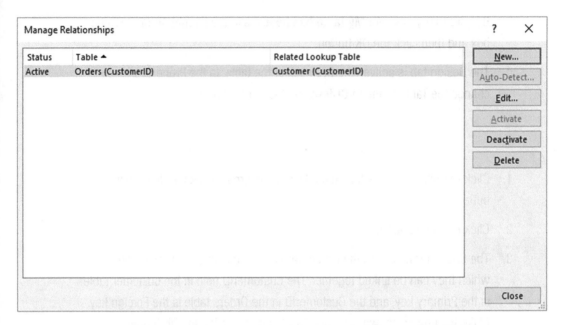

Figure 15-62. *Relationship you created between Orders and Customer*

Defining the Table Relationships in the PivotTable Report Section

1. Click the Ribbon's Insert tab. In the Tables group, click the PivotTable button. The Create PivotTable dialog box displays. See Figure 15-63. The **Table/Range** should be Orders.

2. For the option **Choose whether you want to analyze multiple tables,** select **Add this data to the Data Model**.

3. Click the OK button.

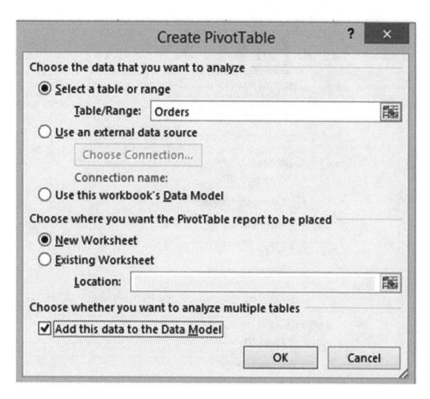

Figure 15-63. *Create PivotTable dialog box*

4. In the PivotTable Fields list on the right side of the worksheet, the word **ACTIVE**
    is highlighted. The word **ACTIVE** is highlighted because there is only one table
    currently active. Click the word **ALL**. See Figure 15-64. Expand the Customer
    and Orders tables so that you can see all of their fields.

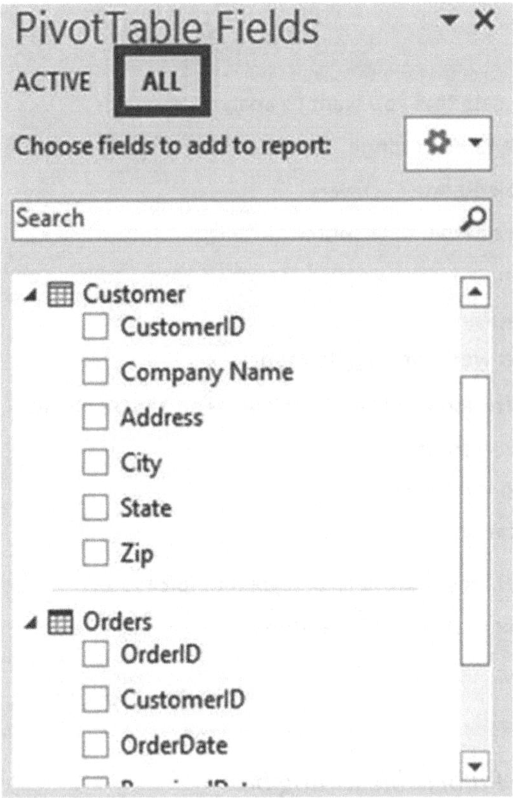

Figure 15-64. *Click the word ALL to see all of the fields for both tables*

We have already created the relationship between the tables, but if we hadn't and you start adding fields in the PivotTables that can't be made sense of without the relationship, you will get the message in Figure 15-65.

Figure 15-65. *Possible message*

Clicking the CREATE button would bring up the same **Create Relationship** dialog box that you already filled out.

5. Now, you can create a PivotTable using both tables. In the PivotTable Fields pane, select the fields CustomerID, City, and State from the Customer table.

6. Select the fields OrderID and Order Amount from the Orders table. The bottom half of the PivotTable Fields list appears as in Figure 15-66.

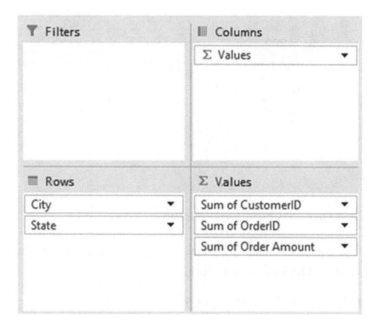

Figure 15-66. *Your areas should look like this*

7. Now, let's rearrange the fields:

   a. Drag State from the Rows area to the Filters area.

   b. Drag Sum of CustomerID from Values to the top of Rows.

   c. Drag Sum of OrderID from Values to the bottom of Rows.

The areas should look like Figure 15-67.

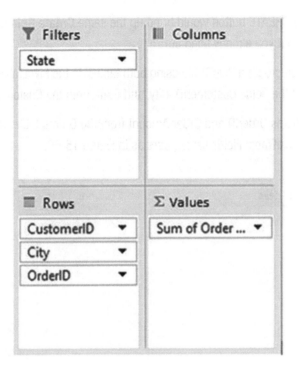

Figure 15-67. *Your areas should look like this*

8. Click the Ribbon's Design tab. In the Layout group, click the Report Layout button and select **Show in Outline Form**.

The resulting PivotTable should look like Figure 15-68.

CustomerID	City	OrderID	Sum of Order Amount
State	**All** ▼		
⊟2			
	⊟Indianapolis		
		1235	3817.5
		1241	3975.75
⊟3			
	⊟Springfield		
		1237	8327.42
		1240	6753.25
⊟4			
	⊟Chicago		
		1233	2750.32
		1236	14975.32
⊟5			
	⊟Fort Wayne		
		1231	3812.64
		1234	7395.5
		1238	4832.17
⊟8			
	⊟Flagstaff		
		1232	4875.85
		1239	10731.15
Grand Total			**72246.87**

Figure 15-68. *Your PivotTable should look like this*

Note   There are no records for those who have a CustomerID of 1, 6, or 7. This is because there are no records for these customers in the Orders table. (They didn't place an order.)

9. You can now filter by State. Click the down arrow for All in cell B1. Click the plus sign to the left of All. See Figure 15-69. You can click the state whose data you want to view.

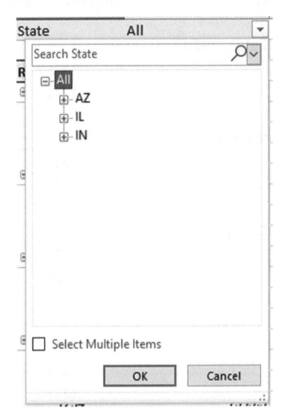

Figure 15-69. *Your PivotTable should look like this*

10. Click AZ. Click the OK button. Only data from Arizona is displayed.

11. Click the Filter icon again. Click IN. Click the OK button. Only data from Indiana is displayed.

12. Click the Filter icon again. Click the check box for **Select Multiple Items**. Check those states whose data you want to view.

Summary

PivotTables provide you with extreme flexibility in what data is to be displayed, how it is to be arranged, and what values you want summarized. It's called a PivotTable because you can pivot your data, switching between rows and columns.

PivotCharts have an advantage over other charts in that they can be filtered.

Small relational databases can be created directly in Excel.

The next chapter deals with using multimedia in a workbook. Multimedia is using different forms of communication. You will use different forms of text, diagrams, pictures, screen captures, audio, and video to communicate in a workbook.

CHAPTER 16

Geography and Stock Data Types

Excel has added two new data types, Geography and Stocks. The way they work is similar. The two data types are located on the Ribbon's Data tab in the Data Type group. See Figure 16-1. Microsoft has plans to add additional data types in the future.

Figure 16-1. *Stocks and Geography Data Types*

Let's first look at Geography. Using the Geography data type, you can gather data about countries, provinces, regions, states, counties, cities, and towns. The information is gathered from Wikipedia and other online sources. The data that you can extract depends on the type of geographic location you have selected. You can gather information from a single geographic location, or you can gather information about a large list of geographic locations all at the same time.

Figure 16-2 displays the name of a small town in Indiana named Hebron in cell A1. When we select the cell and click the Geography button, a question mark will appear to the left of the name. Excel displays this question mark because there are many towns named Hebron and it doesn't know which one you want to use.

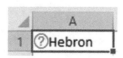

Figure 16-2. *Question mark means the exact location is not known*

© David Slager and Annette Slager 2020
D. Slager and A. Slager, *Essential Excel 2019*, https://doi.org/10.1007/978-1-4842-6209-2_16

At the same time Excel places the question mark in the cell, it shows a Data Selector on the right side of the worksheet. See Figure 16-3. The Data Selector displays the cell address to the left of the name so you can easily identify which cell Excel is questioning. You can click any of these towns to learn more about them. There are more towns named Hebron than what are shown here. To view more towns named Hebron, you can click the magnifying glass or you can click "Show more results" at the bottom of the window. The order in which these appear may be different at different times.

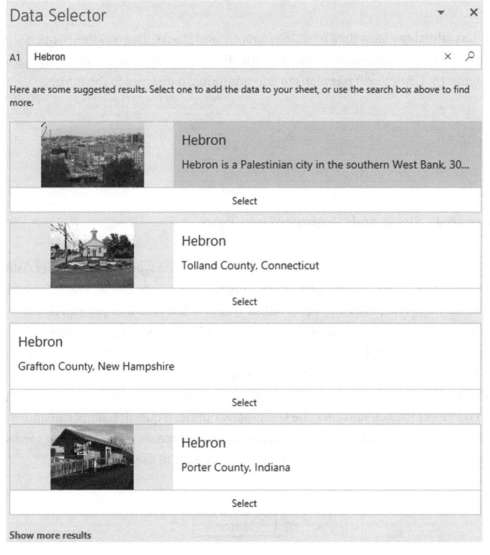

Figure 16-3. *Data Selector*

If you click "Select" below one of the towns, Excel identifies that town as the one you want to gather information about. I will select the Hebron in Porter County, Indiana. If you close the Data Selector without selecting a location, you can click the question mark in the cell at any time to reopen it.

When you have made your selection, Excel places the Geography Data Type Icon 📖 in the cell. When the cell is active, an icon will be displayed to the right of that data. Moving your cursor over the icon, you will see the tool tip "Insert Data Extract data to a cell". See Figure 16-4.

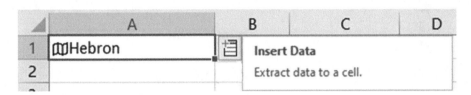

Figure 16-4. *Data Type tool tip*

Clicking the icon will show the options available for the type of location you have entered. See Figure 16-5. The first option is **Admin Division 1**. You may be confused by the word Field at the top of the list. Microsoft has placed it at the top of the list as a heading, not as an option.

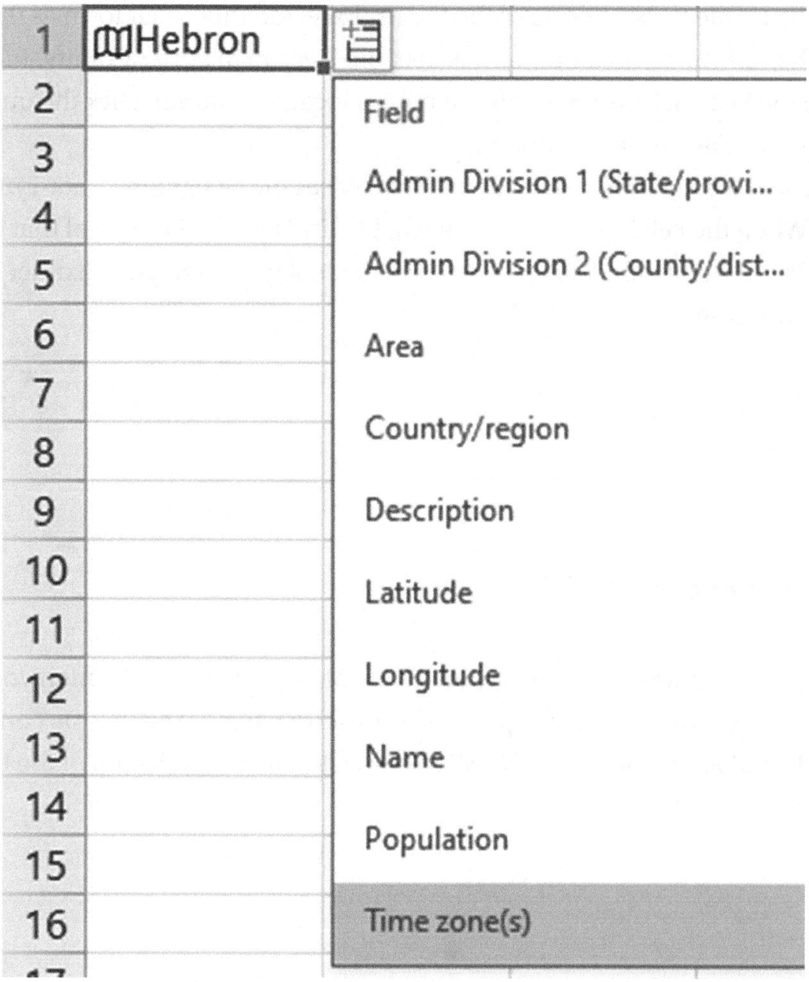

Figure 16-5. Geography Location Options

Selecting the word "population" places the town's population in the cell to the right. Selecting Admin Division 2 places the county name in the first empty cell to the right. See Figure 16-6.

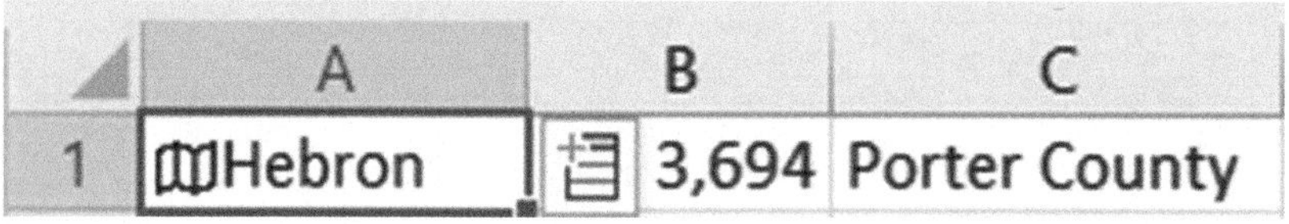

Figure 16-6. Population and Admin Division 2 Options Applied

The Geography Data Type icon shows more information than the Insert Data icon that appears to the right of the cell. Moving your cursor over the Geography Data Type icon will display a tool tip that says "Show Card". Clicking the icon displays the scrollable Card which shows one or more pictures of the location and information about it. Unlike the Insert Data icon, the Card shows the actual data that it is retrieving. For example, the Card doesn't just show Admin Division 1; it shows the actual result—Indiana. If you scroll to the bottom of the Card, you will also see a description of the location. The Card also has clickable links to the location on Wikipedia and additional images. See Figure 16-7. When you move your cursor over the Card's fields, an icon will appear to the right of that field. Clicking the icon will place that information into the first blank cell to the right. For example, in Figure 16-7, clicking the icon to the right of Admin Division, I will place the word Indiana in the first empty cell to the right of Hebron.

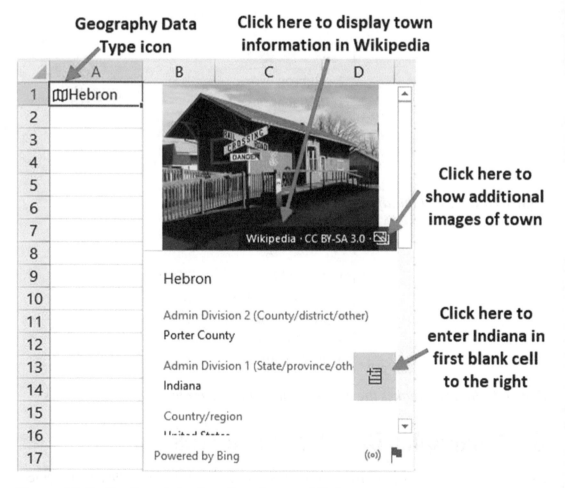

Figure 16-7. *Card contains location data and links*

You can distinguish a city's location by typing the name of the city followed by a comma and then the State, Province, or Region. For example, typing Hebron, Indiana in a cell distinguishes the city you are referring to. When you click the Geography button, a question mark will not appear as Excel will know exactly which Hebron you are referring to. Excel drops the comma and the word Indiana and only leaves the word Hebron in the cell.

Typing this in a cell $\boxed{\text{Hebron, Indiana}}$ and clicking the Geography button returns

this $\boxed{\text{⬚Hebron}}$.

If you selected a range of Indiana cities and towns such as those shown in Figure 16-8 and then clicked the Geography button, Excel would automatically assume that the Cities and Towns you wanted were the ones in Indiana, and it would enter the Geographic Data Type icon instead of the question mark in the cell. See Figure 16-9.

◢	A
1	Hebron
2	Indianapolis
3	Fort Wayne
4	South Bend
5	Vincennes

Figure 16-8. *List of Indiana Cities/Towns*

◢	A
1	⬚Hebron
2	⬚Indianapolis
3	⬚Fort Wayne
4	⬚South Bend
5	⬚Vincennes

Figure 16-9. *Excel makes the assumption that these are Indiana Cities/Towns*

Use Geographic Data Types in a Table

Let's place the states in Figure 16-10 into a table by selecting the cells, clicking the Quick Analysis button, and selecting Tables.

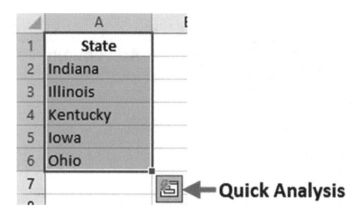

Figure 16-10. *Data to be converted into a Table*

The advantages of using a table are

- A new column of data can be added at one time.

- Column headings are automatically created.

- Because the data is in a table, you can easily filter the data to see only what you want such as only wanting to see states whose median income is greater than 50,000.

- When you add additional locations, they automatically become part of the table.

The states in the table are converted to the Geography Data type by first selecting the cells that contain states and then clicking the Geography button on the Ribbon's Data tab in the Data Types group. Each of the cells will then contain the Geography Data Type icon to the left of the state name. As long as the cursor is in one of the table cells, an icon will appear to the upper right of the table. The tool tip for the icon displays **Add Column**. (When you clicked a Data Type cell that wasn't part of a table, the tool tip for the icon displayed as **Insert Data**.) Clicking the Add Column button displays the options for the location. Notice in Figure 16-11 that the options for States are different than those available for Cities/Towns. Each type of geographic location has its own options.

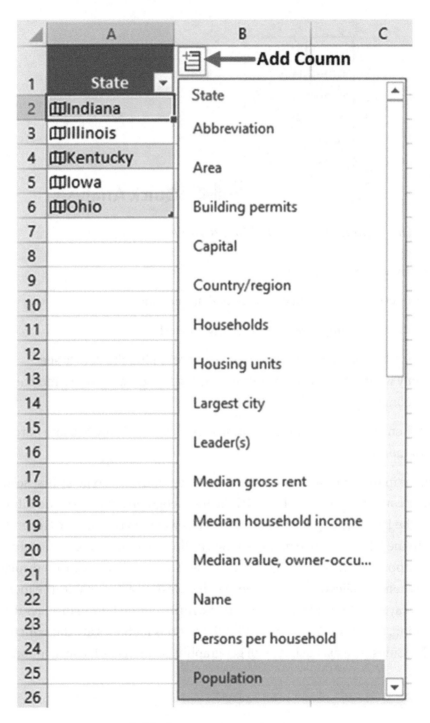

Figure 16-11. *Options available for State Locations*

I will select Population. The Population heading and the population for each of the states appear in the column to the right. See Figure 16-12.

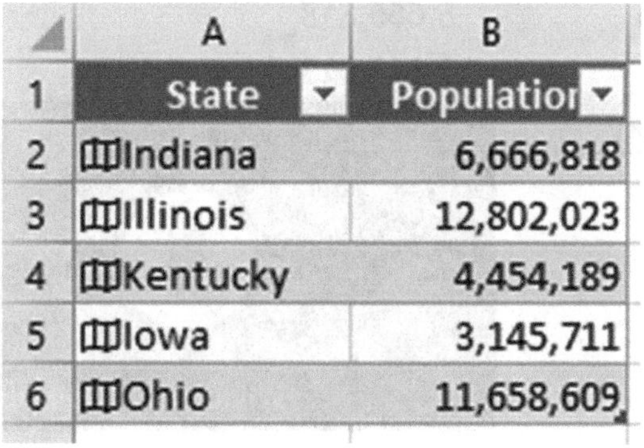

Figure 16-12. *Population for each State*

Click the Geography Data Type icon for Kentucky, and then select capital from the Card. See Figure 16-13.

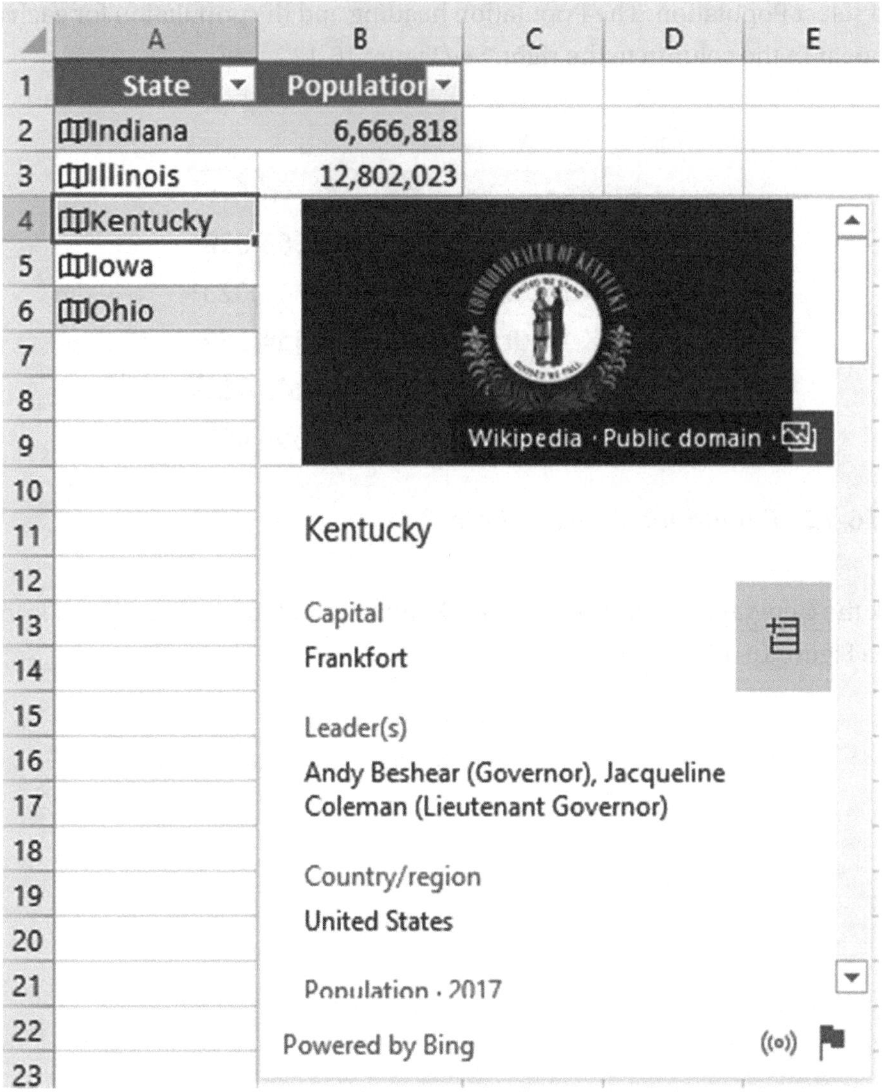

Figure 16-13. *Card for the State of Kentucky*

Even though I selected the capital from the Kentucky Card, Excel entered the capitals from each of the states. See Figure 16-14.

	A	B	C
1	State ▼	Population ▼	Capital ▼
2	🔠Indiana	6,666,818	Indianapolis
3	🔠Illinois	12,802,023	Springfield
4	🔠Kentucky	4,454,189	Frankfort
5	🔠Iowa	3,145,711	Des Moines
6	🔠Ohio	11,658,609	Columbus

Figure 16-14. *Capitals of each State*

If another state is added, it will automatically take on the Geography data type. When Tennessee is entered in row 7, it is automatically assigned a data type of Geography and any column data is automatically filled in. See Figure 16-15.

	A	B	C
1	State ▼	Population ▼	Capital ▼
2	🔠Indiana	6,666,818	Indianapolis
3	🔠Illinois	12,802,023	Springfield
4	🔠Kentucky	4,454,189	Frankfort
5	🔠Iowa	3,145,711	Des Moines
6	🔠Ohio	11,658,609	Columbus
7	🔠Tennessee	6,715,984	Nashville

Figure 16-15. *Data for Tennessee is automatically added*

If the text in cell A7 was changed from Tennessee to Missouri, all the columns of data would automatically change to represent those of Missouri.

Let's say that you want to get the area and population for some cities, states, and countries as shown in Figure 16-16. Since you are using a combination of cities, states, and countries when you click the Add Column button, the field options for all three types display. Cities, states, and countries each have an area and population field, making this possible.

Figure 16-16. *Area and Population for Cities, States, and Countries*

Now you want to know some additional data, such as the Largest City and the percent of Agricultural land. When these fields are added, #FIELD! errors appear in some of the cells. See Figure 16-17. Cities don't provide a field for **Largest City**, and cities and states don't provide a field for **% of Agricultural Land**. This can also occur if there isn't any available data for a particular field in the geographic location you are using. This doesn't necessarily mean there is an error but there just isn't any available data.

Figure 16-17. *#FIELD means no available data*

When I click the Geography Data Type icon to the left of London, the Card is displayed and I can see that Excel assumed that I wanted London, England. What if this isn't the London I wanted? Right-click the cell, and then select Data Type and then Change. See Figure 16-18.

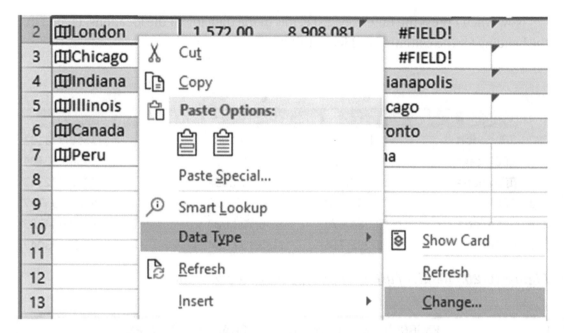

Figure 16-18. *Change the location*

Selecting Change brings up the Data Selector and a different London can be selected.

Using Geographic Location Data in Formulas

Let's say we have these three columns and want to know how many people live in a square km in each geographic location. In cell D2, I typed an equal sign and then clicked cell C2. Excel enters [@Population] to represent the column. I then typed the divide sign and clicked cell B2. See Figure 16-19.

	A	B	C	D
1	Column1 ▼	Area ▼	Population ▼	
2	London	1,572.00	8,908,081	=[@Population]/[@Area]
3	Chicago	589.56	2,704,958	
4	Indiana	94,321	6,666,818	
5	Illinois	149,998	12,802,023	
6	Canada	9,984,670	36,708,083	
7	Peru	1,285,216	32,162,184	

Figure 16-19. *Create formula using Data Type data*

Now when I press the Enter key, Excel fills in the data for all the locations. I typed in the heading for the column. See Figure 16-20.

	A	B	C	D
1	Column1 ▾	Area ▾	Population ▾	# of People in Square KM ▾
2	⊞London	1,572.00	8,908,081	5666.718193
3	⊞Chicago	589.56	2,704,958	4588.10404
4	⊞Indiana	94,321	6,666,818	70.68222347
5	⊞Illinois	149,998	12,802,023	85.34795797
6	⊞Canada	9,984,670	36,708,083	3.676444289
7	⊞Peru	1,285,216	32,162,184	25.02473047

Figure 16-20. *Result of using Data Type data in a formula*

EXERCISE 16-1: USING THE GEOGRAPHY DATA TYPE

Using a Single Data Type Location

1. Create a new workbook. Type Indiana in cell D2 and then press Ctrl + Enter. On the Ribbon's Data tab, in the Data Types group, click Geography. Click the Geography Data Type icon in cell D2. The Data Selector appears to the right of the spreadsheet. There is a town, a county, and a township in Pennsylvania named Indiana. See Figure 16-21.

Figure 16-21. *Data Selector for Indiana*

2. Click the town of Indiana in Indiana County (not the Select under it). You see the data related to this town. Scroll all the way down. You should see a description of the town. This can help you identify what you are looking at. Click the back arrow to return [← | Details] Now, click Select under the town of Indiana in Indiana County. The Geography Data Type icon now appears in the cell. Click the icon. Scroll down until you see Population. Move your cursor over Population. Click the icon to the right. The population shows in cell E2.

3.   Now, let's say that you didn't really mean to get the town; you wanted the county instead. Right-click cell D2 and select Data Type and then click Change. Now, click Select under the County. Notice how the population changed to reflect that the population is now coming from the county.

4.   Select cells D2 and E2. Right-click the selected cells and select Clear Contents.

Type Indiana in cell D3. On the Ribbon's Data tab, in the Data Types group, click Geography. In the Data Selector, type Indiana State. By adding the word State, you have removed any ambiguity and the Data Selector will then only show one result. See Figure 16-22.

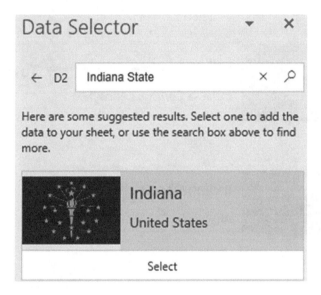

Figure 16-22. *Filter Data Selector results*

There is only one Indiana State and there is only one Indiana County, so if you entered Indiana County, it will only show the one result.

5.   Select cell D2. Right-click the selected cells and select Clear Contents.

Using Data Type Locations in a Table

Enter **State** in cell A1. Type **Indiana** in cell A2. Type **Illinois** in cell A3 and then press Ctrl + Enter. With the cursor still in cell A3, click the Ribbon's Home tab. In the Styles group, click the **Format As Table** command. In the first row of the Medium group, click the second table. Be sure to put a check mark for My table has headers and then click the OK button. See Figure 16-23.

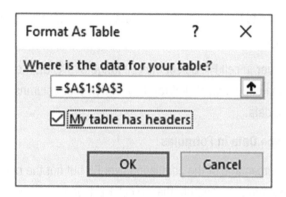

Figure 16-23. *Format As Table with headers*

6. On the Ribbon's Data tab, in the Data Types group, click Geography. The Data Selector doesn't display since both are names of states; Excel assumes that you are referring to states.

7. Type **Kentucky** in cell A4.

8. Click the Geography Data Type icon in cell A2. The Indiana state flag is displayed at the top of the Card. Click the word **Wikipedia**. You are taken to the Wikipedia website for Indiana. Close your browser. Click the Geography icon in cell A2. Click the **More images** icon ▣. Your browser opens with images of Indiana. Close your browser.

9. Click the Geography Data Type icon for one of the three states. It doesn't matter which one you pick. Click Population. Click the Extract icon to the right. The population along with the heading is added for the three states. Expand the column width so that you can see all the data.

10. Click any cell in the table. Click the **Add Column** icon to the right of the Population heading. Click **Median Household Income**.

11. Click the **Add Column** icon. Click **Median Gross Rent**.

12. Click the **Add Column** icon. Click **Population: Age 65+ (%)**.

Adding and Changing Data Types

13. Type Michigan in cell A5. Type Ohio in cell A6. The state data you have selected for the other states is filled in automatically for Michigan and Ohio.

14. Change Michigan to Iowa in cell A5. The data is automatically changed to reflect the state change.

15. Type **65 and Over** in cell F1. Select the cell range C1:F1. On the Ribbon's Home tab in the Alignment group, click Wrap Text. Widen your columns so that you can see all the data.

Using Data Type Data in Formulas

16. You have the percentage of the population over 65 but not the number of people over 65. You can compute that. In cell F2, type the = sign to start the formula. Click cell B2. Type an * for multiplication. Click cell E2. Press Enter.

17. Select the cell range F2:F6. On the Ribbon's Home tab, in the Number group, click the comma. See Figure 16-24.

▲	A	B	C	D	E	F
1	State ▾	Population ▾	Median household income ▾	Median gross rent ▾	Population: Age 65+ (% ▾	65 and Over ▾
2	⊞Indiana	6,666,818	$ 49,255	$ 745	14.6%	973,355.43
3	⊞Illinois	12,802,023	$ 57,574	$ 907	14.2%	1,817,887.27
4	⊞Kentucky	4,454,189	$ 43,740	$ 675	15.2%	677,036.73
5	⊞Iowa	3,145,711	$ 53,183	$ 697	16.1%	506,459.47
6	⊞Ohio	11,658,609	$ 49,429	$ 730	15.9%	1,853,718.83

Figure 16-24. *Computed field 65 and Over with comma formatting*

Let's say you want to know the median % of household income that is being spent on rent in the state of Indiana and Illinois. You don't need to create a new Geography Data type because you can use the ones you already created.

18. Type **Indiana** in cell A14 and **Illinois** in cell A15. Type **% of Income Spent on Rent** in cell B13. Apply Wrap Text to cell B13.

19. Type an = sign in cell B14. You can either click cell A2 or enter its cell address. When you do, you will get all the field options for the state. See Figure 16-25.

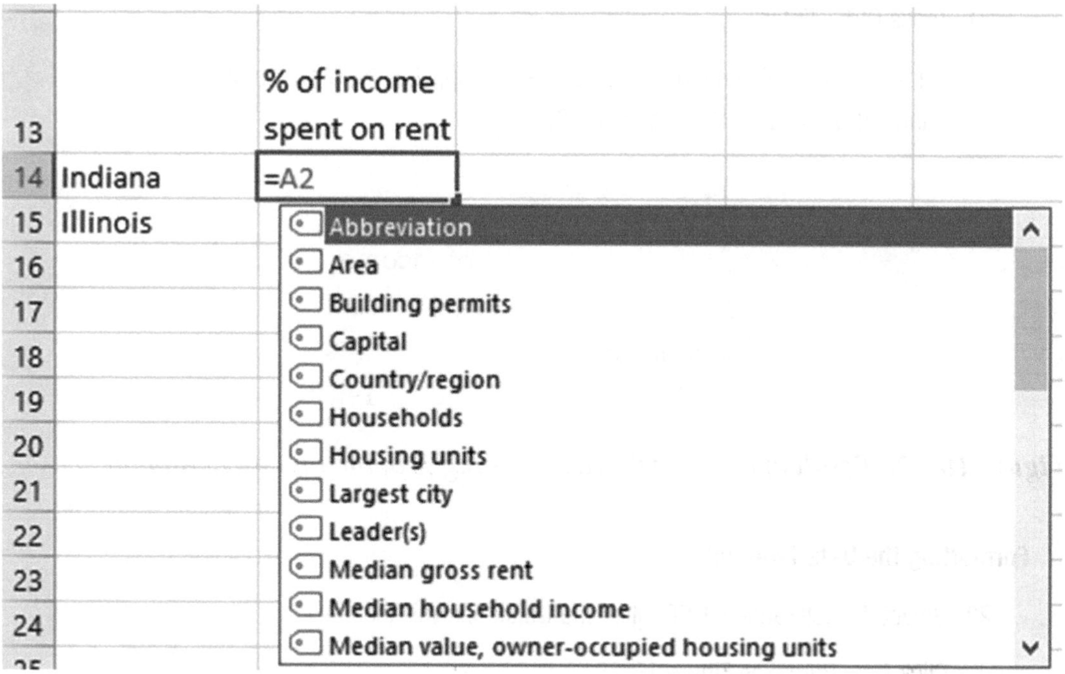

		% of income spent on rent		
13				
14	Indiana	=A2		
15	Illinois	⬭ Abbreviation		⌃
16		⬭ Area		
17		⬭ Building permits		
18		⬭ Capital		
19		⬭ Country/region		
20		⬭ Households		
21		⬭ Housing units		
22		⬭ Largest city		
23		⬭ Leader(s)		
24		⬭ Median gross rent		
		⬭ Median household income		
		⬭ Median value, owner-occupied housing units		⌄

Figure 16-25. *Using available Data Type fields in a formula*

20. Double-click **Median gross rent** to select it.

   Type *** 12** (this is rent per month so you need to multiply it by 12 to get the annual rent).

   Type a divide sign.

   Type **A2** to bring up the field options.

   Double-click **Median household income**. The complete formula is shown in Figure 16-26.

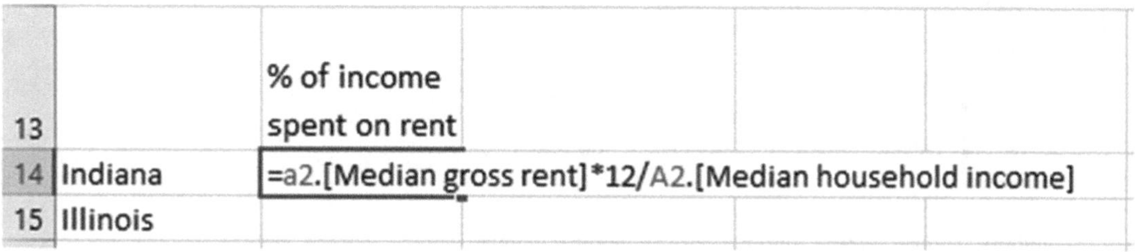

		% of income spent on rent			
13					
14	Indiana	=a2.[Median gross rent]*12/A2.[Median household income]			
15	Illinois				

Figure 16-26. *Formula for determining % of income spent on rent*

21. Press Ctrl + Enter.

22. On the Ribbon's Home tab, in the Number group, click %. Drag the AutoFill handle down to B15. See Figure 16-27.

		% of income
13		spent on rent
14	Indiana	18%
15	Illinois	19%

Figure 16-27. *Result of formula for determining % of income spent on rent*

Formatting the Data Type Data

23. Select the cell range B1:B6. Click the Quick Analysis icon.

Click Data Bars. See Figure 16-28.

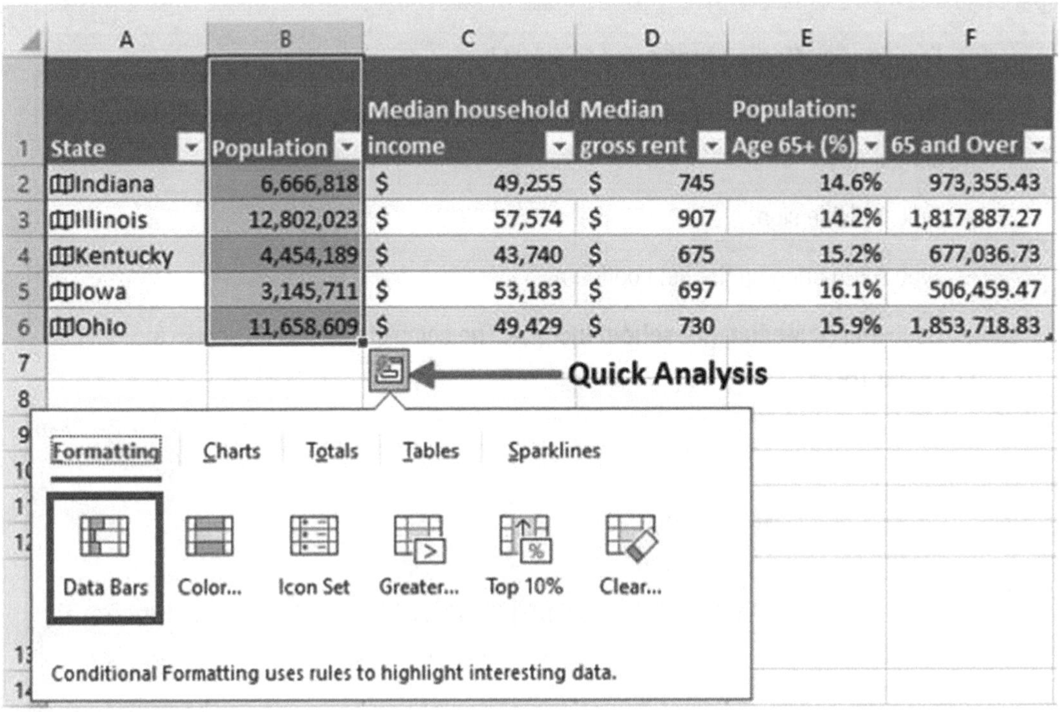

Figure 16-28. *Create Data Bars for Population*

Using the Data Type Data in a Map

24. Select the cell range A1:B6.

    On the Ribbon's Insert tab, in the Charts group, click Maps.

    Click the Filled Map.

    Click Chart Title. Type **Population**.

    Charts have two context menus Design and Format under Chart Tools. Click the Design tab. In the Chart Styles group, click Change Colors. Select whatever color you want. Make any other changes you want. See Figure 16-29.

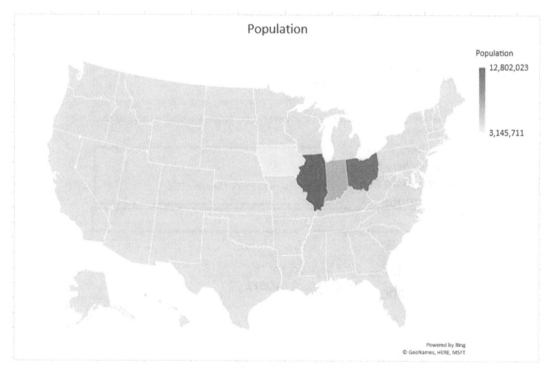

Figure 16-29. *Map for State Population*

Excel's Geography Data Type can be used to enter data into cells by selecting fields that have data pertaining to countries, provinces, regions, states, counties, cities, and towns. Now that you are familiar with the Geography Data Type, you can apply those same skills to the Stocks Data Type.

Stocks

Excel's Stock Data Types work the same way as Geographic Data Types. In order to get information from a Stock data type, you will need to enter the Stock's ticker symbol, company name, or fund name.

The options for a stock are different than those from a fund. Let's first look at the results of entering a company name. If you enter Adobe into a cell, a question mark displays to the left of the company name and the Data Selector appears to the right of the spreadsheet because there are many different Adobe choices. The first Adobe option shows that this stock is on the Nasdaq Stock Exchange and its ticker symbol is ADBE. See Figure 16-30. Just like with the Geography Data Type, you can click the information to view the Data Card or you can click Select to accept it.

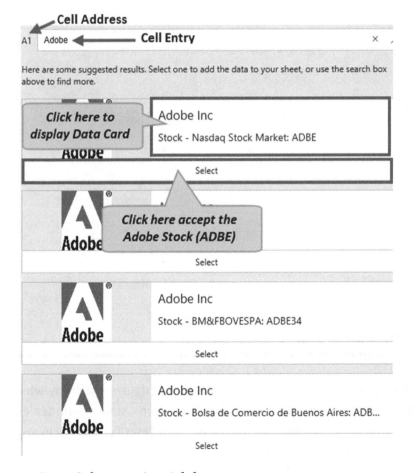

Figure 16-30. *Data Selector using Adobe*

When you click Select, the Stock Data Type icon (🏛) appears to the left of the company name, while the abbreviation for the Exchange and the ticker symbol appear to the right of the company name. See Figure 16-31.

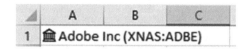

Figure 16-31. *Stock Data Type icon and ticker symbol added to cell*

If you know the ticker symbol, you should enter it rather than the company name. This will probably narrow the number of options shown in the Data Selector. It may even limit the number to one, in which case Excel will know exactly which stock or fund you are referring; therefore, it will not enter a question mark in the cell nor will it display the Data Selector. There are three stocks with the ADBE ticker symbol because they are on three different stock exchanges: one on the Nasdaq Exchange, one on the Swiss Exchange, and the other on the Wiener Boerse Exchange in Vienna, Austria. See Figure 16-32.

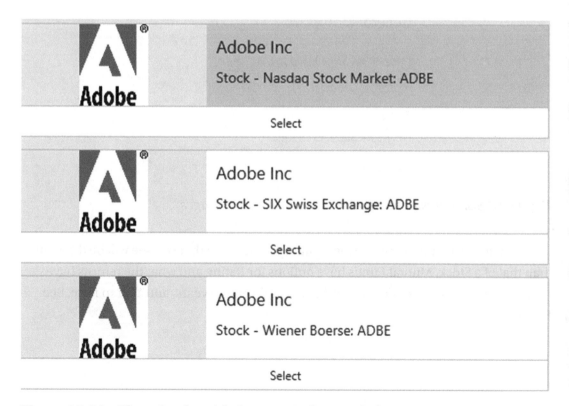

Figure 16-32. *Three Stocks with the same ticker symbol*

The Data Card for Adobe displays their company logo, but Data Cards for other companies may not. Most company Data Cards display the word Wikipedia and have an image icon on their company logo. See Figure 16-33. Clicking the word Wikipedia displays information about that company in Wikipedia. Clicking the image icon displays images of that company.

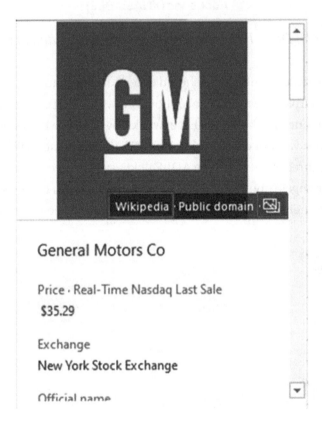

Figure 16-33. *Links to company information and images*

If you make a mutual fund a Stock data type, the fields that you see will be different than that of a Stock. Mutual Funds have options for Rating and what the returns have been for the last week, month, 3 months, 1 year, 3 years, 5 years, and Year to Date. See Figure 16-34.

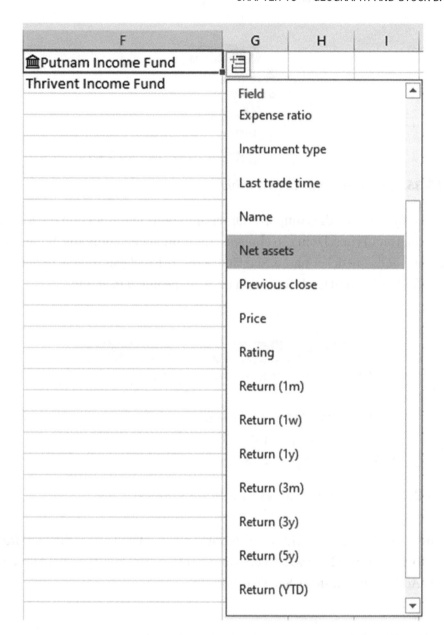

Figure 16-34. *Mutual Fund Fields*

Let's start with some company names in a table. See Figure 16-35.

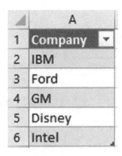

Figure 16-35. *Company names in a table*

I will start by selecting the company names and then click Stocks on the Ribbon's Data tab in the Data Types. Doing so changes the abbreviated company names to their full names and adds the ticker name. See Figure 16-36. Excel was unable to determine what to use for Ford so it placed a question mark to the left of its name.

A
1   Company ▼
2   🏛International Business Machines Corp (XNYS:IBM)
3   ⑦Ford
4   🏛General Motors Co (XNYS:GM)
5   🏛Walt Disney Co (XNYS:DIS)
6   🏛Intel Corp (XNAS:INTC)

Figure 16-36. *Full company names with ticker symbols*

Clicking the question mark displays the Data Selector. I will select the Ford from New York Stock Exchange F. Now, we can start selecting our Data Type field options by clicking the Add Column button. See Figure 16-37.

A		B	C	D	E
Company ▼	📋		Add Column		
🏛International Business Machines Corp (XNYS:IBM)			Extract a field to a table column.		
🏛Ford Motor Co (XNYS:F)					
🏛General Motors Co (XNYS:GM)					
🏛Walt Disney Co (XNYS:DIS)					
🏛Intel Corp (XNAS:INTC)					

Figure 16-37. *Add a field by clicking the Add Column button*

Figure 16-38 displays the field selections that have been made for price, previous close, 52 week high, 52 week low, change, and change %. Some of the values in the Change and Change % columns are in parenthesis. Values in parenthesis are negative values.

	A	B	C	D	E	F	G
			Previous	52 week	52 week		
1	Company	Price	close	high	low	Change	Change (%)
2	🏛International Business Machines Cor	$150.86	$ 151.10	$158.75	$126.85	$ (0.24)	$ (0.00)
3	🏛Ford Motor Co (XNYS:F)	$ 8.00	$ 8.06	$ 10.56	$ 8.00	$ (0.06)	$ (0.01)
4	🏛General Motors Co (XNYS:GM)	$ 34.92	$ 34.41	$ 41.90	$ 32.97	$ 0.51	$ 0.01
5	🏛Walt Disney Co (XNYS:DIS)	$141.30	$ 139.14	$153.41	$107.32	$ 2.16	$ 0.02
6	🏛Intel Corp (XNAS:INTC)	$ 67.11	$ 66.14	$ 69.29	$ 42.86	$ 0.97	$ 0.01

Figure 16-38. *Field selections made after clicking the Add Column button*

Now we can make some computations to compute the current value of each stock, the total current value of our stocks, the value gained or lost on each stock, and the total gain or loss for the day. We could place these computations at the end of the table, but let's place them in a separate area starting in cell A11. In the first two columns, we have the company names and the number of shares we have in each of those companies. See Figure 16-39.

	Company	Shares	Previous Close	Current Value	Gain/Loss
11					
12	IBM	400			
13	Ford	800			
14	GM	1000			
15	Disney	200			
16	Intel	200			
17	**** Totals				

Figure 16-39. *Setup for working with Stock computations*

Our first computation will be to determine the value of each of our stocks at the previous close. We will start the formula in cell C12 with the equal sign and then click cell C2. We will add the multiplication sign and then click cell B12. See Figure 16-40.

Company	Shares	Previous Close	Current Value	Gain/Loss
11				
12 IBM	400	=C2*B12		
13 Ford	800			
14 GM	1000			
15 Disney	200			
16 Intel	200			
17 **** Totals				

Figure 16-40. *Formula for computing Stock values at the Previous Close*

We'll drag the AutoFill handle down to compute the Previous Close for the other companies. See Figure 16-41.

Company	Shares	Previous Close	Current Value	Gain/Loss
11				
12 IBM	400	$ 60,440.00		
13 Ford	800	$ 6,448.00		
14 GM	1000	$ 34,410.00		
15 Disney	200	$ 27,828.00		
16 Intel	200	$ 13,228.00		
17 **** Totals				

Figure 16-41. *Drag AutoFill handle down to copy formula to other Stocks*

The Current Value is the **Price times the Number of Shares** so we will enter the formula =B2*B12 in cell D12. We'll then drag the AutoFill handle down to cell D16.

The Gain/Loss is the Change times the Number of Shares so we will enter the formula =F2*B12 in cell E12. We'll then drag the AutoFill handle down to cell E16. The results are shown in Figure 16-42.

	Company	Shares	Previous Close	Current Value	Gain/Loss
11					
12	IBM	400	$ 60,440.00	$60,344.00	$ (96.00)
13	Ford	800	$ 6,448.00	$ 6,400.00	$ (48.00)
14	GM	1000	$ 34,410.00	$34,920.00	$ 510.00
15	Disney	200	$ 27,828.00	$28,260.00	$ 432.00
16	Intel	200	$ 13,228.00	$13,422.00	$ 194.00
17	**** Totals				

Figure 16-42. *Results for Previous Close, Current Value, and Gain/Loss*

Now the totals can be added by selecting the range B17:E17 and then clicking AutoSum on the Home tab. Apply formatting to the totals and the result is shown in Figure 16-43. We can see that the gain for the day was $992.

	Company	Shares	Previous Close	Current Value	Gain/Loss
11					
12	IBM	400	$ 60,440.00	$ 60,344.00	$ (96.00)
13	Ford	800	$ 6,448.00	$ 6,400.00	$ (48.00)
14	GM	1000	$ 34,410.00	$ 34,920.00	$ 510.00
15	Disney	200	$ 27,828.00	$ 28,260.00	$ 432.00
16	Intel	200	$ 13,228.00	$ 13,422.00	$ 194.00
17	**** Totals	2600	$142,354.00	$143,346.00	$ 992.00

Figure 16-43. *AutoSum used to compute Totals in row 17*

The stock option values in our table are not static. If the stock market is open, some of the values will change throughout the day such as the price, the dollar amount change, and the percent amount change. If the stock market is open, data can be refreshed to get current values (there is a 15-minute delay). To update the data, click the Ribbon's Data tab and then click **Refresh All** in the Queries & Connections group. When the values in the table are updated, then any computations that use that data are also updated.

Figure 16-44 shows the results of clicking Refresh All the following day after entering the table.

Company	Price	Previous close	52 week hi	52 week low	Change	Change (%)
🏛International Bus	$ 151.22	$ 150.86	$ 158.75	$ 126.85	$ 0.36	$ 0.00
🏛Ford Motor Co (XI	$ 8.03	$ 8.00	$ 10.56	$ 7.99	$ 0.03	$ 0.00
🏛General Motors C	$ 35.29	$ 34.92	$ 41.90	$ 32.97	$ 0.37	$ 0.01
🏛Walt Disney Co (X	$ 140.37	$ 141.30	$ 153.41	$ 107.32	$ (0.93)	$ (0.01)
🏛Intel Corp (XNAS:	$ 65.45	$ 67.11	$ 69.29	$ 42.86	$ (1.66)	$ (0.02)

Company	Shares	Previous Close	Current Value	Gain/Loss		
IBM	400	$ 60,344.00	$ 60,488.00	$ 144.00		
Ford	800	$ 6,400.00	$ 6,424.00	$ 24.00		
GM	1000	$ 34,920.00	$ 35,290.00	$ 370.00		
Disney	200	$ 28,260.00	$ 28,074.00	$ (186.00)		
Intel	200	$ 13,422.00	$ 13,090.00	$ (332.00)		
**** Totals	2600	$143,346.00	$143,366.00	$ 20.00		

Figure 16-44. *Stock values automatically updated after clicking Refresh All*

EXERCISE 16-2: USING THE STOCK DATA TYPE

Create a Table for Mutual Funds

In this exercise, you will create a table of mutual fund data using the Stock Data Type.

1. Create a new worksheet.

2. Enter the following data that contains the names of mutual funds or their ticker symbol. Bold the heading name Company. See Figure 16-45.

⬛	A
1	**Company**
2	Putnam Income Fund
3	Thrivent Income Fund;S
4	TIBDX
5	TIGRX
6	TRBCX

Figure 16-45. *Enter this data into a worksheet*

3. Select the cell range A1:A6 and then click the Quick Analysis button. Select the tab Tables and then click the Table icon. See Figure 16-46.

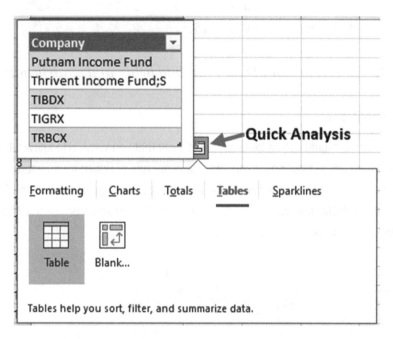

Figure 16-46. *Convert data to a Table*

4. Click the Ribbon's Data Type, and in the Data Types group, click Stocks. The ticker symbols are converted to the mutual fund names, and the Stock Data Type symbols appear to the left of the funds.

5. Click the Add Column button that appears to the upper right of your table. Select Return (1y). Click the Add Column button again and select Return (3y). Click the Add Column button once more and select Return (5y).

6. You will need to apply wrap text to make your column headings readable.
    Select cell range A1:E1. You will add data to column E later. Right-click one of
    the selected cells. Select Format Cells. In the Format Cells window, click the
    Alignment tab and select Wrap Text. Click the OK button. See Figure 16-47.

	A	B	C	D
1	Company ▾	Return (1y) ▾	Return (3y) ▾	Return (5y) ▾
2	🏛Putnam Income I	11.48%	5.55%	3.48%
3	🏛Thrivent Income	13.09%	6.26%	4.70%
4	🏛TIAA-CREF Bond	9.81%	4.92%	3.84%
5	🏛TIAA-CREF Growt	23.55%	14.16%	11.39%
6	🏛T Rowe Price Blu	25.96%	22.33%	16.21%

Figure 16-47. *Convert to Stock Data Type, add fields, and wrap heading text*

Create a Computation

7. There is a 1-month option for returns. Let's say that you wanted to know what
    the rate of return would be if you had that same rate of return for an entire
    year. Type an equal sign in cell E2. Click the Stock Data Type symbol in cell A2.
    Double-click **Return (1m)**. Enter the * for multiplication and then enter 12. Your
    formula should be as shown in Figure 16-48.

	A	B	C	D	E	F	G
1	Company ▾	Return (1y) ▾	Return (3y) ▾	Return (5y) ▾			
2	🏛Putnam Income I	11.48%	5.55%	3.48%	=[@Company].[Return (1m)]*12		
3	🏛Thrivent Income	13.09%	6.26%	4.70%			
4	🏛TIAA-CREF Bond	9.81%	4.92%	3.84%			
5	🏛TIAA-CREF Growt	23.55%	14.16%	11.39%			
6	🏛T Rowe Price Blu	25.96%	22.33%	16.21%			

Figure 16-48. *Formula for determining 1 Year of Return based on the same rate of
the Current 1 Month Return*

8. Press the Enter key, and the data for the other mutual funds is automatically filled in. You can see that this month has been a great month. If the fund would stay at this rate for the entire year, the T Rowe Price Blue Chip would have an increase of 62.75% for the year. Change the column heading for column E to **Adjusted 1 Month to Annual**. See Figure 16-49.

	A	B	C	D	E
1	Company ▼	Return (1y) ▼	Return (3y) ▼	Return (5y) ▼	Adjusted 1 Month to Annual ▼
2	🏛Putnam Income I	11.48%	5.55%	3.48%	18.33%
3	🏛Thrivent Income	13.09%	6.26%	4.70%	20.58%
4	🏛TIAA-CREF Bond	9.81%	4.92%	3.84%	16.32%
5	🏛TIAA-CREF Growt	23.55%	14.16%	11.39%	33.44%
6	🏛T Rowe Price Blu	25.96%	22.33%	16.21%	62.75%

Figure 16-49. *Column E with copied down formula and added heading*

Note   You can change a Data Type back to text by right-clicking the cell and then selecting Data Type ➤ Convert to Text. You would only do this if you wanted to have static data. If you do this, you are removing the online connection to the data. You can't refresh the data, and any cells that contained formulas that used the data will contain a #FIELD! error.

Summary

Microsoft has created two new Data Types: Geography and Stocks, and there will probably be more data types to come. The Geography Data Type has field data pertaining to countries, provinces, regions, states, counties, cities, and towns. The Stock Data Type has field data pertaining to Stocks, Bonds, and Mutual Funds. The two Data Types work the same way. You can display just the data you want from a list of options, or the data can be used in formulas. If Excel can't recognize the geographic location or stock you entered, a question mark will appear in the cell and the Data Selector will display with a message stating that there are no results for that location. A question mark will also appear in the cell if there are multiple locations or stocks with the same name. You will then have to select the geographic location or stock that you are referring to. Clicking a location in the Data Selector or clicking the Data Type icon in the cell will display a data card that contains not only the type of data that is available but also the data itself. The Add Column feature displays the type of data that is available but not the data itself. You can create formulas that use the data type by clicking the cell with the data type and selecting the field option you want to use.

CHAPTER 17

Enhancing Workbooks with Multimedia

Multimedia includes all the different types of media. You have been using text media, but you can add images, shapes, clip art, videos, and audio to your spreadsheets. Multimedia can help you capture your audience's attention by increasing the appeal of your spreadsheets. SmartArt provides a variety of organizational and other charts that you can alter to fit your needs. You can grab a screen capture from a website or an Excel example and then place that capture in the current worksheet.

After reading and working through this chapter, you should be able to

- Add and edit images

- Use WordArt

- Add and change shapes

- Grab an entire window or a portion of it and make it an image

- Use SmartArt

- Add video

- Use audio

799

© David Slager and Annette Slager 2020
D. Slager and A. Slager, *Essential Excel 2019*, https://doi.org/10.1007/978-1-4842-6209-2_17

Adding Pictures to the Worksheet

You can add pictures to your worksheet from either your computer or the Internet. There are two buttons in the Insert tab's Illustrations group for accessing pictures. The Pictures button is for getting pictures from your computer, and the Online Pictures button is for getting pictures from the Internet. See Figure 17-1.

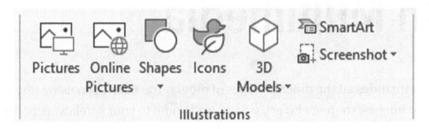

Figure 17-1. *Ribbon's Illustrations group*

If you click the Pictures button, Excel will display the Insert Picture dialog box from which you can browse to locate images on any drives directly connected to your computer or through a network.

If you click the Online Pictures button, you will see a window that displays categories of images. See Figure 17-2. You can click one of the categories and then select an image from the category, or you can enter a search word in the Search Bing text box. Bing is Microsoft's search engine. If you want to access an image from your OneDrive, you can click the OneDrive button.

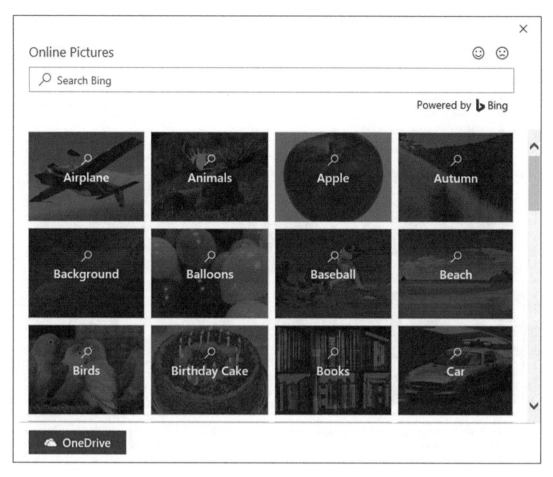

Figure 17-2. Insert Picture window

Dog was entered for the search word in Figure 17-3. If you want to use an image that is licensed under Create Commons, be sure the check box is selected. If you remove the check mark, you will see more images, but you have to pay to use some of those images. Click an image and then click the Insert button to paste it into your worksheet.

Figure 17-3. *The word dog was entered for the search*

You can filter what images appear by clicking the filter button ▽. Figure 17-4 shows the available filters.

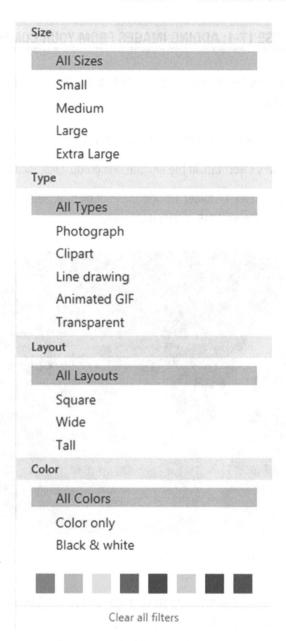

Figure 17-4. *Available filters*

The filter button doesn't apply a filter to the pictures you see; rather, it applies a filter to available images on the Internet. If you apply a filter such as red, you will see images of dogs that are red such as Clifford. You can combine filters. If you wanted to see medium-sized black and white photographs, you could select those three filters.

EXERCISE 17-1: ADDING IMAGES FROM YOUR COMPUTER

In this exercise, you will explore various ways of altering images, adding borders to an image, and applying picture styles.

1. Open workbook Chapter 17. Create a new worksheet.

2. Click the Ribbon's Insert tab. In the Illustrations group, click Pictures.

3. Browse to the Chapter 17 folder. Locate the Pilot picture. Select it and then click the Insert button. Figure 17-5 shows the picture.

Figure 17-5. *Pilot picture*

4. When you have an image selected on your worksheet, the Ribbon will have a Format tab. The Format tab has all the options for altering, sizing, coloring, rotating, cropping, applying effects, adding borders, and so on to your image. Click the Image. Click the Ribbon's Format tab.

5. In the Adjust group, click Corrections. The corrections gallery shows thumbnails of your image with various levels of sharpness, brightness, and contrast. See Figure 17-6. The image in the center of the Sharpen/Soften group and in the center of the Brightness/Contrast group has a border around it. The images with borders around them represent the way your image looks currently.

Move your cursor over the thumbnails while observing the effects they have on the selected image on the worksheet. As you move farther to the right in the Brightness/Contrast group, the brightness increases. As you move farther down in the group, the contrast increases. If you want to apply one of the effects to your selected image, click the thumbnail. If you decide you don't like the effect, you can undo it by pressing Ctrl + Z.

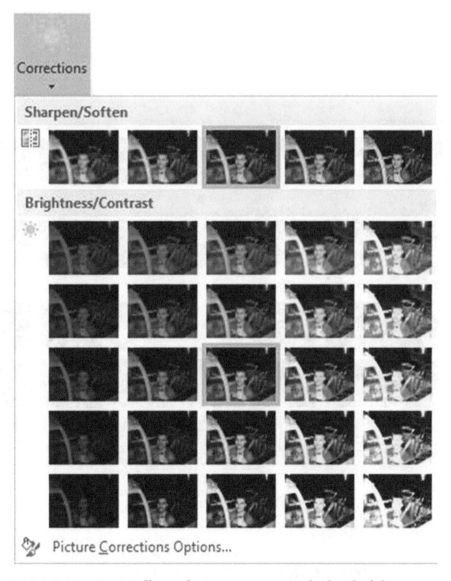

Figure 17-6. *Corrections gallery where you can vary the level of sharpness, brightness, and contrast*

In the Adjust group, there is a Reset Picture button that lets you reset either just the picture or both the picture and its size back to the way it was when you inserted it.

6. In the Adjust group, click the Color button. See Figure 17-7. The adjustments here affect the saturation, tone, and overall color of the image. Move your cursor over the thumbnails while observing the effects they have on the selected image on the worksheet.

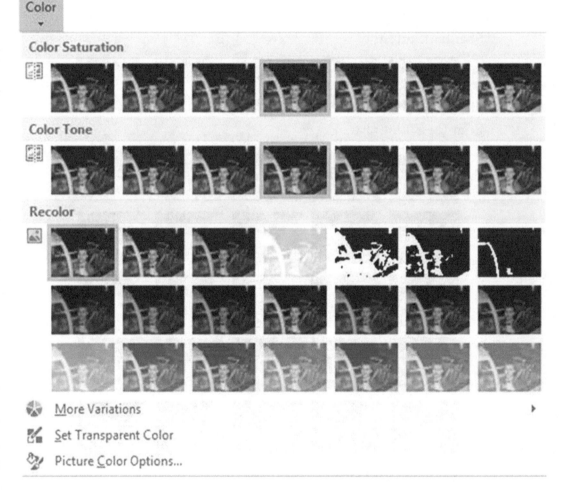

Figure 17-7. *Adjust overall color, saturation, and tone of the image*

7. In the Adjust group, click the Artistic Effects button. Move your cursor over the thumbnails while observing the effects they have on the selected image on the worksheet. Figure 17-8 shows the special effects, pencil sketch and photocopy, applied to the image.

Figure 17-8. *Special effects: pencil sketch and photocopy*

8. The best way to create a dramatic effect without altering the appearance of your picture is to use one of the styles from the Picture Styles Gallery. See Figure 17-9.

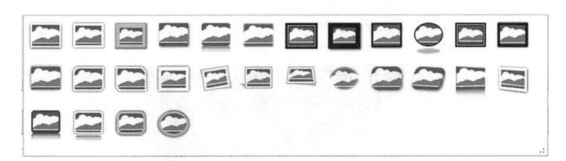

Figure 17-9. *Picture Styles Gallery*

9. Move your cursor over the styles while observing the effects they have on the selected image on the worksheet.

    Figure 17-10 shows some of the various ways that Picture Styles can display your image.

Figure 17-10. *Various ways that Picture Styles can display your image*

10. In the Picture Styles group, click the down arrow for Picture Border button.
    Select a red color and change the weight of the border to 3 pt. See Figure 17-11.
    Press Ctrl + Z to remove the border.

Figure 17-11. *Red border with weight of 3 pt*

11. Click Picture Effects. There are seven groups of various effects. Move your cursor over the various Picture Effects while observing their effect on the selected image on the worksheet. The picture in Figure 17-12 has a glow effect.

Figure 17-12. *Glow effect*

You have learned how to add images to your worksheet from your saved image files on your computer. What if you wanted to add an image that you see on a website or what if you wanted to write training documentation and needed an image from Excel or Word or some other application? Excel has a feature called **Screenshot** that allows you to grab those images.

Using Screenshot

Use Screenshot to grab an entire window or a portion of it and turn it into an image. The Screenshot button is located on the Insert tab in the Illustrations group. Clicking the Screenshot button opens a thumbnail view of all available windows. See Figure 17-13. Clicking one of these thumbnails captures a full-size image of that window.

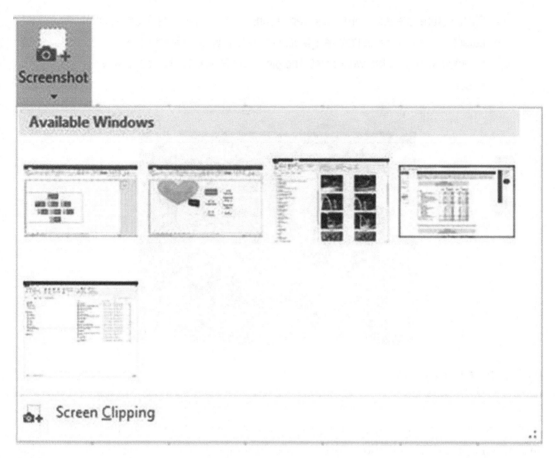

Figure 17-13. *Screenshot can capture one of the open windows on your computer*

If you only want to grab a portion of a window, you will need to open that window prior to clicking Screen Clipping. After clicking Screen Clipping, the Excel program minimizes, the window that you are capturing from dims, and the cursor turns into a crosshair. As you drag your mouse, the area that will be included in your image brightens. When you have finished selecting the area you want, the image appears in your worksheet.

EXERCISE 17-2: GRABBING AN IMAGE WITH SCREENSHOT

In this exercise, you will grab a portion of a website which will then be stored in your worksheet. You will apply a picture style to the captured image.

1. Click the Screenshot worksheet.

2. Close or minimize all open windows except Excel.

3. Click the hyperlink in cell A1.

4. Scroll down on the web page until the entire Income Statement shown in Figure 17-14 is visible.

	A	B	C	D	E
1		Professor Office			
2		Vertical Analysis INCOME STATEMENT			
3		For Years Ended December 31, 2014 and 2013			
4					
5			2014		2013
6		Amount	% of Net Sales	Amount	% of Net Sales
7	Revenue				
8	Sales	$ 225,000	101.74% $	195,000	101.43%
9	Less: Sales returns and allowances	3,840	1.74%	2,750	1.43%
10	Net Sales	$ 221,160	100.00% $	192,250	100.00%
11	Cost of goods sold				
12	Merchandise inventory, January 1	$ 66,000	29.84% $	72,000	37.45%
13	Purchases less returns and allowances	98,000	44.31%	80,000	41.61%
14	Total Cost of merchandise available for sale	$ 164,000	74.15% $	152,000	79.06%
15	Less: Merchandise inventory, March 31	60,225	27.23%	61,000	31.73%
16	Cost of goods sold	$ 103,775	46.92% $	91,000	47.33%
17	Gross Profit	$ 117,385	53.08% $	101,250	52.67%
18	Operating expenses				
19	Delivery expenses	$ 467	0.21% $	200	0.10%
20	Depreciation expense-equipment	3,600	1.63%	3,600	1.87%
21	Payroll taxes expense	388	0.18%	295	0.15%
22	Salary expense	10,739	4.86%	10,500	5.46%
23	Supplies expense	204	0.09%	300	0.16%
24	Telephone expense	175	0.08%	210	0.11%
25	Utilities expense	1,907	0.86%	1,500	0.78%
26	Total operating expenses	$ 17,480	7.90% $	16,605	8.64%
27	Net Income before federal income tax	$ 99,905	45.17% $	84,645	44.03%
28	Federal income tax	$ 12,123	5.48% $	11,800	6.14%
29	Net income after federal income tax	$ 87,782	39.69% $	72,845	37.89%

Figure 17-14. *Window that will be captured with Screenshot*

5. If the web page is maximized, shrink the size of the window so that you can see the Income Statement in your web browser and the Insert tab in Excel at the same time.

6. Click the Ribbon's Insert tab. In the Illustrations group, click the Screenshot button.

7. Click Screen Clipping at the bottom of the window. The Excel program becomes hidden while getting the screen clipping.

8. Using the crosshair, drag across the income statement. As you drag across the income statement, the area that will be included in the captured image becomes highlighted. See Figure 17-15. When you release the mouse, the highlighted area becomes an image that is stored on your worksheet.

	A	B	C	D	E
1		Professor Office			
2		Vertical Analysis INCOME STATEMENT			
3		For Years Ended December 31, 2014 and 2013			
4					
5			2014		2013
6		Amount	% of Net Sales	Amount	% of Net Sales
7	Revenue				
8	Sales	$ 225,000	101.74%	$ 195,000	101.43%
9	Less: Sales returns and allowances	3,840	1.74%	2,750	1.43%
10	Net Sales	$ 221,160	100.00%	$ 192,250	100.00%
11	Cost of goods sold				
12	Merchandise inventory, January 1	$ 66,000	29.84%	$ 72,000	37.45%
13	Purchases less returns and allowances	98,000	44.31%	80,000	41.61%
14	Total Cost of merchandise available for sale	$ 164,000	74.15%	$ 152,000	79.06%
15	Less: Merchandise inventory, March 31	60,225	27.23%	61,000	31.73%
16	Cost of goods sold	$ 103,775	46.92%	$ 91,000	47.33%
17	Gross Profit	$ 117,385	53.08%	$ 101,250	52.67%
18	Operating expenses				
19	Delivery expenses	$ 467	0.21%	$ 200	0.10%
20	Depreciation expense-equipment	3,600	1.63%	3,600	1.87%
21	Payroll taxes expense	388	0.18%	295	0.15%
22	Salary expense	10,739	4.86%	10,500	5.46%
23	Supplies expense	204	0.09%	300	0.16%
24	Telephone expense	175	0.08%	210	0.11%
25	Utilities expense	1,907	0.86%	1,500	0.78%
26	Total operating expenses	$ 17,480	7.90%	$ 16,605	8.64%
27	Net Income before federal income tax	$ 99,905	45.17%	$ 84,645	44.03%
28	Federal income tax	$ 12,123	5.48%	$ 11,800	6.14%
29	Net income after federal income tax	$ 87,782	39.69%	$ 72,845	37.89%

Figure 17-15. As you drag across the income statement, the area that will be included in the captured image becomes highlighted

The screen clipping is an image, and therefore all the functions that can be applied to an image can be applied to the screen clipping.

9.   Maximize your Excel window. Move your captured image to the center of your worksheet window.

10. Click the Ribbon's Format tab. In the Picture's Styles group, select **Drop Shadow Rectangle** from the Styles gallery. This will add a shadow to the bottom and right side of the image.

Working with WordArt

WordArt is decorative text that you have far more control over than the normal text. You can warp WordArt text, make it follow a path, change its shape, give it a 3-D look, change its font, change its background, and so on.

EXERCISE 17-3: USING WORDART

In this exercise, you will explore various ways of creating and altering WordArt. Create a new worksheet.

1.   Create a new worksheet.

2.   Click the Ribbon's Insert tab. In the Text group, click the WordArt button. See Figure 17-16.

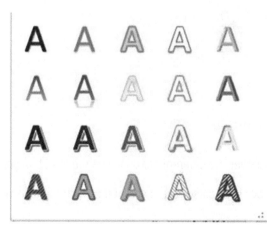

Figure 17-16. *WordArt*

813

3. Select the first A in the first row.

4. Type **Happy Note Music**. Select the text.

    WordArt uses the font type that is selected in the Font group on the Home tab.

5. Click the Ribbon's Home tab. In the Font group, click the down arrow for the Font Type. Move your cursor over the font types as you watch its effect on your WordArt.

6. Click the Ribbon's Format tab. In the WordArt Styles group, click the down arrow for Text Fill. See Figure 17-17.

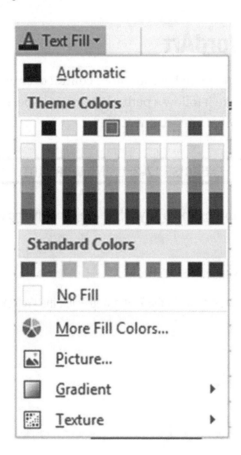

Figure 17-17. *Text Fill options*

7. Move your cursor over the different colors to see the effect on your WordArt.

8. Click Texture. Move your cursor over each of the Textures to see the effect on your WordArt.

9. Notice that you can even fill your text with a Picture.

10.   The Gradient depends upon the color that you are currently using.

We won't make any changes.

11.   In the WordArt Styles group, click the down arrow for Text Outline. You can pick a color for the outline and then change the size of the outline by changing the weight. Select green for the color. Click back on the down arrow for Text Outline and select a weight of 1 pt.

12.   In the WordArt Styles group, click Text Effects. Go through the Shadow, Reflection, and Glow options as you observe their effect on the WordArt.

13.   Click Text Effects again. Click 3-D rotation. You might not be able to see the effect by moving your cursor over the effects. This seems strange, but you first have to apply a 3-D effect before you can view different 3-D effects when moving your cursor over them:

a. Click one of the 3-D effects.

b. Click back on Text Effects on the Ribbon, and select 3-D rotation again. Now, you should be able to see the effects as you move your cursor over each one.

c. Select the **Perspective Relaxed Moderately** effect. See Figure 17-18.

Happy Note Music

Figure 17-18. *Text using the Perspective Relaxed Moderately effect*

We have changed the text, but we can also change the background behind the text. We looked at the Text Fill, Text Outline, and Text Effects. These same options are available for the background. These options are available in the Shape Styles group and are called Shape Fill, Shape Outline, and Shape Effects.

14.   Go through each of the shape options. You might create something similar to Figure 17-19.

Figure 17-19. *See if you can create a similar effect*

15.   Right-click your WordArt. Select **Format Text Effects**. This brings up the
       Format Shape pane with Text Options selected at the top of the pane. Many
       of the options here are the same ones you have used already except that
       some options provide more control over the effect applied. You can see that
       by clicking the Text Fill & Outline button and then expanding Text Fill. See
       Figure 17-20. Click the option buttons on this pane and you will see a lot more
       controls for each effect.

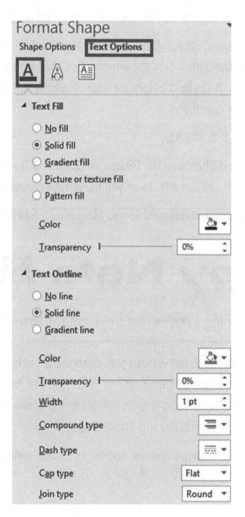

Figure 17-20. *Text Options dialog box*

16.   Click the Text Effects button. There are additional effects here for Material and
      Lighting. See Figure 17-21. Go through each of the options for these effects.
      You will need to click each one. These options do not show the changes as you
      move your mouse across them.

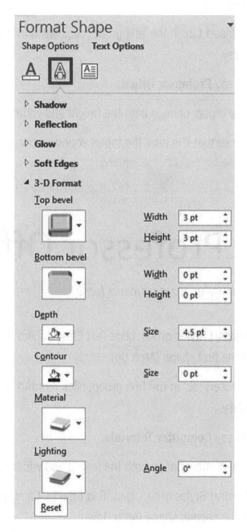

Figure 17-21.   *Text effects*

WordArt can be transformed to follow paths or to be warped into real attention
grabbers.

EXERCISE 17-4: MAKING WORDART FOLLOW A PATH

In this exercise, you will make text follow a path around a circle.

1. Create a new worksheet.

2. Click the Ribbon's Insert tab. In the Text group, click WordArt. Select the first letter **A** in the gallery.

3. Change the text to say **Professor Office**.

4. In the Ribbon's Size group, change both the height and width to 3.94.

5. Drag the border up so that it is near the top of your window. See Figure 17-22.

Figure 17-22. *Drag the border up as shown here*

6. In the Ribbon's WordArt Styles group, click Text Effects. Click Transform. Under Follow Path, click the first shape (Arch Up).

7. Click the Ribbon's Insert tab. In the Text group, click WordArt. Select the first letter **A** in the gallery.

8. Change the text to say **Computer Tutorials**.

9. In the Ribbon's Size group, change both the height and width to 3.94.

10. In the Ribbon's WordArt Styles group, click Text Effects. Click Transform. Under Follow Path, click the second shape (Arch Down).

11. Drag the border up so that the text aligns underneath the **Professor Office** text as shown in Figure 17-23.

12. Click the Ribbon's Insert tab. In the Illustrations group, click Shapes and select the oval.

13. Hold down the Shift key and click down on the left mouse button as you drag it out so that it makes a perfect circle within the text as shown in Figure 17-23.

    This could be used for a logo. You could add more text within the circle. You could jazz it up by applying a 3-D or other effect to the circle.

Figure 17-23. *Make a circle within the text*

14. Click the Ribbon's View tab. In the Show group, uncheck Gridlines.

Adding and Modifying Shapes

Excel provides a large gallery of eight groups of shapes. See Figure 17-24. There is also a group that contains the shapes that you have used most recently. Shapes are located on the Ribbon's Insert tab, in the Illustrations group. To use a shape, click where you want it, and then drag it out to the size you need. Holding down the Shift key while you drag limits the movement of some shapes; for example, lines will be kept straight, a rectangle shape will be perfectly squared, and an oval shape will be a perfect circle. You can add, alter, and delete shapes.

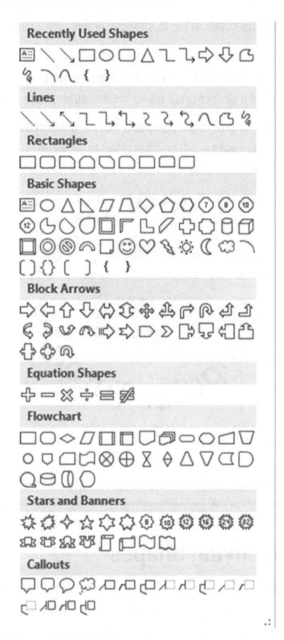

Figure 17-24. *Gallery of eight groups of shapes*

EXERCISE 17-5: USING SHAPES

For this practice, we will use Shapes to create a portion of a flowchart. See Figure 17-25.

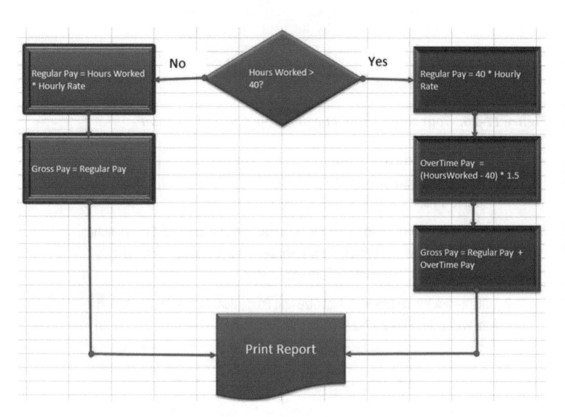

Figure 17-25. *This will be your final result from this exercise*

1. Create a new worksheet. Name the worksheet Shapes.

2. Click the Ribbon's Insert tab. In the Illustrations group, click the Shapes button.

3. Select the Diamond shape in the Flowchart group. The cursor turns into a crosshair.

4. Click the gridline below row 12 and on the gridline between columns F and G. Hold down the Shift key as you drag down and to the right. Expand the shape until it almost takes four columns. See Figure 17-26.

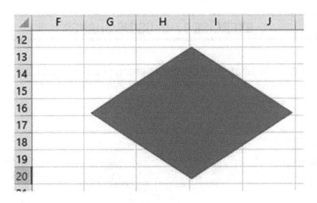

Figure 17-26. *Your shape should be the same size and location*

5. Click the Ribbon's Format tab. In the Shape Styles group, click the More button for the Styles gallery. In the last row of the Theme Styles group, click the blue-colored style.

6. Click the Ribbon's Insert tab. In the Illustrations group, click the Shapes button. In the Flowchart group, click the first shape. It is a rectangle. Click the left corner of where you want the rectangle to start.

7. Right-click the rectangle. Select Size and Properties. Enter 1 for Height and 2 for Width. See Figure 17-27.

Figure 17-27. *Enter the height and width*

8. Move the rectangle so that it aligns as shown in Figure 17-28.

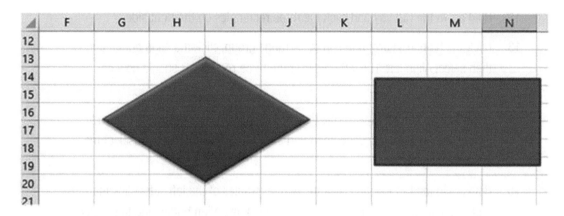

Figure 17-28. *Align the rectangle with the diamond as shown here*

9. With the rectangle selected, click the Ribbon's Format tab. In the Shape Styles group, click the More button for the Styles gallery. In the last row of the Theme Styles group, click the blue-colored style.

10. Press Ctrl + C to copy the rectangle. Press Ctrl + V four times to create four copies of the rectangle.

11. Drag the rectangles so they are positioned as shown in Figure 17-29. Don't worry about exact placement; we will align them next.

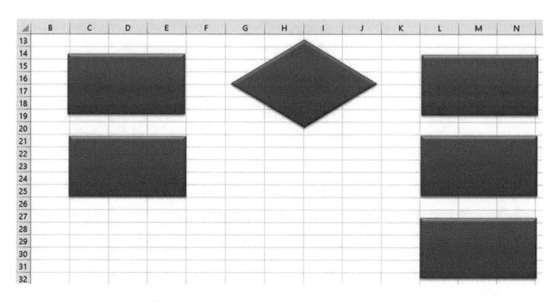

Figure 17-29. *Drag the rectangles so they are positioned as shown*

12.   Hold down the Ctrl key and click the three rectangles on the right side.

13.   Click the Ribbon's Format tab. In the Arrange group, click the Align button and then select Align left. The left border of the three rectangles should now be aligned.

14.   Click the Align button again. This time select Distribute Vertically. This makes the rectangles evenly spaced apart vertically. Click a blank cell to deselect the three rectangles.

15.   Hold down the Ctrl key and click the two rectangles of the left side. Click the Ribbon's Format tab. In the Arrange group, click the Align button and then select Align left. The left border of the two rectangles should now be aligned. Click a blank cell to deselect the two rectangles.

16.   Hold down the Ctrl key and click the second rectangle on the left and the second rectangle on the right. Click the Ribbon's Format tab. In the Arrange group, click the Align button and then select Align bottom.

17.   Click the Ribbon's Insert tab. In the Illustrations group, click Shapes. In the Flowchart group, click the Document icon. Place the cursor where the top left corner of the document shape is to start and then drag it out so that it is in the same location and size as shown in Figure 17-30.

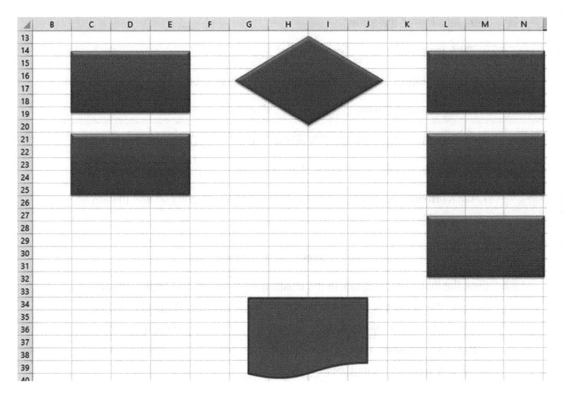

Figure 17-30. *Drag the Document icon out so that it is the same size and in the same location as shown here*

18. With the document shape selected, click the Ribbon's Format tab. Click the More button for the Styles gallery. In the Theme styles group, click the blue color in the last row.

19. Now we can tie our shapes together with arrows:

    a. Click the Ribbon's Insert tab.

    b. In the Illustrations group, click Shapes.

    c. In the lines group, click the second icon (the arrow).

    d. Hold down the Shift key and draw the arrow from the right corner of the decision symbol to the rectangle to its right.

    e. Holding down the Shift key makes the arrow stay in a straight line.

20. Next, let's format the arrow:

    a. Click the arrow.

    b. Click the Ribbon's Format tab. In the Shape Styles group, click Shape Outline.

    c. In the Standard Colors, click red.

    d. In the Shape Styles group, click Shape Outline. Click Weight, and select 2 ¼.

    e. In the Shape Style group, click Shape Outline.

    f. Click Arrows, and select the arrow with the circle on one end and the arrow on the other.

21. Right-click the arrow and then select **Set as Default Line**.

22. Click the Ribbon's Insert tab. In the Illustrations group, click Shapes. The first group in the Shapes is **Recently used Shapes**. The arrow should be in this group. Since you made changes to the look of the arrow and set it as the default line, the arrow line that you select from here will look exactly like the arrow line you just created.

Select the arrow from the Recently used Shapes.

23. Draw the arrow lines as shown in Figure 17-31. You can also speed up the process of doing this without having to go back to the menu every time by copying and pasting the arrow lines and then moving them where you want them. You can rotate the lines and resize them. If they aren't in the exact location you want, you can use your keyboard arrow keys to nudge them in to place.

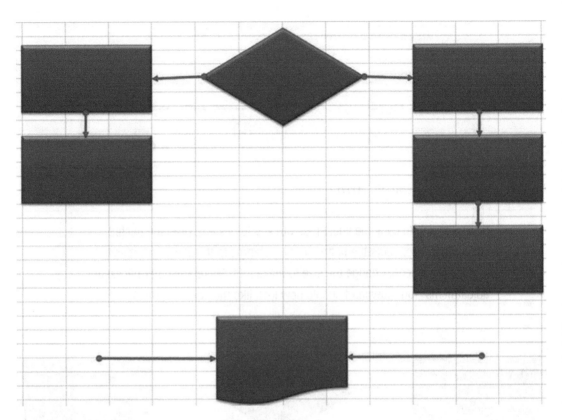

Figure 17-31. *Draw the arrow lines as shown here*

24. Next, we will draw the straight lines that connect the left and right rectangles to the arrows pointing to the document icon:

   a. Click the Ribbon's Insert tab. In the Illustrations group, click Shapes.

   b. In the Lines group, click the first line (the straight line).

   c. Draw a line down from the second rectangle on the left and connect it to the arrow line going to the document icon. Since you created a default line that has an arrow, we are going to have to remove the arrow.

   d. Right-click the line you just created and then select Format Shape. The Format Shape pane appears to the right of the worksheet. Click the Fill & Line button.

   e. Click End Arrow type and then select No Arrow. See Figure 17-32.

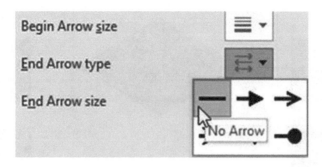

Figure 17-32. *Select the No Arrow type*

25. Press Ctrl + C to copy the line. Press Ctrl + V to paste it. Drag the pasted line so that it connects the bottom right rectangle to the line connecting to the document icon. Change the size to make it fit. Make any other necessary adjustments. Your flowchart should now look like the one in Figure 17-33.

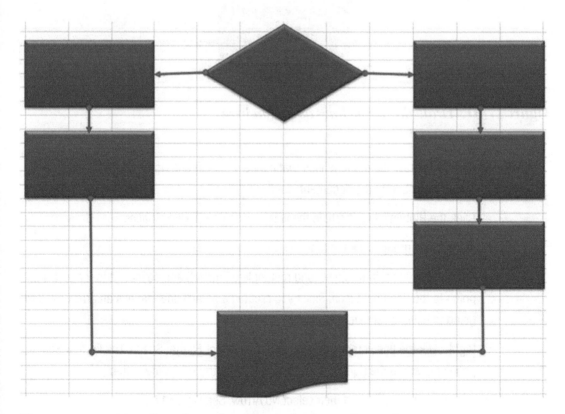

Figure 17-33. *Your flowchart should now look similar to this*

26. Now you can add the text:

    a. Click the top left rectangle.

    b. Enter the text **Regular Pay = Hours Worked * Hourly Rate**.

    c. Click the Ribbon's Home tab. In the Alignment group, click the Middle Align button.

    d. Enter and align the text in the same manner for the other shapes. Enter the text as follows:

       - Second rectangle on left: **Gross Pay = Regular Pay**

       - Decision shape: **Hours Worked > 40?**

       - First rectangle on right: **Regular Pay = 40 * Hourly Rate**

       - Second rectangle on right: **OverTime Pay = (HoursWorked – 40) * 1.5**

       - Third rectangle on right: **Gross Pay = Regular Pay + OverTime Pay**

       - Document shape: Print Report

    e. Increase the size of the font in the Document Shape. Your flowchart should look like the one in Figure 17-34.

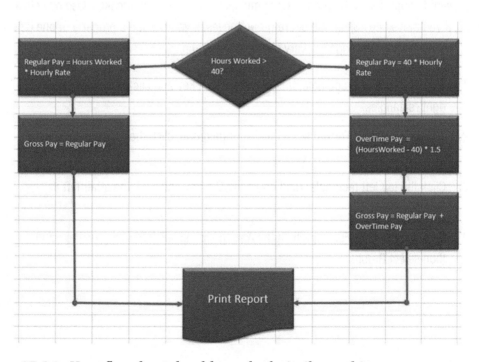

Figure 17-34. *Your flowchart should now look similar to this*

27. We will next add the text **No** to the left of the decision symbol and the text **Yes** to its right:

   a. Click the Ribbon's Insert tab. In the Text group, click Text Box. The cursor changes to a down arrow.

   b. Drag out a rectangle to the left of the decision symbol. Type **No** in the text box.

   c. Right-click in the text box and then select Format Shape. Under Line, select the option **No line**. This removes the text box border. Increase the font size for the text to 16.

   d. Click the outside border of the text box. Press Ctrl + C to copy it. Press Ctrl + V to paste it. Move the pasted text to the right of the decision symbol.

   e. Change the text to **Yes**.

   f. Make any necessary movements.

Grouping and Ungrouping Shapes

If you want to apply the same properties to multiple objects, you can make those objects a group. If you move one object, the entire group moves. If you change a property of one object, Excel makes the same change to every object in the group.

1. Click the top right rectangle. Hold down the Ctrl key and click the other two right rectangles.

2. Click the Ribbon's Format tab. In the Arrange group, click the group button and then select Group.

3. Click the top left rectangle. Hold down the Ctrl key and click the other left rectangle.

4. Click the Ribbon's Format tab. In the Arrange group, click the group button and then select Group.

5.   Click the Ribbon's Format tab. In the Arrange group, click the Selection pane button. The pane shows every object with an eyeball to the right of it. See Figure 17-35. Clicking an eyeball hides the object associated with it. If you click an eyeball to the right of a group, it hides the entire group. The hidden object or group displays a horizontal line. Clicking the horizontal line redisplays the object.

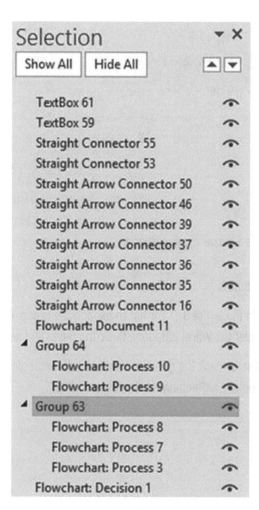

Figure 17-35. *Every object has an eye to the right of it*

6.   Click one of the right rectangles. The group will be selected. Your group name will be different than that shown in Figure 17-35. Click the group name for the right three rectangles. Click the name again. Change the name to RightRectangles.

7. Click the eyeball for the first right rectangle. Click the horizontal line to bring it back. Click the eyeball for the RightRectangles group. Click the horizontal line to bring the group back. Click the word RightRectangles.

8. Click the Ribbon's Format tab. In the Shape Styles group, click the Format Shape button which launches the Format Shape pane. See Figure 17-36.

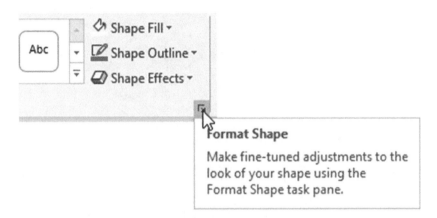

Figure 17-36. *Ribbon's Shape Styles group*

9. Click the Effects button ⬠.

10. Click the arrow to the left of 3-D Format to expand the group. Click Lighting and select Sunset under the Warm category. Click Top bevel and select Round.

11. In the Selection pane, click the group for the two left rectangles. Change the name of the group to LeftRectangles. See Figure 17-37.

Figure 17-37. *Change the name of the group to LeftRectangles*

12. In the Format Shape pane:

   a. Click Contour and select purple.

   b. Change the size of the Contour to 5 pt.

   c. Click Top Bevel and then select Cutout.

   d. Change the width to 7 and the height to 5.

13. Click the Ribbon's Format tab. In the Arrange group, click the Group button and then select Ungroup. Click one of the rectangles on the right. Click the Group button and then select Ungroup.

14. Close the Selection pane.

15. Click the Ribbon's Insert tab. In the Illustrations group, click Shapes. Click the **Thought Bubble Cloud** shape in the Callout group. Draw the callout to the upper left of the decision shape.

    The callout has a yellow circle for one of its handles. You can drag this circle toward the object you want the Callout to be coming from.

16. Drag the yellow circle toward the decision symbol. Type the text **Cloud Application** in the Callout shape. Align the text.

17. You can change any shape into another shape. Let's give the cloud application a cloud shape:

    a. In the Insert Shapes group, click the Edit Shape button, then select Change Shape, and then select the **Thought Bubble Cloud** callout.

    b. Drag the yellow circle from the Cloud Callout to the decision shape.

    c. Make any necessary changes to the size and shape of the callout. See Figure 17-38.

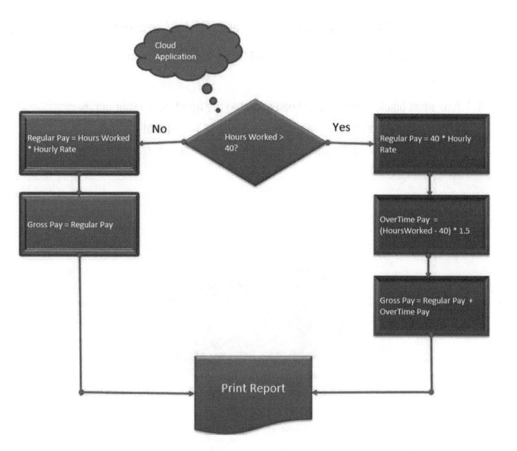

Figure 17-38. *Create a Cloud shape*

**EXERCISE 17-6: CHANGING A SHAPE'S EDIT POINTS AND
PUTTING ONE SHAPE IN FRONT OF ANOTHER**

In this exercise, you will practice changing the shape of an object and moving one object in front of another.

1. Create a new worksheet.

2. Click the Ribbon's Insert tab. In the Illustrations group, click Shapes.

3. In the Basic Shapes group, click the heart. Drag out the heart in the middle of your worksheet.

4. In the Shape Styles group, click Shape Fill and select the color red.

5. In the Insert Shapes group, click the Edit Shape button and then select Edit Points.

6. Two small black handles have been added to the heart. Move your cursor over the top one and drag it up. Two lines appear with a square at the end of each line. Practice moving the squares to change the shape and size of the heart. See Figure 17-39.

Figure 17-39. *Move the squares to change the shape and size of the heart*

7. Click the Ribbon's Insert tab. In the Illustrations group, click Shapes. In the Stars and Banners group, click the third Ribbon (curved and tilted up).

8. Drag out the banner below the heart.

9. Move the heart on top of the banner. The heart moves behind the Ribbon because the heart was added first. See Figure 17-40.

Figure 17-40. *The banner should appear in front of the heart*

10. In the Arrange group, click the top half of the Bring Forward button. Now the heart is front of the Ribbon. Click the top half of the Send Backward button. The banner should now be in front of the heart.

EXERCISE 17-7: CONVERTING AN IMAGE INTO A SHAPE

In this exercise, you will make an image take the shape of a circle.

1. Create a new worksheet.

2. Click the Ribbon's Insert tab. In the Illustrations group, click the Shapes button.

3. In the Basic Shapes group, click the oval.

4. Hold down the Shift key while you drag out the oval; this will make a perfect circle.

5. Click the Format tab. In the Shape Styles group, click the Shape Fill down arrow.

6. Select Picture.

7. Click **From a File**. Browse to the image ShapeImage. Click the Insert button. The picture takes the shape of the circle. See Figure 17-41.

Figure 17-41. *Picture takes the shape of the circle*

In the last two sections, you have learned how to create decorative text using WordArt, which gives your text much more impact. You have seen that Excel provides numerous predefined shapes such as arrows, stars, banners, callouts, and so on, whose shape and color you can modify. Next, you will learn how to use SmartArt. SmartArt takes shapes to another level by combining multiple shapes with text to create various customizable diagrams. There are nine categories of diagrams, and each one contains many layouts to choose from. You can also apply styles to the layouts.

Using SmartArt

SmartArt is a collection of customizable diagrams that are categorized into groups such as Processes, Hierarchy, Matrixes, Pyramids, and so on. The All category shows all of the diagrams from the other categories. You access Choose a SmartArt Graphic dialog box, and all the diagrams appear via the SmartArt button in the Insert tab's Illustrations group. See Figure 17-42.

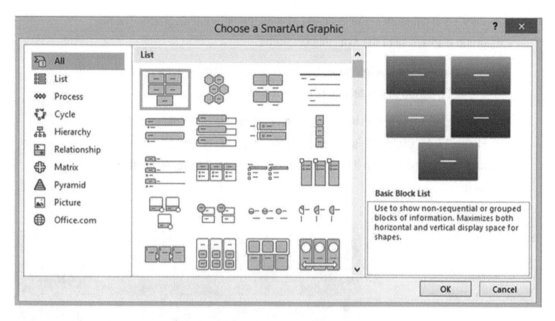

Figure 17-42. *The All category shows all of the diagrams from the other categories*

EXERCISE 17-8: CREATING AN ORGANIZATIONAL CHART

It is best if you draw out how you want your diagram to look before you start creating it in Excel. In this practice exercise, you will create the organizational chart shown in Figure 17-43.

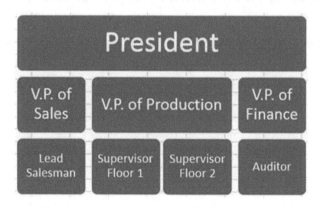

Figure 17-43. *You will create this organizational chart*

1. Create a new worksheet.

2. Click the Ribbon's Insert tab. In the Illustrations group, click SmartArt.

3. Click each category in the left pane to view the different diagrams. The All group displays all of the available diagrams.

4. Click Hierarchy in the left pane. Click the Table Hierarchy in the right pane. It is the last diagram in the second row. Then click OK.

   There is an arrow button that appears on the left side of the diagram. See Figure 17-44. Clicking this button turns the Text pane on and off. We will first change the Text pane so it matches the way we want the organizational chart to look.

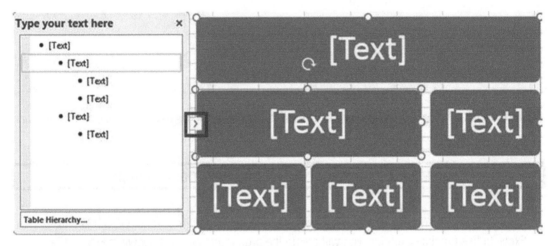

Figure 17-44. *The arrow button turns the Text pane on and off*

5. Click the third text line in the Text pane. See Figure 17-45. Click the Ribbon's Design tab. In the Create Graphic group, click the Promote button.

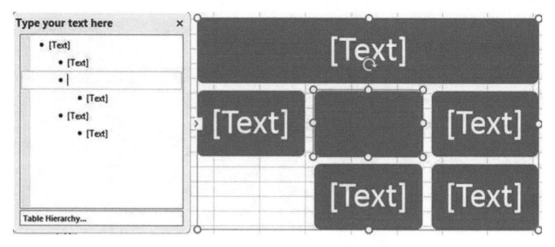

Figure 17-45. *The text block from the bottom row is moved up to the second row*

6. In the Create Graphic group, click the Demote button. The text block moves back down to the third row.

7. Let's perform the same function using a different method. Hold down the Shift key while pressing the Tab key to promote the text block.

8. Press the Tab key to demote the text block.

9. Hold down the Shift key while pressing the Tab key to promote the text block back again.

10. We need to have two text blocks below the second row's second text box. In the Create Graphic group, click the Add Shape's down arrow button. Select **Add Shape Below**.

11. Click the second Text block in the Text pane. In the Create Graphic group, click the Add Shape's down arrow button. Select **Add Shape Below**.

    Now that we have the Text blocks in the order that we want, we can enter the titles for the positions.

12. Enter the title for each position by entering the title either in the chart or in the Text pane. Entering the title in either one will enter it in the other. See Figure 17-46.

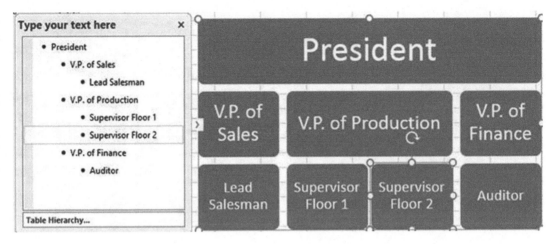

Figure 17-46. *Enter text in the pane and it will appear in the chart and vice versa*

13. Click the Ribbon's Design tab. In the Layouts group, click the More button ⬇ to see the entire Layout gallery.

14. Click the third diagram in the first row. This is the Name and Title Organizational Chart. The benefit of this chart is that besides entering the title for each position, you can enter the name of the person in that position. Just click each text box and enter the names. See Figure 17-47. Add your own names.

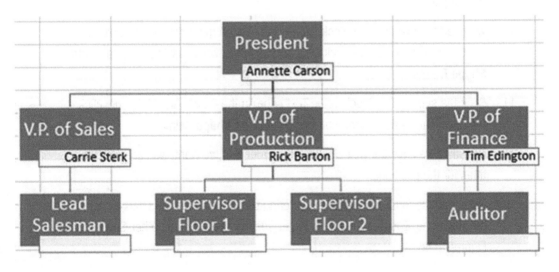

Figure 17-47. *Name and Title organizational chart*

All the gridlines can be very distracting. There are two ways you can remove the lines. One is by turning off the gridlines. The other is by using fill.

Removing the Gridlines by Turning Them Off

1. Click the Ribbon's View tab.

2. In the Show group, remove the check for Gridlines.

   This removes the gridlines for the entire worksheet. If you aren't going to be entering other data in the worksheet cells, this is the best method.

3. In the Show group, check the Gridlines to restore them for the next exercise.

Removing the Gridlines by Using Fill

1. Select all the cells around the chart.

2. Click the Ribbon's Home tab.

3. In the Font group, click the Fill Color button and then select White.

4. Click the down arrow for the border and select **Thick Outside Border**. See Figure 17-48.

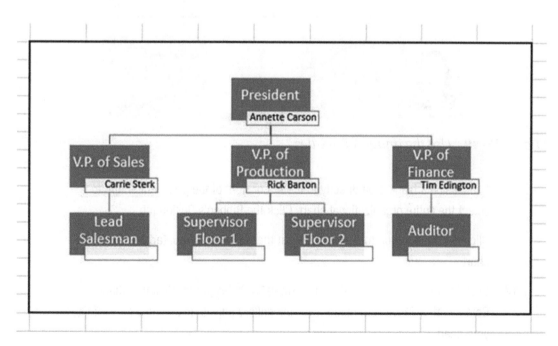

Figure 17-48. *Name and Title organizational chart with thick outside border*

Now that we have the organizational chart the way we want, we can select one of many different layout choices.

5.   Click anywhere on the organizational chart.

6.   Click the Ribbon's Design tab. In the Layouts group, click the second diagram in the first row. This is the Picture Organization Chart.

7.   Click the picture icon ⊞ for the President.

This brings up the Insert Pictures window where you can select a picture from your computer or from the Internet.

8.   Click **From a File**. Browse to the picture President. Click the Insert button.

9.   Follow the same process for the V.P. of Sales. Use the Image VPofSales. See Figure 17-49.

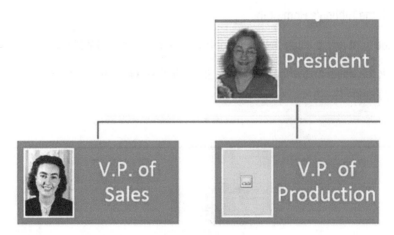

Figure 17-49. *Use the image VPofSales*

10.   Click inside a blank area directly to the left or right of the president shape to select the entire organizational chart. Click the Ribbon's Design tab.

11.   Click the More button for Layouts. Select the **Horizontal Labeled Hierarchy** chart.

12.   Click the Ribbon's Design tab. In the SmartArt Styles group, click the Change Colors button. Move your cursor over the gallery objects to see how they affect your chart.

13.   Select the first object under Primary Theme colors.

14.   Right-click the President shape in your chart. Select Change Shape.

15.   In the Stars and Banners group, select Horizontal Scroll.

16.   Click the Ribbon's Format tab. In the Shape Styles group, click Shape Fill.
       Click whatever color you like. See Figure 17-50.

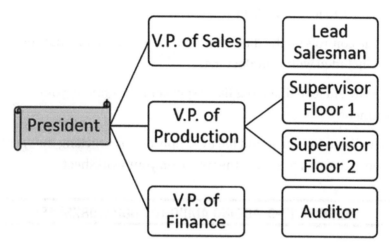

Figure 17-50. *Select a color for Shape Fill*

You have learned how to create impressive worksheets by using images, shapes, and layouts. Next, you will learn how to add audio and video that you or someone else has created to a worksheet.

Inserting Sound into a Worksheet

It is a very simple process to create your own audio sounds. You can buy a good inexpensive microphone, and if you have a laptop, it probably has a built-in microphone. You can add audio to your workbook to provide instruction or to provide information about the data in the workbook or an explanation of why things were done the way they were.

You can either embed the audio file into your workbook or create a link to the audio file. The same is true for video files. If you create a link to the audio file, the file doesn't exist as a part of your workbook. When you play the audio file, Excel searches for the file in the location (path) that you gave it. If you ever move your file from that location, Excel will not be able to find it.

Add sound to your worksheet by doing the following:

1. Select the cell near which you want the sound inserted.

2. Make sure the Insert tab of the Ribbon is displayed.

3. Click the Object tool in the Text group. Excel displays the Object dialog box.

4. Click the Create from File tab.

5. Use the controls on the dialog box to locate a sound file that you want included with your document.

6. Click OK. An icon that looks like a speaker is inserted in your document.

You can later listen to your sound file by simply double-clicking the speaker icon. You can also move the icon to some other place on your worksheet.

EXERCISE 17-9: ADDING AUDIO TO YOUR WORKSHEET

In this exercise, you will embed an audio file into your worksheet.

1. Add a new worksheet to your workbook.

2. Click the Ribbon's Insert tab. In the Text group, click the Object button. The Object dialog box displays.

3. Click the **Create from File** tab. See Figure 17-51.

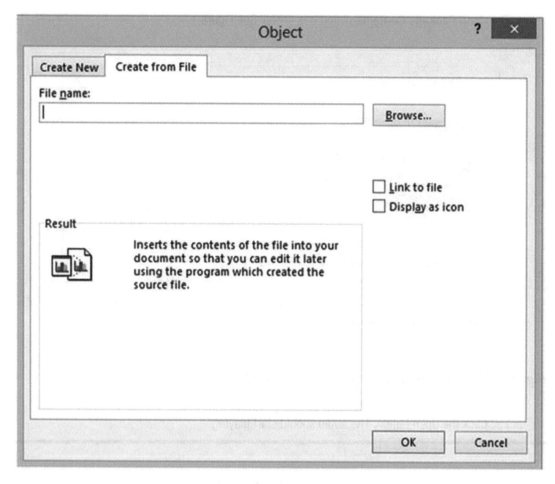

Figure 17-51. *Create from File tab on the Object dialog box*

4. Click the Browse button and then find the file ExcelAudioFile. Click the Insert button.

5. Click the OK button. Since you didn't select Link to file, the file will be embedded into your workbook.

6. Double-click the Audio icon. See Figure 17-52.

Figure 17-52. *Double-click this icon if you want to play the audio file*

The **Open Package Contents** dialog box opens. See Figure 17-53.

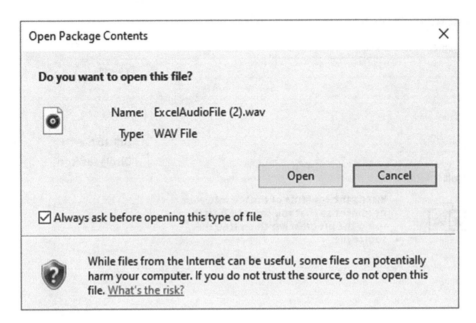

Figure 17-53. *Open Package Contents dialog box*

7.   Click the Open button. The audio should start playing.

Inserting Video into a Worksheet

In order to play a video file in Excel, you will need to have a software program that plays video, such as Windows Media Player.

Just like audio files, video files can be either embedded into your worksheet or linked to it. As mentioned previously, the problem with linking a file is that if you move that file after you have created the link or distribute it to others, Excel will not be able to find it. If you do not choose to link the file, the file will become a part of your workbook. If the video file is large, your workbook of course will take up more space.

Video and Audio files are added the same way to a worksheet.

EXERCISE 17-10: ADDING VIDEO TO YOUR WORKSHEET

In this exercise, you will embed a video file into your worksheet.

1. Place your cursor where you want the icon for the video file to be.

2. Click the Ribbon's Insert tab. In the Text group, click the Object button. The Object dialog box displays.

3. Click the **Create from File** tab.

4. Click the Browse button and then find the file Transpose. Click the OK button.

5. Double-click the Video icon. See Figure 17-54.

Figure 17-54. *Double-click the Video icon to start the video*

6. The **Open Package Contents** dialog box opens. Click the Open button. The video file starts playing in your computer's default video player.

Summary

You have seen how multimedia can have a visual impact as well as improving how quickly the user can grasp what you want him or her to know. Multimedia makes your worksheets look more polished and professional as well as giving you much more power to convey your message.

You can add pictures to your worksheet from either your computer or the Internet. You have seen how you can even create a logo with WordArt by making the text follow a path.

Screenshots can be used to capture an entire screen or a selected portion from a website or other application and then place that capture in the current worksheet. The capture is added to your worksheet as an Image. You can alter it in the same way you can alter any other image.

You have learned how to use Shapes and SmartArt to create flowcharts, organizational charts, and other types of layouts. You can add, alter, and delete a wide variety of shapes.

It is very simple to add audio and video. You can either embed the files into your workbook or link to them. You can buy a good inexpensive microphone to record your audio notes. In order to play a video file in Excel, you will need to have a software program that plays video, such as Windows Media Player.

In the next chapter, you will see all of the various ways of manipulating icons. You will learn how to use the new 3D Images and how to use Object Groups that will allow you to group and ungroup combinations of different objects.

CHAPTER 18

Icons, 3D Images, and Object Grouping

Excel has hundreds of icons for you to use. Icons and SVG files can be converted into shapes. Each separate shape in the icon or SVG file can then be moved, rotated, formatted, copied, or deleted independently or together. Excel 2019 also introduces 3D Images you can rotate 360 degrees.

After reading and working through this chapter, you should be able to

- Convert Icons and SVG files to Shapes

- Work with individual pieces of an icon or SVG File

- Use 3D objects

- Change the visibility of objects

- Change the order of objects

- Group, Ungroup, and Regroup objects

Excel's icons are located on the Ribbon's Insert tab in the Illustrations group. See Figure 18-1.

© David Slager and Annette Slager 2020
D. Slager and A. Slager, *Essential Excel 2019*, https://doi.org/10.1007/978-1-4842-6209-2_18

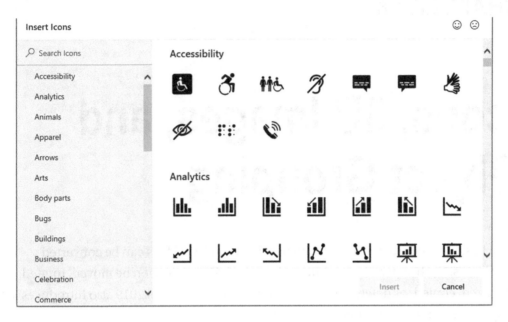

Figure 18-1. *Icons*

To find the icon you want, click the different categories. You can also find the icon you want by scrolling through them. Another way to find an icon is to enter a search word. I will enter the word phone. See Figure 18-2.

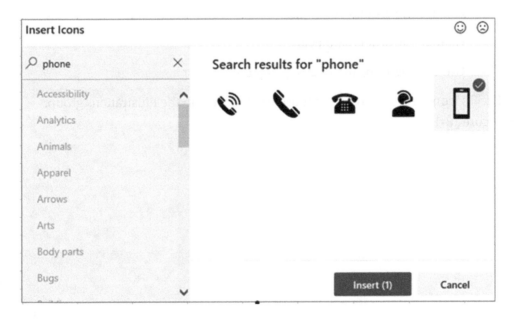

Figure 18-2. *Search for Phone*

When you select an icon, a check mark appears to the upper right of it. The Insert button shows the number of icons you will be inserting. Clicking the Insert button places the selected icons on your spreadsheet. See Figure 18-3. You can increase the size of the phone while keeping the image proportional by dragging out the bottom right handle.

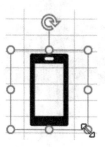

Figure 18-3. *Changing Icon size*

We want to create an image that looks like Figure 18-4.

Figure 18-4. *Image to be replicated*

We need to make the inside of the phone gray. To do that, we need to add a rectangle shape that fills the inside of the phone and then fill that with a gray color. We can click the Ribbon's Insert tab, click Shapes, and select the rectangular shape. Drag that shape so that it fills the inside of the phone. Now we can go to the Home tab and use the Fill color to fill the rectangle with gray. Next, we will click the Ribbon's Insert tab, click Pictures, and select an image. The Image can then be resized to fit in the phone. The name Rhianna was created by using WordArt. WordArt is located on the Ribbon's Insert tab in the Text group. We just pick the style we want, type the name, drag it to where we want it, select the text, and change the font size so that it fits. See Figure 18-5.

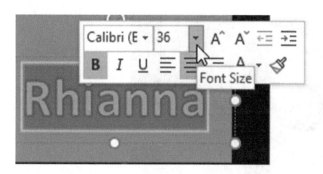

Figure 18-5. *Change font size for WordArt*

Now, we can add the sun at the bottom. The sun is another icon. It can be created using the same process as used for the phone. After clicking the Ribbon's Icons button and then entering the word sun in the search box, the sun icon we want to use can be clicked and then the Insert button. The icon can be resized and dragged to wherever you want it to be. The color of the sun can be changed to gold; to do this, we can right-click the sun and click the Fill color icon and then select a gold color.

Grouping and Ungrouping Icons, Images, Shapes, and WordArt

The image is now complete, but if you move any of the objects we created, they we will move separately from the other objects. We want all the objects to be treated as a single phone. To group all the objects, click one of the objects and then hold down the Ctrl key while you click the other objects. Right-click and select Group and then Group. See Figure 18-6.

Figure 18-6. *Group the objects*

Now when you move or rotate the phone, all the objects move together. See Figure 18-7.

Figure 18-7. *Grouped objects move together*

The individual objects are still each selectable and can be formatted individually or together. If I select the phone Icon and the rectangular shape, the Ribbon will have a Format tab under the contextual tab for Drawing Tools. You can then apply any of the Shape Styles or Shape Effects. Figure 18-8 shows that a blue fill color was applied to the phone and the rectangular shape.

Figure 18-8. *Blue fill color applied to phone and rectangular shape*

You could apply a picture style to the entire phone, or you can click the image and apply a picture style to only it. You can ungroup some or all the objects by selecting the objects you want to ungroup and then right-clicking Group and then Ungroup.

EXERCISE 18-1: CONVERTING ICONS TO SHAPES

In this exercise, you will create a farmer icon that will look like Figure 18-9.

Figure 18-9. *Image to be created in exercise*

Start a new worksheet. Click the Ribbon's Insert tab. In the Illustrations group, click Icons. Spend some time scrolling through the Icons. Click Search Icons in the upper left corner. Type the word farmer. Excel displays three icons related to farmer. See Figure 18-10.

Insert Icons

Figure 18-10. *Icons available when entering farmer for search*

Click on the farmer. Click the Insert button at the bottom right corner of the Insert Icons window to place the icon on your spreadsheet. You can drag the icon anywhere you want and use the resize handles to resize it. You can apply any graphic style. You can apply any color you want. You can apply any graphic effect.

Convert the Icon to a Shape

1. In order to separate the icon into individual shapes, you must select the icon and then click the Ribbon's Format tab and select Convert to Shape in the Change group. Excel will ask you to confirm that you want to turn the icon into a drawing object. Click the Yes button.

2. The farmer is now six separate shapes. Click his hat to select it. Whenever you click a shape, a rotate button is displayed. You can use this button to rotate the shape in either direction. Use this button to tilt the farmer's hat. With the hat still selected, click the down arrow for Fill Color and select a blue color.

3. Select the pitchfork. On the Ribbon, click Format under the Drawing Tools heading. In the Shape Styles, select a Gold gradient fill. Go to Shape Effects and then select one of the 3-D rotations.

4. Add text to the farmer by clicking the shape that forms his body and then select a white font color from the Home tab. Type Farmer.

5. The pitchfork isn't needed. Click the pitchfork and then press the Delete key on your keyboard.

Add Additional Shapes

1. The farmer really needs some eyes. On the Ribbon's Insert tab, in the Illustrations group, click Shapes. Select the oval shape. Drag out a circle on the farmer's face for the left eye. Create the right eye by holding down the Ctrl key (your cursor should show a plus sign) and dragging the left eye to the right.

Group Objects

1. Drag the farmer over a bit. Notice his eyes don't move with him. Grouping the eyes with the farmer will solve this problem. With the farmer selected, hold down the Ctrl key and click each of the eyes. Now, right-click and select Group and then select Group. This groups the shapes together. Now move the farmer; his eyes should move with him.

Using a Shape in Another Location

1. Let's say that you wanted to use only the hat in another area. Double-click the hat to select it. Right-click the hat and select Copy. Go to another location on the spreadsheet, right-click, and select a paste option.

EXERCISE 18-2: GROUPING, UNGROUPING, AND REGROUPING

In this lesson, we will look at grouping, ungrouping, and regrouping multiple icons.

1. Use a new worksheet. On the Ribbon's Insert tab, in the Illustrations group, click the Icons button. Type farmer in the search area.

    The icons that meet the search criteria are displayed. Click the barn and the farmer.

    Excel places a check mark next to each of these icons so that you can easily identify which icons you have selected. The Insert button in the bottom right corner has a 2 in parenthesis to let you know that you will be inserting two icons into your spreadsheet. Click the Insert button. The icons overlap each other. Click a different area to deselect them. Drag them to the center of the spreadsheet but away from each other.

2. Click the barn and resize it. Click the farmer and resize him. The size and placement of your icons should look like Figure 18-11.

Figure 18-11. *Size and placement of icons*

3. With the farmer already selected, hold down the Ctrl key and click the barn. Now that both icons are selected, change the icons to shapes by clicking the Ribbon's Format tab under the Graphics Tools heading, and then in the Change group, click the Convert to Shape button. Excel asks for confirmation to convert these icons to a drawing object. Click the Yes button.

4. You will want to treat both these icons as one. In order to do that, they will need to be grouped together. Click the Ribbon's Format tab and click the Group button in the Arrange group, and then select Group or you could just right-click and then select Group.

5. Since they are now treated as one, they can be moved together. If you change the size of one, you will change the size for both. Move the farmer or the barn; both should move together. Any formatting applied will be applied to both of them. Click the Ribbon's Format tab. Click the down arrow for the Shape Styles. Drag your cursor across them and notice the effect it has on both the farmer and the barn. Try doing the same for Shape Effects.

6. Click the Group button and select Ungroup. They are still grouped together; you must click a different area to deselect them.

7. Next, you will change the color of the barn and its roof. Double-click the barn and then hold down the Ctrl key and click the roof. To change the color, click the Shape Fill Color down arrow 🖌 ▾ and select red.

8. Next, you will change the sun and its rays to yellow. Click one of the rays and then hold down the Ctrl key and click the sun and the other rays. Change the Shape Fill Color to yellow. If you missed any of the rays, select them and then select the Shape Fill Color.

9. To make the image stand out, you can color the background. Click a cell above and to the left of the sun and then drag across the other cells so that you have a rectangle shape that covers both images. Click the down arrow for the Fill Color on the Home tab and select the gold color. See Figure 18-12.

Figure 18-12. *Make image stand out by creating a colored background*

You have now had practice converting icons to shapes and manipulating and coloring those shapes. You can do the exact same thing to SVG files. SVG files are scalable Vector Graphic Files. One of the benefits of SVG files is that they can be enlarged to any size without losing any quality. The file names have an extension of .svg. You can find plenty of free downloadable SVG files from the Internet. Just do a search for **SVG files**.

3D

Excel has added the capability to display rotatable 3D models. 3D models work in Word, Excel, or PowerPoint. The 3D models are located on the Ribbon's Insert tab in the Illustrations group.

You can select a 3D model from a file you have on your computer or OneDrive, or you can grab models from Online Sources. If you click from Online Sources, you will see categorized groups of 3D models. See Figure 18-13. The first three groups are animated. (Yours may look different than the example. You may not have animated groups. This is the way it looked for Office 365 when this book was published.)

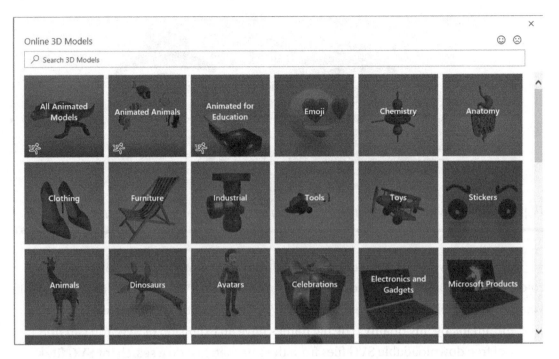

Figure 18-13. *3D image categories*

If you want a model that isn't available in one of these groups, you can enter what you want in the text box and then press Enter. Type the singular of what you want. For example, enter house, not houses, or you may not get any results. You may at times need to enter the plural such as if you wanted legs and not a single leg. There are many programs available that allow you to create 3D images or convert images into a 3D image such as Microsoft's Paint 3D.

Figure 18-14 shows a brain and a heart after being selected from the Medical group.

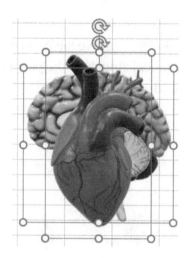

Figure 18-14. *Stacked 3D images after selection*

When you bring in multiple 3D objects, they are placed on top of each other. You will need to deselect them so that you can move them individually. Clicking another location in the spreadsheet will deselect them. You can then drag the objects where you want them located.

When a 3D model is selected, you will see a rotation icon in the center of the model. See Figure 18-15. You can click this rotation icon and spin the heart in any direction, up, down, left, and right.

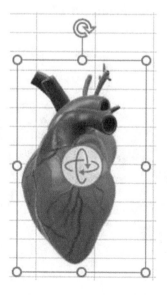

Figure 18-15. *Rotation icon in the center of the heart*

When you have a 3D model selected, you will see the label 3D Model Tools above the Ribbon. Under the label is the context-sensitive tab named Format. Instead of rotating the 3D model, you may find the model already in the position you want in the 3D model views on the Ribbon. See Figure 18-16.

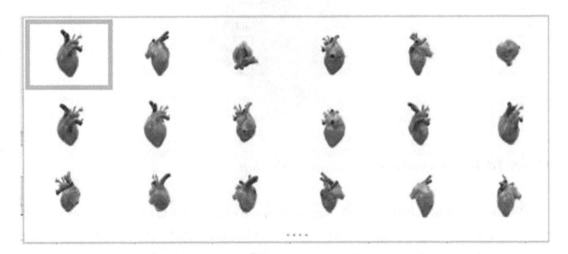

Figure 18-16. *Selectable already rotated 3D model views*

You can group the 3D models by clicking one of them and then holding down the Ctrl key while you click the other one. When both models are selected, you can right-click and then select Group and then Group. When the objects are grouped, there will be a border around the two models. See Figure 18-17.

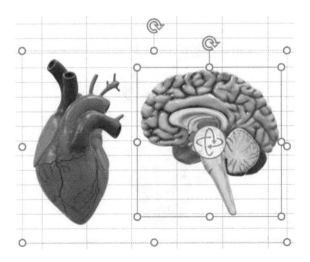

Figure 18-17. *A border appears around grouped objects*

If you want to move both objects together, drag the outside border. If you want to rotate both objects at the same time, use the rotator on the outside border. The 3D model rotator in the center of a model will only work on one model at a time.

You can group together any type of Image object. You can group together Pictures, icons, shapes, 3D images, SmartArt, Charts, Maps, and WordArt. You can also overlay any of these objects on top of each other. So how can you control which object appears on the top of another object? You can use the Bring Forward and Send Backward buttons in the Arrange group, but if you are dealing with numerous objects, you will find it easier to use the Selection Pane button which can control not only the order in which objects are displayed but view how they are grouped and even control if they are visible or not.

EXERCISE 18-3: USING 3D AND OTHER OBJECTS WITH THE SELECTION PANE

In this exercise, you will use icons, shapes, and 3D objects. You will group, ungroup, and regroup the objects. You will learn how to use the Selection Pane to change the visibility and arrangement of your objects.

1. Open workbook Chapter18. Sheet 1 contains a zoo image. On the Ribbon's Insert tab, in the Illustrations group, click Shapes. In the Callouts group, click the Thought Bubble Cloud. The cursor turns into a crosshair. Drag the cloud out so that it appears as though it is coming from the ostrich. Type "What are those strange animals?" See Figure 18-18. You might need to adjust the size of the cloud to get the text to fit the way you want.

Figure 18-18. *Picture with Thought Bubble shape*

2. On the Ribbon's Home tab in the Editing group, click Find & Select and then
    select Selection Pane. The Selection Pane has two objects, the picture and the
    Thought Bubble. The last object you add is placed at the top of the Selection
    Pane. See Figure 18-19. Across from each object is an eye. Click the eye for the
    Thought Bubble. The object disappears and the eye turns into a horizontal line.
    Click the horizontal line; the Thought Bubble reappears and so does the eye.

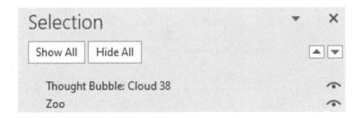

Figure 18-19. *Selection pane with objects whose visibility can be changed*

3. On the Ribbon's Insert tab in the Illustrations group, click the top half of the 3D Models button. Enter zebra in the search box and then press Enter. Click the zebra that looks like the one in Figure 18-20. Click the Insert button at the bottom of the window.

    Drag the zebra to the location shown in Figure 18-20. Try rotating it. Click the different 3D Model Views on the Ribbon.

    Click the name of the object in the Selection Pane. Delete the text and enter Zebra for the name.

4. On the Ribbon's Insert tab in the Illustrations group, click the top half of the 3D Models button. Click the Flowers and Plants Category. Click the pine tree. Click the Insert button.

    Drag the Tree so that it is front of the 3D zebra. Drag the size handles to increase the size of the tree.

    In the Selection Pane, rename the object to 3D Tree.

Figure 18-20. *3D Zebra and 3D tree added to image*

5. On the Ribbon's Insert tab, click Icons. In the Animals group, click the turtle. In the Landscape group, click the double trees. Click the Insert button. Click a blank cell to deselect the two objects. Move them and resize them so that they appear as in Figure 18-21. Change the names of the objects in the Selection Pane to Turtle and Trees.

Figure 18-21. *Turtle and Trees added to image*

6. You can select an object by clicking its name in the Selection Pane. Click the name Turtle. On the Ribbon's contextual Format tab under the heading Graphics Tools in the Change group, click Convert to Shape. If you get a message that asks if you want to convert it to a Microsoft Office drawing object, click the Yes button.

7. Click the name Trees in the Selection Pane. Click Convert to Shape. The Selection pane shows that the Trees and Turtle are each comprised of two separate shapes. See Figure 18-22.

Figure 18-22. *Trees and Turtle are comprised of separate objects*

8. Click the largest tree shape. On the Ribbon's Home tab in the Font group, click the down arrow for the Color Fill and select a green color.

9. Drag the smaller tree shape so that it covers a portion of the larger tree shape. In the Selection pane, click the top shape in the Trees group and then click the down arrow button (Send Backward). This changes the order of the objects. Click on the up arrow button (bring forward) to put the smaller tree back in front. You can also drag an object in the Selection pane to change its order.

10. Click the word Zebra. Drag it to the top or use the up arrow key. The Zebra image moves in front of the trees and 3D Tree. Drag the 3D Tree above the Turtle group. See Figure 18-23.

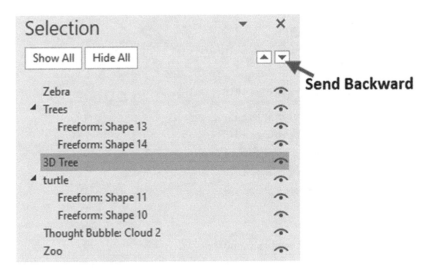

Figure 18-23. *Move the object names to change their order*

The small black trees now appear in front of the green trees. The zebra appears in front of the 3D tree, and the 3D tree appears in front of the turtle. The result is shown in Figure 18-24.

Figure 18-24. *Result of changing object order in the Selection pane*

11. Hold down your Ctrl key and click each object in the Selection pane (when you click a group heading such as Turtle or Trees, you are selecting all the objects in that group). Right-click the picture and select Group and then Group. Move the picture. All the objects move together.

12. Right-click the picture and select Group and then Ungroup. Click a blank cell. Move the picture; the objects don't move with the picture.

13. Right-click the picture and select Group and then Regroup. Move the picture; the objects again move with the picture.

14. Practice removing and bringing back the visibility of the objects. Practice grouping, ungrouping, and regrouping some of the objects. Practice changing the order of the objects.

Summary

This chapter dealt with using various types of objects separately as well as together. You learned how to convert icons to shapes. The same process applies to converting SVG files to shapes. You learned how to group, ungroup, and regroup using right-click or by using the Selections pane. You also learned how to change the visibility and order of objects using the Selection pane. You learned that you can select rotated 3D model views from a 3D object you have added, or you can change the size and freely rotate the 3D objects to look the way you want it to be.

In the next chapter, you will learn how to create macros. A macro is a shortcut that eliminates the need to repeat the same steps over and over for frequently performed tasks.

CHAPTER 19

Automating Tasks with Macros

Macros can be big time-savers for those tasks you find yourself performing repeatedly. Macros enable you to record a series of steps that you normally use to perform a task, store those steps, and then automatically rerun those steps whenever you want. Macros can be quickly activated by assigning shortcut keys to them. If you think you might forget a macro's associated shortcut key, then you can store it in the Quick Access Toolbar. You can also attach macros to buttons and other objects.

After reading and working through this chapter, you should be able to

- Create a macro that uses an absolute cell reference

- Create a macro that uses a relative cell reference

- Assign a macro to a button

- Assign a macro to the Quick Access Toolbar

- Assign a macro to a shape

- Create and update the personal workbook

- Save a macro-enabled workbook

- Use the VBA (Visual Basics for Applications) Development Environment

- Edit a macro

© David Slager and Annette Slager 2020
D. Slager and A. Slager, *Essential Excel 2019*, https://doi.org/10.1007/978-1-4842-6209-2_19

Creating (Recording) a Macro

Simple macros can be created using the Macro Recorder. More complex macros can be created in the Macro Editor. The Macro Recorder records every keystroke and every mouse click from the time you start the recorder to the time you stop it. Since it records everything that you do, don't click any other cells or perform any steps other than those that are directly related to the task you want to perform.

The Macro Recorder can be started from three different locations:

- On the View tab, in the Macros group, click the down arrow for the Macros button and select Record Macro.

- On the Developer tab, in the Code group, click Record Macro.

- On the Status bar, click the Macro Start Recording button 🔢 .

When you start the recorder, a Record Macro dialog box opens where you will need to enter the name you are assigning to the macro, a shortcut key, and where you want the macro stored; you can also provide a description of the macro so that later you can remember what its purpose is. See Figure 19-1.

Figure 19-1. *Entries to make before recording the macro*

There are three places you can store your macro:

- The current workbook

- A new workbook

- The Personal Macro Workbook

If you want to make the macros available for use in other workbooks, select Personal Macro Workbook. The Personal Macro Workbook doesn't exist until you save a macro to it. Excel saves the file as PERSONAL.xlsb. This workbook is hidden and is opened automatically whenever you start Excel.

EXERCISE 19-1: CREATING A SIMPLE MACRO

The macro we will create will apply formatting to a cell. We will use the Start Recording button on the status bar.

Recording the Macro

1. Create a new workbook:

   a. Click the File tab and then select Save As.

   b. Select the location where you want your workbook stored.

   c. Enter **Macro_Practice** for the File name.

   d. Select **Excel Workbook** for the **Save As type**.

   e. Click the Save button.

2. Look on your status bar and see if you have the Macro Start Recording button 📽. If you don't, then right-click the status bar, select the option for **Macro Recording,** and then press the (Esc)ape key to hide the option window.

3. Click inside cell A1.

4. Click the Macro Recording button on the status bar. The Record Macro dialog box opens.

5. The macro name can't contain any spaces. Type **FormatEntry** for the Macro name.

The Shortcut key is currently showing **Ctrl +** and a text box where you are to enter a keystroke that will be used in addition to the Ctrl key. If you enter a lowercase letter, Excel requires you to enter the **Ctrl** key + the letter that you entered for the shortcut key. Unfortunately, just about all the shortcuts that use a Ctrl key and a lowercase letter are already being used by Excel, so we will only be using uppercase letters. If you enter an uppercase letter, Excel adds a Shift key to the shortcut; therefore, you would need to enter Ctrl + Shift + the letter you entered for the shortcut.

6. Click inside the text box and type Shift + F.

7. Click the drop-down arrow for the **Store Macro in** drop-down box and select **This Workbook** because we only want to use the macro in the current workbook.

8. For Description, type **This is my first macro. It will bold, italicize, change the font size to 16, and change the background color to orange.**

9. Click the OK button.

    Every keystroke and mouse click that you make from now until you click the Stop Recording button is recorded, so don't click any other cells or any command buttons other than what you are instructed here to do. You will know when you are recording a macro because the Macro Start Recording button changes to the Macro Stop Recording button ☐. Look at your status bar; the Macro Stop Recording button should be displayed.

10. On the Ribbon's Home tab, in the Font group, click the Bold and Italics buttons. Change the font size to 16. Change the Fill Color to Orange, Accent 2, Lighter 80%. See Figure 19-2.

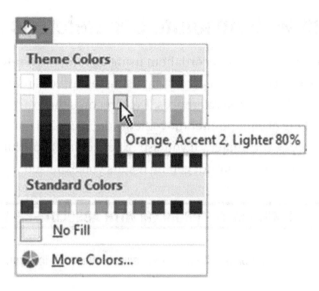

Figure 19-2. *Change the Fill Color to Orange, Accent 2, Lighter 80%*

11. Click the Macro Stop Recording button ☐ on the status bar.

Playing Back the Macro

1. Click cell C2.

2. Type **Hello**.

3. Press Ctrl + Enter.

4. Press the shortcut keys you created for this macro **Ctrl + Shift + F**.

All the keystrokes you used during the recording are now played back and the text in C2 is formatted as defined in the macro.

The Problem with Absolute Cell References

When you record macros, you can record them using **absolute cell references**, which is the default, or you can choose to record them using relative cell references. When you record a macro using absolute cell references, Excel records the exact location where you entered your data. None of your pointer movements are recorded (such as using the up and down arrow keys). Unfortunately, if you use Excel's default absolute cell references when using multiple cells, you may not get the results you expected.

EXERCISE 19-2: LOOKING AT THE PROBLEM WITH ABSOLUTE CELL REFERENCES

This exercise will show the problem with absolute cell references so that you will be able to identify it.

1. Create a new worksheet. Change the name of the worksheet to "Absolute."

2. Click the Macro Start Recording button 🔲 on the status bar.

3. Enter **AbsoluteReference** for the Macro name. (There can be no spaces in a Macro name.)

4. Enter Shift + **A** for the shortcut key.

5. Store the macro in **This Workbook**.

6. For the Description, type **Enter values into multiple cells using Absolute Cell Referencing**.

7. Press the OK button.

8. Click inside cell B4. Type the word **Excel**. Press the Tab key and then type **2019**. Press the Tab key twice and then type **Macro**. Press the Tab key twice and then type **Tutorial**. Press Ctrl + Enter.

9. Click the Macro Stop Recording button 🔲.

10. Click inside cell C6. On the Ribbon's View tab, in the Macros group, the Macros

    button has two parts   .

Clicking the top half of the button displays the **Macro** dialog box which displays all of your macros. Clicking the down arrow on the bottom half of the Macros button brings up a menu from which you can select View Macros. See Figure 19-3. This option also brings up the **Macro** dialog box.

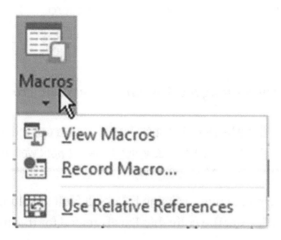

Figure 19-3. *Clicking the bottom half of Macros button brings up a menu*

11. Click the down arrow for the Macros button and then select View Macros.

12. The Macro dialog box opens showing all of the available macros. Select the AbsoluteReference macro. Click the **Run** button. This is another way of running a macro. You could have also used the shortcut you created, which was Ctrl + Shift + A. See Figure 19-4.

The results are probably that cell G4 became the active cell. This is because it repeated the entries that you previously made rather than placing the entries starting at cell C6. This is because Excel uses absolute referencing by default, which doesn't adjust for any changes in the row or columns.

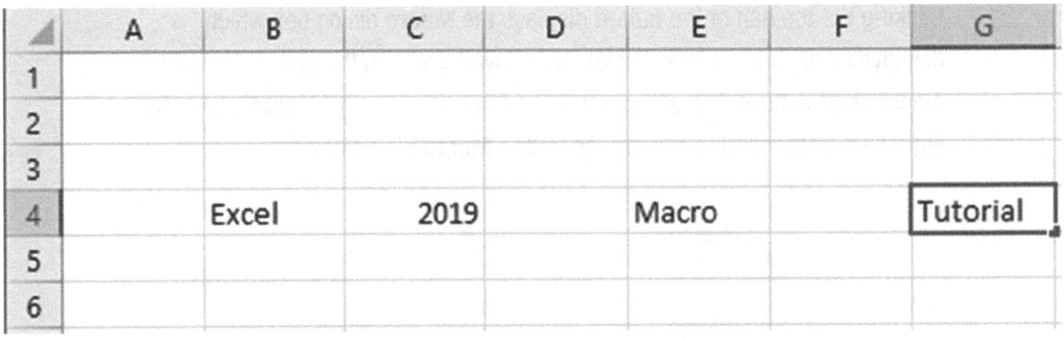

◢	A	B	C	D	E	F	G
1							
2							
3							
4		Excel	2019		Macro		Tutorial
5							
6							

Figure 19-4. Results of running the AbsoluteReference macro

Absolute referencing is fine when you first apply formatting to a cell and then apply that formatting to another cell, as we did previously, but it doesn't work well for multiple entries.

You have learned to create macros using Excel's default absolute cell reference. In Exercise 19-10, you will look at the code that Excel created for this macro and how you can alter it.

Saving a Macro-Enabled Workbook

Workbooks that contain macros must be saved using a different file format than those used for other workbooks. Workbooks that do not contain macros are usually saved with the extension .xlsx. Macro-enabled workbooks need to be stored with the extension .xlsm.

EXERCISE 19-3: SAVING A MACRO-ENABLED WORKBOOK

We originally saved our Macro_Practice workbook as an Excel workbook, but we have added macros to the workbook since we did the save. If we try to save it again as an Excel workbook, we will get a warning message stating that if we continue with this save, we will lose all the macros saved to this workbook.

1. Click the **Save** button 🖫 on the Quick Access Toolbar. See Figure 19-5.

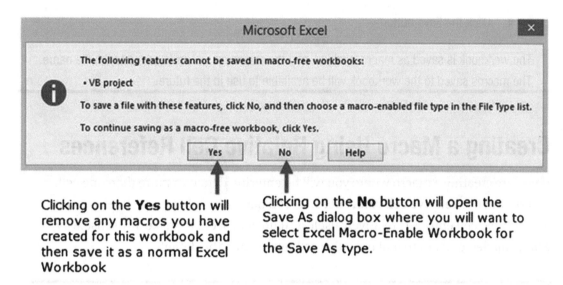

Clicking on the **Yes** button will remove any macros you have created for this workbook and then save it as a normal Excel Workbook

Clicking on the **No** button will open the Save As dialog box where you will want to select Excel Macro-Enable Workbook for the Save As type.

Figure 19-5. *Warning message that is displayed if you try to save a workbook as a normal workbook instead of a macro-enabled workbook*

2. Click the **No** button.

3. Click the down arrow for **Save as type** and then select **Excel Macro-Enabled Workbook**. See Figure 19-6.

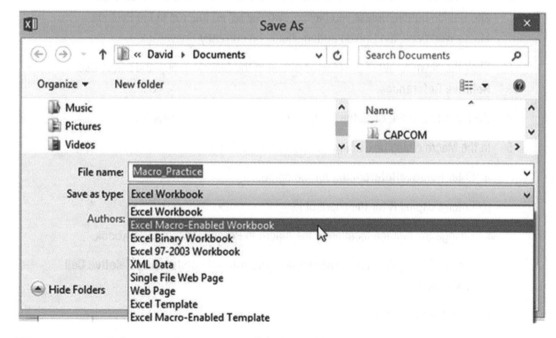

Figure 19-6. *Select Excel Macro-Enabled Workbook*

4.   Click the Save button.

The workbook is saved as macro-enabled, and the extension `.xlsm` is added to the file name. The macros saved to the workbook will be available to use in the future.

Creating a Macro Using Relative Cell References

If you are creating a macro where you will be entering data into more than one cell, you should use **relative cell references**. As you have seen, absolute cell referencing for multiple cell entries is relatively useless. It doesn't seem to make any sense why Microsoft decided not to make relative cell referencing the default.

EXERCISE 19-4: USING RELATIVE CELL REFERENCES

We will create a macro the same way we did for the AbsoluteReferencing macro except that this time we will create it using the relative reference method.

1.   Close all open Excel workbooks. Create a new Excel workbook.

2.   Click the Ribbon's View tab. In the Window group, notice that there is a Hide and Unhide button. Unless you have previously saved macros to the Personal Macro Workbook, the Unhide button should be currently unavailable.

3.   Click the down arrow of the Macros button in the Macros group and select **Use Relative References**.

4.   Click inside cell B4. Click the Record Macro button on the Status bar.

5.   In the Macro dialog box, complete the following steps:

a. Enter **RelativeReferencing** for the name.

b. Enter a capital **R** for the shortcut key.

c. Change the storage location of the macro to **Personal Macro Workbook**.

d. For the Description, type **Enter values into multiple cells using Relative Cell Referencing**.

e. Click the OK button.

6. Type the word **Excel** in cell B4. Press the Tab key and then type **2019**. Press the Tab key twice and then type **Macro**. Press the Tab key twice and then type **Tutorial**. Press Ctrl + Enter.

7. Click the Macro Stop Recording button on the status bar.

8. Click cell C6. Click the top half of the Macros button. You should see the macro you created. Excel has changed the name of the macro to reflect that it is being stored in the PERSONAL.XLSB file.

9. Select the name of your new macro and then click the Run button.

    This time, all the data displays in the new row.

10. Click the View tab. In the Window group, the Unhide button should now be available for use. Click it. See Figure 19-7.

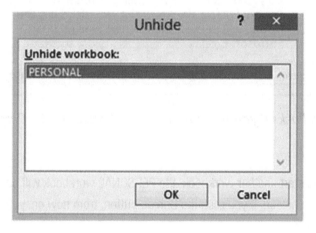

Figure 19-7. *Unhide dialog box*

11. With PERSONAL workbook selected, click the OK button.

    You should now be looking at the PERSONAL workbook. It looks like a normal workbook with one worksheet. Its purpose is just to make its macros available to other workbooks. On the Ribbon on the View tab in the Windows group, click the Hide button.

    The PERSONAL workbook should now be hidden, and you should again be looking at your original workbook.

12. On the Ribbon, click the File tab, and click Save As.

13. Select where you want the file to be saved.

    Because the workbook contains a macro, it must be saved as macro-enabled.

14. Change the Save as type to **Excel Macro-Enabled Workbook**. Name the workbook **Relative**. Click the Save button.

15. Click the 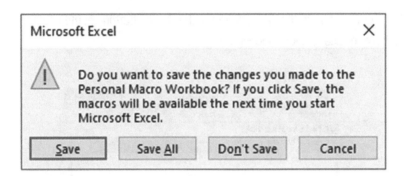 in the upper right corner to exit from Excel.

    If you didn't save the Personal Macro Workbook, you will be asked to save the changes you made to it. See Figure 19-8. Click Save.

Figure 19-8. *Excel asks if you want to save the changes to the Personal Macro Workbook*

Now whenever you start another workbook, the PERSONAL workbook will be automatically loaded, but it will be hidden. If you click the Unhide button, from now on, you will see the PERSONAL workbook.

Adding Macros to the Quick Access Toolbar and Other Objects

So far, we have seen that you can run a macro by typing its shortcut keys and seen how you can run it from the Macros button. For easy and quick availability, you can add a macro to the Quick Access Toolbar, or you can attach it to a button or other object.

EXERCISE 19-5: ASSIGNING INDIVIDUAL MACROS TO THE QUICK ACCESS TOOLBAR

We will assign the RelativeReferencing macro you just created to the Quick Access Toolbar.

1.   Create a new workbook.

2.   Right-click the Quick Access Toolbar. Select **Customize Quick Access Toolbar**. See Figure 19-9.

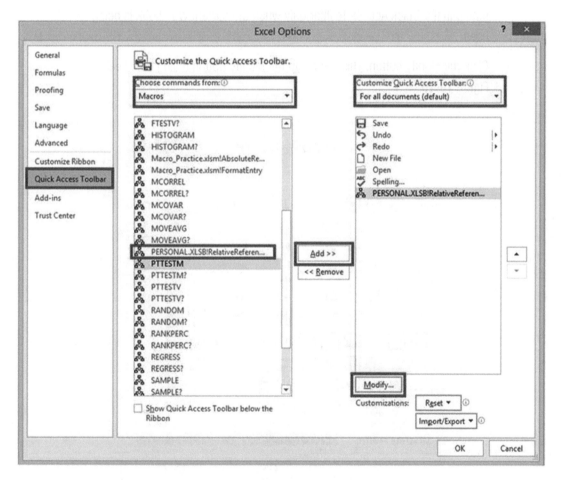

Figure 19-9.   *Excel Options dialog box*

3.   From the left pane, select Quick Access Toolbar.

4.   Under the **Choose commands from** drop-down menu, choose **Macros**.

5. Select **PERSONAL.XLSB!RelativeReferencing**. Click the **Add** button.

Note   Because we saved the macro RelativeReferencing to the Personal
Workbook, it is available to be used in all of our Excel workbooks.

---

6. Under the **Customize Quick Access Toolbar** drop-down menu, choose **For all
    documents (default)**. The other choice for the drop-down menu is to place the
    macro in the Quick Access Toolbar only when the current workbook is being
    used.

7. Click the Modify button. This brings up the Modify Button dialog box. See
    Figure 19-10.

Figure 19-10.   *Modify Button dialog box*

8. Select whatever button you want to associate with the macro. The button that you select will be placed on the Quick Access Toolbar. Click the OK button on the Modify Button dialog box.

9. Click the OK button in the Excel Options dialog box.

10. Click cell A11. Click the button you just created on the Quick Access Toolbar. The RelativeReferencing macro runs.

So far, we have seen that you can run a macro from the Macros button or by typing its shortcut keys. You have also learned how to get quick access to a macro by adding it to the Quick Access Toolbar. In the next exercise, you will assign a macro to a shape that you can click to start the macro.

EXERCISE 19-6: ASSIGNING A MACRO TO A SHAPE

In this exercise, you assign a macro to a shape.

1. Create a new worksheet.

2. Click the Ribbon's Insert tab. In the Illustrations group, click the Shapes button.

3. In the Flowchart group, click the **Flowchart: Punched Tape** icon. See Figure 19-11.

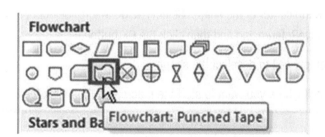

Figure 19-11. *Select the Flowchart: Punched Tape icon*

4. Drag out the shape starting in cell G2 so that its size is similar to Figure 19-13.

5. Click the Ribbon's Home tab. In the Alignment group, click the middle align button and the center button. See Figure 19-12.

Figure 19-12. *Ribbon's Alignment group*

6. Type **Relative Reference** on the shape. Drag across the text to select it. Bold the text and increase its size until it looks similar to the one in Figure 19-13. Click an empty cell to deselect it.

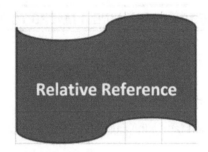

Figure 19-13. *Make your banner look like this*

7. Right-click the center of the shape. Select Assign Macro. Select PERSONAL. XLSB!RelativeReferencing and then click the OK button.

8. Click an empty cell to deselect the shape.

9. Click the cell where you want to start the macro and then click the shape.

The Macros button is on the Ribbon's View tab. This means you have to click the View tab when you want to view the macros or if you need to specify Relative Addressing. To make these features available no matter what the active Ribbon tab is, we will add the Macro button to the Quick Access Toolbar in the next exercise.

EXERCISE 19-7: ASSIGNING THE MACRO BUTTON TO THE QUICK ACCESS TOOLBAR

We will place the Macros button on the toolbar so that you will not need to go to the View tab every time you want to run or edit a macro.

1. Click the File tab on the Ribbon. Click Options.

2. From the left pane, select Quick Access Toolbar.

3. Under the **Choose commands from** drop-down menu, choose **Popular commands**.

4. Scroll to the bottom of the list and find View Macros. Select **View Macros**. Click the **Add** button.

5. Click the OK button.

6. Click the View Macros button on the Quick Access Toolbar. Click the Cancel button.

Adding the Developer Tab to the Ribbon

If you are creating VBA code or assigning a macro to a button, you will need to do this in the Visual Basic Environment which can only be accessed through the Ribbon's Developer tab. If you don't see a Developer tab on your Ribbon, you will need to add it by performing the following steps:

1. Click the **File** tab on the Ribbon. Select **Options** in the left pane to open the Excel Options dialog box.

2. Click **Customize Ribbon** in the left pane. Select **Developer** in the right pane. Click the **OK** button. See Figure 19-14.

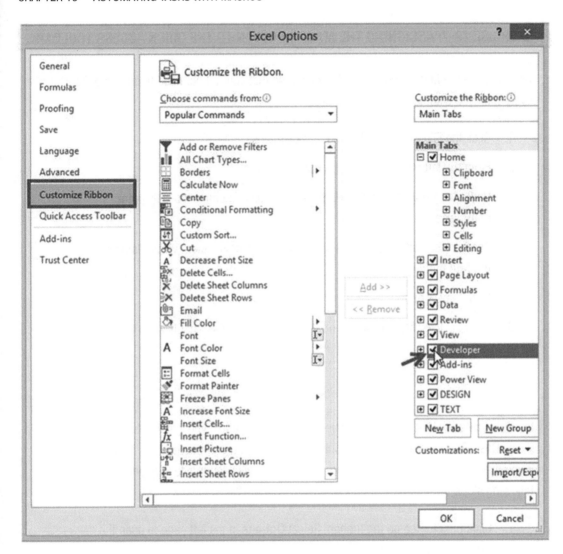

Figure 19-14.   *Select Developer*

You should now see a Developer tab on the Ribbon.

EXERCISE 19-8: ASSIGNING A MACRO TO A BUTTON

We will assign the same macro to a button.

1. Create a new worksheet.

2. On the Ribbon, click the Developer tab. In the Controls group, click the Insert button.

    Excel has two types of controls: Form Controls and ActiveX Controls. Both types have some of the same controls. You can assign macros to Forms Controls, but you cannot assign them to ActiveX Controls. See Figure 19-15.

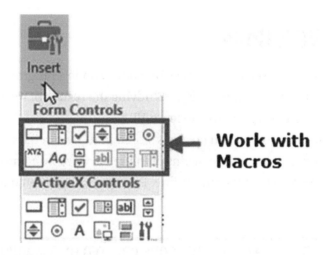

Figure 19-15. *Form Controls*

3. Select the first form control, which is the button. Move the cursor down into the spreadsheet area. The cursor will change to a crosshair.

4. Starting in cell C2, drag down and to the right until you have created enough area for a button.

5. The Assign Macro dialog box appears. Click PERSONAL. XLSB!RelativeReferencing. Click the OK button.

6. Right-click the button and select **Edit Text**. Change the text on the button from Button 1 to Practice.

7. Click inside any empty cell to deselect the button. Click the button you just created.

Sharing the Personal Workbook with Others

You can use the Personal Workbook (`Personal.xlsb`) on other computers or share it with others by emailing them the file or copying it to an external drive and then they could copy the file into the XLSTART folder on their computer.

Looking at VBA Code

When you record a macro, you are creating VBA code. VBA is a programming language. VBA stands for Visual Basic for Applications. Studying the results of the code created when you record a macro is a good way to learn VBA.

The code for each macro that you create is stored in its own separate block of code called a Procedure. A procedure begins with the keyword **Sub** and ends with the keywords **End Sub**. Sub is short for Subroutine. Procedures are often referred to as Subroutines.

EXERCISE 19-9: EDIT THE CODE CREATED BY A MACRO

For this practice, we will alter the code created by the AbsoluteReference macro you created in Exercise 19-3. We will change the code to make the macro repeat the same pattern of text in the row where you start the macro.

Note   Instead of using absolute referencing, we could have used relative addressing to make Exercise 19-3 work, but I want you to see how macros are driven by VBA code and how you can make changes to it. If you want to become a VBA expert, there are entire books devoted to the subject.

---

1. Open the Macro_Practice workbook you saved in Exercise 19-3. You may need to click Enable Content. On the Ribbon, click the View tab. In the Macros group, click the upper half of the Macros button.

2. In the Macro dialog box, select AbsoluteReference.

3. Click the Edit button. You are now in the VBA Editor. Here, you will find the VBA code created by Excel for the FormatEntry and AbsoluteReference macros that you created. The code in Figure 19-16 appears in your VBA Editor window for the AbsoluteReference macro.

```
Sub AbsoluteReference()
'
' AbsoluteReference Macro
' Enter values into multiple cells using Absolute Cell Referencing.
'
' Keyboard Shortcut: Ctrl+Shift+A
'
    Range("B4").Select
    ActiveCell.FormulaR1C1 = "Excel"
    Range("C4").Select
    ActiveCell.FormulaR1C1 = "2019"
    Range("E4").Select
    ActiveCell.FormulaR1C1 = "Macro"
    Range("G4").Select
    Selection.FormulaR1C1 = "Tutorial"
End Sub
```

Figure 19-16. *Code for AbsoluteReference macro*

You can make changes to the code to alter the way the macro works. Notice that the code is placed between Sub and End Sub statements. The Sub statement is given the name you assigned to the macro. After the Sub statement comes several lines that start with a single quote. These lines appear in green. These are comment lines. These lines are ignored when running the macro. They only appear as an aid to help you remember important facts about the macro.

Next comes the code statements that make the macro perform its steps. Notice that cell addresses are placed in Range statements. These statements select these exact cells. That is why, when you ran the macro, the data was entered into the exact same cells as they were originally entered.

We will change the code so that the first cell used is no longer cell B4 but whatever cell you are in when you start the macro.

4. Change the statement Range("B4").Select to ActiveCell.Select.

5. Change the statement Range("C4").Select to ActiveCell.Offset (0, 1).Select.

    The range C4 is one column over from cell B4; therefore, we need our code to reflect the move one column over from the active cell. We can do this using the Offset method. The Offset method uses two values. The first value is the row. The second value is the column. We want all of our data to appear on the same row so we don't want to change the row from the active cell. Therefore, we use a 0 for the row to keep it where it is and we use a 1 to move over one column.

6. Change the statement Range("E4").Select to ActiveCell.Offset (0, 2).Select.

    The range E4 is two columns over from cell C4. We also want our code to move two columns over from the active cell. Therefore, we use 2 for the column value of the Offset method.

7. Change the statement Range("G4").Select to ActiveCell.Offset (0, 2).Select.

    The range G4 is two columns over from cell E4. We also want our code to move two columns over from the active cell. Therefore, we use 2 for the column value of the Offset method. Your code should look like the following:

```
ActiveCell.Select
ActiveCell.FormulaR1C1 = "Excel"
ActiveCell.Offset(0, 1).Select
ActiveCell.FormulaR1C1 = "2019"
ActiveCell.Offset(0, 2).Select
ActiveCell.FormulaR1C1 = "Macro"
ActiveCell.Offset(0, 2).Select
Selection.FormulaR1C1 = "Tutorial"
```

Now that you have finished editing the code, you can return to the worksheet to try it out. You can return to the worksheet by clicking the Excel button on the toolbar or you can press Alt + F11. Pressing Alt + F11 lets you toggle back and forth between the VBA Editor and the worksheet.

8.   Press Alt + F11.

9.   Click inside cell C6. Press the shortcut you created: Ctrl + Shift + A.

Try running the macro in different rows starting in different columns. The pattern should remain the same.

This exercise solved the problem that you observed in Exercise 19-2. You could have, of course, avoided the problem by creating a Relative Cell Address macro rather than an Absolute Cell Address macro.

Creating Macros from Code

A macro doesn't have to be recorded and then played back. Macros can be manually coded into procedures in the Visual Basic Editor's code window. You can still assign shortcut keys and a description to manually coded macros as well as assigning them to objects such as buttons, clip art, images, shapes, and so on.

A procedure is stored within a **module**. A module can contain many procedures. Macros are automatically stored in numbered modules. The first macro you create will be stored in *Module1*. Each macro you create thereafter will also be stored there until you close your workbook. When you reopen the workbook and create a new macro, it will be stored in *Module2*. Whenever you close and reopen the workbook, any new macros that you create will be stored in a new module with the name assigned to the next number in the sequence.

EXERCISE 19-10: CREATING A MACRO FROM CODE

Sometimes just a simple macro can greatly improve the functionality of a worksheet. In this practice, we will create the timeline in Figure 19-17. Entering text at an angle saves space and looks good, but you must bend your head to view it. Wouldn't it be nice if you could view that vertical text in a large cell in a horizontal format without altering the current layout of your data? That is the functionality we are going to add to the timeline. This practice program will allow a user to click one of the cells that contains text displayed at an angle and then either use a shortcut key or click a button to copy the data to a cell that displays the data in a horizontal format.

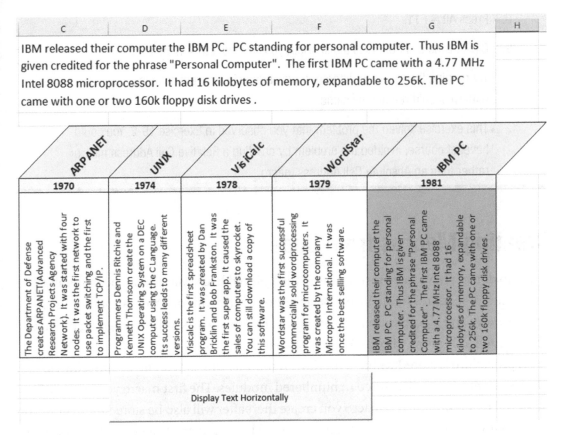

IBM released their computer the IBM PC. PC standing for personal computer. Thus IBM is given credited for the phrase "Personal Computer". The first IBM PC came with a 4.77 MHz Intel 8088 microprocessor. It had 16 kilobytes of memory, expandable to 256k. The PC came with one or two 160k floppy disk drives .

Figure 19-17. *You will create this timeline*

1. Open the workbook Chapter 19.

2. Click the Ribbon's Developer tab. In the Controls group, click the Insert button.

3. Click the Button control in the Form Controls area. Move the cursor below the timeline and drag out a button so that it is in the same location and size as shown in Figure 19-17.

4. The Assign Macro button dialog box appears. Fill in the fields as in Figure 19-18:

    a. Change the **Macro Name** to ConvertToHorizontal.

    b. Change the **Macros in** to This Workbook.

    c. Click the New button.

Figure 19-18. *Click the New button on the Assign Macro window*

You are now in the VBA coding area. The following code is displayed in the Module code window:

```
Sub ConvertToHorizontal()
End Sub
```

5. Place the following code between the two Sub lines:

```
'Make the background color white for the cell range 'C7:G7
Range("C7:G7").Interior.Color = RGB(255, 255, 255)
'Make the background color green for the selected cell
ActiveCell.Interior.Color = RGB(110, 255, 170)
'Take the data from the current cell and display in 'cell C1
Range("C1") = ActiveCell
```

6. Press Alt + F11 to return to the worksheet.

7. Click in front of the Button text. Drag across the text to select it and then type **Display Text Horizontally**. Click an empty cell to deselect the button.

8. Click cell G7. Click the button you created.

    The text from cell G7 should appear in cell C1 in a horizontal format.

9. Click cell C7. Click the button you created.

    The text from cell C7 should appear in cell C1 in a horizontal format.

10. On the Developer tab in the Code group, click the Macros button. Click the Options button on the Macro dialog box. See Figure 19-19.

Figure 19-19. *Select Options from the Macro dialog box*

11. Complete the Macro Options dialog box shown in Figure 19-20:

    a. For the Shortcut key, hold down the Shift key and press T.

    b. For the Description, enter "Display text in horizontal format."

    c. Click the OK button.

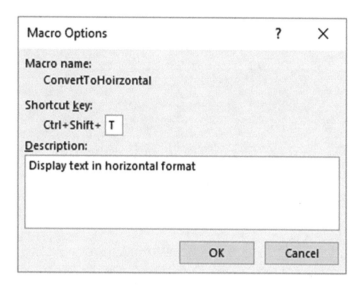

Figure 19-20. *Enter T for the Shortcut key*

12. Click the Cancel button for the Macro window.

13. Click one of the cells in the range C7:G7. Press Ctrl + Shift + T. The
ConvertToHorizontal macro runs and produces the timeline formatted like the
one shown previously in Figure 19-17.

Summary

If you find yourself needing to perform the same steps over and over, then create a macro
to save yourself some time. Macros can be played back by using shortcut keys. Macros
can also be played by clicking buttons or other objects that you assign the macro to.

Macros that you want available to all your workbooks, or to share with others, should
be stored in the Personal.xlsb workbook. This workbook stays hidden until you click
the Unhide button.

Macros can be created using either relative addressing or absolute cell referencing.
Absolute cell referencing works okay when you are creating a macro that affects a single cell,
but if the macro affects multiple cells, you will most likely want to use relative cell addressing.

Excel converts the macros that you create into Visual Basic for Applications code.
This code is editable. You don't have to create a macro to create VBA code. You can
create your own VBA code from scratch. After you have mastered learning VBA, you will
find yourself more often writing the code than letting Excel create it for you.

Index

A

Absolute cell references, 333–337, 880–882

Access database
    import data, 683, 685
    select opened, 683, 684

Account group, 287

Aggregate functions, 489

Aligning data
    context menu, 182–184
    deleting columns/rows, 191, 192
    descriptions, 166
    fitting, 166
    Format Painter, 178–182
    hiding columns/rows, 185, 186
    Home tab, 165
    horizontal alignment, 171, 172
    indenting text, 170, 171
    inserting columns/rows, 190, 191
    inserting/deleting cells, 192–195
    merging and unmerging
        cells, 168–170
    mini-toolbars, 182–184
    rotating text, 172, 173, 175–178
    Shrink to fit, 167, 168
    text wrapping, 166
    unhiding columns/rows, 187
    vertical alignment, 171, 172

AND function, 373

AutoCalculate tools
    AutoSum, 301–305

average, count numbers, max min, 306, 307, 309–311
    viewing, 311, 312

AutoFill
    copy alpha data, 110
    copy cells, 111, 114
    copy date data, 110, 111
    copy numeric data, 110
    custom lists, 112, 113
    duplicate cells, 106, 107
    Fill Formatting Only, 111
    Fill Without Formatting, 111
    options, 109, 114
    serious of values, 107, 108
    teach Excel, 109
    without copy cells, 116
    without Fill Months, 115
    without formatting, 115, 116

AutoFill Handle, 106–110, 113, 476, 477

AutoRecover, 260, 261, 263, 265, 266

B

Backstage
    definition, 233
    navigation group, 234, 235
    overview, 234

Bar charts, 588, 595, 596, 601, 607

Bing Map, 649
    advantages, 644
    disadvantages, 644

D. Slager and A. Slager, *Essential Excel 2019*, https://doi.org/10.1007/978-1-4842-6209-2

Bing Maps (*cont.*)
revenue, 651, 653
use, 649
Built-in microphone, 845

C

Cell background color, 139, 140
Cell borders, 144–146
Cell reference, 53
Cell styles, 565–568
Chart
creation, 28, 587
data labels, 629
elements, 589, 597
fill option, 593
group, 585
horizontal/vertical, 599
modify, 587
More button, 596
options, 590
select data, 30, 601
Smart Lookup, 30–32
suggestions, 29
text option, 593
3-D, 592
Title change, 624
window, 30
Click Charts, 609
Clustered Bar Chart, 601, 602
Clustered Column chart, 609, 622, 746
Column chart, 586
Column sparklines, 657
Column widths
adjust, 85, 86, 90
changing, 86, 89, 90
data, 89
exponential notation, 84, 85

multiple columns, 87
resizing, 87
selection, 86, 87
signs, 84
size, 85
text hidden, 84
text overflow, 83
Combination chart, 620, 621
Comma-delimited file, 666–668
connections, 674
existing connections, 672
imported data, 669
query properties, 671
upload file, 670
Concatenation and Flash Fill, 461, 475, 476
Conditional formatting
apply to table, 577
creating, 569
Data Bars, 571, 572
edit/delete/create rules, 576
formatting rule, 581
highlight duplicate/unique values, 574
highlight values, 575
indicators group, 578
select icon, 579
top/bottom rules, 570, 571
Context menu, 65, 118, 119, 129, 182, 184
Context-sensitive tabs, 13, 866
Custom Views, 228
Add View, 229, 230
delete, 232
dialog box, 228
Hidden Columns, 230
Hidden Rows, 230
show, 231
window, 231

D

Data types
    date/time, 77, 78
    numeric, 76
    text, 76
Data validation
    circling invalid data, 429
    column headings, 417
    custom restriction, creating, 427–429
    error message, create, 420, 421
    LEFT function, 418
    list restriction, creating, 424–427
    order ID, 418, 419
    restriction, 414
    Ribbon's Data Tools group, 415, 416
    rules, 414
    tabs, 421–423
    validation area, 430, 431
    warning style, creating, 422, 423
Date functions
    DAYS, 408, 410
    MONTH, DAY, and YEAR, 405, 407
    NOW, 402, 403
    serial number, 404, 405
    TODAY, 399–401
Delimiter, 666, 667, 679
Deselecting cells, using mouse, 47
Drag-and-drop method, 497–499, 504
Draw Border tool, 147, 148, 150–155

E

Edit Series window, 606
Ellipsis, 67, 184, 351
Entering/correcting data
    AutoComplete, 102, 103
    AutoCorrect
        AutoCorrect Options, 99

        delete items, 101
        Excel Options, 98
        new item, 100
        results, 97
        uses, 96
    AutoFill (*see* AutoFill)
    drop-down list, 103–105
Evaluating formulas
    circular reference, 434–436
    error value messages, 432
    IFERROR, 432, 433
Excel
    current cell identification, 24
    data, 25–27
    definition, 1
    feature, 29
    functions
        AND, 373
        built-in, 348, 349
        date, 398
        IF, 369–373
        logical, 350
        nested, 375–377, 379, 381
        OR, 374
        sum values, 351
    versions, 2, 3
Excel 2019 functions
    IFS, 382–385, 387–389
    MAXIFS and MINIFS, 390–398
Excel versions, 2, 3, 116, 243, 247, 248

F

Fixed-width text file, 679, 682
Flash Fill, 477–479, 481–483, 485
Font group's dialog box launcher, 135
Font size, 134, 135
Format Cells dialog box, 161–165

Format Painter, 178–182

Formatting data

Accounting format, 157

bold, 132

cell, 140–144

cell background color, 139, 140

cell borders, 144–146

cell dialog box, 161–165

colors, 136

Currency format, 158

default formats, 156, 157

double underline, 132

Draw Border tool, 147, 148, 150–155

Font color, 137–139

font group, 132, 135

fonts, 133–135

italics, 132

monetary values, 158, 159

number group, 155, 156

number of decimal places, 160

underline, 132

values to comma style, 160

values to percent style, 159, 160

Formula Auditing

nested function, 443–447

tools, 436

tracing precedents/dependents,
437–439

Watch Window, 440–442

Formulas

arrow keys, 294

Autofill, copying, 298–300

Autofill Handle, 345

bar, 291

cell addresses, 296

column/row headings, range names,
318, 320

computing markup, 297, 298

definition, 289

Excel/results, 292

math operations, 294

math symbols, Excel, 290, 291

named constants, 326–328

named ranges, 328, 329

name manager, 316–318

naming constants, 315

naming ranges, 312–314

noncontiguous cells, naming, 315

reference, 293

selecting named ranges, 320–324, 326

setting scope, name, 330–332

Spreadsheet, 295

typing into cell, 293

Funnel charts, 642

access, 642

stages, 643

G

Game Sales tab, 588

Geography Data Types

data selector, 766

Indiana Cities/Towns, 770

links, 769

location options, 768

population, 768

question mark, 765

tables, 770

advantages, 771

area/population, 776–778

capitals, 775

comma formatting, 782

data bars, 784

Data Selector, 780

formula, 783, 784

headers, 781

map, 785
population, 773
State locations, 772
Tennessee, 775
Grouped objects, 857, 866

H

Help tab, 33
contact support, 36, 37
help button, 37, 38
quick start topics, 35, 36
training options, 34, 35
What's New button, 33, 34
Hierarchical charts
Funnel, 642
Sunburst, 633
Treemap, 625
Hyperlink
address field, 471
AutoFormat, 465, 466
context menu, 467
definition, 461
dialog box, 466, 469, 470
Excel, 462
file, 468
ignore message, 471, 472
Link to pane, 463
notepad, create text, 469
ScreenTip, 468, 473
web page, 464
window, 462, 474
workbook, 472, 473

I, J

Icons, 852
change font size, 855

change size, 853
change to shapes, 859
replicated image, 854
search, 859
size/placement, 861
IF function, 350, 369
IFERROR function, 413, 432–434
IFS functions, 382
Importing data
access, 683
custom column, 699
Excel, 665, 666
ID column, 701
insert step, 698
State columns, 703
tblStudentAddress, 700
tblStudentFees, 693, 705
text files, 666
transform, 690
URL, 691
web address, 691
website, 686, 688, 689
Info group
current workbook, 235, 236
issues
check accessibility, 245, 246
check compatibility, 247, 248
inspect document, 243–245
inspect workbook, 242
options, 243
properties pane
advanced properties, 236, 237
Save As, 237, 238
workbook information, 238
workbook properties, 237
Protect Workbook
Mark as Final, 240, 241
message window, 240

Info group (*cont.*)
    options, 239
    password encryption, 241
    structure, 241, 242
Insert copied cells, 531–534

K

Kentucky Card, 774

L

LEFT function, 418

M

Macro-enabled workbooks, 882, 883
Macro Recorder, 876
Macros, 875
  assign button, 893
  assign shape, 889
  code (*see* Visual Basic for
    Applications (VBA))
  create, 877
  form control, 893
  store, 877
  toolbar, 891
Map Chart, 644
  advantages, 644
  category color, 648
  category data, 647
  coronavirus cases, 646
  disadvantages, 645
  select, 645
MAXIFS and MINIFS
    function, 381, 390, 391
Mini-toolbars, 182, 184, 196
Mixed cell referencing

definition, 338
dollar signs, 338
multiplication table, 339–342
order of precedence, 342, 343
Moving and copying data
  cut/copy buttons, 501
  drag-and-drop method, 498–500
  feature, 498
  Fill Across workbook to another, 516–518
  Fill Across Worksheets, 512, 513, 515
  keyboard, 502, 503
  Paste button gallery, 504–507, 509–511
Moving between cells
  enter values, 43
  Save As, 44
  shortcut keys, 42, 44, 45
Multimedia, 799
  adding images, 804
  adjust color, 806
  corrections, 805
  glow effect, 809
  highlight, 812
  insert pictures, 801
  pictures style, 800, 807, 808
  screenshot, 809–811
  special effects, 807
  wordart (*see* Wordart)
Multiple worksheets, 536, 538–541

N

Navigation basics
  components of workbook, 5, 6
  QAT, 6
  rows/columns, 7
Nested functions, 375, 381, 442
New collaborative comments
  comment buttons, 127

display, 127, 128
enter comment, 124
posted, 124, 125
sign in, different user, 125–127
switch user, 125
Notes, 116
adding, 118
change background, 123
changing name, 119
comments, 121, 122
delete, 119, 121
edit, 119, 121
end of sheet, 122, 123
group, 117
printing, 120, 121
red triangle, 118
viewing, 118–120
NOW function, 402
Number group, 155, 156, 508, 540, 782, 784

O

Office Clipboard, 534–536
Open group
existing workbook
context menu, 253, 254
Navigation pane, 255
pinned, 253
pins, 253
quick access, 255, 256
recent documents, number of, 255
recent list, 252
Open Package Contents, 848
OR function, 374, 418, 428
Orders table, 753
Organizational chart, 839, 841
name/title, 842, 843
VPofSales, 844

P

Page breaks preview
change location, 202
display options, 200
gridlines, 204
horizontal, 200
insert page preak, 202
location, 199
no row/column headers, 203
Print Preview, 203
remove page preak, 202
reset page preak, 202
types, 199
uses, 199
vertical, 200
view tab, 201
Zoom to Page, 205
Page Layout view
add footer, 208
add header, 206, 207
headers/footer tools, 206
horizontal/vertical ruler, 205
options groups, 209, 210
preset headers, 207, 208
print worksheet, 209
remove whitespace, 210, 211
scale/alignment, 209
show options, 211
Panes
creation, 219
freezing, 221, 222
horizontal, 219
scroll bars, 221
split bars, 220, 221
vertical, 219
Paste Special command
calculations, 524–527, 529, 530
definition, 518

Paste Special command (*cont.*)

    dialog box, 519

    operation options, 521, 522

    options, 520

    paste gallery, 518

    transpose rows and columns, 522, 523

Perspective Relaxed Moderately effect, 815

Pie charts, 584, 607

    create, 608

    data, 655

    select, 654

    labels, 608

    standard, 608

    subtype, 612

    title option, 611

    values, 654

PivotTables, 707

    calculations, 721

    collapsed quarters, 729

    Column Labels, 719

    create, 741, 742, 744

    data, 709

    date slicer, 738, 739

    enlarge view, 724

    Fields list, 718

    forms, 735

    handling dates, 725

    inventory data, 725

    manual create, 709

    plus/minus, 713

    product data, 744

    records, 737

    relational database, 752

    report layout, 734

    row labels, 716, 717

    select button, 751

    select fields, 711

    store field, 714

    table format, 726, 727

    timeline, 740

    time slicer, 740

    Wizard, 748, 749

    working, 708

    worksheet, 710

Print group

    options, 272, 273

    printer setting

        collating, 276

        orientation, 276

        paper size, 276, 277

        print options, 275

        scaling options, 279–281

        select margins, 278, 279

        select pages, 275

    select printer, 273, 274

Printing

    different ways, 212

    paste special, 216–218

    print area

        add data, 214, 215

        adding cells, 213

        ignore, 216

        options, 212, 213

        removing, 213, 214, 216

Proof reading cell values, 447, 449

Q

Quick Access Toolbar (QAT), 6, 20, 23, 887

    adding buttons, 22, 23

    autosave, 21

    customizations, 22

    drop-down button, 20

    options, 20

    ribbon setup, 21, 22

    Touch/Mouse mode, 21

R

Random-access memory (RAM), 256
Relational database
    orders vs. customer, 756
    word ALL, 758
Relational databases, 752
Relative cell references, 884, 886
Ribbon
    contextual tabs, 13
    dialog box launchers, 15, 16
    minimizing/hiding, 16, 17
    resizing, 14
    shortcuts, 18, 19
    tabs/groups, 12
Row heights, 91–93

S

Save/ Save As group
    AutoRecover, 257, 261
    backups/limiting changes
        file naming, 268
        General options, 268
        message, 271, 272
        password, 268–270
        Read Only, 271
        save file, 270
        Tools options, 267
    dialog window, 257, 258
    Excel Options window, 260, 261
    file naming, 260
    file types, 259
    select folder, 257, 258
    system crash, 262, 263
    without saving
        closed file, 263, 264
        Manage Workbook, 265
        message, 264–266

        recover, 266
        unsaved files, 264
Screen clipping, 810, 812, 813
Selecting cells
    cell reference
        cell address, 54
        cell selection, 53
        select range, 53
    Go To feature, 55–59
    using keyboard, 50–52
    using mouse, 46
        adjacent cells, 48
        column heads, 47
        deselect/reselect, 49
        nonadjacent cells, 49
        row heads, 48
    using Name Box, 54, 55
Shapes, 820
    arrow lines, 827
    changing, 835
    circle, 838
    convert to image, 837
    flowchart, 821
    height/width, 822
    LeftRectangles, 833
    size/location, 822
    ungrouping, 830
Share group
    email, 285–287
    OneDrive
        edit link, 284
        invite others, 282, 283
        options, 282
        Share window, 285
        sharing link, 283
        view only link, 283, 284
SmartArt, 564, 799, 838, 839, 844,
    850, 867

Sparklines, 656
    column, 657
    create, 657
    marker color, 660
    negative value, 657
    point selected, 662
    spreadsheet, 664
    style, 658
    weight/font, 660
Special characters, 79–83
Spell checking, 449–451
Split Series
    select custom, 618
    select percentage, 618
    select value, 617
Status bar
    accessibility checker, 489
    aggregate functions, 489
    Caps Lock or Num Lock, 487
    Cell Mode, 487
    End Mode, 488
    Flash fill Blank cells/Flash Fill
        changed, 487
    Macro Recording, 488
    options, 486
    Overtype Mode, 488
    Page Number, 489
    Scroll Lock, 488
    Selection Mode, 489
    view shortcuts, 489
    Zoom/Zoom slider, 489–494
Stock Data Type, 786
    add column, 790
    autofill, 792
    autosum, 793
    company name, 790
    computation, 796
    data selector, 786

    field selection, 791
    formula, 792
    gain/loss, 793
    icon, 787
    links, 788
    mutual fund, 789
    refresh, 794
    setup, 791
    ticker, 787
Stocks and Geography Data Types, 765
SUM function
    arguments, 351, 352, 354
    definition, 351
    Insert Function button, 355–360
    adds the cell
        required arguments, 361–363
        optional sum_range
          argument, 363, 364
        syntax, 360
SUMIFS function
    arguments, 365, 366
    cell address, cell value, 367
    definition, 364
    empty cells, 368
    result, 365
    syntax, 364
Sunburst chart, 633, 634
    A2, D15, 637
    create, 636
    removing values, 635
    spreadsheet, 638
    3D effect, 640
Synchronizing Scrolling
    Freeze Panes, 227
    mouse pointer, 226
    scroll bars, 224–226
    select Windows, 223
    split bars, 224

T

Table
  adding to excel
    automatic expansion, 548
    filter, 552–554, 556
    headers, 551
    records, 552
    remove check mark, 549
    result, 558
    worksheet students, 550
  creating and formatting, 544–546
  filtering data, Slicer, 558–560
  sort and filter, 546, 547
  themes, 560–565
Text wrapping, 166
Thesaurus button, 451, 452
3D models, 863
  Bubble shape, 868
  categories, 864
  grouped objects, 866
  object names, 872
  rotated view, 866
  selection pane, 867, 868
  stacked, 865
TODAY() function, 399
Treemap chart, 627
  create, 626
  headings, 628
  label, 629
  label/values, 630
  texture background, 632
  worksheet, 626
Typing mistakes
  clearing contents, 95, 96
  restore cell, 94
  specific characters
    correction, 93

U

Ungrouping, 830, 861

V

Visual Basic for Applications (VBA)
  AbsoluteReference, 895
  create, 897
  edit, 894
  select options, 900

W, X, Y, Z

WordArt, 813
  create, 813
  making, 818
  Perspective Relaxed
    Moderately effect, 815
  text effects, 817
  text fill, 814
Workbook
  creation, 7, 8
  naming, 11, 12
  new, 248, 249, 251
  recently opened, 11
  Save As window, 9, 10
  saving, 8, 9
  Views, 198
Worksheet and cell level protection
  cell level, 457, 458
  locked, default, 454
  message informing, 456
  Password dialog box, 456
  Ribbon's Changes group, 453
  Sheet dialog box, 455
Worksheet, inserting
  video, 848, 849

Worksheets
    activate dialog
        box, 68, 69
    adding/removing, 61–63
    copy, 72
    handling, 67
    hiding/unhiding, 64, 65, 69
    manipulating, 60
    multiple, 64

naming, 60, 70
next, 67
previous, 67
Qtr1 Payroll, 70, 71
Qtr1 Sales, 72
reordering/coping, 65, 66
tab color, 63, 64
tabs, 60
tooltip, 68

DATE DUE

			PRINTED IN U.S.A.